Current Topics in Behavioral Neurosciences

Volume 41

Current Topics in Behavioral Neurosciences provides critical and comprehensive discussions of the most significant areas of behavioral neuroscience research, written by leading international authorities. Each volume in the series represents the most informative and contemporary account of its subject available, making it an unrivalled reference source. Each volume will be made available in both print and electronic form.

With the development of new methodologies for brain imaging, genetic and genomic analyses, molecular engineering of mutant animals, novel routes for drug delivery, and sophisticated cross-species behavioral assessments, it is now possible to study behavior relevant to psychiatric and neurological diseases and disorders on the physiological level. The Behavioral Neurosciences series focuses on translational medicine and cutting-edge technologies. Preclinical and clinical trials for the development of new diagnostics and therapeutics as well as prevention efforts are covered whenever possible. Special attention is also drawn on epigenetical aspects, especially in psychiatric disorders.

CTBN series is indexed in PubMed and Scopus.

More information about this series at http://www.springer.com/series/7854

Timothy Hodgson
Editor

Processes of Visuospatial Attention and Working Memory

Springer

Editor
Timothy Hodgson
School of Psychology
University of Lincoln
Lincoln, UK

ISSN 1866-3370 ISSN 1866-3389 (electronic)
Current Topics in Behavioral Neurosciences
ISBN 978-3-030-31025-7 ISBN 978-3-030-31026-4 (eBook)
https://doi.org/10.1007/978-3-030-31026-4

This Springer imprint is published by the registered company Springer Nature Switzerland AG.
The registered company address is: Gewerbestrasse 11, 6330 Cham, Switzerland

Preface

When I was first approached to edit a volume on visuospatial attention and working memory, I was not only honoured and excited but also daunted by the proposed topic's scope. Consider just the brain structures involved in visuospatial attention and working memory for example: the mid-brain and brainstem centres that control eye movements; the cortical visual pathways in the occipital, parietal and temporal lobes that process object features and locations; through to the lateral and medial frontal lobes of the cerebral cortex and basal ganglia which mediate top-down control over perception and action and maintain information in memory across delays.

To address the challenge, I began by defining the aims of the project in more detail and started to identify features that I thought would make the book distinctive. Firstly, it was decided that the volume should focus on research which addressed processes and mechanisms and that the inclusion of chapters with clinical relevance was also essential. I wanted the volume to bridge a gap I felt existed between research into the control of overt shifts in attention (eye movements) and research into covert attention (paying attention out of the corner of the eye). I also felt no attempt had been made to date to cover evolutionary and developmental perspectives on the topic and so this would be good to attempt. Finally, the book needed to be as international as possible and not just include people whose research I already knew. Following some more thought, many literature searches and much reading, I came up with a "wish list" of authors and chapter topics. In the summer of 2017, I began sending out e-mails and to my surprise the majority of authors responded enthusiastically to my invitation to contribute chapters straight away!

Part I of the book comprises four chapters which focus on *Evolution and Development*. Land's opening chapter outlines an important perspective on how gaze shifting eye movements (saccades) evolved. Surprisingly, cross species comparisons suggest that saccades did not evolve to shift the point of gaze, but to hold the eye stable as the fast phase of the compensatory eye movements that accompany head rotation (nystagmus). That the production and manipulation of stone tools by our ancestors might be key to the evolution of human visuospatial attention and working memory is the implication of the chapter by Bruner and colleagues. The expansion of the medial parietal lobes and precuneus region of the cerebral cortex within the fossil record combined with contemporary neuroimaging evidence is used to argue that enhanced integration of information in working memory was a key factor driving evolution of human cognitive capacities. Following on from these two chapters, it is intriguing to consider how ontology might mimic phylogeny when reading Johnson's contribution, which presents a comprehensive overview of research into how visuospatial attention develops in children. Comparative studies of spatial working memory in humans, children and non-human species are described in De Lillo's chapter. A unique feature of the human capacity to remember spatial locations seems to be the ability to strategically chunk multiple locations in a sequence and encode them as holistic shapes and patterns.

Part II comprises the main body of the book and contains a series of chapters representing a snapshot of contemporary research into *Processes, Mechanisms and Models* in the field. Common themes running through this section are interactions between memory and attention and the extent to which past events influence attention, perception and action in the present. Liesefeld and colleagues outline their dimensional weighting account of trial-by-trial interactions in visual search, whilst Parr and Friston present an active inference framework which successfully models eye movement scan paths by combining memory for what has been seen at a location with predictions regarding what will be seen if one looks there again. Bahami et al. review neurophysiological studies of working memory and selective attention in the prefrontal cortex and the role played by dopamine. Aagten-Murphy and Bays then present a comprehensive review of studies that have examined how information is integrated across saccades and the conditions under which changes to a visual scene are perceived during saccades. The next chapter by Chen and Hutchinson describes how long-term memory influences attention, whilst Laflamme and Enns continue the theme of interactions between memory, perception and attention with new findings from the "superstitious perception" task. Elliot and Coleman's chapter argues that certain types of neural oscillatory activity may be important in selecting objects for action and anticipating future events. Hunt and colleagues address an important long-standing debate in the field, the relationship between covert attention and eye movements. Their chapter suggests that whilst oculomotor programming and covert attention are closely associated, the two processes are not strictly equivalent. Finally in this section, Macaluso's chapter raises the important question of how results obtained in the laboratory apply in the real world, assessing the current state and future opportunities for neuroimaging studies of visuospatial attention in natural visual scenes.

The five chapters in *Part III* present topics of clinical relevance dealing with patient-based research in *Neuropsychology and Neuropsychiatry*. Smith and Archibald consider patients with progressive supranuclear palsy, a condition that can be confused for Parkinson's disease in its early stage, which is characterised by attentional impairments confined to the vertical axis. Zokai and Husain summarise research using delayed reproduction working memory tasks to examine dissociable mechanism underlying deficits in Alzheimer's and Parkinson's patients, whilst Park and Ichinose give an account of working memory impairments in schizophrenia, showing how abnormalities in this group are likely to arise due to attentional and perceptual problems affecting memory encoding. Pennington and Klaus revisit dopamine's role in attention and working memory reviewing recent findings of clinically relevant epigenetic interactions between environmental stressors and genes involved in dopamine transmission. The contribution from my own group on eye movements in neuropsychological tasks comes last and includes an updated version of a model of eye movement control in complex tasks originally presented by Land. Thus, our chapter links back to the opening contribution on the evolution of saccades by Land and therefore seemed a fitting place to end the volume.

I would like to thank all the contributing authors, series editor Thomas Barnes, Alamelu Damodharan and rest of the production team at Springer for all their time and patience with the project. I believe the resulting volume represents a very personal and original survey of a wide and complex field, and I hope you will agree the finished book has been well worth the wait!

Lincoln, UK
July 2019

Timothy Hodgson

Contents

Part I
Evolution and Development

The Evolution of Gaze Shifting Eye Movements

Michael F. Land

Contents

Abstract In animals with good eyesight most eye movements consist of saccades, which rapidly shift the direction of the eye's axis, and intervals between the saccades (fixations) in which gaze is kept stationary relative to the surroundings. This stability is needed to prevent motion blur, and it is achieved by reflexes which counter-rotate the eye when the head moves. This saccade-and-fixate strategy arose early in fish evolution, when the original function of saccades was to re-centre the eye as the fish turned. In primates, and other foveate vertebrates, saccades took on the new function of directing the fovea to objects of interest in the surroundings. Among invertebrates the same saccade-and-fixate pattern is seen, especially in insects, crustaceans and cephalopod molluscs.

Keywords Fixation · Invertebrates · Saccade · Stabilising reflexes · Vertebrates

1 Saccades and Fixations: A Universal Pairing

In humans, all other vertebrates, and in invertebrates with good eyesight, there is one pattern of eye movements that is almost universal: the pairing of fast gaze-shifting movements (saccades) with periods of stable gaze (fixations). In humans there are other types of eye movement – pursuit and vergence – but our usual method of

M. F. Land (✉)
School of Life Sciences, University of Sussex, Brighton, UK
e-mail: m.f.land@sussex.ac.uk

Curr Topics Behav Neurosci (2019) 41: 3–12
DOI 10.1007/7854_2018_60
Published Online: 18 August 2018

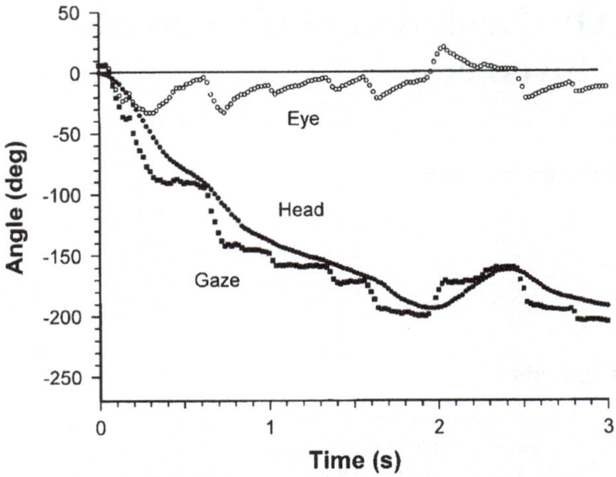

Fig. 1 The author's left eye, looking round a room. The eyes perform saccades and between these compensate for head movements. Gaze (eye + head) moves in a series of discrete fixations, around the direction of the moving head. From Land and Tatler (2009). Note the similarity to Fig. 2

interrogating the surroundings is by using saccades to move our high-resolution foveas around the scene in a series of fixations (Fig. 1).

It seems natural for us to think of saccades, primarily, as movements that shift our direction of acute vision, but this is not how they began. Gordon Walls, in a landmark article, provided an outline of the way that fixations and saccades came about in vertebrates (Walls 1962). As he puts it, the origin of eye movements '. . .lies in the need to keep an image fixed on the retina, not in the need to scan the surroundings'. By the Ordovician period, at least 450 mya, the first fishes already had a reflex in which rotations of the head during swimming evoked compensatory movements of the eyes. This, the vestibulo-ocular reflex (VOR), was brought about by signals from the ampullae or semicircular canals of the vestibular system, and its function was to keep the image still with respect to the surroundings, in spite of movements of the head. At the same time, or slightly later, a second reflex evolved, which took retinal motion signals and fed them back negatively to the eye muscles via the optic tectum. This optokinetic reflex (OKR) also functioned to clamp the image to the surroundings. The need for this double image-stabilising system can be attributed to the fact that vertebrate photoreceptors are slow, cones taking 20 ms or more to respond fully to a change in intensity (Friedburg et al. 2004), so that if the eyes moved passively with the head, the image would be degraded by motion blur. We know from studies on human vision that blurring starts to occur when the image moves across the retina at speeds greater than about 1 degree per second, so compensatory eye movements are essential for clear vision, especially at higher spatial frequencies. These eye movements are continuous, and equal or at least proportional to the disturbance caused by the natural motion of the head, or imposed by an experimenter.

What then happens when a fish makes a turn? If the reflexes are operating as they should, the eyes will become trapped at the limit of their range, and a mechanism is needed to return them to a central point in the orbit. In practice, fish tend to make

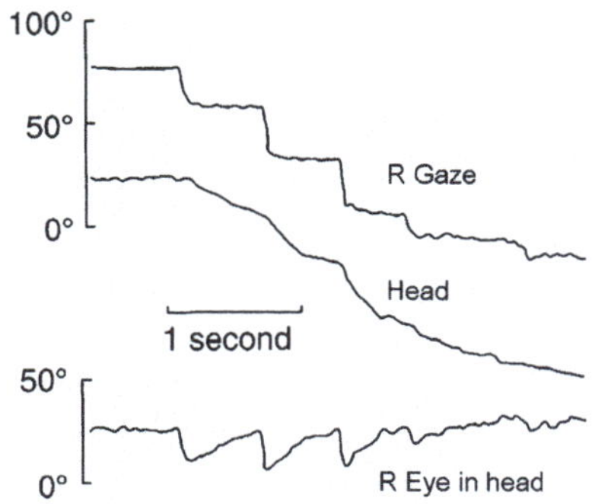

Fig. 2 Records of the eye, head and gaze movements of a goldfish making a turn. The eyes make saccades, but between these their movements are equal and opposite to those of the head. This results in gaze moving in a series of rotationally stable fixations. Goldfish lack a fovea, so the gaze shifts are more a necessary consequence of making a turn than a need to redirect a region of higher resolution. Redrawn from Easter et al. (1974)

these recentring movements early in a turn rather than when the eyes hit a backstop (Fig. 2). These movements need to be fast, since vision will not be possible or desirable during the reset, and this seems to be where the need for saccades originated. To quote Walls again: 'even in the lowest fishes, we see a good reason for saccadic eye movements to be as quick as they notoriously are'. In clinical practice, these reflex movements in humans have become known as 'slow phases' and 'flick backs' or 'fast phases' and together referred to as 'nystagmus'. But they are actually the ancient raw material from which almost all normal movements are derived.

2 Variations in Vertebrates

In some species of most vertebrate classes, this 'saccade and fixate' system has been adapted for a second use, namely, for targeting particular objects in the surroundings for more detailed scrutiny. Where this has evolved, it is always associated with a region of high resolution on the retina, either an area of elevated retinal ganglion cell density, as in a cat, or a smaller distinct fovea, as in a pipefish, chameleon, hawk or primate (Land 2015). These targeting movements do not occur in vertebrates with more uniform retinae, such as goldfish, toads and rabbits: here the appearance of a novel object does not provoke a saccade. The ability to target objects also requires the ability to hold them on a high- resolution area or fovea, and this has led, in humans and other primates, to further oculomotor refinements. These include the ability to fixate and to track moving objects smoothly. Tracking usually means that the reflexes which keep the overall image stationary (VOR and OKN) have to be suspended or modified, so that the eye can move with the target and allow the background to drift (e.g. if you track your moving finger, while observing the background, you can appreciate the destructiveness of motion blur). In primates

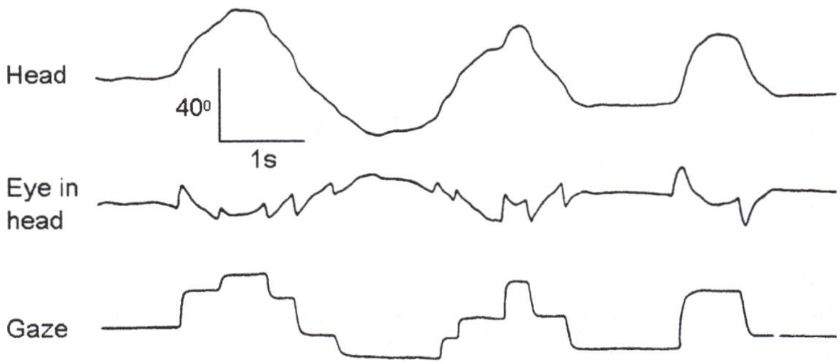

Fig. 3 Head, eye and gaze movements of a rabbit tracking the hand of an observer. Unlike primates, the target is tracked with the head, while the eyes make nystagmus-like saccades and compensatory movements, resulting in steplike gaze shifts. Modified from Collewijn (1977)

tracking extends to the third dimension, so that the foveas of the two eyes can converge on a single object independent of its distance. In mammals with lateral eyes, objects are tracked, but with the head rather than the eyes. Figure 3 shows a rabbit tracking the experimenter's hand with head movements. The eyes remain locked to the background, making a series of saccades and compensatory movements, so that gaze follows the head in a series of discrete fixations that do not directly follow the target, as they would in humans.

In humans, a saccade may be attracted to objects that are novel or otherwise 'eye-catching', but more commonly they are made to objects from which information is needed for the execution of a task, for example, to the spoon needed to stir the coffee or the nail that will be hit with a hammer (Land and Tatler 2009). This implies that the saccadic system has swift access to memories of where things are in the surroundings and information about their identities. Even when the eyes are not involved in information collection, they show the same pattern of saccades and fixations, at a rate of about three per second (Fig. 1).

In vertebrates with necks, head movements add to eye movements in determining the sizes of saccades. Heads have more inertia than eyes, so the contributions of the head need to be managed so that they do not slow up the gaze shifts. In small saccades (<40°) in primates, eye movement contributes most to the gaze shift, with head movement adding to both the speed and amplitude of the saccade, while gaze-stabilising reflexes are temporarily turned off (Fig. 4a). When the predetermined end-point of the gaze change is reached, VOR is turned on, and the eyes move in the opposite direction, exactly counteracting the ongoing head movement and establishing fixation. For smaller saccades the head contributes about 30% to the gaze change. For larger saccades the eyes reach a 'backstop', and, with VOR still turned off, gaze is carried entirely by the head until the gaze end-point is reached, when VOR is re-established (rectangle, Fig. 4b).

The situation in birds is different. Having light heads and flexible necks, the contribution of head movement to gaze shifts is much higher than in mammals, and it is these head movements that make smaller birds seem so busy and vigilant. In

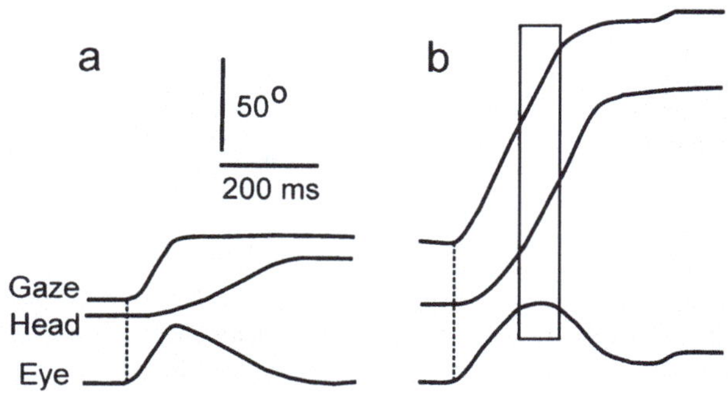

Fig. 4 Eye and head movements during a small (**a**) and large (**b**) combined saccade. Note that in (**b**) VOR remains turned off as the head completes the movement (rectangle) and recommences when the end-point is reached. Redrawn from Guitton (1992)

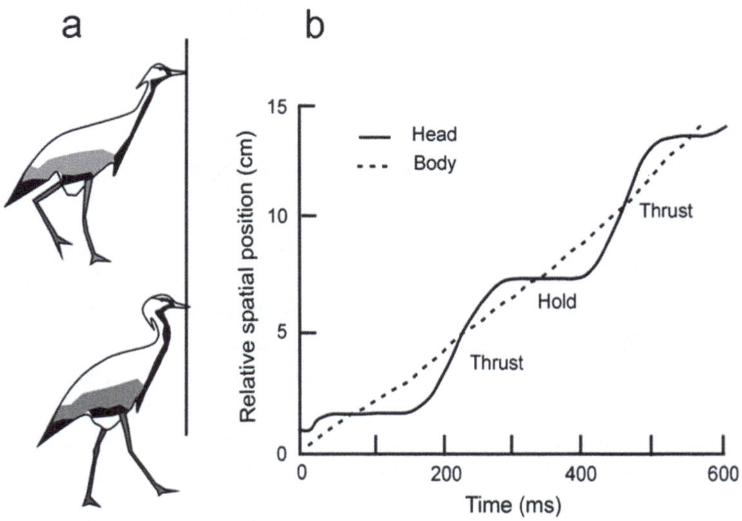

Fig. 5 Translational 'saccades' made by a demoiselle crane (**a**) and a pigeon (**b**). During the 'hold' phases, the head remains still in space (held by an optokinetic mechanism) and then moves forward during the 'thrust' phase. The body motion continues throughout. (**a**) Redrawn from photographs by Necker (2007) and (**b**) records from Frost (1978)

peahens, the head contribution to large gaze shifts (>40°) is almost 90%; although with smaller gaze shifts, this reduces to about 60% (Yorzinski et al. 2015).

Many ground-feeding birds also make what may be called 'translational saccades' while walking. The head moves forward, and then stabilises, as the body continues to walk under the stationary head (Fig. 5). The function of this 'head-bobbing' seems to be similar to that of rotational saccade and fixate behaviour: to allow the view to the side to be held temporarily stationary on the retina while not preventing forward motion.

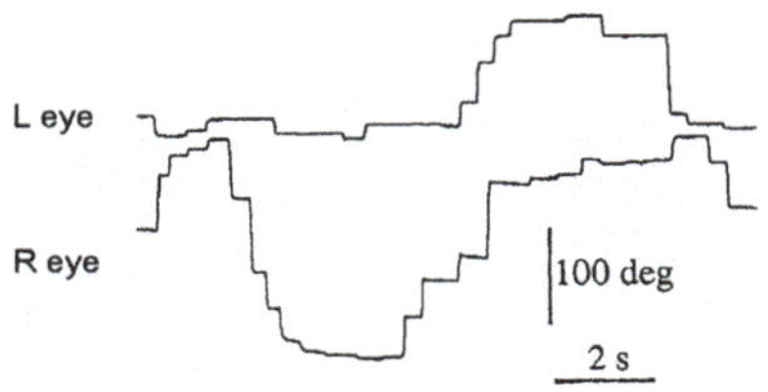

Fig. 6 Asynchronous saccades made by the two eyes of a chameleon, recorded with search coils. Adapted from Ott (2001)

The extent to which the two eyes are yoked during saccades varies considerably across vertebrates. In primates, the eyes are synchronised in timing, and their movements have the same amplitude; in other mammals the eyes are synchronised, but the movement amplitudes are more variable. At the other extreme, chameleon eyes are entirely independent in both the timing and amplitude of saccades (Fig. 6), until the moment preceding a strike, when they both point forwards. Seahorses and pipefish, which, like chameleons, have distinct foveas, have similarly asynchronous saccades. One particularly remarkable fish, the sandlance (*Limnicthyes fasciatus*), differs from all other vertebrates in that, following each saccade, the eye is not held still but drifts back towards a central position (Pettigrew et al. 1999).

3 Saccades and Fixations Outside the Vertebrates

The logic of Walls' argument – that the main function of eye movements is to keep the image still on the retina – should apply equally to those invertebrates that have good eyesight. This means the arthropods, especially crustaceans and insects, and cephalopod molluscs. Figure 7 shows a record of a rock crab, moving in a curvilinear path. Like the goldfish (Fig. 2), the eyes make fast saccadic eye movements with slow movements between them that compensate for body rotations. These slow movements result in periods of stationary gaze. Like vertebrates, crabs have a powerful optokinetic reflex and also the equivalent of a vestibular system that measures body rotation. Clearly, the saccade and fixate strategy is not just an idiosyncrasy of the vertebrate oculomotor system.

Insects too employ saccades and fixations. The situation here is different as the eyes are part of the head, so gaze stability has to be achieved by neck movements to compensate for rotations of the body. Figure 8a shows the head and body of a walking stalk-eyed fly, turning through a right angle. The head makes two saccades, at 120 ms and just before 400 ms, but before and between these movements, the head angle does not change. Flying flies behave rather similarly. In an impressive study, Schilstra and van Hateren (1998) recorded from both the head and thorax of flying blowflies with miniature search coils. Figure 8b shows that both the body and head rotate jerkily, but the head moves faster, and there is compensation for the slower body rotation (head on thorax), resulting in crisp changes of gaze (head).

Hoverflies have such fine control of their body angle during flight that they dispense with neck movements. The result is that 'saccades' appear as rapid changes

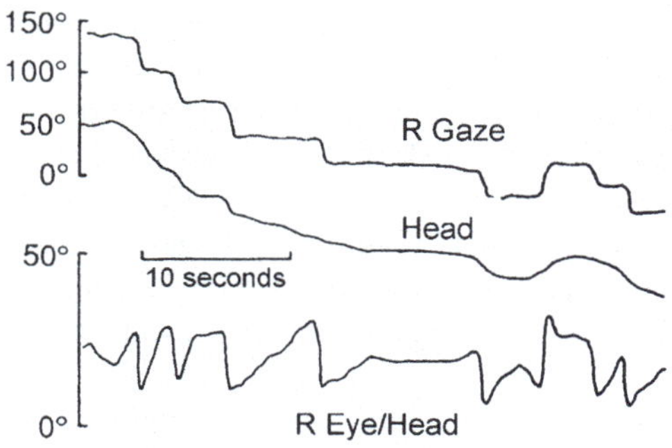

Fig. 7 Gaze, head and eye movements of a rock crab walking on a curved path. As with the goldfish (Fig. 2), the role of eye movements is twofold: to make gaze-shifting saccades and to compensate for movements of the head. Note that the eye/head scale is magnified. Redrawn from Paul et al. (1990)

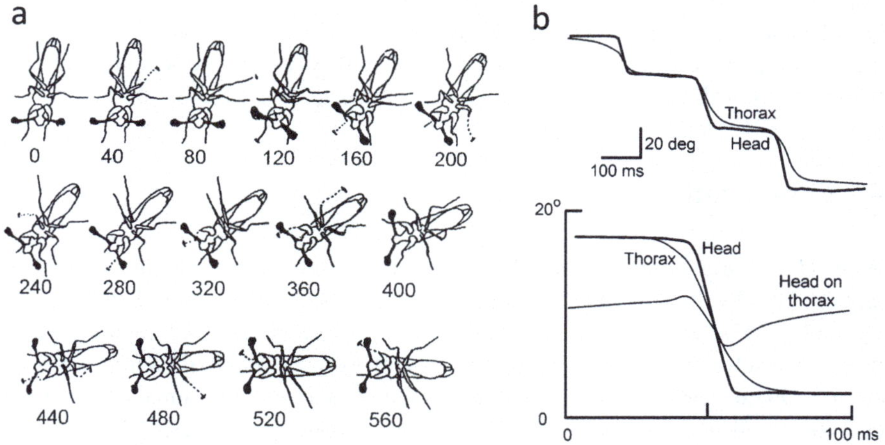

Fig. 8 (**a**) Stalk-eyed fly walking through a right angle on a glass plate. It makes two saccades (at 120 and 380 ms), with fixations in between (redrawn from an unpublished film by W Wickler and U Seibt). (**b**) Three saccadic turns made by a flying blowfly (top) showing the steplike rotations of the head in space and slower movements of the body. Below is a detail of a single saccade, showing the counterrotation of the head on the thorax. Both adapted from Schilstra and van Hateren (1998)

in orientation of the whole body, with 'fixations' between them in which translational flight continues but body angle is held constant (Collett and Land 1975). Interestingly, males have a frontal region of higher resolution, not present in females, and they use this to track the females while remaining out of sight by keeping at a distance of 10 cm. Males, in other words, behave like primates; females behave more like goldfish. Praying mantids also track their prey, with their heads. Against textured backgrounds, the tracking is saccadic, but against plain backgrounds, it is smooth, presumably because the optokinetic signal is too weak to prevent head rotation (Rossel 1980).

Outside the arthropods, the only other major group of animals with mobile eyes are the cephalopod molluscs (*octopus*, cuttlefish and squid). They show the same kind of nystagmus as vertebrates when placed in a rotating striped drum (Hanlon and Messenger 1996), and in the cuttlefish *Sepia*, saccade and fixation movements occur during spontaneous swimming (Collewijn 1970). These compensate for yaw and roll. *Octopus* has seven eye muscles, but cuttlefish and squid have an additional six or seven. These extra muscles are involved in the convergent eye movements which provide binocular vision prior to prey capture, movements that are absent in *octopus* (Budelmann 2009).

There are a few invertebrates that break all Walls' rules by making slow scanning eye movements during which they take in information. Jumping spiders scan images with slow torsional movements and faster side-to-side movements whose functions are to determine identity: specifically, whether a newly detected object is a conspecific or potential prey. Mantis shrimps (Stomatopoda) use scanning movements to extract information about colour and polarisation. Heteropod sea snails scan the water beneath them to detect plankton to feed on. In all these cases, the retinas (or relevant parts of the retina) are one-dimensional strips a few receptors wide, and the scanning movements are sufficiently slow that they do not interfere significantly with resolution (Land and Nilsson 2012).

4 Conclusions

Walls (1962) argued that vertebrate eye movements originated in the need to avoid motion blur by maintaining a stationary image and that this was achieved via vestibulo-ocular and optokinetic fixation reflexes, which evolved in the earliest fishes. Saccades evolved initially as movements to recentre eye direction when an animal turned. This saccade and fixate strategy is found in animals with good eyesight in all major phyla. In man, and other foveate vertebrates, this pattern of eye movements became adapted for a second function: the targeting of particular objects by the fovea, to obtain the benefits of improved resolution.

In primates two other types of eye movement supplement this targeting role. Smooth pursuit allows a moving object to be kept in central vision, but this inevitably means that the background is allowed to blur, with VOR and OKN temporarily suspended. Primates, and some other animals with forward facing eyes, use vergence to direct both foveas to targets at different distances, incidentally allowing stereoscopic range finding.

Head movements augment eye movements in many vertebrates and in birds almost replace them. In insects, with immovable eyes, neck movements, and sometimes whole-body movements, perform saccades and stabilise fixations. Cephalopods have a range of eye movements remarkably similar to those of fish, including vergence movements in cuttlefish and squid when they are about to strike prey.

References

Budelmann BU (2009) Active marine predators: the sensory world of cephalopods. Mar Fresh Behav Physiol 27:59–75

Collett TS, Land MF (1975) Visual control of flight behaviour in the *hoverfly Syritta pipiens* L. J Comp Physiol 99:1–66

Collewijn H (1970) Oculomotor reactions of cuttlefish, *Sepia officinalis*. J Exp Biol 52:369–384

Collewijn H (1977) Eye and head movements in freely moving rabbits. J Physiol 266:471–498

Easter SS, Johns PR, Heckenlively D (1974) Horizontal compensatory eye movements in goldfish (*Carassius auratus*). J Comp Physiol 92:23–35

Friedburg C, Allen CP, Mason PJ, Lamb TD (2004) Contribution of cone receptors and post-receptoral mechanisms to the human photopic electroretinogram. J Physiol 556:819–843

Frost BJ (1978) The optokinetic basis of head-bobbing in pigeons. J Exp Biol 74:187–195

Guitton D (1992) Control of eye-head coordination during orienting gaze shifts. Trends Neurosci 15:174–179

Hanlon RT, Messenger JB (1996) Cephalopod behaviour. Cambridge University Press, Cambridge

Land MF (2015) Eye movements of vertebrates and their relation to eye form and function. J Comp Physiol A 201:195–214

Land MF, Nilsson D-E (2012) Animal eyes. Oxford University Press, Oxford

Land MF, Tatler BW (2009) Looking and acting: vision and eye movements in natural behaviour. Oxford University Press, Oxford

Necker R (2007) Head-bobbing of walking birds. J Comp Physiol A 193:1177–1183

Ott M (2001) Chameleons have independent eye movements but synchronise both eyes during saccadic prey tracking. Exp Brain Res 139:173–179

Paul H, Nalbach H-O, Varjú D (1990) Eye movements of the rock crab *Pachygrapsus marmoratus* walking along straight and curved paths. J Exp Biol 154:81–97

Pettigrew JD, Collin SP, Ott M (1999) Convergence of specialized behaviour, eye movements and visual optics in the sandlance (Teleostei) and the chameleon (Reptilia). Curr Biol 9:421–424

Rossel S (1980) Foveal fixation and tracking in the praying mantis. J Comp Physiol A 139:307–331

Schilstra C, van Hateren JH (1998) Stabilizing gaze in flying blowflies. Nature 395:654

Walls GL (1962) The evolutionary history of eye movements. Vision Res 2:69–80

Yorzinski JL, Patricelli GL, Platt ML, Land MF (2015) Eye and head movements shape gaze shifts in Indian peafowl. J Exp Biol 218:3771–3776

Visuospatial Integration and Hand-Tool Interaction in Cognitive Archaeology

Emiliano Bruner, Annapaola Fedato, María Silva-Gago, Rodrigo Alonso-Alcalde, Marcos Terradillos-Bernal, María Ángeles Fernández-Durantes, and Elena Martín-Guerra

Contents

Abstract Testing cognitive hypotheses in extinct species can be challenging, but it can be done through the integration of independent sources of information (e.g., anatomy, archaeology, neurobiology, psychology), and validated with quantitative and experimental approaches. The parietal cortex has undergone changes and specializations in humans, probably in regions involved in visuospatial integration. Visual imagery and hand-eye coordination are crucial for a species with a remarkable technological and symbolic capacity. Hand-tool relationships are not only a

E. Bruner (✉), A. Fedato, and M. Silva-Gago
Centro Nacional de Investigación sobre la Evolución Humana, Burgos, Spain
e-mail: emiliano.bruner@cenieh.es; annapaola.fedato@cenieh.es; maria.silva@cenieh.es

R. Alonso-Alcalde
Museo de la Evolución Humana, Burgos, Spain
e-mail: ralonso@museoevolucionhumana.com

M. Terradillos-Bernal
Universidad Internacional Isabel I de Castilla, Burgos, Spain
e-mail: marcos.terradillos@ui1.es

M. Á. Fernández-Durantes and E. Martín-Guerra
Sociograph Marketing Science Consulting, Valladolid, Spain
e-mail: mafernandez@sociograph.es; emartin@sociograph.es

Curr Topics Behav Neurosci (2019) 41: 13–36
DOI 10.1007/7854_2018_71
Published Online: 15 December 2018

matter of spatial planning but involve deeper cognitive levels that concern body cognition, self-awareness, and the ability to integrate tools into body schemes, extending the body's functional and structural range. Therefore, a co-evolution between body and technology is to be expected not only in terms of anatomical correspondence but also in terms of cognitive integration. In prehistory, lithic tools are crucial in the interpretation of the cognitive abilities of extinct human species. The shape of tools and the grasping patterns associated with the corresponding haptic experience can supply some basic quantitative approaches to evaluate changes in the archaeological record. At the physiological level, electrodermal activity can be used as proxy to investigate the cognitive response during haptic experiences, revealing differences between tools and between subjects. These approaches can be also useful to evaluate whether and to what extent our complex cognitive resources are based on the capacity to export and delegate functions to external technological components.

Keywords Electrodermal activity · Grasping pattern · Human evolution · Neuroarchaeology · Parietal lobes · Tool shape · Visuospatial integration

1 Prehistory and Neuroscience

Cognitive inferences in prehistoric archaeology have often been provided on the grounds of general terms and processes, rather than specific cognitive theories. The main framework has been a gross and generalized assumption that relies on anatomical and cultural complexity as a proxy for behavioral and cognitive complexity. That is, complex brains are supposed to generate complex behaviors, and complex behaviors are supposed to be necessary to produce complex tools. In the last decade, however, there has been an increasing exchange between anthropologists, archaeologists, neurobiologists, and cognitive scientists, and these research areas have undergone a stimulating multidisciplinary development. Thanks to technical improvements (from digital anatomy to numerical modeling) and the increase in the archaeological record, prehistoric and cognitive sciences have stepped into a more intense and reciprocal process of integration. Some fields have been enhanced, and some others have been introduced as brand-new methodological perspectives. *Paleoneurology* deals with the anatomical study of the endocranial cavity in fossil species and has been improved by the introduction and development of biomedical imaging (Bruner 2017). *Neuroarchaeology* concerns the study of prehistory-related behaviors through physiological and neurobiological approaches, such as functional imaging (Stout and Hecht 2015). *Cognitive archaeology* integrates the archaeological evidence with theories in cognitive science, through neuropsychological perspectives (Coolidge et al. 2015).

Despite the noticeable advantage of mixing archaeological and cognitive knowledge, the limitations are also clear: prehistoric studies are based on indirect traces of structures or processes, and not on the actual targets of interest. In terms of fossil anatomy, paleoanthropology generally works with the fragmented bones of few individuals. Instead of a brain, there is a mold of the endocranial cavity or *endocast*. An endocranial cast can provide information on brain size, some gross cortical proportions, brain geometry and spatial organization, sulcal patterns, and meningeal vascular morphology. All this information is extremely valuable, because it is the only direct evidence we have on the brain anatomy of extinct species. Nonetheless, an endocast is not a brain and should be interpreted with this limitation in mind.

In the case of neuroarchaeology and cognitive archaeology, a main drawback is due to the fact that cognitive processes are investigated and simulated according to the information we have on modern humans (*Homo sapiens*), and not on extinct species. This is of course an intrinsic limitation of these fields. Nonetheless, we often use other species as models when investigating our own biology and evolution (mice, macaques, or chimpanzees), and the differences among species of the same genus (*Homo*) are supposed to be plausibly smaller. The fact that we use cognitive information on modern humans to make cognitive inference on extinct humans must be taken into account, but it should not be taken as a reason to reject the field as a whole. The aim of disciplines that integrate prehistory and neuroscience is to provide consistent hypotheses according to the available information, which can be tested against parallel and independent evidence. Testing hypotheses may be more difficult in extinct species than in living organisms, but the methods and rules are, after all, exactly the same as in any other scientific context.

2 Working Memory and Visuospatial Integration

Early steps in cognitive archaeology were particularly focused on working memory, following the model proposed by Baddeley (see Baddeley 2000, 2001), attempting to trace its components back to archaeological evidence (Coolidge and Wynn 2005; Wynn and Coolidge 2016). Frederick Coolidge and Thomas Wynn, integrating archaeology and neuropsychology, investigated the appearance of behaviors associated with a central executive system, a visuospatial sketchpad, and a phonological loop, in order to evaluate whether our species, *Homo sapiens*, could have enhanced its working memory capacity through a process of selection and adaptation. They suggested, for example, that, according to the technological evidence, Neanderthals' long-term working memory was similar to modern humans, while their working memory capacity was less developed, possibly because of a smaller phonological store or reduced attention levels (Wynn and Coolidge 2004). This conclusion, based on archaeological information, can be used for making behavioral predictions than can be contrasted against the ecological, cultural, and social evidence we have on Neanderthals. Following a similar principle, they also investigated specific behaviors like those associated with managed foraging, as a proxy for cognitive capacities linked to working memory, response inhibition, or space-time integration (Wynn and Coolidge 2003).

The Baddeley model (Baddeley and Hitch 1974) is assumed to rely mainly on a frontoparietal cortical network, and, according to the principles of cognitive archaeology, its functional units can be tentatively tracked back in the cultural remnants of human behaviors, looking for specific aspects associated, for example, with tool use and production, food storage, navigation, art, or social and economic dynamics. The executive system works through inhibition of emotional and spontaneous behaviors, which is probably a crucial hallmark of modernity. The phonological store influences speech and cognitive capacity associated with recursion and hierarchical cognitive organization. The visuospatial sketchpad deals with an egocentric perspective based on imagery (visual) and relational (spatial) capacity.

This last component was relatively neglected in many working memory analyses, but nonetheless it could have been crucial in human evolution. In fact, if we consider the paleoneurological evidence, a major morphological change along the human lineage has been precisely described for the dorsal parietal cortex (Fig. 1), a brain region which is crucial to visuospatial functions (Bruner 2018). Neanderthals display wider superior parietal lobules when compared with more archaic human species, and modern humans show an even larger parietal lobe expansion, which causes a bulging of the parietal profile and their classic "rounded head" (Bruner et al. 2003, 2011; Bruner 2004). Ontogenetic changes suggest that only modern humans have a specific morphogenetic stage of "brain globularization," expressed very early during ontogeny, which is lacking in Neanderthals or chimpanzees (Gunz et al. 2010).

It is interesting, therefore, that the two human species with more complex technological levels display a cortical expansion of areas dedicated to brain-body-environment management and integration, especially when considering that for the human genus "environment" also means "tools." Spatially, the lateral dorsal enlargement of Neanderthals can be tentatively associated with the intraparietal sulcus and superior parietal lobules, while the longitudinal enlargement in modern humans matches the position of superior parietal lobules and precuneus (Bruner 2010; Bruner et al. 2014a; Pereira-Pedro and Bruner 2016). The intraparietal sulcus is more complex in humans than in other primates, and it is largely involved in eye-hand coordination and tool use (Grefkes and Fink 2005; Choi et al. 2006; Tunik et al. 2007; Martin et al. 2011; Verhagen et al. 2012; Zlatkina and Petrides 2014; Kastner et al. 2017). Human specializations of this region are supposed to be directly associated with the evolution of our unique technological skills (Peeters et al. 2009; Goldring and Krubitzer 2017). The precuneus is extremely variable among adult humans, and it is much larger in humans than in chimpanzees (Bruner et al. 2014b, 2017a). It is considered crucial for processes based on integration between somatic (body) and visual cognition, like spatial coordination, visual imagery, mental simulation, autonoesis, and egocentric memory (Fletcher et al. 1995; Cavanna and Trimble 2006; Margulies et al. 2009; Zhang and Li 2012; Freton et al. 2014; Land 2014). The precuneus can be seen as a bridge between the external environment (vision), body cognition, and self-perception, with imagery and inner levels of consciousness. The correspondences

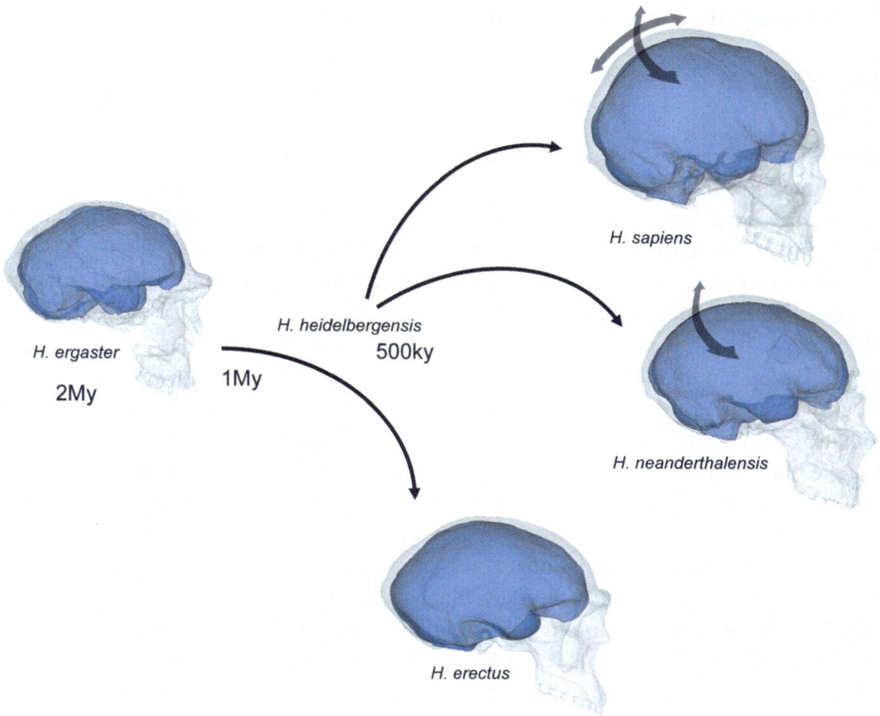

Fig. 1 The earliest fossils of the human genus (*Homo*) are dated to around 2 million years (My). In terms of endocranial morphology, the difference between these early African representatives and later *H. erectus* is apparently a matter of brain size, larger in the latter species. *H. sapiens* and *H. neanderthalensis* also evolved a larger brain size, but in these two cases, there was also evidence for changes in cortical proportions. Some of these changes are related to parietal cortex. Neanderthals display wider parietal lobes, and modern humans have wider and longer parietal lobes. Both lineages probably diverged after 500,000 years (ky) and derived from *H. heidelbergensis* which, as with *H. erectus*, had larger brain size then *H. ergaster* but no noticeable changes of the brain form

between humans and nonhuman areas are not completely clear, although some areas on the primate intraparietal sulcus may have outfolded in humans and the superior parietal lobule might be largely an outer extension of the precuneus (Scheperjans et al. 2008). These cortical areas are also very sensitive to environmental influences, including training and sensorial/somatic stimulation (Quallo et al. 2009; Iriki and Taoka 2012). Furthermore, they are all naturally crucial to specific conceptual and technical skills which range from imagination to tool use. Together, they have all the features of a very powerful visuospatial sketchpad.

3 Visuospatial Integration and Human Evolution

3.1 Body and Space

Most mammals possess homologous areas to the posterior parietal cortex, associated with functions that involve aspects of the body management, but this region is particularly developed in primates, and in particular in humans, due to manipulation skills and technological capacity (Goldring and Krubitzer 2017). The posterior parietal cortex is also crucial to processes aimed at filtering the sensorial information to coordinate attentional and intentional mechanisms (e.g., Posner et al. 1984; Mountcastle 1995; Rushworth et al. 2001; Yantis et al. 2002; Andersen and Buneo 2002; Bisley and Goldberg 2003; Corbetta et al. 2005; Wardak et al. 2004; Freedman and Assad 2006). Such filters are based on experience, as well as on somatic and visual feedbacks. Vision is used to coordinate body and environment, and the body is used as a metric unit of such an environment, in terms of space, time, and even social perspective (Land 2014; Hills et al. 2015; Maister et al. 2015; Peer et al. 2015). Our body perception is largely based on the hands, and the same areas involved in eye-hand coordination are also recruited in decision-making (Tunik et al. 2007). Namely, we can probably say that we often "think with our body," particularly with our hands, planning and simulating actions by using our own body as reference and taking decisions according to simulated or expected body experiences, feedbacks, and capacities. In a behavioral and even neurobiological perspective, the somatosensorial experience is therefore intermingled with the motor experience, generating a blurred separation between "feeling" and "acting" (Ackerley and Kavounoudias 2015).

This framework between body and action becomes further entangled when the body interfaces with technological extensions, namely, during tool use (Bruner and Iriki 2016). Tools are intended as extension of the body schemes, through a functional distinction between *personal space* (the body), *peri-personal space* (within the range of the body), and *extra-personal space* (out of the range of the body) (Maravita et al. 2003; Maravita and Iriki 2004; Farnè et al. 2005; Cléry et al. 2015). The relationship between personal, peri-personal, and extra-personal spaces is particularly relevant when dealing with our evolutionary capacity to extend our body and cognitive functions into technology. The frontoparietal system, in fact, reacts differently to objects positioned in the three spaces, which map to different cortical areas as a function of distance from the body (see Cléry et al. 2015 for a detailed review). A crucial cognitive change takes place when an object is included into the peri-personal space, becoming a potential tool, reachable in terms of body contact and extension. The own body is the metric unit that defines the peri-personal range, and vision supplies the feedback to establish its frontiers, mainly centered on the position of the whole body, of the head, and of the hands. This peri-personal space is updated according to both dynamic changes (momentary and punctual variations) and plastic changes (neural changes after training), and tools have a special role in this sense, artificially altering the extension and capacity of the arms.

Even in simple physical terms, the contact between the body and the tool influences the muscular and sensorial perception of the body itself and, accordingly, all the cognitive mechanisms that use the body as a functional and structural reference (Turvey and Carello 2011). Therefore, visuospatial integration not only concerns gross spatial and mechanical adjustments, but it is also central to fine cognitive functions that deal with self-properties.

In sum, the posterior parietal cortex is involved in cognitive integration between the brain, body, and environment, between body and tools, and between vision and body, using the same resources to coordinate space and time, egocentric perspectives, imagery, and personal memories. This is particularly interesting in the light of the so-called *extended cognition theory*, which interprets cognition as a process generated by the interaction between the nervous system, body experience, and material culture (e.g., Malafouris 2010, 2013; Bruner et al. 2018a).

3.2 A Case Study in Cognitive Archaeology

Neanderthals represent an interesting case study in body cognition because, although they had a brain size comparable with modern humans, the archaeological and paleontological evidence point to distinct visuospatial behavior (see Bruner and Lozano 2014, 2015; Bruner et al. 2016). In particular, the cut marks on their incisors suggest that Neanderthals – and probably their ancestors – used their teeth and mouth to manipulate their technology much more than any extant or extinct modern human population. The mouth is second to the hands in terms of cortical representation of the somatic territories (the "cortical homunculus"), so it is expected that it can be used to provide an additional manipulative body element when hands do not suffice. However, its involvement in manipulation is indeed a risky choice and should be intended as a suboptimal alternative. The significant involvement of the mouth as a "third hand" in Neanderthals may hence suggest a lack of manipulative specialization, when technology reaches a given degree of complexity. These dental marks would not be sufficient to support such cognitive hypothesis, unless associated with many other independent sources of evidence. In Neanderthals the parietal cortex, crucial for visuospatial integration, was probably not enlarged as in modern humans (Bruner 2018). For this species there is no evidence of projectile tools, a technology which is specific of modern humans and associated with throwing ability and visuospatial capacity (Williams et al. 2014; Gärdenfors and Lombard 2018), and Neanderthal hunting techniques were probably based on physical confrontation with the prey (another risky choice, if you are able to catch a prey by shooting from a distance). Also, for Neanderthals there is no evidence of a noticeable iconographic or visual culture. Their few minor suspected graphic manifestations are extremely simple (Hoffmann et al. 2018) and definitely incomparable with both the early and late artistic expressions of *Homo sapiens*. Paradoxically, many people are surprised to see that Neanderthals could have been the authors of very naïve sketches, but in reality we should ask the opposite question: taking into account their large brain size and high encephalization index, why did they not display more complex behaviors? If brain size really matters, with such a large brain (the same size as *Homo sapiens*), they would be expected to go well beyond a scratch or a colored shell. Although the Neanderthal archaeological record may be incomplete, the discrepancy with modern humans is, even only by grade, enormous, suggesting noticeable cognitive differences between the two groups (Wynn et al. 2016). Taken together, all this information (smaller parietal cortex, manipulation by teeth, no projectile technology, absent or negligible graphic culture) supports the hypothesis of a lack of visuospatial specialization and body cognition in Neanderthals when compared with modern humans. Of course, a less specialized cognitive ability is not necessarily a sentence to extinction, and we should not even discard the possibility that Neanderthals may have had other cognitive skills that we did not evolve.

Despite subtle uncertainties in chronology and definitions, it is worth noting that the morphological expansion of the parietal cortex in our species is probably a late acquisition of our lineage, and it matches the appearance of a definite behavioral modernity, including a noticeable visual and iconographic culture and complex technology. In fact, early modern humans shared similar lithic industries with Neanderthals and display only a partial development of the parietal surface (Bruner and Pearson 2013; Bruner et al. 2018b; Neubauer et al. 2018). However, they already had different hand proportions when compared with coeval Neanderthals, and a distinct use of the mouth when supporting manipulative procedures, more associated with the strength of the grip than with its precision (Niewoehner 2001; Fiorenza and Kullmer 2013).

It remains to be considered whether the neuroanatomical changes of the posterior parietal cortex are due to genetic evolution and selection or else to feedback between biology and culture, including training or epigenetic effects (Bruner and Iriki 2016; Krubitzer and Stolzenberg 2014). According to the traditional parcellation approach after Brodmann (see Zilles and Amunts 2010), it can be hypothesized that specific areas evolved, enlarged, or were reused for new emerging functions. By contrast, if brain organization is the result of gradients between sensorimotor regions (Huntenburg et al. 2017), the specialized posterior parietal cortex in primates – and in particular in humans – must be interpreted as an increase of connections and integration between the sensorimotor elements it bridges: body and vision.

4 Haptics and Body Cognition

Human evolution has been characterized by bio-cultural adaptive feedbacks between hand and tool morphology (Susman 1998; Marzke 1997; Almécija et al. 2015). Force distribution during tool use largely depends on the action performed, and it is likely that some behaviors may have had a major influence on hand shape, mostly when dealing with the thumb and the distal phalanxes (Rolian et al. 2011; Williams-Hatala et al. 2018). Hand size is also relevant for tool use, and it is a major factor of variation also among modern adult humans (Key and Lycett 2011; Bruner et al. 2018c).

We can expect that this coevolution between the brain, body, and technology was not only a matter of biomechanics but involved specific cognitive functions associated with hand-tool integration. In general, most studies in this sense are interested in those cognitive abilities that concern planning, decision-making, and the executive functions of the brains. Nonetheless, additionally, we should also consider whether the hand-tool relationship may also require some cognitive process that enhances the integration of the tools into the body schemes.

Visuospatial functions are indeed necessary when planning tools or tool use (the visual imagery functions associated with the precuneus and the intraparietal sulcus). However, beyond these aspects, taking into consideration the importance of the neural management of the personal, peri-personal, and extra-personal spaces, it should be expected that the capacity to integrate tools as body extensions (e.g., in terms of neural plasticity) could be a crucial target of adaptive processes. Although modern humans (*Homo sapiens*) evolved a very specialized tool-based functional extension, the whole human genus (*Homo*) is characterized by a culture and behavior which make us *dependent* on technology (Plummer 2004). In the last 2.5 million years, our ecological, economic, and cultural niches have depended on tools, as essential elements of our behavioral abilities (Key et al. 2016). Such "prosthetic capacity" (Overmann 2015) can therefore not only be an important part of our cognitive system but also a specific ability influenced by natural selection. Interestingly, functional specialization within human brain areas has been shown to be less constrained by genetic factors compared with other living apes and so may be more plastic and sensitive to external influence (Gómez Robles et al. 2015). Such capacity to export cognitive functions to technological (extra-neural) extensions would depend on neural mechanisms, on body experience, and also on the properties of the tools themselves. It is hence mandatory, in cognitive archaeology, to investigate all these three elements, as well as their interactions.

This target is not easy, because of the many factors involved (individual cognitive and sensorial differences, multiple cognitive tasks involved, physical and functional tool parameters and variables, etc.). At experimental levels, simplistic paradigms can be easier to analyze, but scarcely informative. Moreover, many processes involved in behavior and cognition follow complex networks in which the final mechanism is not the simple sum of its parts, and there are emergent properties that can be observed only when analyzing the system as a whole. Actually, the network underlying cognitive extension is supposed to be complex itself, in the sense that, according to the extended cognition hypothesis, there are processes that are activated specifically by the interaction between the brain, body, and culture. Finally, there are major difficulties when trying to quantify specific behavioral resources emerging from body-tool integration because, at present, we still do not know what kind of ability is directly involved in such prosthetic capacity and how to measure it. All these limitations mean that this research area is still in a preliminary methodological stage, in which distinct targets and techniques are investigated so as to evaluate their applications and potentialities. Quantitative methods are, of course, necessary, to step into a full experimental perspective. Three basic components behind body-tool interactions are tool shape, grasping patterns, and cognitive response to hand-tool integration. In the next sections, we show some applications in this sense.

5 Tools, Hands, and Attention: A Synthetic Analysis

5.1 Shape and Technology

Tool physical and geometrical properties influence the interaction with our body by virtue of both visual and haptic information. Beyond affordances associated with possible functional employment of the tool (purpose), the haptic experience is essential to generate an ergonomic spatial and physical integration between body and tool (Turvey and Carello 2011), which is ultimately projected into the newly emerging body schemes. The term "cyborg" has been popularized in a context of science fiction, but technically it refers to any functional integration between a body and a technological element, and humans are a special evolutionary case study of prosthetic extension (Clark 2004). Such "hybrid bodies," in which external components come in contact with the body generating new emerging functions, can be traced back at least 2.5 million years to Africa, when we found the most ancient human technology, the Oldowan. Actually, there is preliminary evidence of older tools (see, e.g., Harmand et al. 2015), but Oldowan is the first technology for which we have a robust and consistent archaeological record (Semaw et al. 1997; Braun et al. 2008; Stout et al. 2010). It was essentially composed by flaked stones with a cutting edge, like the typical "choppers." The raw materials were collected, prepared, and used locally, probably for a quick and momentary utilization. This technology was initially associated with *Homo habilis*, although at present this species is not regarded as a real and homogeneous taxonomic unit, and the hominid (or hominids) associated with this industry remains yet to be determined.

The earliest species undoubtedly assigned to the human genus (*H. ergaster* and *H. erectus*) are associated with stone tools which are much more elaborate and generally labeled as Acheulean technology (Lycett and von Cramon-Taubadel 2008; Hodgson 2015). Acheulean archaeological record begins after 1.7 million years, and the most typical tool in this case is the handaxe, a stone flaked for a larger part of its border, elongated and roughly symmetrical, probably used for multiple tasks. Some features of this industry can be due to stone geometrical constraints (Moore and Perston 2016), but nonetheless it is generally assumed that the complexity of handaxes, when compared with choppers, reveals a cognitive change, because of their design, preparation, and geometry (Wynn 2010; Gowlett 2013). Although Oldowan is more archaic and simpler, it was not substituted by Acheulean, and the two different technologies coexisted independently for at least 600,000 years (Clark and Schick 2000).

Because of the differences between their design and structure, Oldowan and Acheulean technological modes could be particularly interesting when evaluating changes of the human prosthetic capacity to integrate tools as body extensions. It is also worth noting that the same visuospatial processes involved in body cognition and visual imagery are involved as well in integrating self-perception and social context (Coward and Gamble 2008; Hills et al. 2015; Maister et al. 2015; Peer et al. 2015) and that the volume of the cortex in primates is correlated to social group size (Dunbar 1998; Dunbar and Shultz 2007). Tool making and tool use necessarily rely on social learning and imitation (Mesoudi and O'Brien 2008; Schillinger et al. 2015; Gärdenfors and Högberg 2017; Lombao et al. 2017), and technological complexity represents therefore an additional bridge between visuospatial processes and social behavior.

Shape analysis based on coordinates is nowadays a standard in morphometrics, and it can provide a quantitative background to investigate geometric variation and its underlying schemes (e.g., Mitteroecker and Gunz 2009; Adams et al. 2013). In Fig. 2, we have analyzed the shape of some representative lithic tools (seven Oldowan choppers and six Acheulean handaxes obtained through experimental knapping) through landmark analysis of their outline. Outlines were sampled with 30 equally spaced landmarks in two dimensions, after photographs. We also included two natural (not knapped) stones of similar size and of similar composition. Coordinates were superimposed by Procrustes registration to minimize shape differences and then analyzed through Principal Component Analysis (see Zelditch et al. 2004). In this sample there is only one consistent component of shape variation (explaining 76% of the variance) that separates narrower tools (mostly Acheulean) from wider tools (mostly Oldowan). Narrowing/widening appears to be specially

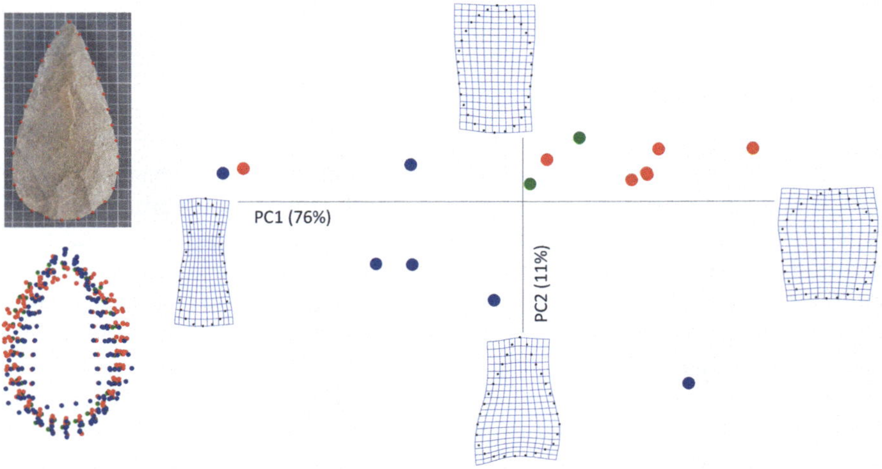

Fig. 2 Shape analysis of the outline of natural stones (green), Oldowan choppers (red), and Acheulean handaxes (blue) after Procrustes superimposition and Principal Component Analysis

pronounced at 2/3 of the tool length, that is, before the tip. Although this shape component largely distinguishes Oldowan and Acheulean tools, there are some exceptions. Natural stones are largely similar to choppers. A second component (11%) separates tools with wide bases and narrow tips (mostly Acheulean) from tools with narrow bases and wider tips (Oldowan, natural stones, and some handaxes). Following principal components explain less that 5%. Tool shape (PC1, PC2, or the overall shape coordinates) is not correlated to tool size (computed as centroid size, namely, the sum of squared distances of all the points from the centroid). Therefore, according to this descriptive tool sample, the main and dominant tool variation deals with elongation (narrow vs wide), while a second minor pattern concerns the inverse proportions between base and tip. When compared with choppers, handaxes are characterized by elongation, wider base and sharper tip. There is only a very minor overlap between these two technological typologies, and apparently their geometric variation does not depend on size. Elongation is a main feature of Acheulean tools, and it is generally interpreted as a proxy of cognitive complexity because it denotes technical skills in knapping procedures (Gowlett 2013). Nonetheless, elongation also influences affordance, extension of the peripersonal space, and dynamic body responses, triggering crucial haptic processes (Turvey and Carello 2011). Accordingly, elongation is expected to have a specific role in inducing alterations of body schema, sensu Maravita and Iriki (2004). This specific pattern of narrowing is really determinant in the variation of these two archaic stone tool typologies, and therefore we should evaluate whether and to what extent it is associated with specific haptic changes, and eventually with relevant body-tool cognitive feedbacks. The same can be proposed for the second shape component (tip-base inverse proportions), although in this case the pattern explains only 10% of the difference and a larger sample is probably necessary to investigate further its reliability.

5.2 *Grasp and Tools*

We asked 46 adults (24 males and 22 females), blindfolded, to handle the same sample of tools until they found a comfortable hand-tool position. Participants were right-handed and not trained in archaeology. They were not allowed to see the tools before the survey. For each tool, when participants reached a stable hand-tool position, the final grasping modality was classified according to Feix et al. (2009). Grasping categories include two power grips (types A and B, with the thumb opposed or else aligned to the other finger) and three precision grips (types C, D, and E, with the index finger on the edge, pinching, or surrounding the tool, respectively). Type C is labeled as precision grip because the thumb approaches the fingertips, but actually it is a sort of intermediate position between power an precision grip. Figure 3 shows the different grasping types and the frequency of each category (A–E) for each tool type (natural stones, Oldowan, Acheulean). All tools were spontaneously grasped primarily with a power grip. Natural stones are predominantly grasped through opponent thumb. Oldowan and Acheulean tools

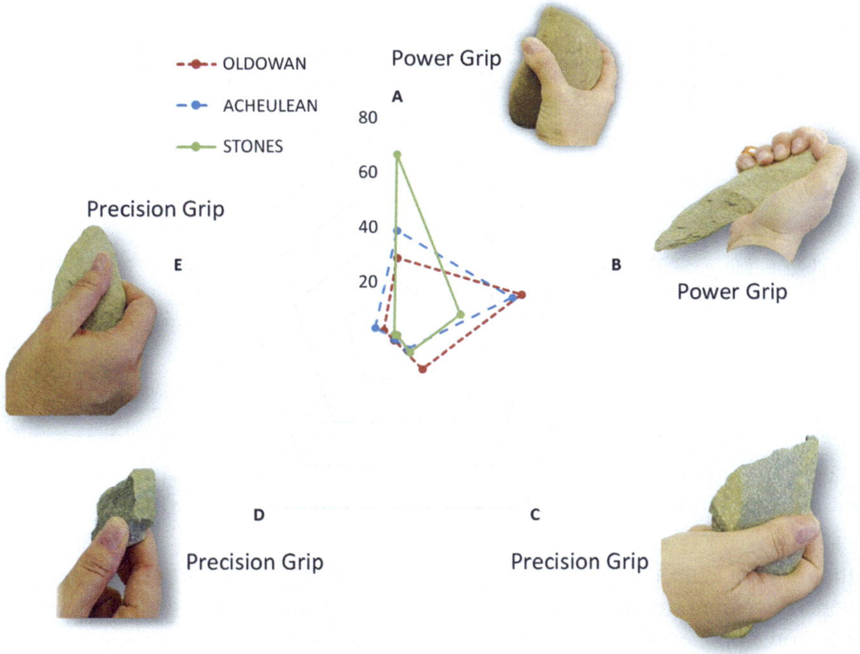

Fig. 3 Percentage of the five grasp types in natural stones, Oldowan choppers, and Acheulean handaxes. Subjects were blindfolded and, without previous archaeological knowledge, were asked to grasp each tools in a way that felt comfortable

display a similar grasping pattern, different from that of natural stones, although the thumb is slightly more lateral (type C) for the choppers and opponent (type A) for the handaxes. In type C grip, the thumb is closer to the fingertips if compared with classic power grips but is also less involved in holding the tool. In this aspect, it is more similar to a power grip that to a real precision grip. Actually, the involvement of the thumb is less relevant in Oldowan than in later technology (Williams et al. 2012), and this can explain the minor difference we detect in this survey between the two tool typologies.

Figure 4 shows a cluster analysis displaying the similarity between tools according to grasp frequencies, basic tool size (maximum length, width, and thickness), and shape (Procrustes coordinates). Similarity is based on Euclidean distances (raw metric differences), and clustering is computed according to an UPGMA (Unweighted Pair Group Method with Arithmetic Mean) criterion, which pairs the most similar figures and then clusters the value most similar to their average. This quantitative comparison suggests that geometry is more efficient in separating choppers from handaxes than their general dimensions or than the grasping modality with which they are handled. Tool affinity based on grasping frequencies looks less determined by tool typology, suggesting that there are factors other than the gross

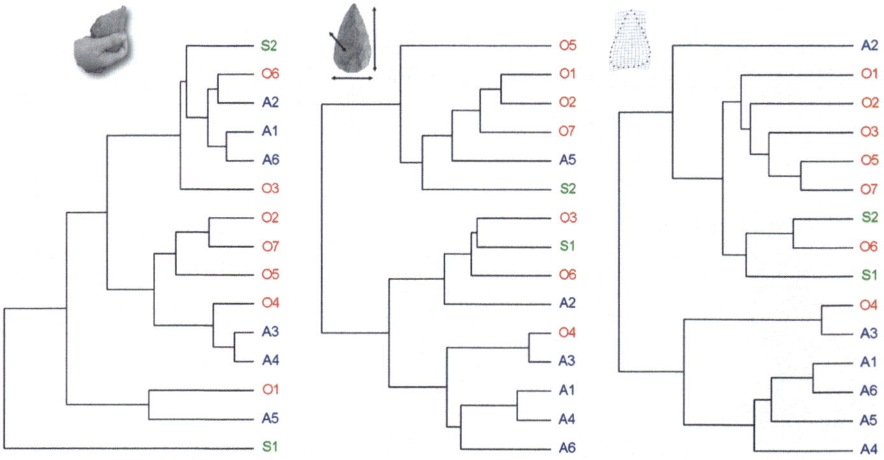

Fig. 4 Cluster analysis based on grasp frequency (left), tool main diameters (center), and tool shape (right), for natural stones (S, green), Oldowan choppers (O, red), and Acheulean handaxes (A, blue)

outline of the object influencing the choice of the hand position. For example, choppers are more similar to natural stones in shape, but the latter involve apparently much more opponent thumb. Acheulean tools are more different from Oldowan tools in shape, but showing a more similar grasping pattern, although with larger frequency of opponent thumb as well. Thus, spontaneous grasping of a tool seems only partially determined by the different degree of technological complexity of these two kinds of industry. This result is partially expected, because shape is actually a major factor when classifying tool typology, and we use to classify stone tools by their shape more than by their haptic properties. But this result also suggests that tool shape and grasping, although sharing reciprocal influences, involve independent additional aspects. Tools are generally categorized according to their form, to the underlying knapping procedure, and to their supposed functions (e.g., Akoshima and Kanomata 2015; Key and Lycett 2017; Ollé et al. 2017). We may wonder whether features associated with grasping and handling should also be taken into account in typological classifications of tools, because of the functional relevance (ergonomics) but also because of possible cognitive involvement (dynamic touch, affordance, prosthetic capacity, etc.).

5.3 *Haptic Stimulation and Attention*

Neural activity during stone-tool manipulation can be visualized through neuroimaging, namely, investigating structural and functional cortical changes associated with archaeologically relevant behaviors (Stout and Chaminade 2007). From earlier to later lithic technologies, we can observe an increase of the hierarchical complexity of the knapping procedures, which is interpreted as a proxy for cognitive complexity (Muller et al. 2017). Functional imaging revealed that Oldowan toolmaking largely relies on parietal cortex activation, while Acheulean toolmaking also implies prefrontal activation (Stout et al. 2015). This suggests a larger cognitive demand for the latter industry, which likely involves a central executive system aimed at evaluating strategies and alternatives to deal with bifacial shaping. Therefore, it looks like Acheulean toolmaking required/induced a relevant response of the frontoparietal network and of the associated working memory processes. At the same time, Acheulean tools also suggest enhanced communication capacity (Gärdenfors and Högberg 2017).

It is important however to note that toolmaking and tool handling do not necessarily rely on the same cognitive capacities and perceptual resources (Goodale et al. 1994). Toolmaking has an important problem-solving component, while haptic feedback and functional extension are more centered in sensorial mechanisms. Similarly, specific abilities associated with dexterity (e.g., the precision of the grip) are not inevitably associated with the capacity of brain-artefact functional integration. Dexterity is largely a mechanical issue, while prosthetic competence involves more crucial cognitive aspects. We can assume that these two characteristics have important reciprocal influences, but not necessarily a stringent correspondence. Namely, body extension, embodiment, and technological integration should not be confounded with specific spatial skills of the osteo-muscular mechanical system.

A quick and practical method to analyze a basic cognitive response to hand-tool interaction is electrodermal activity (Critchley 2002; de Houwer and Hermans 2010). Changes of impedance/conductivity, as recorded at the fingertips, are associated with *attention* (electrodermal level, EDL) and *emotional engagement* (electrodermal response, EDR) (Martin and Venables 1966; Boucsein 2012). Electrodermal variations can be detected during stone tool manipulation (Bruner et al. 2018c). We analyzed EDL and EDR in all the 46 participants handling the tools described previously, with a portable remote electrodermal device SOCIOGRAPH (Fig. 5). The absolute EDR value was used in this comparison. EDL, in contrast, is a variable that depends on individual characteristics. Therefore, we compute EDL %, as the ratio between the peak of maximum attention (corresponding to minimum EDL value) and the average individual value, in order to quantify the maximum shift of attention according to the performance-specific mean figure. Both EDR and EDL show differences for different tools, in terms of mean values and degree of variation (Fig. 5). Therefore, different tools may exert different demands on emotion and attention, during a novel tactile experience. We must assume these different emotional responses depend on the physical features of the tool, mostly size, shape, weight, and texture. In this case, the tools were experimentally prepared, and texture is reasonably homogeneous. Individual variation is large, and this means that, beyond different tool-specific mean values, a tool can exert distinct emotional responses in different persons. The fact that also the degree of variation is not the same for all the tools suggests that some tools can exert more diverse personal responses than others. Both mean value and degree of variation are parameters that

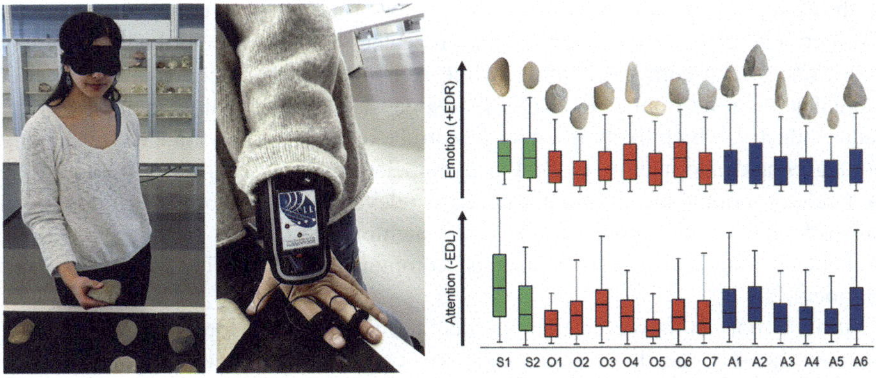

Fig. 5 Subjects were asked to handle the tools (left) while a remote recorder (center) was detecting the variation of electrodermal response and electrodermal level (right: mean, interquartile and range for each tool)

can be investigated through the archeological record to evaluate changes or discontinuity in the patterns of cultural evolution. In this specific case, on average, Oldowan tools induce more emotional response and less attention response than the Acheulean tools, and females display higher attention levels than males (Fig. 6). Both EDL and EDR can be analyzed using also other parameters, like the individual range (difference between minimum and maximum) for each tool, the value at rest (when reaching the comfort position), the value during some specific action, or the whole curve during the entire haptic experience. What is important is that these kinds of interactions can be quantified and used to investigate changes in the cultural record or else specific cognitive hypotheses.

In sum, we have shown some basic methods than can be useful to consider the cognitive interaction between haptic experience and lithic technology. These examples indicate that, when compared with choppers, Acheulean tools are characterized by elongation and further involvement of the thumb, two features that can enhance the cognitive and behavioral extension of the personal and peripersonal space. Taking into account the reciprocal influence between brain and culture, and the possible role of technology as extension of our cognitive system, polarities and causal networks in terms of evolutionary changes are of course difficult to sort out. Generally, it has been assumed that the evolution of a complex brain induces the evolution of a complex culture, but probably this view is too simplistic. Cultural changes can influence behavioral changes, and even channel successive selective pressures (e.g., Crispo 2007; Krubitzer and Stolzenberg 2014). Whatever the causal mechanism behind the transition from Oldowan to Acheulean, we can hypothesize that features like elongation and the involvement of the thumb reveal an increase of the brain-body-tool functional extension and spatial integration. If this is the case, the properties of the Acheulean tools disclose an enhancement of the prosthetic capacity and technical embodiment of the human genus.

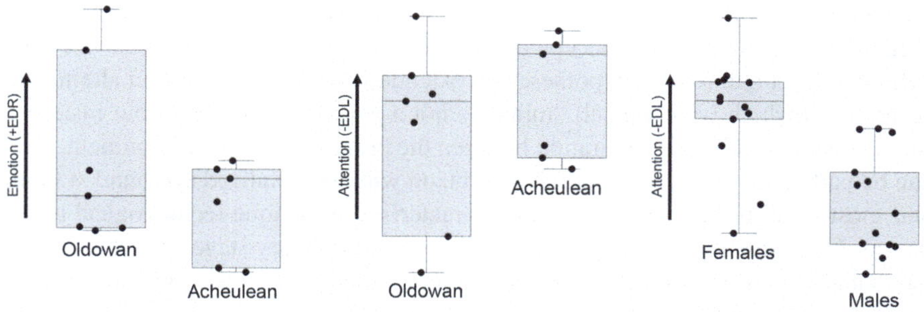

Fig. 6 There are differences in the distribution of mean EDR and EDL for Oldowan and Acheulean tools and for males and females

Although it may be difficult to disentangle causes and effects in such composite system, we can nonetheless use the archaeological record to recognize specific cognitive trends or shifts, according to specific biological signals. We have also seen that different tools can exert different emotional and attention responses, and our Acheulean tools evoked, on average, less emotion but more attention. However, object affordances (both visual and haptic) are based on many distinct factors, and variability (among objects as among subjects) is extremely large. Therefore, these relationships must be investigated through more targeted and specialized samples and, when possible, corroborated by robust statistical validations. Electrodermal activity may represent a useful technique to analyze psychological responses associated with behaviors specifically relevant in the archaeological context and to detect discontinuities in our cultural evolution that may indicate changes in the brain-body-environment relationships. A detailed analysis of these variables and samples is currently in progress.

6 Exploring Body Cognition and Human Evolution

Attention is most often considered in the context of visual selection, but we can also consider attentional processes associated with bodily experience. In terms of technological extension and hand-tool relationships, we can talk of *haptic attention*, when dealing with mechanisms that channel and orient behavioral responses during manipulative experiences. Visual and haptic responses are both crucial to shaping body cognition and visuospatial capacity, and a key question concerns their reciprocal influences. These topics are investigated in psychology, neurobiology, medicine, or even robotics, but they can also be examined by anthropology to provide a comprehensive interpretation of behavior in extinct species (prehistory), in past populations (history) or even between distinct cultures (ethnology).

Although recognizing the importance of environmental influences, a neurocentric paradigm has generally been accepted within which the brain is a computer-like self-sufficient machine. Recent hypotheses on extended cognition suggest an alternative scenario, in which what we call "mind" is not a *product* of the brain but instead a *process* generated by the integration between the brain, body, and environment. This can be particularly relevant in primates, a taxon with a specialized eye-hand system, and most of all in humans, which are characterized by unique technological adaptations. Of course, such a perspective is still in its preliminary stages, in which many terms and definitions are still not clear, interpretations are provisional, and methods are still under construction.

The first step is to evaluate whether technology is a product of our cognitive process, implementing our capacities, or else a constitutive element of the cognitive system. This aim can be hard to achieve, although different scholars are trying to define proper criteria than can be used to validate the hypotheses through quantitative and experimental designs (see Kaplan 2012). For example, the *mutual manipulability criterion* suggests that two entities that alter their reciprocal states pertain to the same system, and the *bandwidth criterion* uses the amount of connectivity to localize relational boundaries between different functional units. All these approaches, bridging biology, philosophy, system theory, and computational neuroscience, although difficult to apply in specific empirical conditions, may supply new perspectives on cognition and behavior and, ultimately, in the interpretation of the archaeological and paleontological records. Of course, we still have some major methodological limitations, because most of our conceptual and experimental tools are based on reductionist perspectives that isolate single elements or single functions, interpreting results through linear causal associations. These cognitive paradigms are relatively recent in our culture, and we probably have to deal with many conceptual and empirical issues before we can adapt our experimental toolkits to different points of view.

An excellent analogy in this sense is the spider's web, in which the threads are an outer extension of the spider's nervous system, produced and shaped by the spider itself (Japyassú and Laland 2017). The web outsources information processing beyond the body of the spider, being a crucial part of the spider's sensorial and cognitive toolkit, essential to its ecological survival. The web, as a peripheral processing element, is involved in spider's memory and attention cognitive machinery and coevolves with the spider's body. Our technology represents an analogous situation, extending our cognitive capacities through external elements which have become necessary to sustain our behavioral abilities as well as our ecological niche. In this sense, the capacity of generate such connection between internal (brain) and external (tools) components is a key feature for selection and adaptation. If the brain, body, and tools are actually part of a single cognitive system, then we have to investigate their respective roles and importance, at both the individual and evolutionary levels.

Acknowledgments We are grateful to Timothy Hodgson for the invitation to participate in this volume and for his detailed comments and suggestions throughout the preparation of this manuscript. We thank the Center for Experimental Archaeology of Atapuerca (CAREX) for its collaboration. Borja Mozo-Grau provided an important technical support during the electrodermal analysis. This study is funded by the Spanish Government (Atapuerca Project; CGL2015-65387-C3-3-P/1-P).

References

Ackerley R, Kavounoudias A (2015) The role of tactile afference in shaping motor behaviour and implications for prosthetic innovation. Neuropsychologia 79:192–205

Adams DC, Rohlf FJ, Slice DE (2013) A field comes of age: geometric morphometrics in the 21st century. Hystrix 24:7

Akoshima K, Kanomata Y (2015) Technological organization and lithic microwear analysis: an alternative methodology. J Anthropol Archaeol 38:17–24

Almécija S, Smaers JB, Jungers WL (2015) The evolution of human and ape hand proportions. Nat Commun 6:7717

Andersen RA, Buneo CA (2002) Intentional maps in posterior parietal cortex. Annu Rev Neurosci 25:189–220

Baddeley AD (2000) The episodic buffer: a new component of working memory? Trends Cogn Sci 4:417–423

Baddeley AD (2001) Is working memory still working? Am Psychol 11:851–864

Baddeley AD, Hitch GJ (1974) Working memory. In: Bower GA (ed) Recent advances in learning and motivation. Academic Press, New York, pp 851–864

Bisley JW, Goldberg ME (2003) Neuronal activity in the lateral intraparietal area and spatial attention. Science 299:81–86

Boucsein W (2012) Electrodermal activity. Springer, Berlin

Braun DR, Tactikos JC, Ferraro JV, Arnow SL, Harris JW (2008) Oldowan reduction sequences: methodological considerations. J Archaeol Sci 3:2153–2163

Bruner E (2004) Geometric morphometrics and paleoneurology: brain shape evolution in the genus *Homo*. J Hum Evol 47:279–303

Bruner E (2010) Morphological differences in the parietal lobes within the human genus: a neurofunctional perspective. Curr Anthropol 51:S77–S88

Bruner E (2017) The fossil evidence of human brain evolution. In: Kaas J (ed) Evolution of nervous systems, vol 4, 2nd edn. Elsevier, Oxford, pp 63–92

Bruner E (2018) Human paleoneurology and the evolution of the parietal cortex. Brain Behav Evol 91:136–147

Bruner E, Iriki A (2016) Extending mind visuospatial integration and the evolution of the parietal lobes in the human genus. Quat Int 405:98–110

Bruner E, Lozano M (2014) Extended mind and visuo-spatial integration: three hands for the Neandertal lineage. J Anthropol Sci 92:273–280

Bruner E, Lozano M (2015) Three hands: one year later. J Anthropol Sci 93:191–195

Bruner E, Pearson O (2013) Neurocranial evolution in modern humans: the case of Jebel Irhoud 1. Anthropol Sci 121:31–41

Bruner E, Manzi G, Arsuaga JL (2003) Encephalization and allometric trajectories in the genus *Homo*: evidence from the Neandertal and modern lineages. Proc Natl Acad Sci U S A 100: 15335–15340

Bruner E, La Cuétara D, Manuel J, Holloway R (2011) A bivariate approach to the variation of the parietal curvature in the genus Homo. Anat Rec 29:1548–1556

Bruner E, de la Cuetara JM, Masters M, Amano H, Ogihara N (2014a) Functional craniology and brain evolution: from paleontology to biomedicine. Front Neuroanat 8:19

Bruner E, Rangel de Lázaro G, Cuétara JM, Martín Loeches M, Colom R, Jacobs HI (2014b) Midsagittal brain variation and MRI shape analysis of the precuneus in adult individuals. J Anat 224:367–376

Bruner E, Lozano M, Lorenzo C (2016) Visuospatial integration and human evolution: the fossil evidence. J Anthropol Sci 94:81–97

Bruner E, Pereira-Pedro AS, Chen X, Rilling JK (2017a) Precuneus proportions and cortical folding: a morphometric evaluation on a racially diverse human sample. Ann Anat 211:120–128

Bruner E, Preuss TM, Chen X, Rilling JK (2017b) Evidence for expansion of the precuneus in human evolution. Brain Struct Funct 222:1053–1060

Bruner E, Spinapolice E, Burke A, Overmann K (2018a) Visuospatial integration: paleoanthropological and archaeological perspectives. In: di Paolo LD, di Vincenzo F, D'Almeida AF (eds) Evolution of primate social cognition. Springer, Cham, pp 299–326

Bruner E, Amano H, Pereira-Pedro AS, Ogihara N (2018b) The evolution of the parietal lobes in the genus *Homo*. In: Bruner E, Ogihara N, Tanabe HC (eds) Digital endocasts. From skulls to brains. Springer, Tokyo, pp 219–237

Bruner E, Fedato A, Silva-Gago M, Alonso-Alcalde R, Terradillos-Bernal M, Fernández-Durantes MA, Martín-Guerra E (2018c) Cognitive archaeology, body cognition and hand-tool interaction. Prog Brain Res 238:325–345

Cavanna AE, Trimble MR (2006) The precuneus: a review of its functional anatomy and behavioural correlates. Brain 129:564–583

Choi HJ, Zilles K, Mohlberg H, Schleicher A, Fink GR, Armstrong E, Amunts K (2006) Cytoarchitectonic identification and probabilistic mapping of two distinct areas within the anterior ventral bank of the human intraparietal sulcus. J Comp Neurol 495:53–69

Clark A (2004) Natural-Born Cyborgs: minds, technologies, and the future of human intelligence. Oxford University Press, Oxford

Clark JD, Schick K (2000) Acheulean archaeology of the eastern Middle Awash. In: Heinzelin J, Clark D, Schick K, Gilbert H (eds) The acheulean and the plio-pleistocene deposits of the Middle Awash valley Ethiopia. Musée Royal de l'Afrique Centrale, Tervuren, pp 51–121

Cléry J, Guipponi O, Wardak C, Hamed SB (2015) Neuronal bases of peripersonal and extrapersonal spaces, their plasticity and their dynamics: knowns and unknowns. Neuropsychologia 70:313–326

Coolidge FL, Wynn T (2005) Working memory its executive functions and the emergence of modern thinking. Camb Archaeol J 15:5–26

Coolidge FL, Wynn T, Overmann KA, Hicks JM (2015) Cognitive archaeology and the cognitive sciences. In: Bruner E (ed) Human paleoneurology. Springer, Cham, pp 177–208

Corbetta M, Kincade MJ, Lewis C, Snyder AZ, Sapir A (2005) Neural basis and recovery of spatial attention deficits in spatial neglect. Nat Neurosci 8:1603

Coward F, Gamble C (2008) Big brains small worlds: material culture and the evolution of the mind. Philos Trans R Soc Lond B 363:1969–1979

Crispo E (2007) The Baldwin effect and genetic assimilation: revisiting two mechanisms of evolutionary change mediated by phenotypic plasticity. Evolution 61:2469–2479

Critchley HD (2002) Electrodermal responses: what happens in the brain. Neuroscientist 8:132–142

de Houwer J, Hermans D (eds) (2010) Cognition and emotion: reviews of current research and theories. Psychology Press, Hove

Dunbar RI (1998) The social brain hypothesis. Brain 9:178–190

Dunbar RI, Shultz S (2007) Evolution in the social brain. Science 317:1344–1347

Farnè A, Iriki A, Làdavas E (2005) Shaping multisensory action–space with tools: evidence from patients with cross-modal extinction. Neuropsychology 43:238–248

Feix T, Pawlik R, Schmiedmayer HB, Romero J, Kragic D (2009) A comprehensive grasp taxonomy. In: Robotics science and systems: workshop on understanding the human hand for advancing robotic manipulation, vol 2, pp 2–3

Fiorenza L, Kullmer O (2013) Dental wear and cultural behavior in Middle Paleolithic humans from the Near East. Am J Phys Anthropol 152:107–117

Fletcher PC, Happe F, Frith U, Baker SC, Dolan RJ, Frackowiak RS, Frith CD (1995) Other minds in the brain: a functional imaging study of "theory of mind" in story comprehension. Cognition 57:109–128

Freedman DJ, Assad JA (2006) Experience-dependent representation of visual categories in parietal cortex. Nature 443:85

Freton M, Lemogne C, Bergouignan L, Delaveau P, Lehéricy S, Fossati P (2014) The eye of the self: precuneus volume and visual perspective during autobiographical memory retrieval. Brain Struct Funct 219:959–968

Gärdenfors P, Högberg A (2017) The archaeology of teaching and the evolution of Homo docens. Curr Anthropol 58:188–208

Gärdenfors P, Lombard M (2018) Causal cognition force dynamics and early hunting technologies. Front Psych 9:87

Goldring AB, Krubitzer LA (2017) Evolution of the parietal cortex in mammals: from manipulation to tool use. In: Kaas J (ed) Evolution of the nervous system, 2nd edn. Elsevier, Amsterdam, pp 259–286

Gómez Robles A, Hopkins WD, Schapiro SJ, Sherwood CC (2015) Relaxed genetic control of cortical organization in human brains compared with chimpanzees. Proc Natl Acad Sci U S A 112:14799–14804

Goodale MA, Meenan JP, Bülthoff HH, Nicolle DA, Murphy KJ, Racicot CI (1994) Separate neural pathways for the visual analysis of object shape in perception and prehension. Curr Biol 4:604–610

Gowlett JAJ (2013) Elongation as a factor in artefacts of humans and other animals: an Acheulean example in comparative context. Philos Trans R Soc Lond B 368:20130114

Grefkes C, Fink GR (2005) The functional organization of the intraparietal sulcus in humans and monkeys. J Anat 207:3–17

Gunz P, Neubauer S, Maureille B, Hublin JJ (2010) Brain development after birth differs between Neanderthals and modern humans. Curr Biol 20:921–922

Harmand S, Lewis JE, Feibel CS, Lepre CJ, Prat S, Lenoble A, Taylor N (2015) 3.3-million-year-old stone tools from Lomekwi 3 West Turkana Kenya. Nature 521:310–315

Hills TT, Todd PM, Lazer D, Redish AD, Couzin ID, Cognitive Search Research Group (2015) Exploration versus exploitation in space mind and society. Trends Cogn Sci 19:46–54

Hodgson D (2015) The symmetry of Acheulean handaxes and cognitive evolution. J Archaeol Sci 2:204–208

Hoffmann DL, Standish CD, García Diez M, Pettitt PB, Milton JA, Zilhão J, Lorblanchet M (2018) U-Th dating of carbonate crusts reveals Neandertal origin of Iberian cave art. Science 359:912–915

Huntenburg JM, Bazin PL, Margulies DS (2017) Large-scale gradients in human cortical organization. Trends Cogn Sci 22:21–31

Iriki A, Taoka M (2012) Triadic (ecological neural cognitive) niche construction: a scenario of human brain evolution extrapolating tool use and language from the control of reaching actions. Philos Trans R Soc Lond B 367:10–23

Japyassú HF, Laland KN (2017) Extended spider cognition. Anim Cogn 20:375–395

Kaplan DM (2012) How to demarcate the boundaries of cognition. Biol Philos 27:545–570

Kastner S, Chen Q, Jeong SK, Mruczek REB (2017) A brief comparative review of primate posterior parietal cortex: a novel hypothesis on the human toolmaker. Neuropsychology 105: 123–134

Key AJM, Lycett SJ (2011) Technology based evolution? A biometric test of the effects of handsize versus tool form on efficiency in an experimental cutting task. J Archaeol Sci 38:1663–1670

Key AJM, Lycett SJ (2017) Form and function in the Lower Palaeolithic: history, progress, and continued relevance. J Anthropol Sci 95:67–108

Key AJM, Proffitt T, Stefani E, Lycett SJ (2016) Looking at handaxes from another angle: assessing the ergonomic and functional importance of edge form in Acheulean bifaces. J Anthropol Archaeol 44:43–55

Krubitzer L, Stolzenberg DS (2014) The evolutionary masquerade: genetic and epigenetic contribution to neocortex. Curr Opin Neurobiol 24:157–165

Land MF (2014) Do we have an internal model of the outside world? Philos Trans R Soc Lond B 369:1–6

Lombao D, Guardiola M, Mosquera M (2017) Teaching to make stone tools: new experimental evidence supporting a technological hypothesis for the origins of language. Sci Rep 7:14394

Lycett SJ, von Cramon-Taubadel N (2008) Acheulean variability and hominin dispersals: a model-bound approach. J Archaeol Sci 35:553–562

Maister L, Slater M, Sanchez Vives MV, Tsakiris M (2015) Changing bodies changes minds: owning another body affects social cognition. Trends Cogn Sci 19:6–12

Malafouris L (2010) The brain – artefact interface (BAI): a challenge for archaeology and cultural neuroscience. Soc Cogn Affect Neurosci 5:264–273

Malafouris L (2013) How things shape the mind: a theory of material engagement. MIT Press, Cambridge

Maravita A, Iriki A (2004) Tools for the body (schema). Trends Cogn Sci 8:79–86

Maravita A, Spence C, Driver J (2003) Multisensory integration and the body schema: close to hand and within reach. Curr Biol 13:531–539

Margulies DS, Vincent JL, Kelly C, Lohmann G, Uddin LQ, Biswal BB, Villringer A, Castellanos FX, Milham MP, Petrides M (2009) Precuneus shares intrinsic functional architecture in humans and monkeys. Proc Natl Acad Sci U S A 106:20069–20074

Martin I, Venables PH (1966) Mechanisms of palmar skin resistance and skin potential. Psychol Bull 65:347

Martin K, Jacobs S, Frey SH (2011) Handedness-dependent and-independent cerebral asymmetries in the anterior intraparietal sulcus and ventral premotor cortex during grasp planning. Neuroimage 57:502–512

Marzke MW (1997) Precision grips hand morphology and tools. Am J Phys Anthropol 102:91–110

Mesoudi A, O'Brien MJ (2008) The cultural transmission of Great Basin projectile-point technology I: an experimental simulation. Am Antiquity 73:3–28

Mitteroecker P, Gunz P (2009) Advances in geometric morphometrics. Evol Biol 36:235–247

Moore MW, Perston Y (2016) Experimental insights into the cognitive significance of early stone tools. PLoS One 11:e0158803

Mountcastle VB (1995) The parietal system and some higher brain functions. Cereb Cortex 5:377–390

Muller A, Clarkson C, Shipton C (2017) Measuring behavioural and cognitive complexity in lithic technology throughout human evolution. J Anthropol Archaeol 48:166–180

Neubauer S, Hublin JJ, Gunz P (2018) The evolution of modern human brain shape. Sci Adv 4:5961

Niewoehner WA (2001) Behavioral inferences from the Skhul/Qafzeh early modern human hand remains. Proc Natl Acad Sci U S A 98:2979–2984

Ollé A, Sala R, Pawlik A, Longo L, Skakun N, Gibaja JF (2017) New contributions to the functional analysis of prehistoric tools. Quat Int 427:2–5

Overmann KA (2015) Teeth tools and human becoming. J Anthropol Sci 93:163–167

Peer M, Salomon R, Goldberg I, Blanke O, Arzy S (2015) Brain system for mental orientation in space time and person. Proc Natl Acad Sci U S A 112:11072–11077

Peeters R, Simone L, Nelissen K, Fabbri-Destro M, Vanduffel W, Rizzolatti G, Orban GA (2009) The representation of tool use in humans and monkeys: common and uniquely human features. J Neurosci 29:11523–11539

Pereira-Pedro AS, Bruner E (2016) Sulcal pattern extension and morphology of the precuneus in adult humans. Ann Anat 208:85–93

Plummer T (2004) Flaked stones and old bones: biological and cultural evolution at the dawn of technology. Am J Phys Anthropol 125:118–164

Posner MI, Walker JA, Friedrich FJ, Rafal RD (1984) Effects of parietal injuries on covert orienting of attention. J Neurosci 4:1863–1874
Quallo MM, Price CJ, Ueno K, Asamizuya T, Cheng K, Lemon RN, Iriki A (2009) Gray and white matter changes associated with tool-use learning in macaque monkeys. Proc Natl Acad Sci U S A 106:18379–18384
Rolian C, Lieberman DE, Zermeno JP (2011) Hand biomechanics during simulated stone tool use. J Hum Evol 61:26–41
Rushworth MF, Paus T, Sipila PK (2001) Attention systems and the organization of the human parietal cortex. J Neurosci 21:5262–5271
Scheperjans F, Eickhoff SB, Hömke L, Mohlberg H, Hermann K, Amunts K, Zilles K (2008) Probabilistic maps morphometry and variability of cytoarchitectonic areas in the human superior parietal cortex. Cereb Cortex 18:2141–2157
Schillinger K, Mesoudi A, Lycett SJ (2015) The impact of imitative versus emulative learning mechanisms on artifactual variation: implications for the evolution of material culture. Evol Hum Behav 36:446–455
Semaw S, Renne P, Harris JWK, Feibel CS, Bernor RL, Fesseha N, Mowbray K (1997) 2.5-million-year-old stone tools from Gona, Ethiopia. Nature 385:333–336
Stout D, Chaminade T (2007) The evolutionary neuroscience of tool making. Neuropsychology 45:1091–1100
Stout D, Hecht E (2015) Neuroarchaeology. In: Bruner E (ed) Human paleoneurology. Springer, Cham, pp 145–175
Stout D, Semaw S, Rogers MJ, Cauche D (2010) Technological variation in the earliest Oldowan from Gona Afar Ethiopia. J Hum Evol 58:474–491
Stout D, Hecht E, Khreisheh N, Bradley B, Chaminade T (2015) Cognitive demands of lower paleolithic toolmaking. PLoS One 10:e0121804
Susman RL (1998) Hand function and tool behavior in early hominids. J Hum Evol 35:23–46
Tunik E, Rice NJ, Hamilton A, Grafton ST (2007) Beyond grasping: representation of action in human anterior intraparietal sulcus. Neuroimage 36:77–86
Turvey MT, Carello C (2011) Obtaining information by dynamic (effortful) touching. Philos Trans R Soc Lond B 366:3123–3132
Verhagen L, Dijkerman HC, Medendorp WP, Toni I (2012) Cortical dynamics of sensorimotor integration during grasp planning. J Neurosci 32:4508–4519
Wardak C, Olivier E, Duhamel JR (2004) A deficit in covert attention after parietal cortex inactivation in the monkey. Neuron 42:501–508
Williams EM, Gordon AD, Richmond BG (2012) Hand pressure distribution during Oldowan stone tool production. J Hum Evol 62:520–532
Williams VM, Burke A, Lombard M (2014) Throwing spears and shooting arrows: preliminary results of a pilot neuroarchaeological study. S Afr Archaeol Bull 69:199–207
Williams-Hatala EM, Hatala KG, Gordon M, Key A, Kasper M, Kivell TL (2018) The manual pressures of stone tool behaviors and their implications for the evolution of the human hand. J Hum Evol 119:14–26
Wynn T (2010) The evolution of human spatial cognition. In: Dolins FL, Mitchell RW (eds) Spatial cognition, spatial perception. Cambridge University Press, Cambridge, pp 213–236
Wynn T, Coolidge F (2003) The role of working memory in the evolution of managed foraging. Before Farming 2:1–16
Wynn T, Coolidge F (2004) The expert Neandertal mind. J Hum Evol 46:467–487
Wynn T, Coolidge F (2016) Archeological insights into hominin cognitive evolution. Evol Anthropol 25:200–213
Wynn T, Overmann KA, Coolidge FL (2016) The false dichotomy: a refutation of the Neandertal indistinguishability claim. J Anthropol Sci 94:201–221
Yantis S, Schwarzbach J, Serences JT, Carlson RL, Steinmetz MA, Pekar JJ, Courtney SM (2002) Transient neural activity in human parietal cortex during spatial attention shifts. Nat Neurosci 5:995–1002

Zelditch M, Swiderski D, Sheets DH, Fink W (2004) Geometric morphometrics for biologists: a primer. Elsevier Academic Press, Waltham
Zhang S, Li CSR (2012) Functional networks for cognitive control in a stop signal task: independent component analysis. Hum Brain Mapp 33:89–104
Zilles K, Amunts K (2010) Centenary of Brodmann's map – conception and fate. Nat Rev Neurosci 11:139–145
Zlatkina V, Petrides M (2014) Morphological patterns of the intraparietal sulcus and the anterior intermediate parietal sulcus of Jensen in the human brain. Proc Soc Biol 281:20141493

Development of Visual-Spatial Attention

Scott P. Johnson

Contents

Abstract This chapter reviews literature on development of visual-spatial attention. A brief overview of brain mechanisms of visual perception is provided, followed by discussion of neural maturation in the prenatal period, infancy, and childhood. This is followed by sections on gaze control, eye movement systems, and orienting. The chapter concludes with consideration of development of space, objects, and scenes. Visual-spatial attention reflects an intricate set of motor, perceptual, and cognitive systems that work jointly and all develop in tandem.

Keywords Attention · Human development · Infancy · Neural development · Object perception · Spatial perception · Visual perception

> …without selective interest, experience is an utter chaos—James (1890).

S. P. Johnson (✉)
Department of Psychology, UCLA, Los Angeles, CA, USA
e-mail: scott.johnson@ucla.edu

Curr Topics Behav Neurosci (2019) 41: 37–58
DOI 10.1007/7854_2019_96
Published Online: 30 April 2019

Attention, the "focalization" or "taking possession" of environmental stimuli or lines of thought (James 1890), can refer to a general level of alertness, focus on specific object, or a shift in the point of gaze (Colombo 2001). Eye movements allow us to quickly form an impression of our surroundings and to make moment-to-moment, context-appropriate decisions about our actions. Eye movements are directly observable, and thus they are good targets for investigations of how infants and children learn about the visual world.

Infants inhabit the same world as do adults and encounter the same visual scenes. How do they meet the challenges of seeing and interpreting the visual environment? Is the infant's visual system sufficiently functional and organized to make sense of the world and able to bind shapes, colors, and textures into coherent forms and to perceive objects, people, and places? Or does the infant's visual system require a period of maturation and experience within which to observe and learn, to coordinate visual and manual skills, to recognize and utilize individual visual cues, and to integrate auditory, haptic, and visual information? These questions are rooted in the perennial "nature-nurture" debate (Johnson 2010; Spelke 1990), and they continue to motivate research investigations of the development of visual-spatial attention. The studies reviewed in this chapter reveal that development of visual-spatial attention is a function of growth, maturation, and experience from learning and from action; all happen at the same time, and all influence one another. The chapter will emphasize the neural foundations of attention and developmental psychology research that shows clear links between attention and learning.

1 Neural Foundations of Visual-Spatial Attention

The purpose of the visual system is to transduce light reflected from surfaces in the environment into neural signals that are relayed to the brain for processing and action planning. Light is first transmitted through the cornea, the outer protective covering, and the lens, which provides focus of near and far objects, and then falls on the retina, a thin film of tissue covering the back of the eyeball. The retina comprises layers of photoreceptors and a rich network of connections and nonsensory neurons that provide initial processing of visual information. Different kinds of photoreceptor accomplish different tasks: There are specialized cells and circuits in the retina for color and contrast, for example, and these help determine how information is subsequently routed to appropriate channels up the visual hierarchy in the brain.

The visual system in the brain consists of a richly interconnected yet functionally segregated network of areas, specializing in processing different aspects of visual scenes and visually guided behavior: contours, motion, luminance, color, objects, faces, approach vs. avoidance, and so forth. Information flows from the retina to the lateral geniculate nucleus (a subcortical structure) and then to the primary cortical visual area (area V1). Reciprocal connections carry information to secondary visual areas (e.g., V2, V3, V4, and the medial temporal area). From these primary and secondary visual areas, visual information diverges to two partly segregated streams, ventral and dorsal (Goodale and Milner 1992; Stiles 2017; Schiller 1996). The ventral stream is connected to temporal cortex and is specialized for object recognition, mostly in inferotemporal cortex or IT (Tanaka 1997). IT connects to perirhinal cortex and other areas involved in categorization of visual stimuli and formation of visual memories (e.g., entorhinal cortex and hippocampus) as well as the lateral prefrontal cortex, involved in learning and planning (Miller 2000; Miyashita and Hayashi 2000). The dorsal stream codes for spatial information (object location and object-oriented action) and connects primarily to parietal structures that are important for voluntary action planning and coordinating somatosensory, proprioceptive, and visual inputs.

1.1 Development of Structure in the Visual System

The visual system, like other sensory and cortical systems, takes shape early in prenatal development. In humans, the retina starts to form around 40 days after conception and cells proliferate until about 160 days. The growth and organization of cells and connections continues well past birth (Finlay et al. 2003). The distinction between foveal and extrafoveal regions is present early, and the topology and patterning of receptors and neurons continue to change throughout prenatal development and the first year after birth. Foveal receptors are over-represented in the cortical visual system, and detailed information about different parts of the scene is enabled by moving the eyes to different points (more on this later). The musculature responsible for eye movements develops before birth in humans, as do subcortical systems (e.g., superior colliculus and brainstem) to control these muscles (Johnson 2001; Prechtl 2001). Different brain areas are present in rudimentary form during the first trimester, but the final forms continue to take shape well after birth; like synaptic pruning, developmental processes are partly the result of experience. Some kinds of experience are intrinsic to the visual system, as opposed to outside stimulation. Spontaneous prenatal activity in visual pathways, for example, contributes to retinotopic mapping (Sperry 1963; Wong 1999).

Many developmental mechanisms are common across mammalian species, including humans, though the timing of developmental events varies (Clancy et al. 2000; Finlay and Darlington 1995). As noted, many major structures (neurons, areas, and layers) in visual cortical and subcortical areas are in place by the end of the second trimester in utero. Later developments consist of the physical growth of neurons and the proliferation and pruning of synapses, which is, in part, activity-dependent.

As soon as neurons are formed, find their place in cortex, and grow, they begin to connect to other neurons. There is a surge in synaptogenesis in visual areas around the time of birth and then a more protracted period in which synapses are eliminated, reaching adult-like levels at puberty (Bourgeois et al. 2000). Synapses are preserved in active cortical circuits and lost in inactive circuits. In contrast to visual areas, auditory cortex experiences a synaptogenesis surge several months earlier, which corresponds to its earlier functionality relative to visual cortex (viz., prenatally). Here, too, pruning of synapses extends across the next several years. (In other cortical areas, such as frontal cortex, there is a more gradual accrual of synapses without extensive pruning.) For the visual system, the addition and elimination of synapses, the onset of which coincides with the start of visual experience, provides an important mechanism by which the cortex tunes itself to environmental demands and the structure of sensory input.

2 Visual Attention in Infancy

Much of what we know about development of visual-spatial attention comes from studies of neural mechanisms in animals and by extensive observations of human infants (Kiorpes and Movshon 2004; Teller and Movhson 1986). Even in infancy, vision is not passive; infants are active perceivers and active participants in their own development (von Hofsten 2004). At no point in development does it appear that humans are simply inactive recipients of stimulation.

2.1 Visual Attention in Newborns

Prior to birth, human fetuses orient their attention preferentially to some stimuli over others (Marx and Nagy 2015; Piontelli 2015). Recordings of cortical activity in utero have demonstrated that fetuses as young as 28 weeks gestation respond to bright light directed at the mother's abdomen (Eswaran et al. 2004; Fulford et al. 2003); moreover, fetuses in the third month of gestation turn preferentially toward light patterns in a face-like configuration vs. control stimuli (Reid et al. 2017).

Human infants are thus born with a functional visual system, and if motivated (i.e., awake and alert), the baby may react to visual stimulation with head and eye movements. Vision is relatively poor; however, acuity (detection of fine detail), contrast sensitivity (detection of different shades of luminance), color sensitivity, and sensitivity to direction of motion all undergo improvements after birth (Banks and Salapatek 1983). The field of view is also relatively small, so that newborns often fail to detect targets too far distant or too far in the periphery. In addition, neonates lack stereopsis, perception of depth from binocular disparity (differences in the input to the two eyes). Maturation of the eye and cortical structures (some of the mechanisms discussed previously) support developments in these visual functions, and learning plays an important role as well, discussed in greater detail in subsequent sections.

Observations of neonates have revealed that despite relatively poor vision, they actively scan the visual environment. Early studies, summarized by Haith (1980), revealed systematic oculomotor behaviors that provided clear evidence of visual organization at birth. Newborns, for example, will search for patterned visual stimulation, tending to scan broadly until encountering an edge, at which point scanning narrows so that the edge can be explored. Such behaviors are likely adaptive for investigating and learning about the visual world.

In addition, newborn infants show consistent visual preferences. Fantz (1961) presented newborns with pairs of pictures and other two-dimensional patterns and recorded which member of the pair which attracted the infant's visual attention, which he scored as proportion of looking times per exposure. Infants often showed longer looking at one member of the pair: bull's-eyes vs. stripes or checkerboards vs. solid forms, for example. Visual preferences have served as a method of choice ever since, in older infants as well as neonates. Slater (1995) described a number of newborns' preferences: patterned vs. unpatterned stimuli, curvature vs. rectilinear patterns, moving vs. static patterns, three-dimensional vs. two-dimensional forms, and high- vs. low-contrast patterns, among others. In addition, perhaps due to the relatively poor visual acuity of the newborn visual system, there is a preference for "global" form vs. "local" detail in newborns (Macchi Cassia et al. 2002).

Fantz (1964) also reported that repeated exposure to a single stimulus led to a decline of visual attention and increased attention to a new stimulus, in 2- to 6-month-olds. Subsequent investigations examined infants' preferences for familiar and novel stimuli as a function of increasing exposure, and these in turn led to standardized methods for testing infant perception and cognition, such as familiarization and habituation paradigms (Cohen 1976), as well as a deeper understanding of infants' information processing (Aslin 2007; Hunter and Ames 1988; Sirois and Mareschal 2002). For example, infants generally show preferences for novel vs. familiar stimuli after habituation (a decline of looking times across trials), implying both discrimination of novel and familiar stimuli and memory for the stimulus shown during habituation. Neonates and older infants also recognize visual constancies or invariants, the identification of common features of a stimulus across some transformation, for instance, shape, size, slant, and form (Slater et al. 1983). Recognition of invariants, in turn, forms the basis for visual categorization (Mareschal and Quinn 2001).

Finally, some kinds of functional visual development have been explained in terms of visual maturation (Atkinson 2000; Johnson 1990, 2005). Acuity, for example, improves in infancy with a number of developments, all taking place in parallel: migration of receptor cells in the retina toward the center of the eye, elongation of the receptors to catch more incoming light, growth of the eyeball to augment the resolving power of the lens, myelination of the optic nerve and cortical neurons, and synaptogenesis and pruning. And as described in the next section, development of gaze control in infancy has been thought to reflect changes in visual-spatial attention stemming from brain development.

3 Development of Gaze Control

Both infants and adults scan visual scenes actively – on the order of 2–4 eye movements per second in general (Johnson et al. 2004; Melcher and Kowler 2001). Gaze control in adults is accomplished with a coordinated system comprising both subcortical and cortical components. Six muscles are connected to the eyeball, each under direct control by brainstem. Eye movements are initiated in areas such as the frontal eye fields, in the cortex, and the superior colliculus, a subcortical area that receives inputs from several cortical areas. Both these areas connect to the brainstem, from which the actual signals to drive eye movements originate. Research on development of eye movements has often been viewed as an indirect means to examine cortical development, on the assumption that oculomotor behaviors can serve as "markers" to specific brain systems (e.g., Johnson 1990). In young infants, cortical influences on gaze control are less prominent.

Several types of eye movements, which develop over the first 6 postnatal months, are made during visual tracking: reflexive and voluntary saccades (quick eye movements from location to location) and smooth pursuit eye movements (slower eye movements that track moving targets; Richards 2001). Voluntary and smooth pursuit eye movements require attention, whereas reflexive eye movements are responses to sudden onset stimuli. Visual pathways that support reflexive eye movements include the retina, lateral geniculate nucleus, superior colliculus, and potentially primary visual cortex (Schiller 1985, 1988). Voluntary eye movements are planned attention-driven saccades that involve early cortical visual areas, parietal cortex, and frontal eye fields (Schiller 1985, 1988). Smooth pursuit eye movements are supported by visual pathways involving medial temporal lobe, middle superior temporal lobe, and perhaps parietal cortex (Schiller 1985, 1988). Johnson (1990, 1995) suggested that the pathways for reflexive eye movements are mature at birth, whereas primary visual area layers supporting voluntary saccadic eye movements mature rapidly during the first 6 postnatal months in conjunction with behavioral changes. Smooth pursuit, however, continues to improve over the first 2 years (Richards 1998).

Early arguments (Bronson 1974) claimed that the development of visual orienting involved a shift from solely subcortical to a mixture of subcortical and cortical processing. However, recent evidence suggests that the development proceeds in a graded manner, with some limited cortical activity likely present in newborns (Johnson 2005). In addition, development of smooth and saccadic eye movements in infants (described in detail in the next section) has often been interpreted as revealing development of distinct cortical systems that control them. Another early and influential proposal held that there are two discrete visual systems, a relatively primitive and phylogenetically older "secondary" system and a relatively sophisticated "primary" system that is more recent to humans in evolutionary time (Schneider 1969). In the neonate, visual behavior was held to be guided principally by the secondary system, which is characterized by poor foveal vision. The secondary system is restricted to reflexive or reactive eye movements to peripheral stimuli and does not participate in analysis of complex visual patterns. The primary system was thought to develop across the first several postnatal months, accompanies improvements in acuity and contrast sensitivity, and is responsible for the emergence of endogenous or internal control of saccades so as to support inspection of visual scenes (Bronson 1974). More recent interpretations of the two-system model have suggested that visual attention is largely under subcortical control until the first few months after birth, after which there is increasing cortical control (Atkinson 1984; Colombo 2001; Johnson 1990).

An example of how oculomotor control may emanate from cortical development is found in Johnson (1990). There are rapid improvements between 6 and 10 weeks in smooth pursuit, the ability to track a small moving target against a featureless background, and in motion direction discrimination, the ability to discriminate dynamic random dot patterns (Aslin 1981; Wattam-Bell 1996). Johnson (1990) suggested that a common developmental path underlies emergence of both smooth pursuit and motion sensitivity: maturation of pathways to and from the medial temporal area. That is, the perception of motion in the visual environment and the programming of eye movements to follow motion are thought to be supported by the same cortical networks (Thier and Ilg 2005). This suggestion was tested empirically by Johnson et al. (2008), who observed infants between 58 and 97 days of age in both a smooth pursuit and a motion direction discrimination task. Individual differences in performance on the two tasks were strongly correlated and were also positively correlated with age, consistent with a maturational model (though not necessarily uniquely predicted by it).

Other visual functions in infancy that have been linked to cortical maturation include development of form and motion perception, stemming from maturation of parvocellular and magnocellular processing streams, respectively (Atkinson 2000), and development of visual memory for object features and object locations, stemming from maturation of ventral and dorsal processing streams (Mareschal and Johnson 2002).

4 Eye Movements

As noted previously, humans move their eyes from location to location even before birth, a behavior that presumably reflects some kind of decision made somewhere in the visual system to find locations in the visual scene for closer inspection. This is known as foveation, the bringing of an image in the environment to the fovea, the center of the visual field and the location on the retina producing the highest acuity inputs to the brain. Foveation is an ingenious mechanism to balance the need to derive detailed visual information from the world and the need to reduce as much as possible the metabolic demands of a large brain necessary to process the information. The fovea has the highest concentration of photoreceptors, and these are preferentially mapped onto visual cortical tissue. Acuity is best at the point of gaze and drops off abruptly with increasing visual eccentricity into a low-resolution visual surround; this drop-off is reflected in the distribution of photoreceptors on the retina

(Winkler et al. 1990). Detailed representation of a visual scene, therefore, which entails extensive processing by visual cortex, takes place only for a region within about 2° visual angle of the viewed scene (approximately the size of a thumbnail at arm's length). To avoid the need to build the large brain that would be needed to process detail about the entire scene, the visual system compromises by periodically shifting the point of gaze with saccadic eye movements and thus reorienting the specific location in the scene that is best represented and processed.

4.1 Primary Eye Movement Systems

There are four primary eye movement systems, each of which is produced by separate neural circuitry, all channeled through the brainstem, which innervates the ocular musculature. Saccades are the most common form of eye movement, consisting of quick changes of fixation – that is, holding the gaze stable – whose function is to place the retinal "image" of an object of interest on the fovea. Smooth pursuit eye movements maintain foveation on a moving item of interest. Optokinetic nystagmus (OKN) is a semi-reflexive eye movement pattern driven by a large visual array that moves with respect to a stationary observer, and the vestibulo-ocular response (or VOR) is a semi-reflexive eye movement pattern driven by a stationary large visual array placed in front of, or surrounding, a moving observer. The relation of eye movements to the head is similar in both OKN and VOR, a relatively slow, smooth track followed by a quick saccade-like movement, but the stimulus conditions, developmental progression, and neural systems controlling each are different. OKN, VOR, and smooth pursuit work in tandem to yield visual stability, the ability to stabilize the retinal image despite perturbations due to eye, head, and body movements and motion in the observer's surroundings. Visual stability is vital for the effective extraction of detailed information about the visual environment as observers move about, and it improves markedly after 2 months with the onset of consistent smooth pursuit and suppression of OKN and VOR (Aslin and Johnson 1996). This is described in greater detail in the subsequent section.

4.1.1 Scanning

Visual attention is a combination of saccades and fixations, working in tandem to move the point of gaze across the visual field. Sequencing of saccades and fixations is known as scanning, and it constitutes a form of overt (observable) visual attention. During a saccade, the point of gaze for both eyes sweeps rapidly across the scene, and during a fixation, the point of gaze is relatively stationary. Information about the scene is acquired during the fixations. Analysis of the scene cannot be performed during a saccade, whose purpose is to direct attention to a different part of the scene for subsequent processing. Scanning can be interspersed with smooth eye movements, as when the head translates or rotates as the point of gaze remains stabilized on a single point in space (the eyes move to compensate for head movement) or when following a moving target.

Newborn infants, if awake and alert, examine their surroundings with a series of fixations, indicating that some of the neural circuitry for saccade generation is in place at birth. Young infants' fixations, however, often do not extend beyond areas of high contrast, such as edges (Bronson 1994), or remain centered around a limited set of stimulus features (Bronson 1990; Johnson and Johnson 2000). Infants older than 3 months will more often scan in what appears to be an exploratory fashion. Older infants will also scan between individual stimuli more readily than will younger infants (Bronson 1994). This pattern of development has been interpreted as a shift from reflexive to more purposive scanning, consistent with the maturation of cortical pathways, as noted previously.

4.1.2 Sticky Fixation

One- and 2-month-old infants have been found to exhibit much longer looking times than either neonates or 4-month-old infants in studies that use habituation (Hood et al. 1996; Johnson 1996; Slater et al. 1996a) or other looking time methods (e.g., Johnson et al. 1991). This so-called sticky fixation has been interpreted as difficulty disengaging attention and has been tied to tonic inhibition of the superior colliculus by the substantia nigra and basal ganglia, later controlled by cortical mechanisms subserving endogenous attentional control and peripheral expansion of the visual field (Johnson 1990; see also Hood et al. 1998; Maurer and Lewis 1998). (However, there are other, more cognitive explanations for longer looking times in 1- and 2-month-olds; Johnson 1996.) A decrease in fixation durations with age has also been found in studies of infants' natural scene viewing (Helo et al. 2016; Wass and Smith 2014). Attentional disengagement continues to improve throughout childhood (Gregory et al. 2016) and into adolescence (Luna et al. 2008). Interestingly, individual differences in fixation durations are linked to attentional and behavioral control in childhood (Papageorgiou et al. 2014).

4.1.3 Smooth Pursuit

In contrast to saccades, smooth pursuit is limited in very young infants. When presented with a small, moving target, infants younger than 2 months will often attempt to track it with "catch-up" saccades, rather than smooth eye movements (Aslin 1981). Younger infants may engage in short bouts of smooth pursuit if target speed is not too high, but smooth pursuit is not robust (Kremenitzer et al. 1979; Roucoux et al. 1983). The limitation in pursuit is not an inability to move the eyes smoothly: OKN, which contains a slow-movement component, can be observed in neonates. Rather, very young infants may be incapable of engaging in predictive eye movements, such that the future location of a moving target cannot be computed, or they may be unable to track due to limitations in motion processing (a function of the medial temporal area; Komatsu and Wurtz 1989). Alternatively, immaturity of retinal photoreceptors may prohibit firm registration of the target on the fovea, such that a series of saccades is necessary to recenter gaze and maintain fixation.

5 Orienting

Many studies have examined orienting – engagement of visual attention – as infants are presented with a limited number of small static or moving targets (Johnson 2005; Richards 1998). Oculomotor behaviors that have been examined include detection of targets in the periphery, saccade planning, oculomotor anticipations, sustained vs. transient attention, effects of spatial cuing, and eye/head movement integration; other tasks examined inhibition of eye movements, such as disengagement of attention, inhibition of return, and spatial negative priming. Bronson (1990, 1994) explored developmental changes in scanning patterns as infants viewed simple geometric forms. The youngest infants tested (2 weeks) were reported to attend primarily to a single prominent feature, whereas older infants (3 months) were more likely to scan between features and to direct saccades with greater accuracy, again perhaps reflecting a transition from reflexive to "volitional" scanning.

5.1 Overt vs. Covert Attention

Along with these developments in overt orienting, infants become capable of covert orienting – an internal shift of attention that can facilitate saccades to particular spatial locations. Two related phenomena are inhibition of return (IOR), a delay in eye movements toward a previously cued location, and spatial negative priming, a delay in eye movements toward a separate location presented alongside the fixated one (i.e., subsequent to the first fixation).

5.1.1 Inhibition

In a typical spatial cueing paradigm, an infant is shown a central stimulus as a stimulus flashes in one of the two possible peripheral locations; the stimulus is too brief to elicit a saccade. After a delay, a target is presented in one or both of the peripheral locations; evidence for covert orienting is provided by effects of the cue's location. In a study by Clohessy et al. (1991), for example, infants sat in front of three screens. A dynamic stimulus first appeared in the center for fixation, followed by a cue on one of the peripheral screens and then a presentation of a target on both peripheral screens. Infants 6 months and older made more saccades opposite the location where the original cue was flashed, taken as evidence of IOR. Using similar methods, Johnson et al. (1994) presented 4-month-old infants with stimulus sequences in which a 100 ms cue flashed contralateral to the location of target appearing 400 ms later. After 12 such "training" trials, "test" trials were presented intermixed with more training trials; in test trials the target appeared in the cued location after a 100 ms delay. Despite the training (to make a saccade to the side contralateral to the cue), infants showed faster saccadic RTs to orient to the target on test trials, consistent with the possibility that covert attention facilitated saccades to the cue's location. Johnson and Tucker (1996) subsequently investigated effects of cue-target timing intervals on the speed and direction of orienting in 2-, 4-, and 6-month-olds. Two-month-olds showed only weak effects of the cue, but 4-month-olds showed facilitation to the cued location when the target appeared 200 ms after cue onset but inhibition if it appeared 700 ms after cue onset. Six-month-olds showed evidence of inhibition only (cf. Clohessy et al. 1991). Researchers using paradigms that rely on overt orienting have reported IOR in newborns (Simion et al. 1995; Valenza et al. 1994), whereas those that require covert orienting have found evidence of IOR only in infants 4 months of age and older (Johnson et al. 1994; Johnson and Tucker 1996; Richards 2001).

Scanning efficiency is assessed in the IOR paradigm by the inhibition of eye movements to previously processed locations. In a negative priming paradigm, when an object or location that had previously been ignored or inhibited becomes the target to be selected, responses to it are slowed in comparison to a control stimulus (Neill 1977; Tipper 1985), suggesting that the ignored object or location was attended covertly and inhibited in favor of selecting the attended stimulus. In spatial negative priming (SNP) tasks, competition between the target and distractor must be resolved by suppression of the distractor and simultaneous selection of the target, a demand unnecessary in IOR tasks because the cue generally precedes the target. Impairment during target selection, therefore, can be seen as a validation that the distractor location was attended and inhibited during target selection in the prime. IOR, therefore, is a measure of preference for novel unattended locations, whereas SNP provides a measure of the ability to inhibit a location, while selection of another is underway. These are two distinct processing demands, and behavioral results from the two methods shed light in unique ways on the development of inhibition and its role in learning in infancy.

A study with children ranging from 5 to 12 years, for example, reported negative priming of identity, location, and conceptual task features (Pritchard and Neumann 2004; Simone and Mccormick 1999). The magnitude of spatial negative priming (SNP) effects (i.e., inhibition of location) has been found to remain fairly constant from about age 6 years into adulthood (Davidson et al. 2006). Amso and Johnson (2005) used a spatial negative priming task to assess the ability of 9-month-old infants and adults to select between simultaneously presented locations when inhibiting a distractor location in favor of a target location. Interstimulus intervals (ISIs) between presentation of cues and targets were manipulated to examine temporal effects on inhibition efficiency. Each trial began with simultaneous presentation of a salient cue and a nonsalient distracter, followed (after varying ISIs) by a target in the location previously occupied by the distracter or another location. Both infants and adults exhibited the SNP effect at the longest (550 ms) and intermediate (200 ms) ISIs – that is, saccades were slower to targets when they appeared at the distracter location. In a follow-up study with 3-, 6-, and 9-month-olds, the youngest infants provided no evidence of inhibition; rather, target locations were facilitated (evinced by faster saccadic RTs) (Amso and Johnson 2008). These studies provide evidence, therefore, for covert attention in young infants, but not fully mature inhibitory processes.

5.1.2 Selective Attention

Visual selective attention is the ability to select relevant stimuli for processing and to ignore or inhibit competing alternatives. Efficient allocation of visual attention is critical to learning in infancy. To organize the world into coherent percepts, a child must attend to certain environmental features while simultaneously ignoring others. This requires selection of relevant stimuli for processing and inhibition of those that are irrelevant.

Developmental investigations into visual selective attention have made use of visual search paradigms, which have origins in work by Treisman and colleagues (Treisman 1988; Treisman and Gelade 1980). In a visual search paradigm, a unique target element is placed in an array of distracters. For example, Treisman and Gormican (1988) presented an oblique target among vertically oriented distracters. Participants quickly detected the oblique target, and saccadic RTs to the target were largely unaffected by the number of distracters in the display. The authors concluded that the fact that the increase in elements of one feature did not interfere with the processing of the other suggests that oblique and vertical must be coded as separate features in the visual module that codes for orientation. The "pop-out" effect has been attributed to a parallel processing mechanism that directs attention to the location of the unique item during the preattentive early stages of visual processing (Treisman 1988; Treisman and Gelade 1980).

Research using the novelty preference method (Colombo et al. 1995; Quinn and Bhatt 1998) and mobile conjugate reinforcement techniques (Adler et al. 1998; Rovee-Collier et al. 1996; Goodale and Milner 1992) has reported evidence of pop-out as early as 3 months of age. Also consistent with adult findings, this pop-out effect has been found to be unaffected by the number of distractors (Rovee-Collier et al. 1996).

Visual search tasks have also been used to test infants' sensitivity to competition between parts of a visual scene (Dannemiller 1998, 2000, 2002). In these studies, a moving target was embedded in an array of 27 static red and green distracters, evenly distributed on the left and right sides of a monitor; each trial commenced when the infant looked at an attention getter in the center of the screen. Attention to the target was influenced by the spatial distribution of the distracters, such that orienting was weakened when higher salience bars were contralateral to the moving target. (Salience was determined by the contrast of static bars with their background; Ross and Dannemiller 1999; Zenger et al. 2000.) Thus, attention was divided when salient static features competed for orienting with the moving probe, and sensitivity to this competition, indexed as poorer performance with increased competition, may reflect maturing selective attentional mechanisms.

Finally, Amso and Johnson (2006) used a visual search task to examine relations between visual selection and object perception in infancy. Three-month-old infants were shown displays in which single targets of varying levels of salience were presented among homogeneous static vertical distractors. Infants also completed a "perceptual completion" task in which they were habituated to a partly occluded moving rod and subsequently presented with unoccluded broken and complete rod test stimuli. Infants were divided into two groups. "Perceivers" were those infants whose posthabituation preference indicated perception of object unity in the perceptual completion task (i.e., a novelty preference for the unoccluded broken rod stimulus), whereas the "non-perceivers" looked longer at the complete rod test stimulus. Perceivers tended to scan the rod parts more than did non-perceivers; in addition, perceivers detected more targets during the visual search task. That is, infants who provided evidence of perceiving the unity of disjoint surfaces also provided evidence of efficient visual selective attention in the search task. With the emergence of selective attention, therefore, infants become better able to detect and learn about relevant parts of the visual scene. This is discussed further subsequently.

6 Perception of Space, Objects, and Scenes

6.1 Development of Spatial Perception

The gaze moves frequently from place to place in the visual scene, and our heads and bodies move around in space. Despite near-continual transformations, disruptions, and interruptions in visual input, we usually experience the visual world as an inherently stable place. This can be demonstrated vividly by trying to read while shaking the head back and forth (reading is not much compromised) vs. shaking the reading material back and forth (reading can be quite difficult). When an observer rotates the head, the VOR (vestibuloocular response) provides compensatory eye movements which allow the point of gaze to remain fixed or to continue moving volitionally as desired (as when reading). When the page moves, there is no such compensatory response.

Evidence from three paradigms suggests that visual stability emerges gradually across the first year after birth. First, young infants have difficulty discriminating optic flow patterns that simulate different directions of self-motion (Gilmore et al. 2004). Infants viewed a pair of random dot displays in which the dots repeatedly expanded and contracted around a central point to simulate the effect of moving forward and backward under real-world conditions. On one side, the location of this point shifted periodically, which for adults specifies a change in heading direction; the location on the other side remained stationary. Under these circumstances adults detected a shift simulating a 5° change in heading, but infants were insensitive to all shifts below 22°, and sensitivity was unchanged between 3 and 6 months. Gilmore et al. speculated that optic flow sensitivity may be improved by self-produced locomotion after 6 months of age or by maturation of the ventral visual stream.

Second, young infants' saccade patterns tend to be retinocentric, rather than body-centered, when studied in a "double-step" tracking paradigm (Gilmore and Johnson 1997). Retinocentric saccades are programmed without taking into account previous eye movements. Body-centered eye movements, in contrast, are programmed while updating the spatial frame of reference or coordinate system in which the behaviors occur. Infants first viewed a fixation point that then disappeared, followed in succession by the appearance and extinguishing of two targets on either side of the display. The fixation point was located at the top center of the display, and targets were located below it at the extreme left and right sides. As the infant viewed the fixation point and targets in sequence, there was an age-related transition in saccade patterns. Three-month-olds tended to direct their gaze *downward* from the first target, as if directed toward a target below the current point of gaze. In reality the second target was below the first location – the original fixation point – not the current point of gaze. Seven-month-old infants, in contrast, were more likely to direct gaze directly toward the second target. These findings imply that the young infants' visual-spatial coordinate system, necessary to support perception of a stable visual world, may be insensitive to extra-retinal information, such as eye and head position, in planning eye movements. In addition, the child's field of view transforms with the advent of the transition to walking, which yields new kinds of visual experiences (Kretch et al. 2014).

Third, there are limits in the ability of infants younger than 2 months to switch attention flexibly and volitionally to consistently maintain a stable gaze. Movement of one's body through the visual environment can produce an optic flow pattern, as can head movement while stationary (recall the head-shaking example). The two scenarios may produce similar visual inputs from optic flow, yet we readily distinguish between them. In addition, adult observers can generally direct attention to either moving or stationary targets, nearby or in the background, as desired. These are key features of visual stability, and the four eye movement systems described previously work in concert to produce it. OKN (optokinetic nystagmus) stabilizes the visual field on the retina as the observer moves through the environment. OKN is triggered by a large moving field, as when gazing out the window of a train: The eyes catch a feature, follow it with a smooth movement, and saccade in the opposite direction to catch another feature, repeating the cycle. The VOR, described previously, helps maintain a stable gaze to compensate for head movement (OKN and the VOR are present and functional at birth, largely reflexive or obligatory, and are likely mediated by subcortical pathways; Atkinson and Braddick 1981; Preston and Finocchio 1983). The others are the saccadic eye movement system and smooth pursuit, to compensate for or cancel the VOR or OKN as appropriate. Aslin and Johnson (1994) observed suppression (cancellation) of the VOR to fixate a small moving target in 2- and 4-month-olds, but not 1-month-olds, and Aslin and Johnson (1996) observed suppression of OKN to fixate a stationary target in 2-month-olds, but not in a younger group.

6.2 Development of Object Perception

Studies of object perception in young infants have revealed a developmental pattern from "piecemeal" to coherent perception of the visual environment that extends from birth through the first several months afterward, implying a fundamental shift in the infant's perceptual experience (Johnson 1997; Bremner et al. 2015). Experiments that examined perceptual completion, for example, revealed that newborn infants appear to construe a partly occluded rod behind a box as consisting of two disjoint parts (Slater et al. 1990; Slater et al. 1996b; but see Valenza and Bulf 2011 for counterevidence). By 4 months, perceptual completion has become more robust and reliable under different conditions (see Bremner et al. 2015, for review). As noted previously, an important means by which infants may come to perceive occlusion is by improvements in selective attention (Amso and Johnson 2006; Johnson et al. 2008; Schlesinger et al. 2012).

Several studies have reported that young infants can maintain active representations of the solidity and location of fully hidden objects across brief delays (e.g., Aguiar and Baillargeon 1999; Spelke et al. 1992), and some interpretations of these reports have appealed to innate object concepts (e.g., Baillargeon 1993; Spelke and Kinzler 2007). However, as noted in the previous paragraph, newborns provide little evidence of perceiving partly occluded objects, begging the question of how perception of *complete* occlusion, or simple object persistence, emerges in infancy. Experiments addressing this question have examined infants' responses to objects that move on a trajectory, disappear behind an occluder, reappear on the far side, and reverse direction, repeating the cycle (see Bremner et al. 2015 for review). These experiments have tended to support a "perceptual" account of object persistence, as opposed to an account based on object persistence as an innate principle through which events are interpreted. Object persistence is specified by perceptual information such as deletion and accretion of objects at an edge, surface appearance, rate and orientation of object motion, and other visual cues. The developmental question thus concerns infants' changing ability to perceive object persistence on the basis of these cues. Four-month-olds perceive persistence across shorter spatial and temporal gaps than 6-month-olds, for example, and they require more cues to specify occlusion (and hence persistence) than adults (Johnson et al. 2003).

Other advances in object perception come from coordinated visual attention and manual exploration, which can help infants understand objects as solid in 3D space. For example, 4-month-olds who exhibited high levels of spontaneous engagement with complex objects, operationalized as manual manipulation of the object accompanied by visual attention to the object, appeared to be better at mental rotation, the ability to imagine the appearance of objects from different viewpoints and to discriminate objects from different views (Slone et al. 2018). In addition, 4- to 7-month-olds' visual-manual object exploratory skills predicted 3D object completion, the ability to perceive objects as solid in 3D space (Soska et al. 2010).

6.3 Development of Scene Perception

In our everyday visual environment, visual scenes are complex and generally characterized by a number of objects at different distances, overlapping one another from the observer's perspective. Two types of scene characteristics have been extensively studied in adults: perceptual salience (low-level features such as edges, color, and contrast) and semantic relevance (objects). Low-level scene characteristics can be measured with perceptual salience maps (Itti et al. 1998), and there is evidence that visual attention is driven toward salient regions based on these low-level properties (Borji et al. 2013; Itti and Koch 2000; Itti et al. 1998). Other studies have showed, in contrast, that attention allocation can be better explained as looking at objects or semantically relevant characteristics of a scene, rather than salience (Einhaüser et al. 2008; Nuthmann and Henderson 2010; Stoll et al. 2015). Moreover, fixations often fall on the centers of objects (Foulsham and Kingstone 2013; Nuthmann and Henderson 2010; Xu et al. 2014), which is difficult to explain from a saliency point of view, because saliency models typically highlight object contours and edges. In addition, effects of low-level salience on visual attention are minimal when task demands are made more challenging, such as engagement in visual-motor tasks (Tatler et al. 2011), or when viewing scenes with familiar objects in everyday settings, which can help the viewer establish areas of the scene with the highest semantic content (Henderson and Hayes 2017). Thus in adults, the role of salience in scene viewing is limited, and eye movements are more likely to be driven by meaning.

As noted in previous sections, development of visual-spatial attention in infancy has been thought of as a shift from relying on exogenous features (e.g., perceptual salience) toward more endogenous (e.g., knowledge) (Johnson 1990, 2001), but findings from studies of young infants' free viewing of complex scenes are mixed with respect to the role salience plays. For instance, Frank et al. (2009) found that salience was a better predictor of attention in younger infants (3-month-olds) than faces, whereas in older (6- and 9-month-olds) infants faces were a better predictor than salience. Frank et al. (2014) replicated this finding and showed that looking at faces in the youngest age group was correlated with performance on a visual search task (i.e., number of targets detected). This implies that detecting semantically relevant information is partly a function of emerging selective attention skills. However, other studies have reported that the predictive value of perceptual salience for children's visual attention actually increases with age (Amso et al. 2014; Franchak et al. 2016), and a recent comparison of eight saliency models found that most were better able to predict fixation locations for adults than for infants (Mahdi et al. 2017), which would not be expected if the role of saliency in driving attention was reduced over developmental time. Part of the discrepancy among findings may stem from the fact that objects are salient (Elazary and Itti 2008), and so some of the reported age-related effects of perceptual salience on visual-spatial attention may actually be an increase in attention toward meaningful objects.

In a recent study of the development of scene perception, 3- to 15-month-old infants were presented with a series of everyday photographs (outdoor and indoor scenes, no faces) as their eye movements were recorded (van Renswoude et al. 2019); infants also participated in visual search and spatial negative priming tasks as described previously. In addition, a group of adults viewed the photos, and their gaze patterns were compared to those of infants. Infants' visual attention to the scenes was then modeled as a function of perceptual salience, adult fixation locations, age, and selective and inhibitory attentional mechanisms while accounting for general biases (e.g., to produce horizontal eye movements; van Renswoude et al. 2016). Interestingly, all of these factors play a role in guiding eye movements as infants viewed complex scenes. Perceptual salience and adult fixation locations both made unique predictions to infants' attention allocation. Older infants were more likely to fixate parts of scenes more frequently fixated by adults relative to younger infants, and fixation durations were longer on regions more frequently fixated by adults. In addition, there was a stronger decrease in fixation durations with age in those infants with better orienting skills. Taken together, these results suggest that development of scene perception is characterized by a growing tendency to look at semantically meaningful regions in scenes and that this process is gated in part by the emergence of gaze control.

7 Conclusions

Development of visual-spatial attention is a complex interplay between multiple systems: innate propensities, neural maturation, various kinds of experiences, motor development, and learning. Many of the insights into development of visual-spatial attention described in this chapter have resulted from remarkable advances in technology to record infants' and children's eye movements (Franchak et al. 2011) and brain activity (Johnson 2005; Richards 2001), yet there are many aspects of attentional development that remain unknown, in particular the precise roles of different kinds of experiences and the impact of developmental disabilities.

References

Adler SA, Inslicht S, Rovee-Collier C, Gerhardstein PC (1998) Perceptual asymmetry and memory retrieval in 3-month-old infants. Infant Behav Dev 21:253–272

Aguiar A, Baillargeon R (1999) 2.5-month-old infants' reasoning about when objects should and should not be occluded. Cogn Psychol 39:116–157

Amso D, Johnson SP (2005) Selection and inhibition in infancy: evidence from the spatial negative priming paradigm. Cognition 95:B27–B36

Amso D, Haas S, Markant J (2014) An eye tracking investigation of developmental change in bottom-up attention orienting to faces in cluttered natural scenes. PLoS One 9:e85701

Amso D, Johnson SP (2006) Learning by selection: visual search and object perception in young infants. Dev Psychol 42:1236–1245
Amso D, Johnson SP (2008) Development of visual selection in 3- to 9-month-olds: evidence from saccades to previously ignored locations. Infancy 13:675–686
Aslin RN (1981) Development of smooth pursuit in human infants. In: Fisher DF, Monty RA, Senders JW (eds) Eye movements: cognition and visual perception. Erlbaum, Hillsdale, pp 31–51
Aslin RN (2007) What's in a look? Dev Sci 10:48–53
Aslin RN, Johnson SP (1994) Suppression of VOR by human infants. Poster presented at the association for research in vision and ophthalmology conference, Sarasota, FL, May 1994
Aslin RN, Johnson SP (1996) Suppression of the optokinetic reflex in human infants: implications for stable fixation and shifts of attention. Infant Behav Dev 19:233–240
Atkinson J (1984) Human visual development over the first 6 months of life: a review and a hypothesis. Hum Neurobiol 3:l–74
Atkinson J (2000) The developing visual brain. Oxford University Press, New York
Atkinson J, Braddick O (1981) Development of optokinetic nystagmus in infants: an indicator of cortical binocularity? In: Fisher DF, Monty RA, Senders JW (eds) Eye movements: cognition and visual perception. Erlbaum, Hillsdale, pp 53–64
Baillargeon R (1993) The object concept revisited: new directions in the investigations of infants' physical knowledge. In: Granrud CE (ed) Visual perception in infancy. Carnegie-Mellon symposia on cognition, vol 23. Erlbaum, Hillsdale, pp 265–316
Banks M, Salapatek P (1983) Infant visual perception. In: Haith MM, Campos JJ (ed) Infancy and developmental psychobiology. Handbook of child psychology (ser ed: Mussen PH), vol 2, 4th edn. Wiley, New York, pp 435–572
Borji A, Sihite DN, Itti L (2013) What stands out in a scene? A study of human explicit saliency judgment. Vis Res 91:62–77
Bourgeois J-P, Goldman-Rakic PS, Rakic P (2000) Formation, elimination, and stabilization of synapses in the primate cerebral cortex. In: Gazzaniga MS (ed) The new cognitive neurosciences. MIT Press, Cambridge, pp 45–53
Bremner JG, Slater AM, Johnson SP (2015) Perception of object persistence: the origins of object permanence in infancy. Child Dev Perspect 9:7–13
Bronson G (1974) The postnatal growth of visual capacity. Child Dev 45:873–890
Bronson G (1990) Changes in infants' visual scanning across the 2- to 14-week age period. J Exp Child Psychol 49:101–125
Bronson G (1994) Infants' transitions toward adult-like scanning. Child Dev 65:1243–1261
Clancy B, Darlington RB, Finlay BL (2000) The course of human events: predicting the timing of primate neural development. Dev Sci 3:57–66
Clohessy AB, Posner MI, Rothbart MK, Vecera S (1991) The development of inhibition of return in early infancy. J Cogn Neurosci 3:345–350
Cohen LB (1976) Habituation of infant visual attention. In: Tighe RJ, Leaton RN (eds) Habituation: perspectives from child development, animal behavior, and neurophysiology. Erlbaum, Hillsdale, pp 207–238
Colombo J (2001) The development of visual attention in infancy. Annu Rev Psychol 52:337–367
Colombo J, Ryther JS, Frick JE, Gifford JJ (1995) Visual pop-out in infants: evidence for preattentive search in 3- and 4-month olds. Psychon B Rev 2:266–268
Dannemiller JL (1998) A competition model of exogenous orienting in 3.5-month-old infants. J Exp Child Psychol 68:169–201
Dannemiller JL (2000) Competition in early exogenous orienting between 7 and 21 weeks. J Exp Child Psychol 76:253–274
Dannemiller JL (2002) Relative color contrast drives competition in early exogenous orienting. Infancy 3:275–301
Davidson M, Amso D, Cruess L, Diamond A (2006) Development of cognitive control and executive functions from 4–13 years: evidence from manipulations of memory, inhibition, and task switching. Neuropsychologia 44:2037–2078

Einhaüser W, Spain M, Perona P (2008) Objects predict fixations better than early saliency. J Vis 8:18
Elazary L, Itti L (2008) Interesting objects are visually salient. J Vis 8:3
Eswaran H, Lowery CL, Wilson JD, Murphy P, Preissi H (2004) Functional development of the visual system in human fetus using magnetoencephalography. Exp Neurol 190:52–58
Fantz RL (1961) The origin of form perception. Sci Am 204:66–72
Fantz RL (1964) Visual experience in infants: decreased attention to familiar patterns relative to novel ones. Science 146:668–670
Finlay BL, Clancy B, Kingsbury MA (2003) The developmental neurobiology of early vision. In: Hopkins B, Johnson SP (eds) Neurobiology of infant vision. Praeger, Westport, pp 1–41
Finlay BL, Darlington RB (1995) Linked regularities in the development and evolution of mammalian brains. Science 268:1578–1584
Foulsham T, Kingstone A (2013) Optimal and preferred eye landing positions in objects and scenes. Q J Exp Psychol 66:1707–1728
Franchak JM, Kretch KS, Soska KC, Adolph KE (2011) Head-mounted eye-tracking: a new method to describe infant looking. Child Dev 82:1738–1750
Franchak JM, Heeger DJ, Hasson U, Adolph KE (2016) Free viewing gaze behavior in infants and adults. Infancy 21:262–287
Frank MC, Vul E, Johnson SP (2009) Development of infants' attention to faces during the first year. Cognition 110:160–170
Frank MC, Amso D, Johnson SP (2014) Visual search and attention to faces in early infancy. J Exp Child Psychol 118:13–26
Fulford J, Vadeyar SH, Dodampahala SH, Moore RJ, Young P, Baker PN et al (2003) Fetal brain activity in response to a visual stimulus. Hum Brain Mapp 20:239–245
Gilmore RO, Johnson MH (1997) Body-centered representations for visually-guided action emerge during early infancy. Cognition 65:B1–B9
Gilmore RO, Baker TJ, Grobman KH (2004) Stability in young infants' discrimination of optic flow. Dev Psychol 40:259–270
Goodale MA, Milner AD (1992) Separate visual pathways for perception and action. Trends Neurosci 15:20–25
Gregory NJ, Hermens F, Facey R, Hodgson TL (2016) The developmental trajectory of attentional orienting to socio-biological cues. Exp Brain Res 234:1351–1362
Haith MM (1980) Rules that babies look by: the organization of newborn visual activity. Erlbaum, Hillsdale
Helo A, Rämä P, Pannasch S, Meary D (2016) Eye movement patterns and visual attention during scene viewing in 3-to 12-month-olds. Vis Neurosci 33:E014
Henderson JM, Hayes TR (2017) Meaning-based guidance of attention in scenes as revealed by meaning maps. Nat Hum Behav 1:743–747
Hood BM, Murray L, King F, Hooper R, Atkinson J, Braddick O (1996) Habituation changes in early infancy: longitudinal measures from birth to 6 months. J Reprod Infant Psychol 14:177–185
Hood BM, Atkinson J, Braddick O (1998) Selection-for-action and the development of orienting and visual attention. In: Richards JE (ed) Cognitive neuroscience of attention: a developmental perspective. Erlbaum, Mahwah, pp 219–250
Hunter MA, Ames EW (1988) A multifactor model of infant preferences for novel and familiar stimuli. In: Rovee-Collier C, Lipsitt LP (eds) Advances in child development and behavior, vol 5. Ablex, Norwood, pp 69–93
Itti L, Koch C (2000) A saliency-based search mechanism for overt and covert shifts of visual attention. Vis Res 40:1489–1506
Itti L, Koch C, Niebur E (1998) A model of saliency-based visual attention for rapid scene analysis. IEEE Trans Pattern Anal Mach Intell 20:1254–1259
James W (1890) Principles of psychology. Holt, New York

Johnson MH (1990) Cortical maturation and the development of visual attention in early infancy. J Cogn Neurosci 2:81–95
Johnson MH (1995) The development of visual attention: a cognitive neuroscience perspective. In: Gazzaniga MS (ed) The cognitive neurosciences. MIT Press, Cambridge, pp 735–747
Johnson SP (1996) Habituation patterns and object perception in young infants. J Reprod Infant Psychol 14:207–218
Johnson SP (1997) Young infants' perception of object unity: implications for development of attentional and cognitive skills. Curr Dir Psychol Sci 6:5–11
Johnson SP (2001) Neurophysiological and psychophysical approaches to visual development. In: Kalverboer AF, Gramsbergen A (eds) Handbook of brain and behaviour in human development. Elsevier, Amsterdam, pp 653–675
Johnson MH (2005) Developmental cognitive neuroscience, 2nd edn. Blackwell, Oxford
Johnson SP (2010) How infants learn about the visual world. Cogn Sci 34:1158–1184
Johnson SP, Johnson KL (2000) Early perception-action coupling: eye movements and the development of object perception. Infant Behav Dev 23:461–483
Johnson MH, Tucker LA (1996) The development and temporal dynamics of spatial orienting in infants. J Exp Child Psychol 63:171–188
Johnson MH, Posner MI, Rothbart MK (1991) Components of visual orienting in early infancy: contingency learning, anticipatory looking, and disengaging. J Cogn Neurosci 3:335–344
Johnson MH, Posner MI, Rothbart MK (1994) Facilitation of saccades toward a covertly attended location in early infancy. Psychol Sci 5:90–93
Johnson SP, Bremner JG, Slater A, Mason U, Foster K, Cheshire A (2003) Infants' perception of object trajectories. Child Dev 74:94–108
Johnson SP, Slemmer JA, Amso D (2004) Where infants look determines how they see: eye movements and object perception performance in 3-month-olds. Infancy 6:185–201
Johnson SP, Davidow J, Hall-Haro C, Frank MC (2008) Development of perceptual completion originates in information acquisition. Dev Psychol 44:1214–1224
Kiorpes L, Movshon JA (2004) Neural limitations on visual development in primates. In: Chalupa LM, Werner JS (eds) The visual neurosciences. MIT Press, Cambridge, pp 159–188
Komatsu H, Wurtz RH (1989) Relation of cortical areas MT and MST to smooth pursuit eye movements. In: Ito M (ed) Neural programming. Japan Scientific Societies Press, Tokyo, pp 137–148
Kremenitzer JP, Vaughan HG Jr, Kurtzberg D, Dowling K (1979) Smooth-pursuit eye movements in the newborn infant. Child Dev 50:442–448
Kretch KS, Franchak JM, Adolph KE (2014) Crawling and walking infants see the world differently. Child Dev 85:1503–1518
Luna B, Velanova K, Geier CF (2008) Development of eye-movement control. Brain Cogn 68:293–308
Macchi Cassia V, Simion F, Milani I, Umiltà C (2002) Dominance of global visual properties at birth. J Exp Psychol Gen 131:398–411
Mahdi A, Su M, Schlesinger M, Qin J (2017) A comparison study of saliency models for fixation prediction on infants and adults. IEEE Trans Cogn Dev Syst 10:485–498
Mareschal D, Johnson SP (2002) Learning to perceive object unity: a connectionist account. Dev Sci 5:151–185
Mareschal D, Quinn PC (2001) Categorization in infancy. Trends Cogn Sci 5:443–450
Marx V, Nagy E (2015) Fetal behavioural responses to maternal voice and touch. PLoS One 10: e0129118
Maurer D, Lewis TL (1998) Overt orienting toward peripheral stimuli: normal development and underlying mechanisms. In: Richards JE (ed) Cognitive neuroscience of attention: a developmental perspective. Erlbaum, Mahwah, pp 51–102
Melcher D, Kowler E (2001) Visual scene memory and the guidance of saccadic eye movements. Vis Res 41:3597–3611
Miller EK (2000) Organization through experience. Nat Neurosci 3:1066–1068

Miyashita Y, Hayashi T (2000) Neural representation of visual objects: encoding and top-down activation. Curr Opin Neurobiol 10:187–194
Neill WT (1977) Inhibitory and facilitatory processes in selective attention. J Exp Psychol Human 3:444–450
Nuthmann A, Henderson JM (2010) Object-based attentional selection in scene viewing. J Vis 10:20
Papageorgiou KA, Smith TJ, Wu R, Johnson MH, Kirkham NZ, Ronald A (2014) Individual differences in infant fixation duration relate to attention and behavioral control in childhood. Psychol Sci 25:1371–1379
Piontelli A (2015) Development of normal fetal movements. Springer, Milan
Prechtl HFR (2001) Prenatal and early postnatal development of human motor behavior. In: Kalverboer AF, Gramsbergen A (eds) Handbook of brain and behaviour in human development. Elsevier, Amsterdam, pp 415–427
Preston KL, Finocchio DV (1983) Development of the vestibulo-ocular and optokinetic reflexes. In: Simons K (ed) Early visual development: normal and abnormal. Oxford University Press, New York, pp 80–88
Pritchard VE, Neumann E (2004) Negative priming effects in children engaged in nonspatial tasks: evidence for early development of an intact inhibitory mechanism. Dev Psychol 40:191–203
Quinn PC, Bhatt RS (1998) Visual pop-out in young infants: convergent evidence and an extension. Infant Behav Dev 21:271–288
Reid VM, Dunn K, Young RJ, Amu J, Donovan T, Reissland N (2017) The human fetus preferentially engages with face-like visual stimuli. Curr Biol 27:1825–1828
Richards JE (1998) Cognitive neuroscience of attention: a developmental perspective. Erlbaum, Mahwah
Richards JE (2001) Cortical indices of saccade planning following covert orienting in 20-week-old infants. Infancy 2:135–157
Ross S, Dannemiller JL (1999) Color contrast, luminance contrast, and competition within exogenous orienting in 3.5-month-old infants. Infant Behav Dev 22:383–404
Roucoux A, Culee C, Roucoux M (1983) Development of fixation and pursuit eye movements in human infants. Behav Brain Res 10:133–139
Rovee-Collier C, Bhatt RS, Chazin C (1996) Set size, novelty, and visual pop-out in infancy. J Exp Psychol Human 22:1178–1187
Schiller PH (1985) A model for the generation of visually guided saccadic eye movements. In: Rose D, Dobson V (eds) Models of the visual cortex. Wiley, New York, pp 62–70
Schiller PH (1988) The neural control of eye movements. In: Richards JE (ed) Cognitive neuroscience of attention: a developmental perspective. Erlbaum, Mahwah, pp 3–50
Schiller PH (1996) On the specificity of neurons and visual areas. Behav Brain Res 76:21–35
Schlesinger M, Amso D, Johnson SP (2012) Simulating the role of visual selective attention during the development of perceptual completion. Dev Sci 15:739–752
Schneider GE (1969) Two visual systems. Science 163:895–902
Simion F, Valenza E, Umiltà C, Dalla Barba B (1995) Inhibition of return in newborns is temporonasal asymmetrical. Infant Behav Dev 18:189–194
Simone PM, Mccormick EB (1999) Effect of a defining feature on negative priming across the life span. Vis Cogn 6:587–606
Sirois S, Mareschal D (2002) Models of habituation in infancy. Trends Cogn Sci 6:293–298
Slater A (1995) Visual perception and memory at birth. In: Rovee-Collier C, Lipsitt LP (eds) Advances in infancy research, vol 9. Ablex, Norwood, pp 107–162
Slater AM, Morison V, Rose DH (1983) Perception of shape by the new-born baby. Br J Dev Psychol 1:135–142
Slater A, Morison V, Somers M, Mattock A, Brown E, Taylor D (1990) Newborn and older infants' perception of partly occluded objects. Infant Behav Dev 13:33–49
Slater AM, Brown E, Mattock A, Bornstein M (1996a) Continuity and change in habituation in the first 4 months from birth. J Reprod Infant Psychol 14:187–194

Slater A, Johnson SP, Brown E, Badenoch M (1996b) Newborn infants' perception of partly occluded objects. Infant Behav Dev 19:145–148

Slone LK, Moore DS, Johnson SP (2018) Object exploration facilitates 4-month-olds' mental rotation performance. PLoS One 13:e0200468

Soska KC, Adolph KE, Johnson SP (2010) Systems in development: motor skills acquisition facilitates 3D object completion. Dev Psychol 46:129–138

Spelke ES (1990) Principles of object perception. Cogn Sci 14:29–56

Spelke ES, Kinzler KD (2007) Core knowledge. Dev Sci 10:89–96

Spelke ER, Breinlinger K, Macomber J, Jacobson K (1992) Origins of knowledge. Psychol Rev 99:605–632

Sperry RW (1963) Chemoaffinity in the orderly growth of nerve fiber patterns and their connections. Proc Natl Acad Sci U S A 50:703–710

Stiles J (2017) Principles of brain development. WIREs Cogn Sci 8:e1402. https://doi.org/10.1002/wcs.1402

Stoll J, Thrun M, Nuthmann A, Einhaüser W (2015) Overt attention in natural scenes: objects dominate features. Vis Res 107:36–48

Tanaka K (1997) Mechanisms of visual object recognition: monkey and human studies. Curr Opin Neurobiol 7:523–529

Tatler BW, Hayhoe MM, Land MF, Ballard DH (2011) Eye guidance in natural vision: reinterpreting salience. J Vis 11:5

Teller DY, Movhson JA (1986) Visual development. Vis Res 26:1483–1506

Thier P, Ilg UJ (2005) The neural basis of smooth-pursuit eye movements. Curr Opin Neurobiol 15:645–652

Tipper SP (1985) The negative priming effect: inhibitory priming by ignored objects. Q J Exp Psychol 37:571–590

Treisman A (1988) Features and objects: the fourteenth Bartlett memorial lecture. Q J Exp Psychol 40:210–237

Treisman A, Gelade G (1980) A feature integration theory of attention. Cogn Psychol 12:97–136

Treisman A, Gormican S (1988) Feature analysis in early vision: evidence from search asymmetries. Psychol Rev 95:15–48

Valenza E, Bulf H (2011) Early development of object unity: evidence for perceptual completion in newborns. Dev Sci 14:799–808

Valenza E, Simion F, Umiltà C (1994) Inhibition of return in newborn infants. Infant Behav Dev 17:293–302

van Renswoude DR, Johnson SP, Raijmakers MEJ, Visser I (2016) Do infants have the horizontal bias? Infant Behav Dev 44:38–48

van Renswoude DR, Visser I, Raijmakers MEJ, Tsang T, Johnson SP (2019) Real-world scene perception in infants: what factors guide attention allocation? Revision in preparation

von Hofsten C (2004) An action perspective on motor development. Trends Cogn Sci 8:266–272

Wass SV, Smith TJ (2014) Individual differences in infant oculomotor behavior during the viewing of complex naturalistic scenes. Infancy 19:352–384

Wattam-Bell J (1996) Visual motion processing in 1-month-old infants: habituation experiments. Vis Res 36:1679–1685

Winkler KC, Williams RW, Rakic P (1990) Photoreceptor mosaic: number and distribution of rods and cones in the rhesus monkey retina. J Comp Neurol 297:499–508

Wong ROL (1999) Retinal waves and visual system development. Annu Rev Neurosci 22:29–47

Xu J, Jiang M, Wang S, Kankanhalli MS, Zhao Q (2014) Predicting human gaze beyond pixels. J Vis 14:28–28

Zenger B, Braun J, Koch C (2000) Attentional effects on contrast detection in the presence of surround masks. Vis Res 40:3717–3724

Variations in the Beneficial Effects of Spatial Structure and Serial Organisation on Working Memory Span in Humans and Other Species

Carlo De Lillo

Contents

Abstract This chapter reviews studies of spontaneous search in large-scale settings and studies featuring variations of the Corsi test in humans and animals. It aims to highlight a synergy of working memory (WM) processes and the use of spatio-temporal structure and explain its underpinnings within a comparative framework. The chapter starts by showing that the degree of organisation of serial search patterns spontaneously deployed by humans and animals in simulated foraging tasks is associated with a reduction of WM errors. Then, by comparing studies conducted on different species, it exposes a parallel between the degree of search organisation and taxonomic relatedness to humans. Such a parallel could indicate that a hallmark of the cognition of humans and closely related species is the ability to offload WM by developing serially organised search patterns that exploit the spatial structure of the environment. However, a causal relationship between serial organisation and search efficiency can only be inferred with serial recall tasks, where the structure of specific sequences can be systematically manipulated. Thus, studies using variations of the Corsi test are considered subsequently, which suggest that humans might enjoy an exceptional aptitude to benefit from the spatio-temporal structure in serial

C. De Lillo (✉)
University of Leicester, Leicester, UK
e-mail: CDL2@LE.AC.UK

Curr Topics Behav Neurosci (2019) 41: 59–84
DOI 10.1007/7854_2019_97
Published Online: 20 July 2019

tasks, despite remarkable memory abilities shown by other primate species as well. The extent to which the benefit of spatial organisation in human WM span must be mediated by perceptual grouping processes is then considered. To clarify this issue, recent experiments using virtual reality to compare serial recall in small visual displays that afford perceptual grouping and in immersive navigational spaces that cannot do so are discussed. The results of these latter experiments indicate that the effects of structure in serial recall emerge in conditions not affording grouping at perceptual level. Thus, it is suggested that more central representational processes play a role in the interaction between spatio-temporal organisation and working memory span in humans.

Keywords Comparative cognition · Corsi test · Evolution · Memory · Search

1 Introduction

Working memory (WM) refers to the temporary storage of information which is necessary for the operation of higher cognitive processes and provides a buffer for the interfacing of perception, long-term memory and action (Baddeley 2003). In humans it is associated with general intelligence and linked to attention and consciousness. Given the close relationship between WM and human higher-level cognitive processes, the systematic study of WM processes in different species at a different taxonomic distance from humans could provide critical information about the steps that led to the emergence of human cognitive sophistication.

Spatial WM refers to the retention of series of spatial locations. According to the most influential model of human WM (Baddeley and Hitch 1974), it is served by dedicated components of WM which do not pertain to verbal information (see also Smith et al. 1996). As the direct comparison of WM in humans and in other animals requires the assessment of WM for information that is not linguistic in nature, spatial WM is ideally suited for comparative studies. Spatial WM is often measured as spatial span, which denotes the number of locations that an organism can retain and monitor over a limited period or until the task at hand requires the memory buffer to be reset. Despite its widespread use, it could be argued that spatial span is the measure of spatial WM that is most prone to be confounded by other factors. For example, memory for multiple locations will be greatly affected by their specific configuration and familiarity. Moreover, algorithms can be developed for the efficient exploration of locations arranged in given configurations, making spatial span likely to be affected by rule learning processes or the development of specific serial search patterns. However, I shall argue in this chapter that it is exactly because it taps processes at the interface between mental capacity, long-term memory and organisational principles that spatial span, if tested under appropriate conditions, is an ideal measure for the characterisation of the idiosyncrasies of human cognition in comparison with the mental capacity of other species.

It is becoming increasingly accepted that understanding how WM interacts with organisational processes is key for the refinement of current models of human working memory (Hurlstone et al. 2014; Vandierendonck 2018). Assessing similarities and differences of such interaction in other species could help us determine what makes it possible. Moreover, by systematically testing species with which humans shared an ancestor at different points of their natural history should allow deductions concerning the steps that led to the architecture of WM in modern humans. Such a comparative approach would require measuring spatial span in humans and animals with similar procedures. However, attempts to do so have been rare.

2 Spatial WM in Animals and Humans

In animals, spatial WM is often determined by assessing memory for locations explored during search for food in simulated foraging tasks. Searching multiple locations is a ubiquitous process that naturally pertains to a number of real-life problems and cognitive tasks. Indeed, it has been claimed that similar processes are deployed when searching real or virtual environments for resources or rewards (Astur et al. 2005), the Internet for information (Pirolli 2009) or the mental lexicon for given words (Wilke et al. 2009). In general, efficient search requires avoiding redundant revisits of locations explored in a given foraging bout. Spatial span in search can be naturalistically determined by the number of locations that an organism can keep track of before starting to make WM errors by revisiting previously explored locations. The specific path taken by an organism through the search space provides information about organisational principles spontaneously deployed to maximise search efficiency.

Search builds on natural behaviours of a variety of species. As such, it is a powerful comparative tool. Nonetheless, the interdependence between spatial structure, serial organisation and WM capacity can be evinced only correlationally with naturalistic search tasks, making it necessary to use other paradigms to establish causal links between these factors.

In humans, the most common test for the assessment of spatial span is the Corsi block-tapping test (CBTT; Corsi 1972). The CBTT is a serial recall task. Participants observe a tester tap with their finger wooden blocks glued on a board, or in computerised versions of the test icons flashing on a monitor, according to sequences that they are required to rehearse and reproduce. Despite its popularity with humans, the CBTT has not been used with animals, apart from very rare exceptions (Botvinick et al. 2009; Fagot and De Lillo 2011). Indeed, the extent to which animals are able to perform a serial recall task in conditions analogous to those used with humans has been questioned (Carruthers 2013). In contrast with search tasks, the ecological validity of the Corsi test may be limited. In fact, aside from some specific forms of imitation (e.g. of a dance routine in humans or perhaps very specific forms of tool use and manufacture), it is difficult to envisage naturalistic contexts where an organism is required to repeat all the steps of a sequence that has just been observed. Nonetheless, in this chapter I shall argue that serial recall tasks such as the Corsi test can be a very powerful tool for assessing experimentally the relationship between serial spatial organisation and WM span.

Thus, it seems that a fertile line of comparative investigation would require the integration of approaches focused on spontaneous search, so far mostly investigated in animals and those focused in serial recall largely investigated in humans. The possibility of pursuing such line of research is becoming more tangible. One of the reasons why different paradigms are used to assess spatial WM in humans and smaller animals is the scale of the environment required by different species. It is difficult to test humans in large unfamiliar environments where key variables can be flexibly manipulated. Modern immersive virtual reality (VR) techniques start to offer the opportunity to test humans in realistic navigational environment where search as a well as serial recall tasks can be implemented in novel and very informative ways.

In the attempt to highlight the importance of considering search behaviour alongside serial recall from a comparative perspective, this chapter is organised as follows. In the next section, I will describe studies which point to differences in the way in which various species spontaneously organise the serial exploration of a set of locations and how search organisation relates to WM. In the section that follows, I shall review evidence gathered with serial recall tasks and variations of the Corsi test applied to a comparative framework. Throughout the chapter, I will present evidence recently gathered using VR for studying search and serial recall in humans to add key evidence that could not be obtained with desktop tasks.

3 The Spontaneous Emergence of Serial Control in Search and Foraging

One of the most established set-up for assessing spatial working memory in animals is the radial maze (Olton and Samuelson 1976). It features a number of arms departing radially from a central platform. In the original procedure (Olton and Samuelson 1976), each of the eight arms in the maze is baited at the beginning of each trial. The subject is then allowed to search, and the number of different arms it visits within the first eight arm choices provides an indication of spatial span. This measure and variations of it are related to span as they take into account WM errors consisting in re-entering arms that have already been searched in a trial.

Olton and Samuelson (1976) carried out one of the first systematic investigations of the ability of animals to keep track of a number of visited locations when exploring a large set of them. This ability taps onto a naturalistic aspect of foraging, which typically requires searching a large number of locations within the home range of the organism and monitoring the availability and depletion of resources in each of them. The radial structure of the maze was originally conceived as an ingenious way of ensuring that the animal, by returning to the central platform after each arm's exploration, is always faced with the same large number of choices at the time of selecting the next arm to visit (Olton and Samuelson 1976). Ergonomically, the radial maze is most suitable for small mammals, but variations of it, consisting in a set of containers arranged in a circle, have been used as an alternative set-up with pigeons (Spetch and Edwards 1986) and children (Foreman et al. 1984). Virtual reality versions of the task have been developed for adult humans and used in fMRI studies (Astur et al. 2005).

A peculiarity of a circular configuration of search locations is that they can be efficiently explored algorithmically by developing systematic patterns of responses. For example, Olton and Samuelson (1976) noted that although the rats they tested in the radial maze did not generally search adjacent arms consecutively, they showed a tendency to choose arms according to a rule such as "always choose an arm that is 90 degrees from the arm just explored" and occasionally observed strings of adjacent arms chosen in succession in clockwise or counter-clockwise order (Olton and Samuelson 1976; see also Dubreuil et al. 2003). Since this type of algorithmic responding can offset memory, a number of procedures have been developed to prevent it, including blocking some arms to prevent access by the animals or altering the regularity of their angular arrangement (see Schenk et al. 1995). Nonetheless, I contend that it is precisely the ability to organise systematic search patterns as a function of the shape of the search space to reduce memory load that is essential to study, if we want to understand the ontogenetic and phylogenetic origins of higher human cognition.

A first line of evidence in support of this claim shows that the emergence of even very simple forms of algorithmic responding in the radial maze parallels cognitive development in humans. Two-year-old children tested in variations of the radial maze do not show evidence of algorithmic responding, and their ability to keep track of previously visited locations is very poor (Foreman et al. 1984; Aadland et al. 1985). By contrast, by the age of 4, children develop algorithmic search patterns consisting in searching adjacent arms consecutively and consistently following a given direction of travel, clockwise or counter-clockwise (Foreman et al. 1984; Aadland et al. 1985). In contrast with 2-year-olds, 4-year-old children rarely make redundant visits to locations already explored within a trial. Moreover, when older children are prevented from using algorithmic strategies, their ability to monitor the arms they have visited collapses, whereas radial maze performance in younger children remains unchanged in these conditions (Foreman et al. 1984; Aadland et al. 1985). This latter observation suggests that there may be a causal relationship between deploying systematic searches and performance.

It could be argued that the peculiar shape of a radial maze does not reflect the layout of naturalistic environments and that the presence of structure is unnaturally exacerbated in this type of maze. It is however important to note that any spatial arrangement of multiple locations can afford some form of structure, which could in principle be exploited by an agent that is left free to select their search path through them. For example, another simple form of structure from which organisms can benefit is the organisation of target locations in spatial clusters. This is the arrangement of patchily distributed resources in naturalistic foraging environments. Spatial clusters can be defined formally on the basis of the relative distance of each location from any other, so that there are groups of locations where the distance between each of them is shorter than the minimum distance of locations within any two groups.

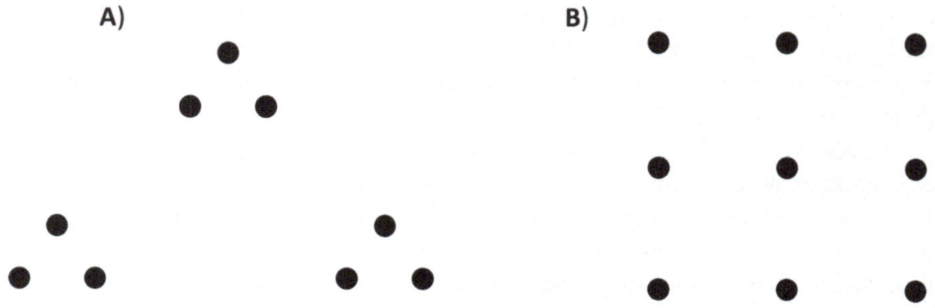

Fig. 1 (**a**) Clustered arrangement of nine locations; (**b**) nine locations arranged as a square matrix

An example of a regular clustered arrangement of nine locations into three groups (clusters) of three locations each is provided in Fig. 1a.

This type of clustered arrays affords principled explorations. An organism could systematically explore each of the clusters of locations exhaustively before moving onto another cluster. By developing this search pattern, it could reduce the number of locations that it needs to retain in memory during search. In fact, efficient search would require the subject to keep track of the locations within a cluster while exploring it. However, once a cluster has been fully explored, it would only be necessary to encode in WM the approximate location of the entire cluster (rather than that of each individual item within it) as an explored area to avoid going back to it. Subsequently, the same could be done for other clusters. By deploying such a systematic search pattern, an efficient exploration of the set of locations could be carried out which would only require monitoring the number of clusters explored rather than the much larger number of individual locations visited during search. As a consequence, the memory load imposed by the task would be dramatically reduced compared to the load imposed by an unsystematic search.

3.1 Effect of Spatial Clustering on Foraging in Monkeys, Rats and Humans

In a study aimed at determining the extent to which monkeys searching for food in a simulated foraging task benefit from spatial clustering in the way just described, we tested a group of capuchin monkeys housed at the primate centre of the National Research Council of Italy in Rome (De Lillo et al. 1997). The set-up comprised nine plastic containers suspended from the ceiling of a large enclosure. Each container was baited with a peanut at the beginning of each trial, before letting the animal enter the enclosure and allowing it to retrieve the peanuts from all the nine containers. When this happened, the trial terminated. The animal was let out of the enclosure, the containers were rebaited out of sight of the animal, and then a new trial was started by letting the animal back into the enclosure. As a dependent measure, we recorded the total number of times that the subject visited any of the containers before all the nine peanuts were retrieved. The set-up with suspended containers was designed so that visiting a container did not alter its position or orientation. Thus, no visible trace was left of the containers that had been explored. Therefore, by looking at the number of total visits, we could judge the ability of the animals to keep track of the locations visited during each trial and ensure search efficiency. In order to measure the efficiency gained by capuchin monkeys when searching a clustered set of locations, compared to a non-clustered set, we tested first the animals with 60 trials of a baseline condition where they searched a square matrix of 9 containers. The configuration of this search space is shown in Fig. 1b. Then, we tested the animals with an experimental condition featuring the clustered search space described above for 60 trials (Fig. 1a). Finally, an additional 60 trials were given as a control condition featuring the matrix of locations (Fig. 1b) to determine whether any increase in search performance in the clustered configuration compared to the baseline condition could be due to task practice alone, rather than to the configuration of the search space.

With the above paradigm, we observed a dramatic improvement in the search performance of capuchin monkeys when they were faced with the clustered configuration compared to both the baseline and the control conditions that featured a matrix of locations (De Lillo et al. 1997). Interestingly a linear trend in the reduction of redundant moves to previously visited locations was observed across the 60 experimental trials. This suggests that the increase in search efficiency was not just immediately induced by the clustered configuration but instead required the animals to learn to exploit the constraints of such a configuration. Of particular importance was to note the strategy that accompanied the increase in search efficiency. This consisted in increasingly avoiding exiting the area of a cluster and starting exploring a different cluster before all the containers within the first cluster were exhaustively explored. In fact, the occasions in which the subjects exited a cluster before having exhaustively explored it were correlated to the number of errors consisting in visiting an already visited container. Moreover, the majority of errors made by the animals consisted in failing to identify which container had already been visited within a previously partially explored cluster. This pattern of results is consistent with the notion that monkeys' search was sustained by hierarchical coding. This means that when all the locations within a cluster had been visited, the area of the cluster could be marked as an explored region rather than having to keep track of each individual location that had been visited within a cluster (see Fig. 2b).

Capuchin monkeys did not show a positive trend in search efficiency in the search space arranged as a matrix. For comparative purposes, I subsequently tested rats and preschool children using the same paradigm (De Lillo 2012). The environments used to test rats and children were appropriately scaled. Each matched the ratio between the size of the environment used to test their search skills in monkeys and the body size of the monkeys. Rats were tested using a search space made up of baited poles that the rats had to explore by rearing to collect fragments of nuts in food wells at the top of the poles. They received the same number of trials used for the monkeys. Preschool children (3–5 years of age) were tested in similar conditions as part of a

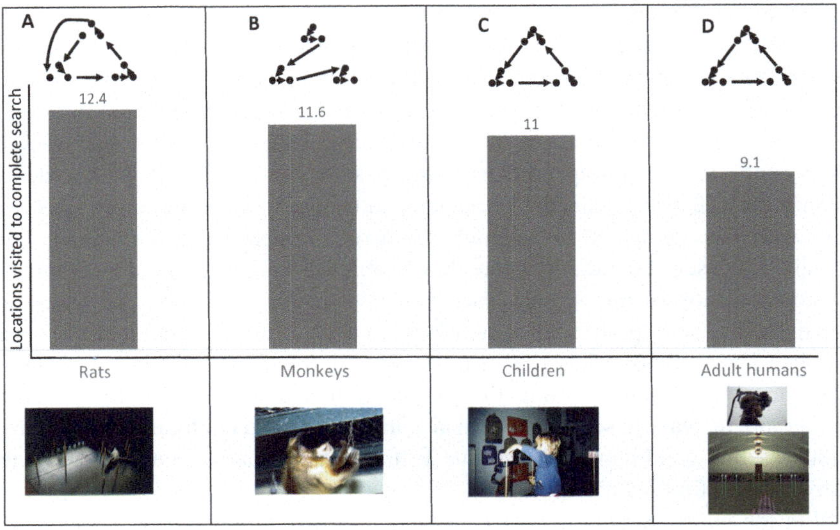

Fig. 2 Interspecies comparison of search efficiency and organisation. Data for rats (**a**) and children (**c**) from De Lillo (2012); data for capuchin monkeys (**b**) from De Lillo et al. (1997); data for adult humans (**d**) from De Lillo et al. (2014). The top row depicts the typical search pattern observed for the given species. The depicted pattern is the observed trajectory used to explore the set of nine target locations in nine total visits (optimal search) that was most highly correlated with all the other trajectories used in optimal searches by the given species (see De Lillo 2012 for a detailed explanation of the analysis). The bar graph in the middle row shows the average number of visits used by the given species to visit all the nine locations in the search space. The bottom line depicts the real-life (**a–c**) or VR search environment (**d**) used to test the given species

game where they had to retrieve toy insects from poles surmounted by coloured containers representing flowers in a field. The poles could be adjusted depending on the height of each child. This ensured that the children could not look into the flower/containers to check whether they still hosted a toy insect. They could only check for the presence of the toy by raising their arms over their head and exploring the inside of a flower/container with their hands (see Fig. 2c, bottom row). This made the set-up similar to that used with rats. Children benefitted from the clustered arrangement of the search space compared to both the baseline and the control condition, suggesting a strong effect of configuration of search targets. Rats showed an improvement of search efficiency in the clustered configuration compared to the baseline condition, but the performance did not drop at baseline levels in the control condition suggesting that the clustering effects may be at least in part confounded by a general effect of practice with the search task in this species. When the three species were compared on the basis of their performance on the clustered configuration, children emerged as the most proficient searchers, followed by monkeys and then by rats.

Crucially, this pattern of search efficiency across species was found to be associated with measures of search organisation. Children used a clear clustering principle whereby all the locations within each cluster were usually exhaustively explored before exiting the cluster. Moreover, children also showed a tendency to follow a particular direction of travel (clockwise or counter-clockwise depending on individual children) when exploring one cluster after the other. The prototypical search pattern shown by children is shown in Fig. 2c.

Rats also showed a tendency to search containers within the same cluster consecutively (see Fig. 2a). However, they exited a cluster before all the containers within it had been explored much more frequently that it was the case for children and monkeys (De Lillo 2012). Thus, the trend in search efficiency observed when comparing rats, monkeys and children is accompanied by a trend in search organisation with children showing a more principled search pattern than monkeys and monkeys showing a more principled search pattern than rats when exploring a clustered set of locations.

Further studies which have assessed search efficiency in variations of the paradigm used by De Lillo et al. (1997) with other species provide additional interesting results. Mice (Valsecchi et al. 2000) fail to benefit from a clustered organisation of containers and so do rats tested in another study (Foti et al. 2007). A study (Bartolomucci et al. 2001) with tree shrews (*Tupaia belangeri*) – a species that has been widely considered the closest taxonomically, among non-primate species, to the order primates – shows that this species too fails to benefit from a clustered search space, compared to a matrix of locations. Nonetheless, the authors (Bartolomucci et al. 2001) report that the tree shrew shows more evidence of clustering than the mice tested by Valsecchi et al. (2000).

As it is impractical to devise a large-scale environment suitable for testing adult humans in conditions similar to those used with children or non-human species, we devised an immersive virtual reality version of the task where we could scale-up the environment used with children to test adult humans in similar conditions (De Lillo et al. 2014). The search space used with adult humans consisted in nine poles surmounted by a sphere within a large hall with richly textured surfaces and a variety of landmarks (see Fig. 2d, bottom line). Thus, this was similar to the real-life environment where children were tested, which was a nursery hall (see De Lillo 2012 for a description and Fig. 2c bottom line). Adult humans were tested using a stereoscopic head-mounted display. A position tracker determined the viewpoint depending on the head and body movement of the participants who operated a wand to interact with the environment. Participants were asked to select the spheres on each of the nine poles by navigating towards them, placing a virtual hand visible within the display and pulling the trigger of the wand to select it. This mimicked the task used with children. In the virtual environment, a transient written message provided feedback that the location had been "visited". Participants then travelled to another pole in the environment, and so on, until each of the poles had been selected. As soon as this happened, the trial ended and a new trial started. As for the studies with other species, no cues were left to mark visited locations during a trial. Environments with poles arranged in a matrix or in a clustered configuration alternated in consecutive trials until six trials were presented. In these conditions, participants proved extremely proficient. On average, they completed a search of nine poles with 9.07 (SD = 0.20) pole selections in the clustered configuration and 9.47 (SD = 0.99) selections in the matrix configuration. This nearly optimal performance was accompanied by highly organised search patterns in both conditions. In the clustered configuration, participants exhibited the same strategy observed in children. They explored each of the clusters exhaustively before moving to another cluster and did so by following a specific clockwise or counter-clockwise direction of travel. However, in contrast with both children and non-human animal species, adult humans deployed a clear search strategy in the matrix configuration too, which consisted in always selecting poles in the same row or column in consecutive selections.

Considered together, the results of the studies reviewed above suggest a progression in the deployment of serially organised search patterns that goes in parallel with taxonomic relatedness to humans for cross-species comparisons and with cognitive growth among humans (see Fig. 2a–d, top row). Moreover, searches that are most organised serially are associated with a reduction of working memory errors consisting in revisiting previously explored locations (see Fig. 2a–d, second row). This suggests a possible causal link between the serial organisation of search and memory efficiency. However, the presence of such causal link is impossible to prove with search studies. The study of search behaviour in relation to WM is essential for detecting the spontaneous emergence of serially organised behaviour and how it interfaces with mental capacity. Moreover, it enables the formulation of plausible hypotheses about what adaptive value serial ordering could have in an ecologically valid task domain such as search and foraging.

As we have seen in the examples above, it is possible in search studies to manipulate the structure of the search space experimentally to see how it affects search organisation and memory performance. Nonetheless, in studies of search and foraging, the relationship between search organisation and memory performance can only be assessed with correlational methods. For this reason, it is of key importance to integrate research on search and foraging, where the organism is left free to select the trajectory to follow throughout the search space, with research using tests where the path taken to explore the search space can be manipulated as part of the experimental design. Serial recall tasks are ideally suited for this purpose. In the next section, I review studies that have used serial recall tasks such as variations of the Corsi block-tapping test (CBTT) to assess experimentally the nature of the relationship between serial organisation and memory performance.

4 The Experimental Control of Serial Organisation: Structure and the Corsi Test

As introduced at the beginning of this chapter, the most commonly used test of visuospatial working memory is the CBTT (Corsi 1972). The CBTT is a serial recall task requiring participants to repeat back an observed sequence of locations in the correct order. An implicit assumption of early studies carried out using the CBTT was that, because the configuration of locations featured in the test is spatially irregular, random sequences with the same number of items should be of a similar level of difficulty to reproduce. However, it soon became clear that this is not the case and that, even if the array of items is spatially irregular, some sequences are more difficult to reproduce than others (Smirni et al. 1983). Thus, attempts have been made at standardising the test so that the same sequences, presented in an irregular array, can be used across different studies and clinical groups (Kessels et al. 2000). Nonetheless, when spatial span is measured with irregular arrays it is always the product of an unspecified combination of memory capacity and organisational processes that can be used to segment the sequence into chunks whenever some elements of structure are detected (e.g. when consecutive elements are perceived to be in a linear path or in close spatial proximity to each other). This makes it difficult to determine which cognitive mechanisms and resources support the recall of sequences in these conditions. This problem can be obviated by devising variations of the CBTT with arrays of items that feature a very regular spatial structure. Then, sequences presented therein can be manipulated, so that some sequences conform to specific organisational factors afforded by the structure of the array and some are not consistent with them (see Bor et al. 2003; De Lillo 2004; De Lillo and Lesk 2010). These variations of the CBTT are the ideal tool for investigating the causal relationship between serial organisation and the exploration of a set of locations. This approach was used to probe further, with adult humans, the nature of the relationship between specific search trajectories in a clustered foraging environment and search efficiency as observed in comparative studies (De Lillo 2004). Participants were asked to repeat back sequences of three types. Sequences of type A replicated the highly organised search pattern deployed by children and consisting in selecting all the locations within a cluster before moving to a different cluster and in following the same direction of travel both within and between clusters. Sequence type B replicated the pattern observed in monkeys. They followed the same constraints of sequences type A, so that all the locations within a cluster were selected before moving on to a different cluster. However, the direction of the trajectory was not kept constant within and between clusters. Finally, sequences type C were designed to exemplify the trajectories spontaneously shown by rats. These latter sequences violated the constraints of both sequences type A or type B, so that consecutive selections were never in the same cluster. The results clearly showed that participants were more accurate in reproducing sequences type A than type B, which in turn were reproduced more accurately than sequences type C.

The pattern of results described above suggest a causal relationship between the type of trajectory used to visit a series of locations and the ability to keep track of them to avoid returning to previously visited ones. However, different hypotheses can be formulated to explain the nature of this relationship in the recall of sequences presented within a clustered array. The first is that principled serial patterns, where each cluster is explored in turn, allow participants to form hierarchical representations of the set of locations. According to this hypothesis, a subordinate level of the hierarchical representation would be used to keep track of individual locations within a cluster, whereas a superordinate level of the representation would keep track of the location of the clusters, once they have been explored.

A hierarchical representation would enable a more efficient use of information in WM during search as the organism would only need to monitor the set of locations within a cluster before the cluster is fully explored and then the set of clusters, as explained above. A second explanation is that the path characteristics of the movement used to reproduce the sequence fully account for the beneficial effects of clustering without having to assume the formation of a hierarchical representation. One prominent path characteristic is the length of the movement necessary to visit the locations in the prescribed order. Path length has been shown to negatively affect performance in humans tested with the CBTT and related tasks. The reason why sequences with a longer path are more difficult to reproduce is still unclear, but it has been proposed that it may be due to the longer time presumably required to rehearse long-path sequences by scanning a mental image of the set of locations (Parmentier et al. 2005). Another path characteristic shown to deteriorate recall is the number of times that the trajectory of the movement connecting the items in the sequence crosses over itself. The contribution of these different factors in determining the advantage for systematic search trajectories often can be difficult to disentangle. For example, in the study mentioned above (De Lillo 2004), sequences type A would on average have a shorter path and less crossings than sequences type C. Thus, effects of path length, path crossings and hierarchical memory organisation can be confounded.

Further studies (De Lillo and Lesk 2010; De Lillo et al. 2014, 2016) clarify the extent to which path characteristics can explain clustering effects without invoking the causal role of a hierarchical representation. It seems that path length in itself cannot fully explain the benefits of serial clustering in spatially clustered arrays. This is demonstrated by an experiment where path length and clustering were experimentally dissociated (De Lillo et al. 2016, Experiment 1). This was done by using large and small clustered CBTT arrays presented on a touchscreen. Examples of these arrays of different sizes and sequences to be reproduced therein are shown in Fig. 3.

The use of small and large arrays made it possible to present sequences not segregated by clusters in the small array that had a much shorter path length than that of sequences segregated by clusters in the large array. The critical comparison of the recall of clustered sequences with a short-path length with that of not clustered sequences with a long-path length made it possible to determine the effects of structure independently of that of path length. The results of the experiment showed clearly that participants found it easier to reproduce the sequences segregated by clusters, even in the large array, despite their longer path length, compared to that of unstructured sequences presented in the small arrays. Interestingly, very similar results were obtained in a version of the task implemented in immersive virtual reality navigational environments where the scale of the environments exacerbated further the differences in path length in small and largely spaced arrays (De Lillo et al. 2014).

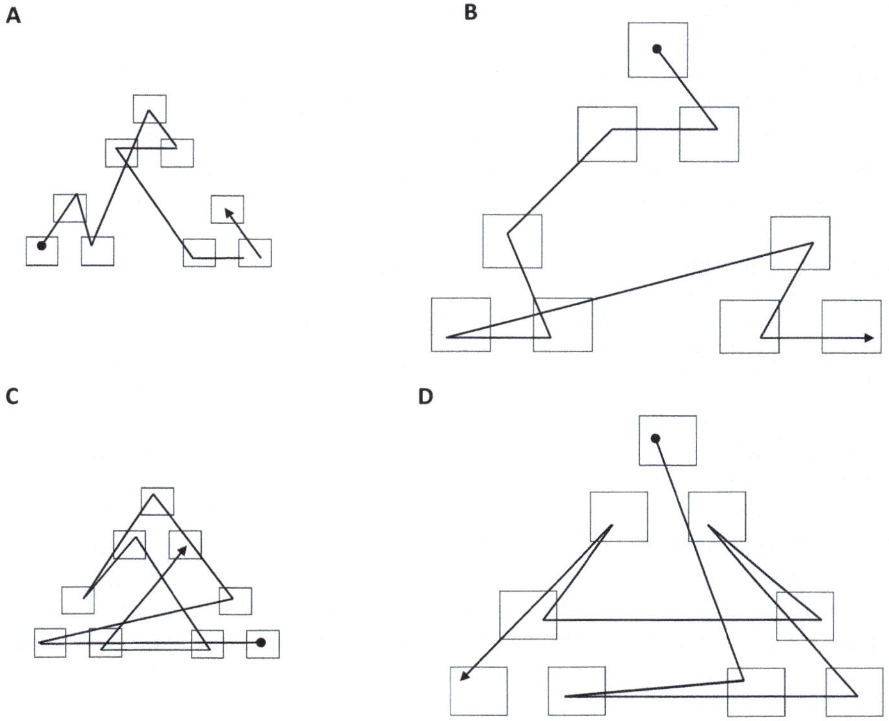

Fig. 3 Examples of sequences and displays used for a Corsi-type task in the study by De Lillo et al. (2016), Experiment 1. Small displays (**a** and **c**) and large displays (**b** and **d**) were used for presenting sequences of nine items that required a short or long movement path, respectively, in order to connect the items in the required serial order. The use of large and small arrays enabled the manipulation of the path length of the sequence independently from the sequence being segregated or not segregated by clusters. (**a**) Example of a sequence segregated by clusters and with a short-path length; (**b**) example of a sequence segregated by clusters and with a long-path length; (**c**) example of a sequence not segregated by clusters and with a short-path length; (**d**) example of a sequence not segregated by clusters and with a long-path length. Redrawn with permission from De Lillo et al. (2016)

4.1 Timing-Based Evidence of Hierarchical Representations in Serial Spatial Recall

Perhaps, the strongest support for the hypothesis that people form hierarchical representations of sequences that are segregated by clusters, compared to sequences that are not, is provided by response time (RT) analyses following manipulations of the composition of clustered CBTT. First, whenever people report sequences segregated by spatial clusters, RT peaks are observed at cluster boundaries (De Lillo 2004; De Lillo and Lesk 2010). This RT pattern is taken as evidence for hierarchical organisation and chunking in a variety of domains (Klahr et al. 1983; Povel and Collard 1982; Sternberg et al. 1988, 1990) and could be an expression of hierarchical coding of information in clustered versions of the CBTT too (De Lillo 2004; De Lillo and Lesk 2010; De Lillo et al. 2016). By contrast, when we exclude the RT for the selection of the first item of the sequence that is very long as it is likely to reflect the planning of the execution of the entire sequence (Fischer 2001; De Lillo and Lesk 2010), sequences not segregated by clusters show an inverted U pattern of RT that resembles the serial position curve observed for non-hierarchical representations of unstructured sequences, with items in the middle of the sequence being the most difficult to recall (De Lillo 2004). These patterns of RT are shown in Fig. 4.

It could be argued that in a spatial task such as the CBTT, RT peaks at cluster boundaries in clustered sequences could be a spurious consequence of the fact that finger movements connecting consecutive touches of items in different clusters are longer than movements connecting touches of items located within the same cluster. This alone could explain the increased RT at cluster boundaries.

A first line of evidence against the above hypothesis comes from an experiment that used a variation of the CBTT procedure. Instead of having to repeat the entire sequences in the recall phase, a single probe item flashed in the display, and participants were required to touch the item that followed it in the previously presented sequence (see De Lillo 2004, Experiment 2). Thus, the same movement length was required for each response irrespectively of whether the item to be selected was at a cluster boundary or within a cluster in the presented sequence. Exactly the same pattern with longer RT at cluster boundaries emerged in this experiment, indicating that this RT pattern is an expression of the representation of the sequence and not just of the time taken to move along longer or shorter distances across the screen during sequence reproduction.

Moreover, compelling evidence in favour of the hierarchical coding hypothesis of the clustering effect in WM is provided by a number of studies featuring the systematic manipulation of clusters' number and size for CBTT sequences of the same number of ordinal steps (De Lillo and Lesk 2010). First, the initial RT for the first touch of the sequence is affected by the number of clusters in which the sequence is segregated. The more clusters there are, the longer the initial RT. Whereas RT at cluster boundaries increases as a function of the number of items within each cluster. These results strongly resemble those used to evidence hierarchical organisation in motor planning for other types of sequences such as typewriting or speech production (Sternberg et al. 1988, 1990). Interestingly, in clustered versions of the CBTT, patterns consistent with hierarchical organisation emerge also when people have to judge whether two sequences are the same and participants do not need to programme a serial motor output. This suggests that hierarchical organisation pertains to the memory coding of the sequence and not the motor plan for the execution of the sequence (De Lillo and Lesk 2010).

Finally, additional support for the notion that clustered sequences in the CBTT are represented hierarchically comes from an experiment where pauses were introduced at critical points during the presentation of the sequences. The results showed that performance deteriorates considerably when pauses are introduced during the

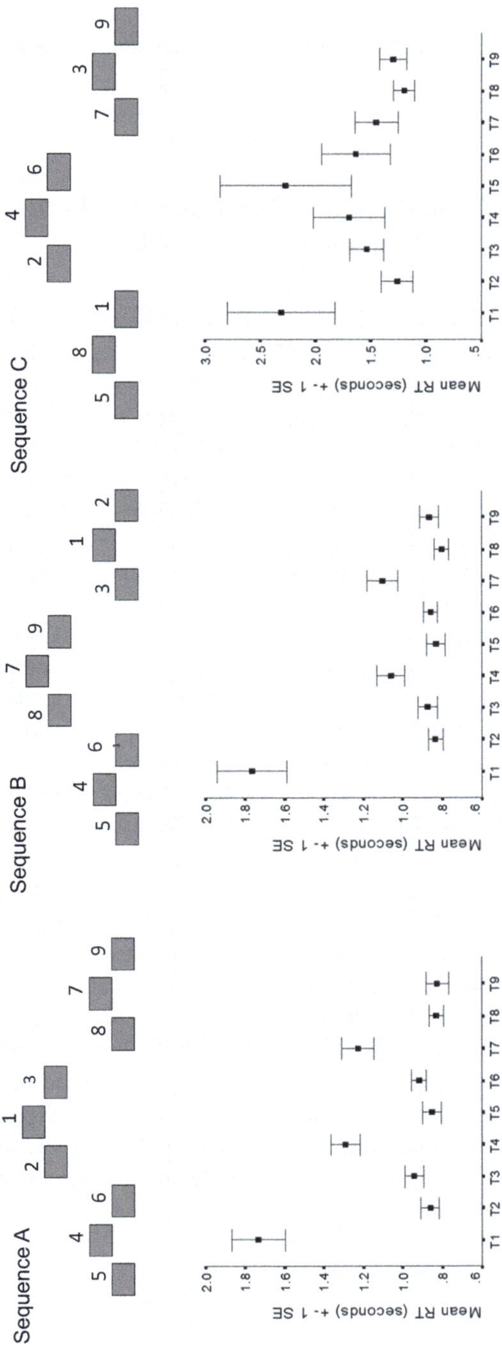

Fig. 4 Sequences and corresponding response times (RT) as observed by De Lillo (2004) (redrawn with permission). The displays featured squares presented on a touch-sensitive monitor (top row). The numbers on the squares are used in the figure to indicate the serial order in which the squares had to be touched and were not shown in the display. The horizontal axis of the graphs (bottom row) indicates the serial order of the touch. *SE* standard error of the mean

presentations of items within a cluster, and as such are not consistent with the segmentation imposed by a hierarchical representation, compared to conditions where the pauses are presented at cluster boundaries and as such are consistent with it (De Lillo et al. 2016).

4.2 The Relative Contribution of Structure and Path Characteristics to Efficiency in Spatial Recall

CBTT displays arranged in spatial clusters are interesting to study because they capture an intuitive and ecologically plausible way in which space can be partitioned in areas and subareas. Also, they mimic the arrangement of patchily distributed natural food resources that human ancestors and other primates with a frugivorous lineage exploited when foraging. Indeed it has been proposed that the evolution of working memory in humans and frugivorous primates stems from specialisations enabling the efficient exploration of resources grouped in space and time (Milton 1993).

However, as we have seen, separating the effects of path characteristics and effects pertaining to the structure of the representation afforded by sequences segregated by clusters requires laborious experimental manipulations. Thus, other highly structured arrays, such as square matrices of items arranged in regular rows and columns, can be extremely useful to disentangle the complex relationship between the processing of spatial structure and spatial span. In arrays arranged as a matrix, it is possible to define structured sequences as those where consecutive items are either within the same row, column or diagonal and unstructured sequences as those that violate this rule (Bor et al. 2003). With this type of arrays, we assessed whether or not path characteristics such as path length and the presence of crossings (another path characteristic that is known to affect spatial WM) can explain all the effects of structure in Corsi-type tasks (De Lillo et al. 2016, Experiment 3). Using a 5×5 square matrix of items, we implemented a fully nested experimental design featuring sequences of seven items with a short- or long-path length, which could contain or could not contain path crossings and that could be either structured or unstructured. We observed a significant effect of all the three factors: path length, crossings and structure. However, when looking at the proportion of variance explained by each of these factors, it emerged clearly that structure explained most of it. This suggests that in humans there is an important effect of structure that cannot be explained by path characteristics such as path length and presence of crossings. These results are presented in Fig. 5.

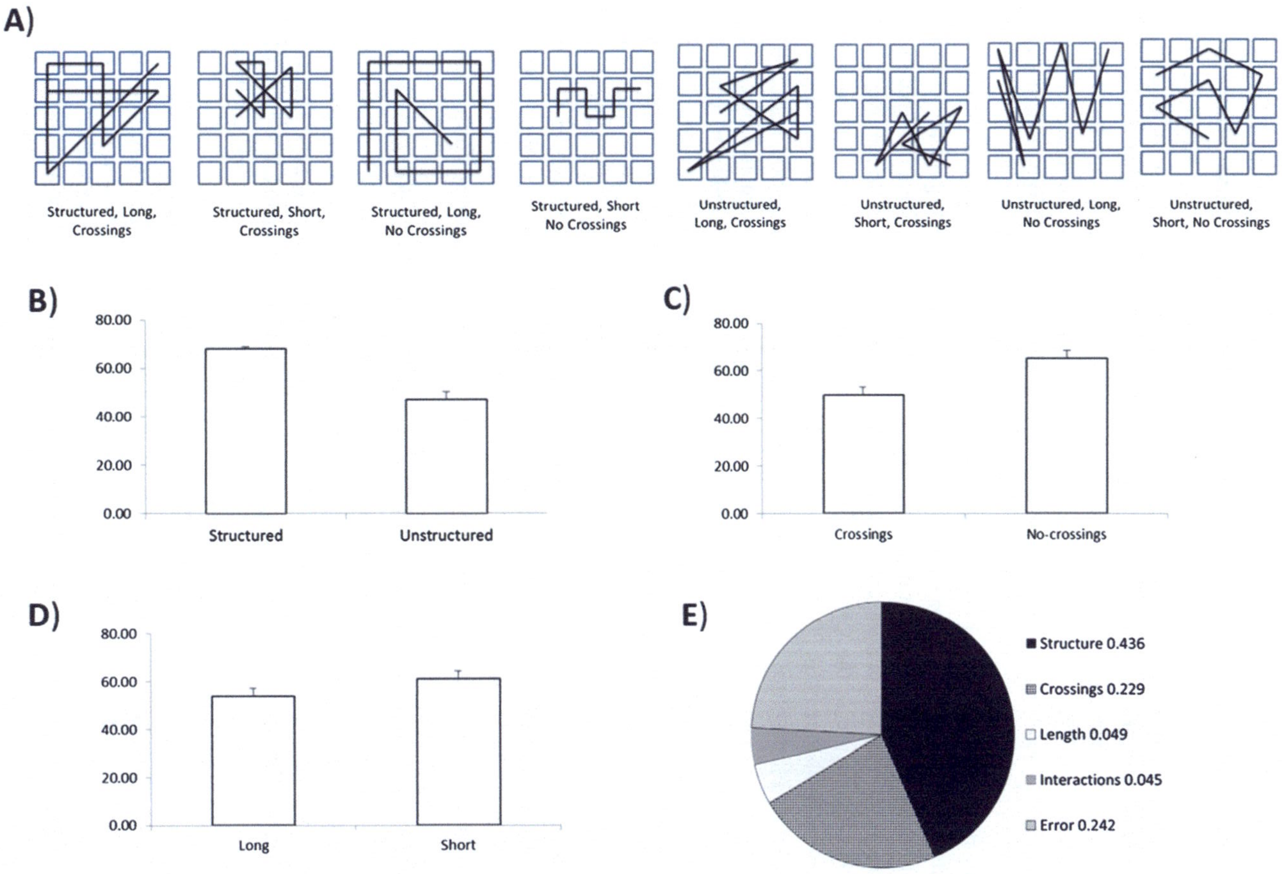

Fig. 5 Display, sequences and main results of Experiment 3, reported in De Lillo et al. (2016). (**a**) Examples of sequences of different types used for the experiment. The design featured three factors: structure (structured/unstructured), presence of crossings in the path (crossings/no crossings), and path length

4.3 Comparative Studies of Perceptual Grouping Effects on Spatial Span

From a comparative perspective, it is worth noting that the beneficial effects on spatial span, which are induced by the detection and encoding of structure in CBTT tasks, may be peculiar to humans. When a direct comparison of humans and non-human primates on a Corsi-type task featuring sequences that could be structured or unstructured, long or short and with or without crossings has been performed, it emerged that whereas humans are sensitive to the structure of the sequence, baboons (*Papio papio*) are sensitive to its path length (Fagot and De Lillo 2011).

An important theoretical issue that requires clarification is the extent to which effects of structure in the tasks just described are mostly due to organisation and grouping at perceptual level or, if by contrast, pertain to the memory encoding of the sequence. The notion that visual grouping and perceptual organisation can explain effects of structure in the CBTT is often proposed, more or less explicitly, in the literature (Hurlstone et al. 2014; Bor et al. 2003). If perceptual organisation explains the benefits of structure in CBTT sequences, then there would be good reason to believe that known differences in perceptual organisation between humans and monkeys could suffice to explain interspecies variability in the ability to benefit from structure in WM and serial recall too. In fact, there is very robust evidence that humans have a much stronger propensity than monkeys to derive shapes formed by grouping local elements into larger global configurations. Pioneered by a study on humans and baboons (Fagot and Deruelle 1997), a large body of literature has accumulated regarding this issue using hierarchical visual stimuli (Navon 1977) in comparative studies of humans and non-human primates (e.g. Spinozzi et al. 2003; De Lillo et al. 2005; Truppa et al. 2017). In most of these studies, hierarchical visual stimuli feature local geometric shapes, which are spatially arranged to form larger global configurations (see Fig. 6).

Using matching-to-sample tasks (e.g. Fagot and Deruelle 1997; Spinozzi et al. 2003; De Lillo et al. 2005) or visual search tasks (Deruelle and Fagot 1998) featuring hierarchical stimuli, it is possible to devise conditions that require the participants to process either the global configurations of the stimuli or their local elements. There is now overwhelming evidence indicating that, in contrast with humans, a variety of species of monkeys are more accurate when processing the local level of these hierarchical stimuli (Fagot and Deruelle 1997; Spinozzi et al. 2003; De Lillo et al.

Fig. 5 (continued) (long/short). (**b**) Main effect of structure. The total number of items correctly recalled in the appropriate order for the structured sequences is higher than that recalled for unstructured sequences. (**c**) Main effect of crossings. The total number of items correctly recalled in the appropriate order for the sequences without crossings is higher than for sequences with crossings. (**d**) Main effect of path length. The total number of items correctly recalled in the appropriate order for the sequences with a short path is higher than that recalled for sequences with a short path. (**e**) Variance explained by each factor, interactions and error. The presence/absence of structure explains most of the variance in the data

Fig. 6 Examples of hierarchical visual stimuli. In these stimuli, elements of a particular shape (local level of stimulus structure) are spatially arranged to form configurations of a given shape (global level of stimulus structure). Humans are typically faster and more accurate when required to process global configurations. By contrast, many studies indicate that monkeys are more accurate when required to process the local levels of the stimuli (see Fagot and Deruelle 1997; De Lillo et al. 2005, 2012)

2005; Truppa et al. 2017). It is likely that the differences are mediated by top-down attentional biases towards one or the other level of stimulus structure (De Lillo et al. 2011). Given this body of literature, it would be plausible that the limited ability of non-human primates to benefit from structure in the CBTT stemmed from their bias towards attending to the local items considered in isolation rather than global properties of their configuration and the spatio-temporal structure of the sequence with which they are presented. Moreover, in addition to this general difference in bias towards attending global or local levels of stimulus structure, comparative studies have highlighted specific differences in how humans and monkeys use visual grouping cues (Spinozzi et al. 2009). Such differences in visual grouping could add plausibility to the notion that interspecies differences in the ability to benefit from structure in the CBTT are explained by perceptual processes.

It is therefore critical to determine whether or not the advantage shown by humans for structured sequences in the CBTT is mediated by perceptual organisation, gestalt formation and grouping processes. In a first attempt to provide more

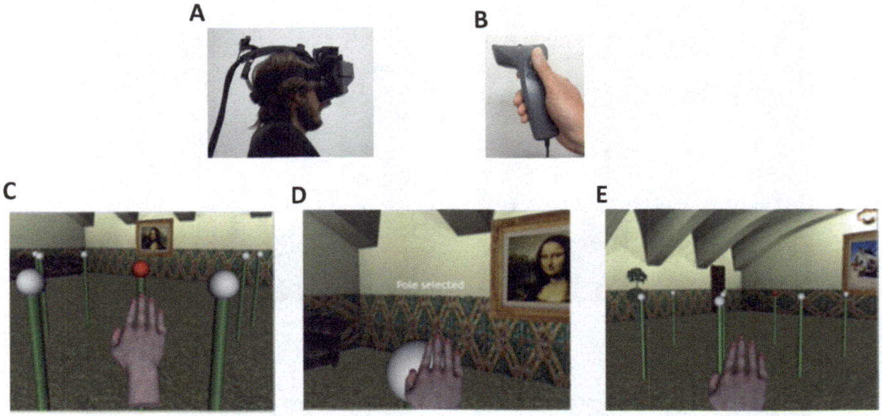

Fig. 7 Example of an immersive virtual reality (VR) implementation of a Corsi test in a navigational space as featured in De Lillo et al. (2014), Experiment 2, and De Lillo et al. (2016), Experiment 5. (**a**) Headset used for the task. (**b**) Wand used for the participants to control their movement and in the VR environment and the selection of items. A view of the environment and the selection of a pole during the presentation phase are presented in (**c–e**). (**c**) During the presentation phase of each trial, a sphere on the top on one of the poles in the environment turns red. (**d**) The participant has to approach and select the pole. A written feedback is provided confirming that the pole has been selected. The pole then turns white. (**e**) The next item in the sequence is presented (another pole in the array turns red). The procedure continues until the entire sequence is presented. In the recall phase, the array only features white poles that the participant has to select in the same order as in the presentation phase. Changes in viewpoint, and the relatively long time taken for participants to move from one pole to another, prevent the participant from benefitting from perceptual grouping processes during the presentation phase. Any effects of organisation in this navigational environment must pertain to memory coding and/or mental images representing the sequence

information about this issue, we carried out an experiment using immersive virtual reality with humans to assess whether effects of structure are also present when people perform a navigational version of the CBTT in a large-scale environment which does not afford grouping at a perceptual level (De Lillo et al. 2016, Experiment 3). The virtual environment used for the experiment (see Fig. 7), consisted in a set of poles surmounted by white spheres within a large textured virtual hall, as it was used for other experiments on search and foraging (De Lillo et al. 2014, Experiment 1).

However, differently from the search task/foraging task (De Lillo et al. 2014, Experiment 1), the navigational version of the CBTT test featured a presentation phase where participants were induced to select the poles in a prescribed order that they subsequently had to reproduce in a recall phase. This novel version of the CBTT was implemented as follows. In the presentation phase, participants saw one of the white spheres turn red. They had to approach it by navigating towards it in the environment and then select it using the trigger on a handheld wand. The red sphere turned white upon selection, and, simultaneously, one of the other white spheres turned red. The participant had now to approach and select this second pole in order to make it revert to white. As soon as this happened, a third sphere turned red, and so on until a sequence of seven pole selections was experienced in this way. On completion of this presentation phase, a recall phase ensued. All the spheres remained white throughout the recall phase, and the participant had to approach and select them in the same order as that followed in the presentation phase. The conditions described above for the touchscreen version of the task were used, so that sequences could be long or short, in terms of the length of the movement necessary to complete it. They could contain three crossings or no crossings, and they could be structured or unstructured. In other words the same 2 (sequence length) × 2 (presence of crossing) × 2 (presence of structure) ANOVA design used for the touchscreen version of the task (see description above for De Lillo et al. 2016, Experiment 3) was implemented in a large-scale navigational environment. Importantly, as illustrated in Fig. 7, in this version of the task the participant was immersed in the environment and navigated throughout the configuration of poles without a bird-eye view of their locations. Moreover, the view point of the participant changed continuously throughout the presentation phase. Finally, time taken by the participants to move between poles far exceeded the time resolution for perceptual grouping to occur in this phase.

In these conditions, which are not compatible with the operation of perceptual grouping processes, we replicated the effects obtained with touchscreen versions of the task. This indicates that, at least in humans, effects of structure can be independent from any effects due to perceptual grouping (De Lillo et al. 2016). Thus, it is plausible that interspecies differences in the ability to detect spatio-temporal structure are independent from the differences in perceptual organisation that have been observed between monkeys and humans. This conclusion would be in line with evidence suggesting that even the observed difference in global and local processing between humans and monkeys may be mediated by relatively high-level functions rather than pre-attentive early stages of visual processing (De Lillo et al. 2011). New data on non-human primates tested in virtual reality or large-scale serial recall tasks such as the one described would be very useful to fully clarify the extent and nature of differences between human and non-human primates in serial organisation.

5 Summary, Final Considerations and Conclusions

This chapter provides a summary of three lines of evidence, which converge in suggesting that an important dimension for the characterisation of higher adult human cognitive processes is the extent of the ability to pick-up and exploit the structure in the environment to reduce the load of the otherwise cognitively demanding coding of serially organised responses. First, search and foraging tasks where individuals are left free to select their trajectory through multiple target locations yielding food resources have been informative in regards to the tendency of different organisms to spontaneously deploy organised search trajectories in large-scale environments. A pattern of results emerged there that suggests a parallel between search efficiency, relative ability to organise principled search in space and taxonomic closeness to humans. Second, serial recall tasks implemented in small manipulative space, such as the CBTT, have enabled the demonstration of a causal relationship between the degree of organisation of serial trajectories and WM span for visited locations. RT (De Lillo 2004; De Lillo and Lesk 2010), fMRI (Bor et al. 2003) and other performance measures (De Lillo et al. 2016) afforded by these tasks strongly suggest that grouping during the encoding of sequences underpins the enhanced WM span so far observed for organised sequences. More recent experiments, which have used virtual reality to implement serial recall task in immersive navigational spaces, suggest that benefitting from structure during the encoding of spatial sequences does not need to be mediated by path characteristics (De Lillo 2004) or visual perceptual grouping (De Lillo et al. 2016).

Several animal species show a large memory capacity (e.g. Balda and Kamil 1992; Conway and Christiansen 2001). Baboons and pigeons have an outstanding long-term memory for arbitrary associations between a visual image and a left-right response (Fagot and Cook 2006). WM for random sequences of numerals has been claimed to be outstanding in chimpanzee and equal in capacity to that of humans (Silberberg and Kearns 2009), if not even superior to it (Inoue and Matsuzawa 2007). On the basis of a detailed analysis of error patterns shown by rhesus monkeys tested in a serial recall task for unstructured sequences, which resembles the pattern of errors made by humans in similar tasks, it has been suggested that a single mechanism for short-term serial order memory may be shared across species (Botvinick et al. 2009). All the above studies converge in claiming that when memory for unrelated material is considered, the analogies between the memory processes of different animal species and those between humans and animals outweigh interspecies differences. The capacity of baboons to memorise 3,500–5,000 arbitrary associations compared to the only marginally less remarkable ability of pigeons to memorise 800–1,200 of them, with similar forgetting rates, reaction time, and serial position observed in the two species, has been taken as an indication that 250 million years of divergent evolution between birds and primates only produced minor changes in memory capacity rather than in memory functions and dynamics (Fagot and Cook 2006). The evidence reviewed in this chapter suggests that the 96 million years of evolution separating humans and rodents (Nei et al. 2001) and even the mere 30 million years or so that separate human and monkeys (Sarich and Wilson 1967) witnessed major changes in the ability to impose spatio-temporal structure on information that needs to be retained in the short term and to benefit from it. It has been claimed that differences in intelligence and higher cognition between humans and animals can be reduced to linguistic, motor and perceptual skills (Macphail 1987).

At odds with such claims, the studies reviewed in this chapter show differences between humans and monkeys and between monkeys and other mammals in the ability to exploit spatio-temporal structure in conditions that suggest a minimal role, if any, for motor planning (De Lillo and Lesk 2010) or perceptual grouping processes (De Lillo et al. 2016) and in a task domain that is largely independent from language in humans (Baddeley and Hitch 1974). The identification and precise characterisation of the functions that mediate this ability and the pressures (both ecological and internal to the cognitive economy of the organism) that lead to its emergence is now of key importance for the following reasons. First, it would help clarify what is unique about human cognition and why. Second, it would shed light on the theoretical conundrum of whether change in memory capacity determines differences in cognitive ability across species (Fagot and Cook 2006) or, *vice versa,* change in attentional and executive skills causes differences in memory span by enabling more sophisticated coding strategies. Finally, it would guide practical choices with vast ethical implications, such as selecting suitable animal models of human higher cognitive and brain functions.

References

Aadland J, Beatty WW, Maki RH (1985) Spatial memory of children and adults assessed in the radial maze. Dev Psychobiol 18:163–172

Astur S, St. Germain SA, Baker EK, Calhuon V, Pearlson GD, Constable RT (2005) fMRI hippocampal activity during a virtual radial arm maze. Appl Psychophysiol Biofeedback 30:307–317

Baddeley AD (2003) Working memory: looking back and looking forward. Nat Rev Neurosci 4:829–839

Baddeley AD, Hitch GJ (1974) Working memory. In: Bower GH (ed) The psychology of learning and motivation, vol 8. Academic Press, London

Balda RP, Kamil AC (1992) Long-term spatial memory in clark's nutcracker, *Nucifraga columbiana*. Anim Behav 44:761–769

Bartolomucci A, de Biurrun G, Fuchs E (2001) How tree shrews (*Tupaia belangeri*) perform in a searching task: evidence for strategy use. J Comp Psychol 115(4):344–350

Bor D, Duncan J, Wiseman RJ, Owen AM (2003) Encoding strategies dissociate prefrontal activity from working memory demand. Neuron 37:361–367

Botvinick MM, Wang J, Cowan E, Roy S, Bestianen C, Mayo JP et al (2009) An analysis of immediate serial recall performance in a macaque. Anim Cogn 12:671–678

Carruthers P (2013) Evolution of working memory. PNAS 110(Supp 2):10371–10378

Conway CM, Christiansen MH (2001) Sequential learning in nonhuman primates. Trends Cogn Sci 5:539–546

Corsi PM (1972) Human memory and the medial temporal region of the brain. Diss Abstr Int 34 (02):891B. (University microfilms No. AA105-77717)

De Lillo C (2004) Imposing structure on a Corsi-type task: evidence for hierarchical organisation based on spatial proximity in serial spatial memory. Brain Cogn 55(3):415–426

De Lillo C (2012) The use of spatial structure in working memory: a comparative standpoint. In: McGonigle M, McFarland D, Stennings K (eds) The complex mind: an interdisciplinary approach. Palgrave Macmillan, Basingstoke, pp 38–54

De Lillo C, Lesk V (2010) Spatial clustering and hierarchical coding in immediate serial recall. Eur J Cogn Psychol 22:216–246

De Lillo C, Visalberghi E, Aversano M (1997) The organisation of exhaustive searches in a “patchy” space by capuchin monkeys (*Cebus apella*). J Comp Psychol 111:82–90

De Lillo C, Spinozzi G, Truppa V, Naylor DM (2005) A comparative analysis of global and local processing of hierarchical visual stimuli in young children and monkeys (*Cebus apella*). J Comp Psychol 119:155–165

De Lillo C, Spinozzi G, Palumbo M, Giustino G (2011) Attention allocation modulates the processing of hierarchical visual patterns: a comparative analysis of capuchin monkeys (*Cebus apella*) and humans. J Exp Psychol Anim Behav Process 37:341–352
De Lillo C, Palumbo M, Spinozzi G, Giustino G (2012) Effects of pattern redundancy and hierarchical grouping on global-local visual processing in monkeys (*Cebus apella*) and humans (*Homo sapiens*). Behav Brain Res 226:445–455
De Lillo C, Kirby M, James FC (2014) Spatial working memory in immersive virtual reality foraging: path organisation, travelling distance and search efficiency in humans (Homo sapiens). Am J Primatol 76:436–446
De Lillo C, Kirby M, Poole D (2016) Spatio-temporal structure, path characteristics, and perceptual grouping in immediate serial spatial recall. Front Psychol 7:1686
Deruelle C, Fagot J (1998) Visual search for global/local stimulus features in humans and baboons. Psychon Bull Rev 5:476–481
Dubreuil D, Tixier C, Dutrieux G, Edeline JM (2003) Does the radial arm maze necessarily test spatial memory? Neurobiol Learn Mem 79:109–117
Fagot J, Cook R (2006) Evidence for large long-term memory capacities in baboons and pigeons and its implications for learning and the evolution of cognition. PNAS 103:17564–17567
Fagot J, De Lillo C (2011) A comparative study of working memory: immediate serial recall in baboons (Papio papio) and humans. Neuropsychologia 49:3870–3880
Fagot J, Deruelle C (1997) Processing of global and local visual information and hemispheric specialisation in humans (*Homo sapiens*) and baboons (*Papio papio*). J Exp Psychol Hum Percept Perform 23:429–442
Fischer MH (2001) Probing spatial working memory with the Corsi blocks task. Brain Cogn 45:143–154
Foreman N, Arber M, Savage J (1984) Spatial memory in preschool infants. Dev Psychobiol 17:129–137
Foti F, Spirito F, Mandolesi L, Aversano M, Petrosini L (2007) Effects of spatial food distribution on search behaviour in rats (*Rattus norvegicus*). J Comp Psychol 212:290–299
Hurlstone MJ, Hitch GJ, Baddeley AD (2014) Memory for serial order across domains: an overview of the literature and directions for future research. Psychol Bull 140(2):339–373
Inoue S, Matsuzawa T (2007) Working memory of numerals in chimpanzees. Curr Biol 17:R1004–R1005
Kessels RP, van Zandvoort MJE, Postma A, Kappelle LJ, de Haan EHF (2000) The Corsi Block-Tapping Task: standardization and normative data. Appl Neuropsychol 7:252–258
Klahr D, Chase WG, Lovelace EA (1983) Structure and process in alphabetic retrieval. J Exp Psychol Learn Mem Cogn 9:462–477
Macphail EM (1987) The comparative psychology of intelligence. Behav Brain Sci 10:645–695
Milton K (1993) Diet and primate evolution. Sci Am 269:86–93
Navon D (1977) Forest before trees: the precedence of global features in visual perception. Cogn Psychol 9:353–383
Nei M, Xu P, Glazco G (2001) Estimation of divergence times from multiprotein sequences for a few mammalian species and several distantly related organisms. PNAS 98:2497–2502
Olton DS, Samuelson RJ (1976) Remembrance of places passed: spatial memory in rats. J Exp Psychol Anim Behav Process 2:97–116
Parmentier FB, Elford G, Maybery MT (2005) Transitional information in spatial serial memory: path characteristics affect recall performance. *J Exp Psychol Learn Mem Cogn* 31:412–427
Pirolli P (2009) Information foraging theory. Oxford University Press, Oxford
Povel DJ, Collard R (1982) Structural factors in patterned finger tapping. Acta Psychol 52:107–123
Sarich VM, Wilson AC (1967) Immunological time scale for hominid evolution. Science 158:1200–1203
Schenk F, Grobety MC, Lavenex P, Lipp HP (1995) Dissociation between basic components of spatial memory in rats. In: Alleva E, Fasolo A, Lipp HP, Nadel L, Ricceri L (eds) Behavioural

brain research in naturalistic and semi-naturalistic settings: possibilities and perspectives, NATO ASI series, series D: behavioural and social sciences, pp 277–300

Silberberg A, Kearns D (2009) Memory for the order of briefly presented numerals in humans as a function of practice. Anim Cogn 12:405–407

Smirni P, Villardita C, Zappala G (1983) Influence of different paths on spatial memory performance in the Block-Tapping Test. J Clin Neuropsychol 5:355–359

Smith EE, Jonides J, Koeppe RA (1996) Dissociating verbal and spatial working memory using PET. Cereb Cortex 6:11–20

Spetch ML, Edwards CA (1986) Spatial memory in pigeons (*Columba livia*) in an open-field feeding environment. J Comp Psychol 100:266–278

Spinozzi G, De Lillo C, Truppa V (2003) Global and local processing of hierarchical visual stimuli in tufted capuchin monkeys (*Cebus apella*). J Comp Psychol 117:15–23

Spinozzi G, De Lillo C, Truppa V, Castorina G (2009) The relative use of proximity, shape similarity and orientation as visual perceptual grouping cues in tufted capuchin monkeys (*Cebus apella*) and humans (*Homo sapiens*). J Comp Psychol 123:56–68

Sternberg S, Knoll RL, Monsell S, Wright CE (1988) Motor programs and hierarchical organisation in the control of rapid speech. Phonetica 45:175–197

Sternberg S, Knoll RL, Turock DL (1990) Hierarchical control in the execution of action sequences: tests of two invariance properties. In: Jeannerod M (ed) Attention and performance XIII: motor representation and control. Lawrence Erlbaum Associates, Hillsdale, pp 3–55

Truppa V, Carducci P, De Simone DA, Bisazza A, De Lillo C (2017) Global/local processing of hierarchical visual stimuli in a conflict-choice task by tufted capuchin monkeys (*Sapajus spp.*). Anim Cogn 20:347–357

Valsecchi P, Bartolomucci A, Aversano M, Visalberghi E (2000) Learning to cope with two different food distributions: the performance of house mice (*Mus musculus*). J Comp Psychol 114:272–280

Vandierendonck A (2018) Working memory benchmarks – a missed opportunity: comment on Oberauer et al. (2018). Psychol Bull 144:963–971

Wilke A, Hutchinson JMC, Todd PM, Czienskowskia U (2009) Fishing for the right words: decision rules for human foraging behaviour in internal search tasks. Cogn Sci 33:497–529

Part II
Processes, Mechanisms and Models

Biasing Allocations of Attention via Selective Weighting of Saliency Signals: Behavioral and Neuroimaging Evidence for the Dimension-Weighting Account

Heinrich René Liesefeld, Anna M. Liesefeld, Stefan Pollmann, and Hermann J. Müller

Contents

Abstract Objects that stand out from the environment tend to be of behavioral relevance, and the visual system is tuned to preferably process these salient objects by allocating focused attention. However, attention is not just passively (bottom-up) driven by stimulus features, but previous experiences and task goals exert strong biases toward attending or actively ignoring salient objects. The core and eponymous assumption of the dimension-weighting account (DWA) is that these top-down biases are not as flexible as one would like them to be; rather, they are subject to

H. R. Liesefeld (✉)
Department Psychologie, Ludwig-Maximilians-Universität, Munich, Germany

Graduate School of Systemic Neurosciences, Ludwig-Maximilians-Universität, Munich, Germany
e-mail: Heinrich.Liesefeld@psy.lmu.de

A. M. Liesefeld
Department Psychologie, Ludwig-Maximilians-Universität, Munich, Germany

S. Pollmann
Institute of Psychology and Center for Behavioral Brain Sciences, Otto von Guericke University, Magdeburg, Germany

H. J. Müller
Department Psychologie, Ludwig-Maximilians-Universität, Munich, Germany

Department of Psychological Sciences, Birkbeck College, University of London, London, UK

Curr Topics Behav Neurosci (2019) 41: 87–114
DOI 10.1007/7854_2018_75
Published Online: 27 December 2018

dimensional constraints. In particular, DWA assumes that people can often not search for objects that have a particular feature but only for objects that stand out from the environment (i.e., that are salient) in a particular *feature dimension*. We review behavioral and neuroimaging evidence for such dimensional constraints in three areas: search history, voluntary target enhancement, and distractor handling. The first two have been the focus of research on DWA since its inception and the latter the subject of our more recent research. Additionally, we discuss various challenges to the DWA and its relation to other prominent theories on top-down influences in visual search.

Keywords Dimension weighting · Priority map · Review · Task history · Visual search

1 Saliency Computations for Guiding Visual Search

The visual environment provides a wealth of information of which only a tiny fraction is processed. A crucial factor determining the information uptake is *focused attention*. Major theories of visual attention assume that only one object is highlighted by focused attention at a time, thereby obtaining a competitive advantage in influencing perception and action. Objects standing out from their surround, such as a black sheep in a flock of white sheep, often receive this competitive advantage. This drawing of attention by "special" stimuli is referred to as stimulus-driven or bottom-up guidance of attention and is explained via the workings of saliency maps. Saliency maps are spatial representations of the visual scene coding for the conspicuity of each object in terms of a single saliency value at each (occupied) location. A common assumption is that for each feature dimension, the distance in feature space between each object and its (immediate) surround (its *saliency*) is calculated before spatial attention is allocated and that these dimension-specific saliency maps are then integrated into a (pre-attentive) priority map that influences the allocation of attentional resources (e.g., Bundesen et al. 2011; also termed "activation map," Wolfe 2007; "saliency map," Itti and Koch 2001 and Li 1999; or "master map," Treisman 1988). This means that instead of the object's features proper, the relation of an object's features to the respective surrounding features influences attention allocations. The more an object differs from its immediate surround, the more salient and the easier to find it is (Duncan and Humphreys 1989; Liesefeld et al. 2016; Nothdurft 1993; Töllner et al. 2011).

To illustrate the concept of saliency more firmly, take a red, vertical bar, for example. This object is salient due to its color only when presented among objects of a different color (e.g., among green objects; Fig. 1a), but not among homogeneous objects with the same or a very similar shade of red (Fig. 1c). In the latter case, the object can still be salient due to other features, such as its orientation, if presented among red objects of another orientation (e.g., tilted to the right, Fig. 1b). In this review, we will refer to objects that stand out in a particular dimension (mostly

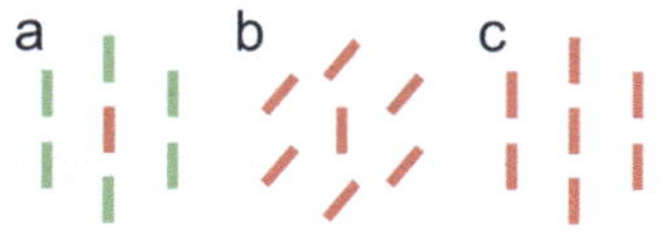

Fig. 1 Saliency depends on local feature contrast. The same object (red, vertical bar) is either salient (**a**) because of its color or (**b**) because of its orientation or (**c**) not salient at all. In other words, the very same red, vertical bar is a color singleton in (**a**), an orientation singleton in (**b**), and not a singleton in (**c**)

targets, but also salient distractors) as *singletons* in that dimension; for example, Fig. 1a features a *color singleton* and Fig. 1b an *orientation singleton*. Thus, it is not any feature of the object per se that is salient, but saliency is defined by the relationship to other features from the same dimension. Saliency signals from the various dimensions are integrated on the priority map. Due to this integration, the priority map does not represent any information on *how* a salient object differs featurally from its surround but only on *how much* it differs: the priority map is feature-and dimension-less (see Fig. 2). One consequence of this is that a strong activation at the priority map can be induced by any singleton, independently of its specific featural relation to its environment, so that knowledge of target and distractor features is not required to guide attention to a singleton target (Müller et al. 2017).

However, several decades of research have firmly established that spatial attention is not only reflexively guided by bottom-up saliency but also by how the observer is "tuned" prior to the start of the search (i.e., before the search display is presented in a laboratory visual search task). This tuning proceeds either voluntarily, because of what the observer intends to find (*search goals*), or involuntarily, because of what the observer has done before (*search history*; sometimes also termed "selection history"; see Awh et al. 2012, or Wolfe and Horowitz 2017, for reviews on these and other potential influences on search guidance). One specific idea of how this tuning might work is that incoming saliency signals are weighted before they are integrated on the priority map (see Fig. 1 and Box 1). In this conceptual framework, "tuning" means changing the weights to the advantage of one type of saliency signal (e.g., target enhancement) or at the disadvantage of another type of saliency signal (e.g., distractor suppression). Biased in this way, the priority map then "guides" attention, ensuring that objects with a high priority are processed first (Wolfe 2007), which higher likelihood (Liesefeld et al. 2016; Moran et al. 2013, 2017) and/or more efficiently (Bundesen et al. 2011; Moran et al. 2016).

Given that such tuning heavily influences perception via controlling attention allocations (e.g., Treisman 2006), one of the most pressing questions is: what are the properties the visual system is tuned to in various situations? One intuitively obvious answer is that the system is tuned toward certain feature values. For example, when searching for a book and knowing that its binder is red, the observer might up-weight the feature "red" during search, thereby conferring an advantage to all red objects in

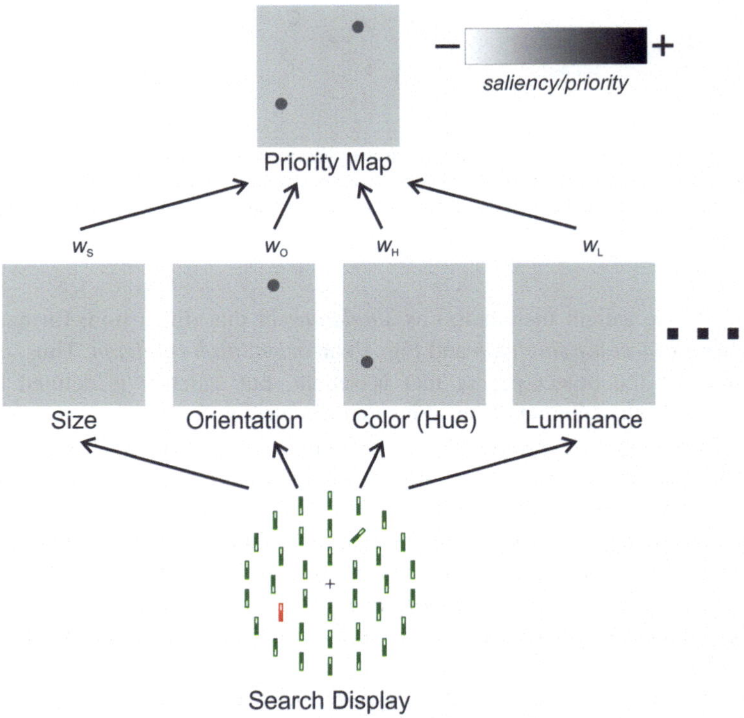

Fig. 2 Simplified sketch of saliency computations in the visual system. From the search display, saliency values are extracted for each feature dimension, reflecting how strongly each object differs from its surround. As there are only color (hue) and orientation singletons in the display, only these two maps contain any (differential) activation. These activations are weighted and integrated at the superordinate priority map, which in turn guides the allocation of focal attention. Note that the saliency maps at the intermediate layer do not preserve any information on feature values and the priority map is even blind to feature dimensions. In other words, for the intermediate saliency maps, it does not matter how an object differs from its surround but only whether and by how much it differs in the coded dimension; the priority map is even more abstract in that it cares only about how much an object differs from its surround in *any* dimension. The influences of the various saliency maps on the priority map depend on the weight settings (w_S, w_O, w_M, and w_L; see text); this example assumes that all weights are set to the same positive value and that the bottom-up saliency of the red and the tilted object are identical, so that their priority map activations are identical as well

the competition for focal attention. The aim of the present review is to summarize the extensive existing evidence that this view might be too simplistic: the dimension-weighting account (DWA) postulates that people in many different situations do not (and sometimes cannot) tune to specific features but to singletons in specific dimensions.

Box 1 Core Assumptions of the DWA

The dimension-weighting account (DWA) can be summarized by a few core assumptions, all of which have received considerable empirical support, as detailed in the text.

Integration of saliency signals on and search guidance by the priority map. The DWA assumes that several saliency maps constitute a subordinate level that inputs to a superordinate (master-saliency or) priority map. Activations at that superordinate map, in turn, guide the allocation of focal attention. Accordingly, DWA is a version/specification of Guided Search (Wolfe 2007).

Dimensional constraints on weighting. Similar to most Guided-Search-type theories, the DWA assumes that saliency signals are not just summed at the priority map but that—depending on the current context—some signals are amplified (up-weighted) and others attenuated (down-weighted). The eponymous (and the core distinguishing) feature of the DWA is its emphasis on dimensional constraints of this weighting.

Involuntary changes in weights. Whereas most theories focus on goal-driven voluntary control of saliency weighting, the DWA—since its inception—emphasizes involuntary mechanisms (due to search history) that influence these weights and that actually are often irrelevant to the task (because they do not improve performance) or even run counter to search goals (because a different set of weights would help perform the task more efficiently). Of note, both voluntary control and search history are top-down influences (see Gaspelin and Luck 2018c) that are assumed to work via the same mechanism (dimension weighting) but for different reasons (search goals vs. experience).

Pre-attentive locus of weighting. Another strong emphasis of the DWA lies on the idea that weights are set before the search display comes up and influence saliency extraction from the display—either voluntarily or involuntarily. Much work has shown that processing a target involuntarily changes dimensional weight settings that then influence saliency computations on the next trial. As saliency processing, in turn, influences attention allocations, this also means that dimension weighting takes effect at a pre-attentive stage (with respect to the current trial) of stimulus processing.

DWA started off with an intriguing observation: search responses to the exact same target item were slower when the target could be either an orientation, size, or color singleton, with the odd-one-out target item changing unpredictably across trials (across-dimension condition), compared to when it was always a color singleton and the specific color changed across trials (within-dimension condition; Egeth 1977, p. 300; Treisman 1988, pp. 207–210). In both cases, it is by design not possible to prepare for a specific target feature, but it is possible to prepare for a specific target dimension in the latter, easier condition (viz., the color dimension). A follow-up study replicated the effect and additionally showed that keeping the target constant at a specific feature value did not improve performance beyond keeping the

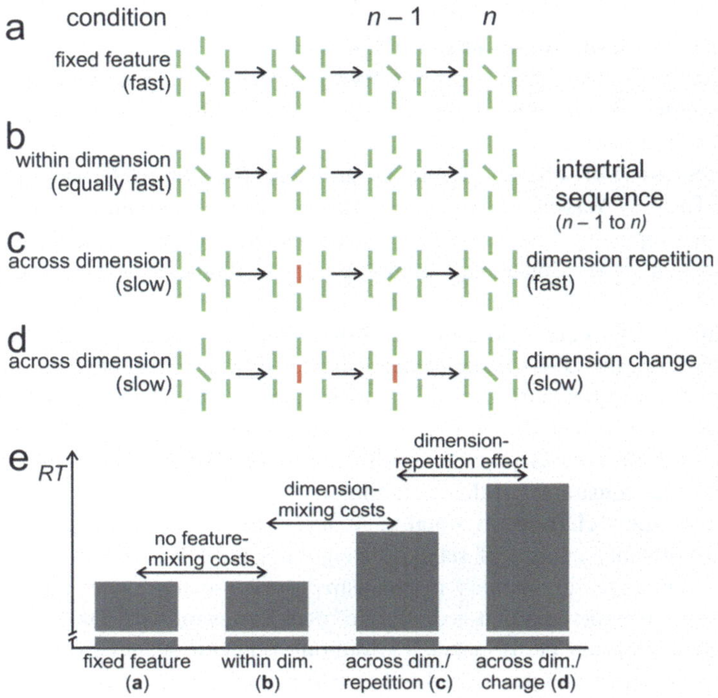

Fig. 3 Empirical findings that have inspired and supported the DWA. Search is faster when the physically identical target (on trial *n*) is defined in a constant dimension within a block (**a** and **b** vs. **c** and **d**), whereas it does not matter whether the target is always defined by the same feature (**a** vs. **b**). With variable targets (across-dimension condition, **c** and **d**), search is (relatively) fast when the target dimension repeats and slow when the dimension changes across trials (**c** vs. **d**). (**a–d**) Only illustrate local contrast; target position is unpredictable across trials. (**e**) Illustrates the various effects on RTs which support the DWA

dimension constant (see Fig. 3; Müller et al. 1995; but see Wolfe et al. 2003). Thus, it seems that observers can prepare for (i.e., up-weight) a specific target dimension, but not a specific target feature (hence dimension-weighting account).

2 The Impact of the Preceding Trial: Search History Effects on Dimensional Weights

The studies reviewed above (Egeth 1977; Müller et al. 1995; Treisman 1988; Wolfe et al. 2003) showed that observers cannot effectively prepare for a target if they do not know in advance in which dimension the target will differ from its surround in the across-dimension condition (Fig. 3c, d); in other words, no goal-driven, voluntary control is possible when the target's singleton dimension is not predictable. Interestingly, however, attention allocations in this condition are not merely bottom-up driven either: examining how the target dimension on the previous trial (trial $n - 1$) influences performance on the current trial (trial n), Müller et al. (Müller et al. 1995; see Fig. 3c–e) found that responses are faster when the dimension repeats (e.g., orientation [on trial $n - 1$] → orientation [on trial n]) compared to when it changes (color → orientation). This effect has been taken to indicate that processing of the target (dimension) on trial $n - 1$ somehow changes the dimensional weight settings and that these changes persist into trial n. Persistence of the weights is likely involuntary, given that the dimension on trial $n - 1$ did not predict the dimension on trial n (i.e., there was no incentive to keep these settings). This indicated a third influence beyond bottom-up saliency and voluntary control that was dubbed *intertrial effects* and constitutes a particularly powerful form of (immediate) *search history* (Allenmark et al. 2018; see also Awh et al. 2012; Theeuwes 2018; Wolfe and Horowitz 2017). Found and Müller (1996) went on to confirm the weight-shifting account and showed that, just like the mixing costs reviewed above (Fig. 3a, b, e), this effect is largely dimension-specific rather than feature-specific in nature: repeating the exact target feature (e.g., left-tilted → left-tilted) did not (typically) improve performance beyond repeating the dimension (e.g., right-tilted → left-tilted). Notably, this *dimension-repetition effect* emerges even though observers are not usually aware of the target dimension on a given trial (Müller et al. 2004), indicating that it reflects a largely implicit memory effect.

A strong claim of the DWA is that dimensional weighting influences (via the priority map) the allocation of attention (Box 1). This means that—with respect to the search display on trial n—dimension-repetition effects must emerge at a pre-attentive stage of processing. Töllner et al. (Töllner et al. 2008; see also Gramann et al. 2010; Töllner et al. 2010) used ERPs to directly test this claim: Observers searched for a singleton target object that was either a blue square or a red circle among blue circles and responded to the orientation of its object-internal grating (vertical vs. horizontal) while their EEG was recorded. The most important finding was that the N2pc component (see Box 2 and Fig. 4) emerged earlier for dimension repetitions than for dimension changes, indicating that dimension-repetition effects influence the very first allocation of covert attention (for a review of converging behavioral evidence, see Krummenacher and Müller 2012).

Box 2 Obtaining Neuroscientific Evidence for the DWA

Various neuroscientific techniques have helped to elucidate the nature of dimension weighting:

EEG. The high temporal resolution of event-related potentials (ERPs) extracted from the electroencephalogram (EEG) allows the examination of cognitive processes that occur before an overt response is executed. This is of considerable importance, because the DWA assumes that the dimensional weights are set before the search display is presented, so that they influence processing already at a pre-attentive stage (see Box 1). Response times are,

(continued)

Box 2 (continued)
however, measured only at the end of a trial when attention has already been allocated to the target. The *N2pc* component (posterior-contralateral negativity in the N2 range, sometimes referred to simply as posterior-contralateral negativity, PCN) is a validated marker of spatial allocation of *covert attention* and therefore particularly useful for examinations of attentional dynamics (Eimer 1996; Luck and Hillyard 1994a, b). In particular, the N2pc can be used to determine (a) whether attention was allocated to a particular object and (b) when in time attention was allocated. Similar (though not identical) information can be gained from eye tracking by measuring *overt attention* allocations (fixations; e.g., Geyer et al. 2008).

fMRI. The high spatial resolution of functional magnetic resonance imaging (fMRI) permits identification of brain areas that are selectively activated by experimental manipulations. This has led to the mapping of a large network of brain areas, including the frontoparietal attention network (e.g., Corbetta and Shulman 2002), that are active during the dimension-weighting process. The consequence of this process (up-weighting of certain dimensions), in turn, is reflected by increased activation in dimension-specific brain areas.

Lesions and TMS. Existing or induced changes of local neuronal function by lesions or transcranial magnetic stimulation (TMS), respectively, have been used to test the functional contribution of brain areas whose activation was observed with fMRI. Both disruption (by lesions) and facilitation (by TMS) supported a functional contribution of various cortical regions to dimension weighting.

To identify the neuronal substrate of dimension weighting, Pollmann et al. (2000) had participants search for a target either defined by color or by movement. In one condition, target identity was kept constant within a block (akin to Egeth 1977; Müller et al. 1995; Treisman 1988). They observed target-dimension-specific activations in areas involved in the processing of the respective aspects of visual information (see Box 2): in the color block, bilateral areas in the fusiform gyrus (concordant with previous reports of V4) were selectively activated, whereas in the motion block, locations in the vicinity of hMT+ were selectively activated. Thus, V4, known to be involved in color processing (e.g., Bartels and Zeki 2000), exhibits increased activation as long as color is up-weighted, and hMT+, known to support motion processing (e.g., Beauchamp et al. 1997), shows increased activation when motion is up-weighted (see also Pollmann et al. 2006b; see Schledde et al. 2017 for converging single-cell evidence).

While dimensional biases would likely be implemented in posterior brain areas that are involved in sensory processing of the respective dimension, (top-down induced) weight changes were expected to activate frontal and parietal areas (e.g., Corbetta and Shulman 2002; Box 2). Indeed, dimension changes elicited widespread

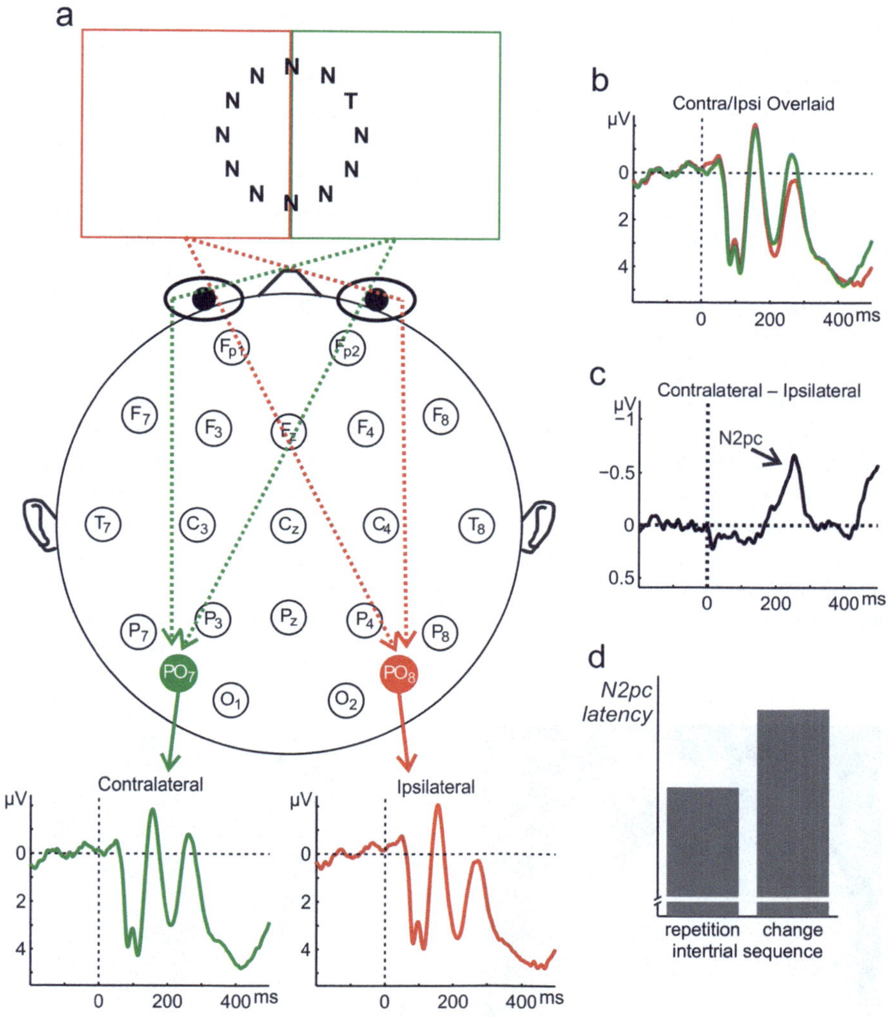

Fig. 4 Generation of the N2pc component. (**a**) The left (red) and right (green) hemifield are initially processed by both eyes, but information on the left hemifield is projected mainly to right cortical regions and vice versa. Consequently, processing of a target (T) placed in the right hemifield is more strongly reflected in EEG activity measured at posterior electrodes contralateral to the target (i.e., left) than at electrodes ipsilateral to the target (i.e., right). Lateralized activity measured at electrode pair PO7/8 is typically used to extract the N2pc component. (**b**) Some 200 ms after stimulus onset, activity is more negative at contralateral than ipsilateral electrodes. This can be seen by overlaying both traces. (**c**) Target-related activity, including the N2pc component, is extracted by subtracting ipsilateral from contralateral activity. (**d**) The relative time at which the N2pc emerges in different experimental conditions is informative with respect to the timing of attention allocations (Töllner et al. 2011)

activation of the dorsal attention network (Pollmann et al. 2000). In addition to this network, anterior prefrontal cortex, including, most notably, the left frontopolar cortex, was activated following dimension changes (Pollmann et al. 2000; Weidner et al. 2002). At the time, this was quite unexpected, because anterior prefrontal cortex was seen as a high-level executive control structure (e.g., Koechlin et al. 1999), which would likely be insensitive to nonvoluntary processes. However, the target-dimension changes in a simple feature search task that elicited frontopolar activation shared an element of ambiguity with other tasks that involved frontopolar cortex (Burgess et al. 2005). More specifically, while attentional weighting of the new target dimension was by no means required by the rules of the search task, it is an adaptive process—facilitating processing of the next target in the new dimension—that might be supported by frontopolar cortex (Pollmann 2004).

The causal role of lateral frontopolar cortex was confirmed in a lesion study (Pollmann et al. 2007; Box 2): During across-dimension search, patients with a lesion centered on left lateral frontopolar cortex (Fig. 5) showed a selective increase of reaction times on dimension-change trials—but not on (within-dimension) feature-change trials, thus indicating a functional contribution of left lateral frontopolar cortex to dimension weighting. This selective increase of dimension-change effects was not observed in a control group with frontomedial lesions. Thus, dimension-change-related frontopolar activations had not been epiphenomenal; rather, left lateral frontopolar cortex indeed supports dimension-weighting processes in across-dimension search. This early evidence for frontopolar involvement in attention weighting fits well with the developing concept of a critical role of anterior

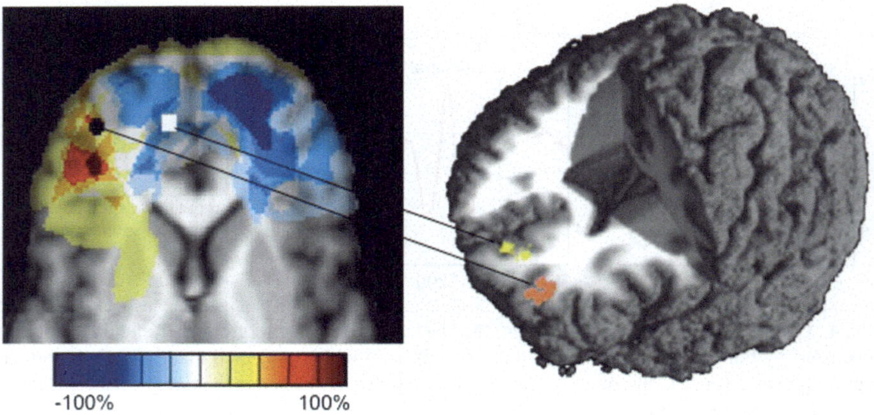

Fig. 5 The involvement of frontal cortex in dimension weighting. The right side shows fMRI data by Weidner et al. (2002). Left lateral frontopolar cortex (marked red) showed selective BOLD increase for dimension changes, but not for (within-dimension) feature changes during feature search. In contrast, left frontomedial cortex (yellow) showed the same activation pattern during conjunction search. Left side: The causal involvement of left lateral frontopolar cortex in dimension weighting during feature search was shown by selectively increased dimension-change effects in patients with left lateral frontopolar lesions (red) as compared to patients with frontomedian lesions (blue) and matched controls. The color scale indicates lesion density in the patient sample. The black disk indicates the center of the left lateral frontopolar activation and the white square the center of the frontomedial activation observed by Weidner et al. (2002)

prefrontal cortex in resource exploration versus exploitation, which is based on recent findings across paradigms and across human as well as nonhuman primates (e.g., Pollmann 2016; Raja Beharelle et al. 2015).

The causal role of right inferior parietal cortex was confirmed as well. Sorting patients according to whether they showed dimension-repetition effects (lesion-symptom mapping) revealed that lesions of right inferior parietal cortex—one of the structures with dimension-change-related fMRI activation (Pollmann et al. 2000)—prevented dimension weighting (Utz et al. 2013). In turn, repetitive TMS over right angular gyrus facilitated subsequent stimulus processing selectively on same-dimension, same-response trials, thus indicating amplified dimension weighting (Bocca et al. 2015).

Furthermore, Pollmann et al. (2006b) found that dimension-change-related activity precedes response-change-related activity, thus providing converging evidence for an early stage of dimension-repetition/change effects. Taken together these findings indicate that prefrontal areas control the weight shifting between visual dimensions and that areas in parietal and temporal cortices implement these control signals by inducing dimension-specific pre-activations in visual input areas.

That across-dimension effects (Egeth 1977; Müller et al. 1995; Treisman 1988) and dimension-repetition effects (Found and Müller 1996) were observed for pop-out searches actually poses an explanatory challenge for theories of saliency computations, such as the DWA (e.g., Fig. 2). This is because pop-out is the subjective phenomenon that the target stands (or pops) out of the display so that it is found *immediately*, without any need to scrutinize any of the nontarget objects (like the black sheep in a flock of white sheep). Thus, one might assume that pop-out search is already maximally efficient, so that no further speedup is possible. Search efficiency is typically measured by varying the number of nontarget objects in the display and determining the slope of the function relating response times to number of objects (search slope). It has been shown that there is a continuum of search slopes (i.e., search efficiencies) across different visual search tasks (Wolfe 1998) and within the same task dependent on a variation of target saliency (Duncan and Humphreys 1989; Liesefeld et al. 2016; Nothdurft 1993). Some might consider pop-out to be one pole of this continuum. Liesefeld et al. (2016), however, showed that search can speed up beyond the point where the target reliably pops out. Their observers searched for a tilted target bar among vertical nontarget bars. Saliency was manipulated by varying the tilt of the target in fine steps from 1.5° to 45°; the resulting search slopes ranged from 70 ms/object (very inefficient) to 0 ms/object (very efficient). Furthermore, and of critical importance for the present point, absolute response times decreased even in the range of very efficient searches (flat search slopes, i.e., ~0 ms/object) with a further increase in saliency. These decreases ranged from 615 ms to 440 ms (6° and 45° targets, respectively), showing that there is indeed quite some leeway (~175 ms) for further speedup even in the efficient (pop-out) range.

Töllner et al. (2011) showed that (some of) this leeway emerges at a pre-attentive stage (see Box 1): They manipulated target saliency while making sure (via pilot studies) that all targets were still within the pop-out range. Measured N2pc latency indeed decreased with increasing saliency, indicating that attention allocations were speeded. Notwithstanding the general speedup with increases in saliency, the Liesefeld et al. (2016) data also showed that the effect of saliency is not linear but approaches an asymptote with higher saliencies, indicating that the probability of finding any effect on search times decreases with the bottom-up saliency of the target object: the more salient a target is, the less leeway there is for further speedups. Indeed, Zehetleitner et al. (2011) observed smaller dimension-repetition effects for very salient objects compared to medium-salient objects (all within the pop-out range), for which an additional increase in priority due to favorable dimensional weight settings would yield only marginal extra benefits. A general take-home message from the Zehetleitner et al. and the Liesefeld et al. studies is that the absence (or nonsignificance) of effects on search times might sometimes simply reflect that target priority is already maximal (at ceiling) in the less efficient condition (e.g., dimension-change condition), a situation that should be avoided if sensitive statistical tests are desired. However, the Liesefeld et al. study also shows that this ceiling is not yet reached when search slopes level off (when the target pops out), thus supporting the interpretation that dimension repetitions do speed up search proper even in this very efficient search range.

Dimension-repetition effects do not only occur for selected dimensions but are observed for almost all dimensions that have so far unambiguously been shown to guide search (Wolfe and Horowitz 2004, 2017). In Guided Search (Wolfe 2007), attributes are considered “basic” or “guiding” when targets defined by that attribute are found efficiently (pop-out; see above). According to Wolfe and Horowitz (2017; see their Box 1), “undoubted” guiding attributes are color, motion, orientation, and size. Indeed, dimension-repetition effects were observed for color (e.g., Found and Müller 1996; Müller et al. 2003; Krummenacher et al. 2009; Töllner et al. 2008; Pollmann et al. 2000; Weidner and Müller 2013; Weidner et al. 2002; Zehetleitner et al. 2011), motion (e.g., Pollmann et al. 2000; Weidner and Müller 2013; Weidner et al. 2002), and orientation (e.g., Found and Müller 1996; Müller et al. 2003; Krummenacher et al. 2009; Zehetleitner et al. 2011). Furthermore, dimension-repetition effects were shown for “probable” (according to Wolfe and Horowitz 2017) guiding attributes such as shape (Töllner et al. 2008) and luminance polarity (Zehetleitner et al. 2011). This renders the DWA a general account of visual search, not limited to a single dimension (see also Sect. 5).

Notwithstanding the strong focus on pop-out search, dimension-repetition effects are assumed to influence saliency computations in general (Box 1). Therefore, they should also occur for inefficient searches. Indeed, Weidner and Müller (2013) observed dimension-repetition effects in a conjunction search task. In this type of task, the target is defined by a conjunction of two features (e.g., a red, vertical bar among green vertical and red horizontal bars), and search is inefficient even if the individual features are very salient (e.g., Wolfe 1998; Wolfe et al. 2010). Weidner and Müller’s participants searched for a target that was (unpredictably across trials) defined by either size (large vs. small) *and* motion direction (diagonal vs. horizontal oscillatory motion) or by size (large vs. small) *and* color (red/blue vs. green). That is, one of the two target-defining dimensions stayed constant across trials (size; primary dimension), whereas the other dimension varied (motion direction vs. color; secondary dimension). Replicating findings from efficient feature search tasks, responses were speeded when the secondary dimension repeated (e.g., size and color → size and color) relative to when it changed (e.g., size and motion direction → size and color).

The neuronal sources of dimension weighting might be somewhat different for conjunction compared to feature search (Fig. 5). In contrast to the Pollmann et al. (2000) study reviewed above, Weidner et al. (2002) employed a conjunction search task comparable to that of Weidner and Müller (2013) and found dimension-change-related activation in pregenual frontomedian cortex (instead of left lateral frontopolar cortex found by Pollmann et al. 2000 during feature search). In a further experiment, Weidner et al. replicated this dissociation between conjunction and feature search within the same participants. In more posterior brain areas, in contrast, dimension-repetition/change-related activation was comparable across studies, matching the known structural specialization of the visual system (see Box 2).

Two or more dimensions can be up-weighted concurrently but at a cost for each individual dimension. In the Weidner and Müller (2013) study just reviewed, orientation (arguably) had to be up-weighted throughout, and still dimension-repetition effects emerged for the "secondary" dimension (size vs. motion direction). Krummenacher et al. (2001, 2002a, b) examined situations in which the target was a feature singleton on each trial (feature search) but differed either in one or in two features from its surround (because each feature on its own already sufficiently defines the target [in contrast to the Weidner and Müller 2013, conjunction task], this is referred to as *redundant-signals paradigm*). Most important for the present review, Krummenacher et al. found that up-weighting one dimension comes at a cost for other dimensions: when a single-feature trial was preceded by a redundant-feature trial (e.g., orientation *and* color → color), responses were slower compared to when the single-feature trial was preceded by a single-feature trial from the same dimension (e.g., color → color). Responses on a single-feature trial were, however, faster when preceded by a redundant-feature trial (e.g., orientation *and* color → color) compared to when preceded by a single-feature trial from the other dimension (e.g., orientation → color). This indicates that processing a redundantly defined target on trial $n - 1$ increases the weights for both feature dimensions, but not as much as a single-feature trial increases the weight for its particular dimension. This pattern of results can be explained by either of two ideas: (a) a limit to the total attentional weight that is shared between dimensions if multiple dimensions are relevant for the search or (b) it is not the absolute weight assigned to a given dimension that determines target priority, but the relative weight with respect to all other feature dimensions (a form of divisive normalization; e.g., Carandini and Heeger 2012). Differentiating these two ideas is an interesting avenue for future research.

The strength of behavioral dimension-repetition effects is somewhat modulated by the type of task. When the task is not to merely detect whether a target is present, but rather to classify the target according to some (other) property not critical for basic detection (e.g., find a tilted bar among vertical distractors and decide whether that bar has a gap on its top or bottom; see Fig. 2), or to localize the target (e.g., left vs. right side of the display), dimension-repetition effects manifest in behavioral, response-time measures are reduced or abolished (Chan and Hayward 2009; Müller and Krummenacher 2006; Kumada 2001; Krummenacher et al. 2002b, 2009; Pollmann et al. 2000, 2006b; Theeuwes et al. 2006; Zehetleitner et al. 2011).

To determine the temporal locus of the dimension-repetition effect in the classification task, Töllner et al. (2008) directly measured attention allocations via the N2pc (see Box 2). As predicted by the DWA, allocation of attention was speeded for dimension repetitions compared with dimension changes. Also dimension-repetition/change-related fMRI activation in dimension-specific cortical areas comparable to that found in a detection task (Pollmann et al. 2000; see Box 2) was observed in a classification task as well (Pollmann et al. 2006a). This indicates that dimension-repetition effects emerge at a pre-attentive stage, independently of the search task.

Töllner et al. (2012a) went on to show that the timing of the N2pc is unaffected by the task demands, whether the task requires simple target detection or localization or classification. However, task demands influence a later, post-selective stage at which response-relevant information is extracted from the attended stimulus to decide on the appropriate motor action (stimulus-response, S-R, mapping). Töllner et al. (2008) provided electrophysiological evidence (based on comparing the timing of the N2pc with the *stimulus*-locked lateralized readiness potential) that this stage is subject to implicit "linked-expectancy" effects: repetition of the target-defining dimension biases S-R mappings toward the same response issued on the previous trial (whereas a target-dimension change may induce a bias toward a different response). Accordingly, response selection is fast when both the target-defining dimension and the response-critical target property repeat; but response selection is slow when one property repeats and the other changes (partial-repetition costs), because this requires the overcoming of a dimension-induced S-R expectancy. Such partial-repetition costs (Hommel 1998; Hommel et al. 2001) seem to cancel out dimension-repetition effects on response-change trials, thus yielding an underestimation of the "true" dimension-repetition (vs. dimension-change) effect in averaged classification-search data (Müller and Krummenacher 2006; Pollmann et al. 2006a).

A complementary approach to measuring response times is to present search displays only briefly followed by a mask and then measure accuracies. The mask serves to erase iconic memory of the display and thereby effectively abort search at an experimentally controlled point in time. The minimum time between search display and mask onset that is needed to achieve a certain accuracy level is a measure of the time taken for visual search proper, excluding processes occurring at the response-selection stage, so that any effects can be less ambiguously attributed to a perceptual stage of processing (see previous paragraph and Box 1). Using this approach, Zehetleitner et al. (2011, Exp. 1) presented search displays for about 65 ms, immediately followed by a mask, and participants had to indicate whether a target bar (luminance or orientation singleton) was presented on the right or the left of the display. Replicating RT results, localization accuracy was indeed higher on dimension-repetition than on dimension-change trials, thus providing additional evidence for a perceptual stage of dimension-repetition effects.

A similar, yet in a crucial aspect distinct, intertrial effect was discovered at about the same time as dimension-repetition effects and was termed "priming of pop-out" (PoP; Lamy et al. 2011; Maljkovic and Nakayama 1994): Similar to most of the experiments reviewed above, the target in a typical PoP task is defined as the only object of a particular feature (a color singleton in most experiments). Crucially, however, only very few other (nontarget) objects are typically present, and target and nontarget features swap randomly across trials. In contrast to the studies reviewed above (where the nontargets were kept constant and numerous), there was a huge effect of whether the target (and therefore also nontarget) *feature* was (a) predictable (constant across all trials; see also Bravo and Nakayama 1992) and (b) repeated or swapped relative to the previous trial, even though the target *dimension* was fixed. That is, in contrast to the dimension-specific effects reviewed above, PoP is a feature-specific effect. The different outcomes of the two designs were explained by Rangelov et al. (Rangelov et al. 2013; see also Krummenacher et al. 2010; Rangelov et al. 2011a, b; Zehetleitner et al. 2012): The feature-specific PoP vanished when more nontargets were presented. They argued (and later showed more directly; see Rangelov et al. 2017) that with a (for PoP studies typical) set size of three objects (one target and two nontargets), targets often do actually not pop-out, that is, often distractors are attended before the target (see also Becker 2008). One interpretation of these findings is that the target is not particularly salient in sparse displays (it lacks local contrast; see Liesefeld et al. 2016; Nothdurft 1993). If the target is not sufficiently salient to reliably draw attention on every trial, more elaborate processing might become necessary before a response is issued. This would be needed to ensure that the attended object is indeed the target, so as to avoid erroneous responses to a nontarget, and (logically) this may require a process of feature comparison of some attended object against the other items. PoP might emerge because repeatedly processing the same target feature speeds up this validation process. On this interpretation, "priming of pop-out" (in contrast to dimension-repetition effects) is not, or only to a limited degree, driven by saliency-computation mechanisms and would be better described as priming of (post-selective) target feature validation (see Huang et al. 2004).

3 Partial Control over Dimensional Weights: Interactions of Search History and Voluntary Control

In the preceding section, we summarized evidence that dimensional weights do shift without the observer's explicit intention in situations where changing the weight settings confers no particular advantage. We will now review how far dimensional weighting can occur when the observer knows that changing the weights is advantageous for an upcoming search, so that weight changes are intended by the observer or at least in line with search goals. In other words, we will address voluntary control over dimensional weights based on task goals.

Also starting from the inception of DWA, it was clear that dimensional weights are, indeed, subject to voluntary control: Müller et al. (1995, Exp. 3) found that when observers knew that one target dimension appeared more likely in an across-dimension search task, performance for these targets improved dramatically. The flexibility of this voluntary preparation was later examined directly by, on each trial, informing participants about the upcoming target in an across-dimension search task (Müller et al. 2003; see also Wolfe et al. 2003; Zehetleitner et al. 2011): The target could be either one of two orientation singletons (20° tilted to the left or right) or one of two color singletons (red or blue), and a symbolic (written) cue was given before onset of the search display that informed participants about the singleton dimension of the upcoming target with 80% validity. For example, when observers saw the word "color," the upcoming target was a color singleton in 80% of cases and an orientation singleton in 20%. Compared to a neutral baseline (the cue word "neutral" was shown), responses were speeded for valid and slowed for invalid trials, indicating that participants were able to voluntarily prepare for the upcoming target. In a second experiment, the specific target feature was cued with a 79% validity using the words "red," "blue," "left," and "right." As predicted by the DWA, cueing a specific feature also expedited responses to another feature from the same dimension on invalid trials (e.g., blue targets were detected faster after the cue "red"), even though it was equally likely to appear as any feature from the respective other dimension (each 7%).

Just like dimension-repetition effects, the strength of the cueing advantage depends on the type of task. Cueing effects are sometimes abolished when the task is not just to detect whether a target is present but to classify one of its features (classification task) or to localize it. This response-dependence of the effect might indicate that cueing effects occur at a post-selective, response-related stage (Mortier et al. 2010; Theeuwes et al. 2006). However, in line with a pre-attentive origin (Box 1), cueing effects were re-established by increasing task difficulty (Zehetleitner et al. 2011) and by increasing the incentive to use the cue (Müller and Krummenacher 2006). Again, ERPs provided a crucial piece of evidence in this debate by demonstrating that the cueing advantage due to dimensional weighting has a pre-attentive origin, even in a classification task (Töllner et al. 2010). Similar to Müller et al. (2003), Töllner et al. (2010) presented a cue word (color, shape, or neutral), followed by a short inter-stimulus interval and a search display featuring either a color- or an orientation-singleton target. This time, the task was to identify the orientation of the target (horizontal or vertical grating), instead of simply detecting the presence of a target (as in Müller et al. 2003). Besides replicating the Müller et al. (2003) cueing effect on RTs with this classification task (cf. Theeuwes et al. 2006; see also Müller and Krummenacher 2006), they also showed that valid cues speeded attention allocations as indicated by expedited (and increased) N2pcs relative to trials with invalid cues and, thus, confirmed a pre-attentive origin of the cueing effect.

These findings clearly show that dimensional weight settings are susceptible to voluntary control. If there are two influences on these weight settings, voluntary control and search history, one might wonder which of the two is more potent when pitted against each other. To examine this question, Müller et al. (Müller et al. 2003;

see also Zehetleitner et al. 2011) compared the dimension-repetition effects on trials with neutral and with informative cues. As expected, strong dimension-repetition effects occurred when cues were uninformative (neutral) and could therefore not influence the weight settings. These effects were reduced, but not completely abolished when cues were informative, both when the informative cue was valid and when it was invalid, indicating that search history effects persist, even if observers try to voluntarily control their weight settings. In fact, dimension-repetition effects remained even when the cue was 100% valid, thus removing any incentive to up-weight the preceding target dimension instead of the cued dimension.

Weidner et al. (2009) examined the neuronal correlates of voluntary weight shifts in a paradigm similar to Müller et al. (2003) and added a manipulation of target saliency. Decreases in saliency (and therefore in search efficiency) yielded activity increases in the frontoparietal attention network and the right anterior middle frontal gyrus and activity decreases in the medial anterior prefrontal cortex. Having prepared for the wrong target dimension (because of an invalid cue) caused activity increases in the left lateral frontopolar cortex, the left supramarginal gyrus, and the cerebellum as well as bilaterally in the posterior orbital gyrus, the inferior frontal gyrus, and the pre-SMA and decreases (with respect to validly cued trials) in the superior frontal gyrus, anterior to the junction with the precentral gyrus (frontal eye fields, FEF). These areas might be involved in reorganizing the weight settings when the current settings are invalid. Furthermore, an interaction between saliency and cue validity was observed in the left temporoparietal junction, rendering this a probable site for integration of saliency signals and dimensional weight settings as would be expected to occur at the priority map (see also Serences et al. 2005).

4 Handling Distraction by Setting Dimensional Weights

The top-down setting of dimensional weights also allows successful handling of salient-but-irrelevant distractors. In his classical studies, Theeuwes (1991, 1992) showed that search for a singleton pop-out target is hampered by the presence of a salient distractor that is a singleton in a different dimension (e.g., a color distractor during search for a shape target; see Fig. 6a). This was interpreted as evidence for attentional capture: attention is involuntarily allocated first toward the salient distractor, before it can be reallocated to the target. The first misallocation (capture) of attention causes a delay of target processing and therefore produces response-time costs. However, more recent evidence suggests that such a distractor does not typically capture attention but is suppressed (at least when target and distractor are constant across trials, so that observers can prepare effectively; Burra and Kerzel 2013; Hickey et al. 2006; Kerzel and Barras 2016; Kiss et al. 2012; McDonald et al. 2013). Jannati et al. (2013), for example, showed that a color distractor during search for a shape target (as displayed in Fig. 5a) does not elicit an N2pc but a P_D indicating suppression of the distractor instead of attentional capture (Hickey et al. 2009; Sawaki et al. 2012; Toffanin et al. 2011).

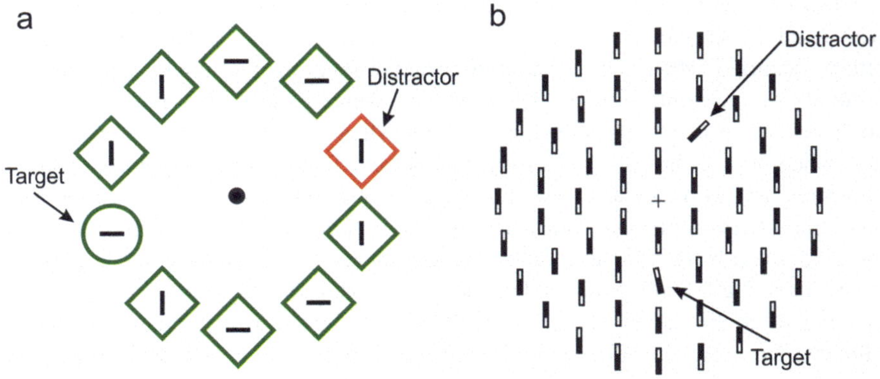

Fig. 6 Search displays used to examine attentional capture. (**a**) An example of a typical additional-singleton task (as used by Jannati et al. 2013). Observers had to determine the orientation of the line within the circle (vertical vs. horizontal). (**b**) The additional-singleton task used by Liesefeld et al. (2017). Observers had to determine the position of the notch in the bar tilted 12° to the left (top vs. bottom). Nontargets (green diamonds and vertical bars, respectively) and singleton distractor (red diamond and bar tilted 45° to the right, respectively) were completely irrelevant in both tasks

That the distractor in Jannati et al. (2013, and many similar studies) does not have a strong potential to capture attention follows from the DWA: In preparation for a search, observers either up-weight the target dimension (shape) so that the target is the object achieving the highest value on the priority map, and/or they down-weight the distractor dimension (color) so that the distractor does not stand out on the priority map. The observed evidence for suppression (Jannati et al. 2013; Gaspar and McDonald 2014; Gaspelin and Luck 2018a, b) might indicate that this weighting is not perfect and the distractor produces some residual activation on the priority map that must be suppressed to pave the way for an attention allocation to the target (i.e., implementing a winner-takes-all mechanism; Desimone and Duncan 1995).

Assigning a low priority to a salient-but-irrelevant distractor would be impossible or counterproductive if distractor and target are coupled to the same weight. According to the DWA, this is the case when both singletons are defined in the same dimension. Any attempt to down-weight the distractor (dimension) would also down-weight the target (dimension) and would therefore obstruct search. As a consequence, such a same-dimension distractor should reliably capture attention. Liesefeld et al. (2017) directly tested this prediction with a 45° distractor and a 12° target, tilted into opposite directions (see Fig. 6b). The notion of attentional capture implies that (a) attention is first allocated to the distractor and only afterward to the target and (b) that attention allocation toward the target is delayed with respect to a search display without a distractor (where attention is allocated directly to the target). Indeed, Liesefeld et al. observed an N2pc to the distractor that preceded an N2pc to the target, and the target N2pc on distractor-present trials was delayed with respect to the target N2pc on distractor-absent trials. Furthermore, distractor presence delayed responses by more than 200 ms! Thus, when the target is constant across trials, a different-dimension distractor is unlikely to capture attention, but a same-dimension distractor reliably does (see also Schubö and Müller 2009; Töllner et al. 2012b; for behavioral evidence, see Liesefeld et al. in press; Sauter et al. 2018, in press; for a focused review on dimension weighting as a mechanism for distractor handling, see Liesefeld and Müller in rev.).

5 Is Color Special?

Even the first studies on DWA already observed an apparent idiosyncrasy for one particular feature dimension: color. As for other dimensions, responses were speeded for a color target preceded by a color target compared to a color target preceded by, say, an orientation target. In contrast to other dimensions such as orientation, however, it mattered whether the exact same color was repeated: There was a further speedup when a red target was preceded by a red target relative to when a red target was preceded by a blue target (Found and Müller 1996). This peculiarity of color was also observed for cueing studies: Whereas cueing the exact feature did not matter for orientation, cueing a specific color speeded up responses when the cue matched the subsequent target (Müller et al. 2003). Furthermore, in contrast to orientation distractors during search for orientation targets (Liesefeld et al. 2017), salient color distractors do not (reliably) capture attention during search for color targets (Gaspar and McDonald 2014; Gaspar et al. 2016; but see Weichselbaum and Ansorge 2018).

Much of the confusion probably arises from treating "color" as one dimension, whereas it is in fact a multidimensional property. Already at early stages of color processing, retinal ganglion cells (as transmitted to the early visual cortical areas via the thalamic lateral geniculate nucleus) represent color in a three-dimensional space (De Valois et al. 1966; Derrington et al. 1984). A similar dimensional structure is implemented in the psychologically meaningful CIE *Lab* color space, where color is defined along the dimensions luminance (L), red-green (a) and blue-yellow (b). It is thus far unknown how this multidimensional space is structured for saliency computations; and this is further complicated by the possibility that it has more than three dimensions (see D'Zmura 1991; Lindsey et al. 2010; but see Martinovic et al. 2018). But let us, for the sake of argument, speculate using the three dimensions of Lab space: even if, in most experiments, one dimension is kept constant (by using equiluminant stimuli), the employed colors still typically vary on (at least) two dimensions. Gaspar and McDonald (2014), for example, had observers search for a yellow target among green distractors and found that a red distractor does not capture attention. It is quite evident that in our hypothetical example search for a yellow target would be supported by up-weighting yellow-blue (b) and a red distractor is best ignored by down-weighting red-green (a). Thus, red distractors can be sufficiently down-weighted to avoid attentional capture during search for yellow targets among green nontargets (Gaspar and McDonald 2014; for a related, though somewhat different, explanation, see the literature on linear separability; Bauer et al. 1996a, b; Daoutis et al. 2006; Kong et al. 2016).

Notwithstanding the above, the various dimensions of color space are certainly not treated fully independently. Separate but nonindependent dimensions are typically referred to as "integral" (Algom and Fitousi 2016; Garner 1974). The empirical criterion for integrality is that during classification according to one dimension, irrelevant variability in the other dimension cannot be (fully) ignored (thus hampering classification). It appears likely that dimension weighting typically spills over to integral dimensions but can be restricted to single dimensions if necessary. This would explain both dimension-repetition and feature-repetition effects (Found and Müller 1996) as well as dimension-cueing and feature-cueing (Müller et al. 2003) effects observed for color.

6 The Scope of the DWA and Its Relation to Other Ideas

The DWA makes the positive claim that the transfer from the dimension-specific saliency maps to the priority map can be biased by dimensional weighting. However, it does not make the negative claim that there are no other nodes in the priority calculation system that are (under certain conditions) influenced by search history and/or voluntary control, including the possibility of feature weighting. Weighting mechanisms specific to a certain feature, to a certain dimension, or even to a certain modality (visual vs. auditory vs. haptic; e.g., Töllner et al. 2009) as well as spatial mechanisms might work in concert, because each mechanism in isolation is less effective and/or less efficient (i.e., would cost more cognitive resources) and because a combination of mechanisms increases the adaptability of the visual system.

Contingent capture (Folk and Remington 1998; Folk et al. 1992) in a DWA interpretation means that not only distractors with target *features* capture attention but also distractors that are singletons in the same *dimension* as the target. Regarding the discussion on search modes (singleton detection vs. feature search; Bacon and Egeth 1994; Leber and Egeth 2006), we would contend that observers can (also) take on a dimension-search mode. There is, indeed, evidence of contingent capture in a general color-search mode (Folk and Anderson 2010; Folk and Remington 1998; Harris et al. 2015). *Relational coding* (Becker 2010; Becker et al. 2010) assumes that observers search for a singleton that differs from its surround in a particular direction (e.g., redder or larger) instead of searching for any singleton in a given dimension (e.g., red-green singleton or size singleton, without a negative or positive sign). It appears plausible that relational coding is possible for some dimensions, while other dimensions are restricted to general dimension weighting (such as orientation; Liesefeld et al. 2017).

Given these numerous interrelations, future research should identify situations in which one and/or the other mechanism takes effect and how the various mechanisms are related, instead of treating these accounts as competitors. Any attempt to disentangle dimension-based and feature-based mechanisms should rather focus on dimensions that are easily characterized such as orientation and luminance and avoid dimensions that are less so, such as color or shape. Notwithstanding the general compatibility of these approaches, such a research agenda might turn out to unify some (or all) of these ideas.

Acknowledgments This work was supported by German Research Foundation (DFG) grants MU 773/14-1 and MU 773/16-1 (to H.J.M.), Open Research Area grant DFG PO 548/16-1, and DFG grant CRC779/A4 (to S.P.), by LMU Munich's Institutional Strategy LMUexcellent within the framework of the German Excellence Initiative (to H.J.M. and H.R.L.) and by the Graduate School of Systemic Neurosciences, Munich Center for Neurosciences – Brain & Mind (to H.R.L.).

References

Algom D, Fitousi D (2016) Half a century of research on Garner interference and the separability-integrality distinction. Psychol Bull 142:1352–1383. https://doi.org/10.1037/bul0000072

Allenmark F, Müller HJ, Shi Z (2018) Inter-trial effects in visual search: factorial comparison of Bayesian updating models. PLoS Comput Biol 14:e1006328. https://doi.org/10.1371/journal.pcbi.1006328

Awh E, Belopolsky AV, Theeuwes J (2012) Top-down versus bottom-up attentional control: a failed theoretical dichotomy. Trends Cogn Sci 16:437–443. https://doi.org/10.1016/j.tics.2012.06.010

Bacon WF, Egeth HE (1994) Overriding stimulus-driven attentional capture. Percept Psychophys 55:485–496. https://doi.org/10.3758/BF03205306

Bartels A, Zeki S (2000) The architecture of the colour centre in the human visual brain: new results and a review. Eur J Neurosci 12:172–193. https://doi.org/10.1046/j.1460-9568.2000.00905.x

Bauer B, Jolicoeur P, Cowan WB (1996a) Distractor heterogeneity versus linear separability in colour visual search. Perception 25:1281–1294. https://doi.org/10.1068/p251281

Bauer B, Jolicoeur P, Cowan WB (1996b) Visual search for colour targets that are or are not linearly separable from distractors. Vis Res 36:1439–1465. https://doi.org/10.1016/0042-6989(95)00207-3

Beauchamp MS, Cox RW, DeYoe EA (1997) Graded effects of spatial and featural attention on human area MT and associated motion processing areas. J Neurophysiol 78:516–520. https://doi.org/10.1152/jn.1997.78.1.516

Becker SI (2008) Can intertrial effects of features and dimensions be explained by a single theory. J Exp Psychol Hum Percept Perform 34:1417–1440. https://doi.org/10.1037/a001138

Becker SI (2010) The role of target–distractor relationships in guiding attention and the eyes in visual search. J Exp Psychol Gen 139:247–265. https://doi.org/10.1037/a0018808

Becker SI, Folk CL, Remington RW (2010) The role of relational information in contingent capture. J Exp Psychol Hum Percept Perform 36:1460–1476. https://doi.org/10.1037/a0020370

Bocca F, Töllner T, Müller HJ, Taylor PC (2015) The right angular gyrus combines perceptual and response-related expectancies in visual search: TMS-EEG evidence. Brain Stimul 8:816–822. https://doi.org/10.1016/j.brs.2015.02.001

Bravo MJ, Nakayama K (1992) The role of attention in different visual-search tasks. Percept Psychophys 51:465–472. https://doi.org/10.3758/BF03211642

Bundesen C, Habekost T, Kyllingsbæk S (2011) A neural theory of visual attention and short-term memory (NTVA). Neuropsychologia 49:1446–1457. https://doi.org/10.1016/j.neuropsychologia.2010.12.006

Burgess PW, Simoins JS, Dumontheil I, Gilbert SJ (2005) The gateway hypothesis of rostral prefrontal cortex (area 10) function. In: Duncan J, Phillips L, McLeod P (eds) Measuring the mind: speed, control, and age. Oxford University Press, Oxford, pp 217–248

Burra N, Kerzel D (2013) Attentional capture during visual search is attenuated by target predictability: evidence from the N2pc, Pd, and topographic segmentation. Psychophysiology 50:422–430. https://doi.org/10.1111/psyp.12019

Carandini M, Heeger DJ (2012) Normalization as a canonical neural computation. Nat Rev Neurosci 13:51–62. https://doi.org/10.1038/nrn3136

Chan LH, Hayward WG (2009) Feature integration theory revisited: dissociating feature detection and attentional guidance in visual search. J Exp Psychol Hum Percept Perform 35:119–132. https://doi.org/10.1037/0096-1523.35.1.119

Corbetta M, Shulman GL (2002) Control of goal-directed and stimulus-driven attention in the brain. Nat Rev Neurosci 3:201–215. https://doi.org/10.1038/nrn755

D'Zmura M (1991) Color in visual search. Vis Res 31:951–966. https://doi.org/10.1016/0042-6989(91)90203-H

Daoutis CA, Pilling M, Davies IL (2006) Categorical effects in visual search for colour. Vis Cogn 14:217–240. https://doi.org/10.1080/13506280500158670

De Valois RL, Abramov I, Jacobs GH (1966) Analysis of response patterns of LGN cells. J Opt Soc Am 56:966–977. https://doi.org/10.1364/JOSA.56.000966

Derrington AM, Krauskopf J, Lennie P (1984) Chromatic mechanisms in lateral geniculate nucleus of macaque. J Physiol 357:241–265. https://doi.org/10.1113/jphysiol.1984.sp015499

Desimone R, Duncan J (1995) Neural mechanisms of selective visual attention. Annu Rev Neurosci 18:193–222. https://doi.org/10.1146/annurev.ne.18.030195.001205

Duncan J, Humphreys GW (1989) Visual search and stimulus similarity. Psychol Rev 96:433–458. https://doi.org/10.1037/0033-295X.96.3.433

Egeth H (1977) Attention and preattention. In: Bower GH (ed) The psychology of learning and motivation, vol 11. Academic Press, New York, pp 277–320

Eimer M (1996) The N2pc component as an indicator of attentional selectivity. Electroencephalogr Clin Neurophysiol 99:225–234. https://doi.org/10.1016/0013-4694(96)95711-9

Folk CL, Anderson BA (2010) Target-uncertainty effects in attentional capture: color-singleton set or multiple attentional control settings? Psychon Bull Rev 17:421–426. https://doi.org/10.3758/PBR.17.3.421

Folk CL, Remington R (1998) Selectivity in distraction by irrelevant featural singletons: evidence for two forms of attentional capture. J Exp Psychol Hum Percept Perform 24:847–858. https://doi.org/10.1037/0096-1523.24.3.847

Folk CL, Remington RW, Johnston JC (1992) Involuntary covert orienting is contingent on attentional control settings. J Exp Psychol Hum Percept Perform 18:1030–1044. https://doi.org/10.1037/0096-1523.18.4.1030

Found A, Müller HJ (1996) Searching for unknown feature targets on more than one dimension: investigating a 'dimension-weighting' account. Percept Psychophys 58:88–101. https://doi.org/10.3758/BF03205479

Garner WR (1974) The processing of information and structure. Erlbaum, Potomac

Gaspar JM, McDonald JJ (2014) Suppression of salient objects prevents distraction in visual search. J Neurosci 34:5658–5666. https://doi.org/10.1523/JNEUROSCI.4161-13.2014

Gaspar JM, Christie GJ, Prime DJ, Jolicoeur P, McDonald JJ (2016) Inability to suppress salient distractors predicts low visual working memory capacity. Proc Natl Acad Sci U S A 113:3693–3698. https://doi.org/10.1073/pnas.1523471113

Gaspelin N, Luck SJ (2018a) Combined electrophysiological and behavioral evidence for the suppression of salient distractors. J Cogn Neurosci 30:1265–1280. https://doi.org/10.1162/jocn_a_01279

Gaspelin N, Luck SJ (2018b) The role of inhibition in avoiding distraction by salient stimuli. Trends Cogn Sci 22:79–92. https://doi.org/10.1016/j.tics.2017.11.001

Gaspelin N, Luck SJ (2018c) "Top-down" does not mean "voluntary". J Cogn 1:25. https://doi.org/10.5334/joc.28
Geyer T, Müller HJ, Krummenacher J (2008) Expectancies modulate attentional capture by salient color singletons. Vis Res 48:1315–1326. https://doi.org/10.1016/j.visres.2008.02.006
Gramann K, Töllner T, Müller HJ (2010) Dimension-based attention modulates early visual processing. Psychophysiology 47:968–978. https://doi.org/10.1111/j.1469-8986.2010.00998.x
Harris AM, Becker SI, Remington RW (2015) Capture by colour: evidence for dimension-specific singleton capture. Atten Percept Psychophys 77:2305–2321. https://doi.org/10.3758/s13414-015-0927-0
Hickey C, McDonald JJ, Theeuwes J (2006) Electrophysiological evidence of the capture of visual attention. J Cogn Neurosci 18:604–613. https://doi.org/10.1162/jocn.2006.18.4.604
Hickey C, Di Lollo V, McDonald JJ (2009) Electrophysiological indices of target and distractor processing in visual search. J Cogn Neurosci 21:760–775. https://doi.org/10.1162/jocn.2009.21039
Hommel B (1998) Event files: evidence for automatic integration of stimulus-response episodes. Vis Cogn 5:183–216. https://doi.org/10.1080/713756773
Hommel B, Müsseler J, Aschersleben G, Prinz W (2001) The theory of event coding (TEC): a framework for perception and action planning. Behav Brain Sci 24:849–937. https://doi.org/10.1017/S0140525X01000103
Huang L, Holcombe AO, Pashler H (2004) Repetition priming in visual search: episodic retrieval, not feature priming. Mem Cogn 32:12–20. https://doi.org/10.3758/BF03195816
Itti L, Koch C (2001) Computational modelling of visual attention. Nat Rev Neurosci 2:194–203. https://doi.org/10.1038/35058500
Jannati A, Gaspar JM, McDonald JJ (2013) Tracking target and distractor processing in fixed-feature visual search: evidence from human electrophysiology. J Exp Psychol Hum Percept Perform 39:1713–1730. https://doi.org/10.1037/a0032251
Kerzel D, Barras C (2016) Distractor rejection in visual search breaks down with more than a single distractor feature. J Exp Psychol Hum Percept Perform 42:648–657. https://doi.org/10.1037/xhp0000180
Kiss M, Grubert A, Petersen A, Eimer M (2012) Attentional capture by salient distractors during visual search is determined by temporal task demands. J Cogn Neurosci 24:749–759. https://doi.org/10.1162/jocn_a_00127
Koechlin E, Basso G, Pietrini P et al (1999) The role of the anterior prefrontal cortex in human cognition. Nature 399:148–151. https://doi.org/10.1038/20178
Kong G, Alais D, van der Burg E (2016) An investigation of linear separability in visual search for color suggests a role of recognizability. J Exp Psychol Hum Percept Perform 42:1724–1738. https://doi.org/10.1037/xhp0000249
Krummenacher J, Müller HJ (2012) Dynamic weighting of feature dimensions in visual search: behavioral and psychophysiological evidence. Front Psychol 3:221. https://doi.org/10.3389/fpsyg.2012.00221
Krummenacher J, Müller HJ, Heller D (2001) Visual search for dimensionally redundant pop-out targets: evidence for parallel-coactive processing of dimensions. Percept Psychophys 63:901–917. https://doi.org/10.3758/BF03194446
Krummenacher J, Müller HJ, Heller D (2002a) Visual search for dimensionally redundant pop-out targets: parallel-coactive processing of dimensions is location specific. J Exp Psychol Hum Percept Perform 28:1303–1322. https://doi.org/10.1037/0096-1523.28.6.1303
Krummenacher J, Müller HJ, Heller D (2002b) Visual search for dimensionally redundant pop-out targets: redundancy gains in compound tasks. Vis Cogn 9:801–837. https://doi.org/10.1080/13506280143000269
Krummenacher J, Müller HJ, Zehetleitner M, Geyer T (2009) Dimension- and space-based intertrial effects in visual pop-out search: modulation by task demands for focal-attentional processing. Psychol Res 73:186–197. https://doi.org/10.1007/s00426-008-0206-y

Krummenacher J, Grubert A, Müller HJ (2010) Inter-trial and redundant-signals effects in visual search and discrimination tasks: separable pre-attentive and post-selective effects. Vis Res 50:1382–1395. https://doi.org/10.1016/j.visres.2010.04.006

Kumada T (2001) Feature-based control of attention: evidence for two forms of dimension weighting. Percept Psychophys 63:698–708. https://doi.org/10.3758/BF03194430

Lamy D, Zivony A, Yashar A (2011) The role of search difficulty in intertrial feature priming. Vis Res 51:2099–2109. https://doi.org/10.1016/j.visres.2011.07.010

Leber AB, Egeth HE (2006) It's under control: top-down search strategies can override attentional capture. Psychon Bull Rev 13:132–138. https://doi.org/10.3758/BF03193824

Li Z (1999) Contextual influences in V1 as a basis for pop out and asymmetry in visual search. Proc Natl Acad Sci U S A 96:10530–10535. https://doi.org/10.1073/pnas.96.18.10530

Liesefeld HR, Müller HJ (in rev.) Distractor handling via dimension weighting

Liesefeld HR, Moran R, Usher M, Müller HJ, Zehetleitner M (2016) Search efficiency as a function of target saliency: the transition from inefficient to efficient search and beyond. J Exp Psychol Hum Percept Perform 42:821–836. https://doi.org/10.1037/xhp0000156

Liesefeld HR, Liesefeld AM, Töllner T, Müller HJ (2017) Attentional capture in visual search: capture and post-capture dynamics revealed by EEG. NeuroImage 156:166–173. https://doi.org/10.1016/j.neuroimage.2017.05.016

Liesefeld HR, Liesefeld AM, Müller HJ (in press). Distractor-interference reduction is dimensionally constrained. Visual Cognition. https://doi.org/10.1080/13506285.2018.1561568

Lindsey DT, Brown AM, Reijnen E et al (2010) Color channels, not color appearance or color categories, guide visual search for desaturated color targets. Psychol Sci 21:1208–1214. https://doi.org/10.1177/0956797610379861

Luck SJ, Hillyard SA (1994a) Electrophysiological correlates of feature analysis during visual search. Psychophysiology 31:291–308. https://doi.org/10.1111/j.1469-8986.1994.tb02218.x

Luck SJ, Hillyard SA (1994b) Spatial filtering during visual search: evidence from human electrophysiology. J Exp Psychol Hum Percept Perform 20:1000–1014. https://doi.org/10.1037/0096-1523.20.5.1000

Maljkovic V, Nakayama K (1994) Priming of pop-out: I. Role of features. Mem Cogn 22:657–672. https://doi.org/10.3758/BF03209251

Martinovic J, Wuerger SM, Hillyard SA, Müller MM, Andersen SK (2018) Neural mechanisms of divided feature-selective attention to color. NeuroImage 181:670–682. https://doi.org/10.1016/j.neuroimage.2018.07.033

McDonald JJ, Green JJ, Jannati A, Di Lollo V (2013) On the electrophysiological evidence for the capture of visual attention. J Exp Psychol Hum Percept Perform 39:849–860. https://doi.org/10.1037/a0030510

Moran R, Zehetleitner M, Müller HJ, Usher M (2013) Competitive guided search: meeting the challenge of benchmark RT distributions. J Vis 13(8):24. https://doi.org/10.1167/13.8.24

Moran R, Zehetleitner M, Liesefeld HR, Müller HJ, Usher M (2016) Serial vs. parallel models of attention in visual search: accounting for benchmark RT-distributions. Psychon Bull Rev 23:1300–1315. https://doi.org/10.3758/s13423-015-0978-1

Moran R, Liesefeld HR, Usher M, Müller HJ (2017) An appeal against the item's death sentence: accounting for diagnostic data patterns with an item-based model of visual search. Behav Brain Sci 40:e148. https://doi.org/10.1017/S0140525X16000182

Mortier K, van Zoest W, Meeter M, Theeuwes J (2010) Word cues affect detection but not localization responses. Atten Percept Psychophys 72:65–75. https://doi.org/10.3758/APP.72.1.65

Müller HJ, Krummenacher J (2006) Locus of dimension weighting: preattentive or postselective? Vis Cogn 14:490–513. https://doi.org/10.1080/13506280500194154

Müller HJ, Heller D, Ziegler J (1995) Visual search for singleton feature targets within and across feature dimensions. Percept Psychophys 57:1–17. https://doi.org/10.3758/BF03211845

Müller HJ, Reimann B, Krummenacher J (2003) Visual search for singleton feature targets across dimensions: stimulus- and expectancy-driven effects in dimensional weighting. J Exp Psychol Hum Percept Perform 29:1021–1035. https://doi.org/10.1037/0096-1523.29.5.1021

Müller HJ, Krummenacher J, Heller D (2004) Dimension-specific intertrial facilitation in visual search for pop-out targets: evidence for a top-down modulable visual short-term memory effect. Vis Cogn 11:577–602. https://doi.org/10.1080/13506280344000419

Müller HJ, Liesefeld HR, Moran R, Usher M (2017) Parallel attentive processing and pre-attentive guidance. Behav Brain Sci 40:e149. https://doi.org/10.1017/S0140525X16000194

Nothdurft H (1993) Saliency effects across dimensions in visual search. Vis Res 33:839–844. https://doi.org/10.1016/0042-6989(93)90202-8

Pollmann S (2004) Anterior prefrontal cortex contributions to attention control. Exp Psychol 51:270–278. https://doi.org/10.1027/1618-3169.51.4.270

Pollmann S (2016) Frontopolar resource allocation in human and nonhuman primates. Trends Cogn Sci 20:84–86. https://doi.org/10.1016/j.tics.2015.11.006

Pollmann S, Weidner R, Müller HJ, von Cramon DY (2000) A fronto-posterior network involved in visual dimension changes. J Cogn Neurosci 12:480–494. https://doi.org/10.1162/089892900562156

Pollmann S, Weidner R, Müller HJ et al (2006a) Selective and interactive neural correlates of visual dimension changes and response changes. NeuroImage 30:254–265. https://doi.org/10.1016/j.neuroimage.2005.09.013

Pollmann S, Weidner R, Müller HJ, von Cramon DY (2006b) Neural correlates of visual dimension weighting. Vis Cogn 14:877–897. https://doi.org/10.1080/13506280500196142

Pollmann S, Mahn K, Reimann B et al (2007) Selective visual dimension weighting deficit after left lateral frontopolar lesions. J Cogn Neurosci 19:365–375. https://doi.org/10.1162/jocn.2007.19.3.365

Raja Beharelle A, Polanía R, Hare TA, Ruff CC (2015) Transcranial stimulation over frontopolar cortex elucidates the choice attributes and neural mechanisms used to resolve exploration–exploitation trade-offs. J Neurosci 35:14544–14556. https://doi.org/10.1523/JNEUROSCI.2322-15.2015

Rangelov D, Müller HJ, Zehetleitner M (2011a) Dimension-specific intertrial priming effects are task-specific: evidence for multiple weighting systems. J Exp Psychol Hum Percept Perform 37:100–114. https://doi.org/10.1037/a0020364

Rangelov D, Müller HJ, Zehetleitner M (2011b) Independent dimension-weighting mechanisms for visual selection and stimulus identification. J Exp Psychol Hum Percept Perform 37:1369–1382. https://doi.org/10.1037/a0024265

Rangelov D, Müller HJ, Zehetleitner M (2013) Visual search for feature singletons: multiple mechanisms produce sequence effects in visual search. J Vis 13(3):22. https://doi.org/10.1167/13.3.22

Rangelov D, Müller HJ, Zehetleitner M (2017) Failure to pop out: feature singletons do not capture attention under low signal-to-noise ratio conditions. J Exp Psychol Gen 146:651–671. https://doi.org/10.1037/xge0000284

Sauter M, Liesefeld HR, Zehetleitner M, Müller HJ (2018) Region-based shielding of visual search from salient distractors: target detection is impaired with same- but not different-dimension distractors. Atten Percept Psychophys 80:622–642. https://doi.org/10.3758/s13414-017-1477-4

Sauter M, Liesefeld HR, Müller HJ (in press). Learning to suppress salient distractors in the target dimension: region-based inhibition is persistent and transfers to distractors in a non-target dimension. J Exp Psychol Learn Mem Cogn. https://doi.org/10.1037/xlm0000691

Sawaki R, Geng JJ, Luck SJ (2012) A common neural mechanism for preventing and terminating the allocation of attention. J Neurosci 32:10725–10736. https://doi.org/10.1523/JNEUROSCI.1864-12.2012

Schledde B, Galashan FO, Przybyla M et al (2017) Task-specific, dimension-based attentional shaping of motion processing in monkey area MT. J Neurophysiol 118:1542–1555. https://doi.org/10.1152/jn.00183.2017

Schubö A, Müller HJ (2009) Selecting and ignoring salient objects within and across dimensions in visual search. Brain Res 1283:84–101. https://doi.org/10.1016/j.brainres.2009.05.077

Serences JT, Shomstein S, Leber AB et al (2005) Coordination of voluntary and stimulus-driven attentional control in human cortex. Psychol Sci 16:114–122. https://doi.org/10.1111/j.0956-7976.2005.00791.x

Theeuwes J (1991) Cross-dimensional perceptual selectivity. Percept Psychophys 50:184–193. https://doi.org/10.3758/BF03212219

Theeuwes J (1992) Perceptual selectivity for color and form. Percept Psychophys 51:599–606. https://doi.org/10.3758/BF03211656

Theeuwes J (2018) Visual selection: usually fast and automatic; seldom slow and volitional. J Cogn 1:29. https://doi.org/10.5334/joc.13

Theeuwes J, Reimann B, Mortier K (2006) Visual search for featural singletons: no top-down modulation, only bottom-up priming. Vis Cogn 14:466–489. https://doi.org/10.1080/13506280500195110

Toffanin P, de Jong R, Johnson A (2011) The P4pc: An electrophysiological marker of attentional disengagement? Int J Psychophysiol 81:72–81. https://doi.org/10.1016/j.ijpsycho.2011.05.010

Töllner T, Gramann K, Müller HJ et al (2008) Electrophysiological markers of visual dimension changes and response changes. J Exp Psychol Hum Percept Perform 34:531–542. https://doi.org/10.1037/0096-1523.34.3.531

Töllner T, Gramann K, Müller HJ, Eimer M (2009) The anterior N1 component as an index of modality shifting. J Cogn Neurosci 21:1653–1669. https://doi.org/10.1162/jocn.2009.21108

Töllner T, Zehetleitner M, Gramann K, Müller HJ (2010) Top-down weighting of visual dimensions: behavioral and electrophysiological evidence. Vis Res 50:1372–1381. https://doi.org/10.1016/j.visres.2009.11.009

Töllner T, Zehetleitner M, Gramann K, Müller HJ (2011) Stimulus saliency modulates pre-attentive processing speed in human visual cortex. PLoS One 6:e16276. https://doi.org/10.1371/journal.pone.0016276

Töllner T, Rangelov D, Müller HJ (2012a) How the speed of motor-response decisions, but not focal-attentional selection, differs as a function of task set and target prevalence. Proc Natl Acad Sci U S A 109:e1990–e1999. https://doi.org/10.1073/pnas.1206382109

Töllner T, Müller HJ, Zehetleitner M (2012b) Top-down dimensional weight set determines the capture of visual attention: evidence from the PCN component. Cereb Cortex 22:1554–1563. https://doi.org/10.1093/cercor/bhr231

Treisman A (1988) Features and objects: the fourteenth Bartlett memorial lecture. Q J Exp Psychol A 40:201–237. https://doi.org/10.1080/02724988843000104

Treisman A (2006) How the deployment of attention determines what we see. Vis Cogn 14:411–443. https://doi.org/10.1080/13506280500195250

Utz S, Humphreys GW, Chechlacz M (2013) Parietal substrates for dimensional effects in visual search: evidence from lesion-symptom mapping. Brain 136:751–760. https://doi.org/10.1093/brain/awt003

Weichselbaum H, Ansorge U (2018) Bottom-up attention capture with distractor and target singletons defined in the same (color) dimension is not a matter of feature uncertainty. Atten Percept Psychophys 80:1350–1361. https://doi.org/10.3758/s13414-018-1538-3

Weidner R, Müller HJ (2013) Dimensional weighting in cross-dimensional singleton conjunction search. J Vis 13(3):25. https://doi.org/10.1167/13.3.25

Weidner R, Pollmann S, Müller HJ, von Cramon DY (2002) Top-down controlled visual dimension weighting: an event-related fMRI study. Cereb Cortex 12:318–328. https://doi.org/10.1093/cercor/12.3.318

Weidner R, Krummenacher J, Reimann B et al (2009) Sources of top–down control in visual search. J Cogn Neurosci 21:2100–2113. https://doi.org/10.1162/jocn.2008.21173

Wolfe JM (1998) What can 1 million trials tell us about visual search? Psychol Sci 9:33–39. https://doi.org/10.1111/1467-9280.00006

Wolfe JM (2007) Guided search 4.0: current progress with a model of visual search. In: Gray W (ed) Integrated models of cognitive systems. Oxford University Press, New York, pp 99–119

Wolfe JM, Horowitz TS (2004) What attributes guide the deployment of visual attention and how do they do it? Nat Rev Neurosci 5:495–501. https://doi.org/10.1038/nrn1411

Wolfe JM, Horowitz TS (2017) Five factors that guide attention in visual search. Nat Hum Behav 1:58. https://doi.org/10.1038/s41562-017-0058

Wolfe JM, Butcher SJ, Lee C, Hyle M (2003) Changing your mind: on the contributions of top-down and bottom-up guidance in visual search for feature singletons. J Exp Psychol Hum Percept Perform 29:483–502. https://doi.org/10.1037/0096-1523.29.2.483

Wolfe JM, Palmer EM, Horowitz TS (2010) Reaction time distributions constrain models of visual search. Vis Res 50:1304–1311. https://doi.org/10.1016/j.visres.2009.11.002

Zehetleitner M, Krummenacher J, Geyer T et al (2011) Dimension intertrial and cueing effects in localization: support for pre-attentively weighted one-route models of saliency. Atten Percept Psychophys 73:349–363. https://doi.org/10.3758/s13414-010-0035-0

Zehetleitner M, Rangelov D, Müller HJ (2012) Partial repetition costs persist in nonsearch compound tasks: evidence for multiple-weighting-systems hypothesis. Atten Percept Psychophys 74:879–890. https://doi.org/10.3758/s13414-012-0287-y

Active Inference, Novelty and Neglect

Thomas Parr and Karl J. Friston

Contents

Abstract In this chapter, we provide an overview of the principles of active inference. We illustrate how different forms of short-term memory are expressed formally (mathematically) through appealing to beliefs about the causes of our sensations and about the actions we pursue. This is used to motivate an approach to active vision that depends upon inferences about the causes of 'what I have seen' and learning about 'what I would see if I were to look there'. The former could manifest as persistent 'delay-period' activity – of the sort associated with working memory, while the latter is better suited to changes in synaptic efficacy – of the sort that underlies short-term learning and adaptation. We review formulations of these ideas in terms of active inference, their role in directing visual exploration and the consequences – for active vision – of their failures. To illustrate the latter, we draw upon some of our recent work on the computational anatomy of visual neglect.

Keywords Active inference · Active vision · Markov decision process · Novelty · Saccades · Visual neglect

---

T. Parr (✉) and K. J. Friston
Wellcome Centre for Human Neuroimaging, Institute of Neurology, University College London, London, UK
e-mail: thomas.parr.12@ucl.ac.uk; k.friston@ucl.ac.uk

Curr Topics Behav Neurosci (2019) 41: 115–128
DOI 10.1007/7854_2018_61
Published Online: 18 August 2018

1 Introduction

Active inference is a principled framework for describing Bayes optimal behaviour. It formalises the notion of perception as hypothesis testing (Gregory 1980; Friston et al. 2012), emphasising the active manipulation of sensory epithelia to perform perceptual experiments. Here, we focus upon active vision (Wurtz et al. 2011; Andreopoulos and Tsotsos 2013; Ognibene and Baldassarre 2014; Mirza et al. 2016). Our formulation implies that perception of space is fundamentally tied to motor representations, as visual input at a point in space is the consequence of a saccade to that location (Zimmermann and Lappe 2016). This enactive view also brings with it perspectival aspect to the way that we sample – and internalise – our lived world: see Rudrauf et al. (2017) for a treatment of the implicit phenomenology. In what follows, we provide an overview of active inference in the context of visual foraging. This takes the form of a generic description in terms of Markov decision processes and draws examples from our recent work to illustrate these ideas in concrete scenarios. We will illustrate visual foraging in an environment with volatile contingencies, to disclose the intimate relationship between past beliefs (i.e. working memory and attention) and future plans (i.e. working memory and intention). We then turn to scenes with multiple stimuli and show that the optimisation of beliefs about the causes of sensory impressions is an efficient way to deal with such environments. To demonstrate the importance of this sort of inference, we illustrate – via simulations – how these mechanisms can fail. This provides us with several plausible mechanisms that could account for visual neglect – a common neuropsychological syndrome that reflects a cardinal failure of active vision.

2 A Note on Terminology

In the following, we call on some technical concepts – e.g. evidence, free energy and generative model. In this section, we briefly introduce the terms that will become important in the later parts of the chapter. Most of these terms reflect the fact that we are trying to understand behaviour in terms of a principled sampling of the world to make perceptual inferences. Formally speaking, this (active) inference can be described in terms of probability distributions or 'beliefs'. Priors and posteriors are the probabilities before and after making an observation (respectively). Likelihoods quantify the probability of an observation given we knew how it was caused. A generative model is a probabilistic specification of beliefs about how a certain kind of sensory data is generated. It comprises priors and likelihoods and can be used to make predictions about the most probable cause of sensations, namely, the posterior. Model evidence is the probability that a model could account for some given data. This is sometimes referred to as a marginal likelihood, or negative surprise. Self-evidence depends upon a 'self-model' that specifies the kinds of data that will be observed by an organism given the sort of creature it is. For example, given that a creature is a fish, it might expect to be surrounded by water. Observing this carries high self-evidence, while finding itself on dry land carries little evidence for its continued existence. From this, it is intuitively sensible that self-evidence should not drop too low. Negative free energy (sometimes referred to as an 'evidence lower bound') is a quantity that is always less than the evidence for a model. Minimising free energy thus maximises the lower bound on evidence and ensures that self-evidence is maintained above a certain value. This is just a mathematical way of describing behaviour that minimises surprise (i.e. maximises model evidence) and thereby resolves uncertainty about sensory exchanges with the world.

3 Active Inference

Bayes optimal behaviour requires that action maximises self-evidence (Hohwy 2016) or, equivalently, minimises surprise (Friston et al. 2010). This is not always possible (or efficient) to compute directly, but can be approximated by a free energy bound (Beal 2003), as illustrated by Jensen's inequality:

$$\underbrace{F}_{\text{Free Energy}} = \underbrace{-E_Q\left[\ln \frac{P(\tilde{o},\tilde{s},\pi)}{Q(\tilde{s},\pi)}\right] \geq -\ln E_Q\left[\frac{P(\tilde{o},\tilde{s},\pi)}{Q(\tilde{s},\pi)}\right]}_{\text{Jensen's inequality}} = \underbrace{-\ln P(\tilde{o})}_{\text{Surprise}}$$

In this equation, P is the probability distribution that represents a generative model. This probabilistic model expresses the beliefs an animal has about the way in which its sensations ($\tilde{o}$) are generated from hidden (unobservable) states ($\tilde{s}$) and the policy (π) or action sequence that it is pursuing. The tilde (~) notation indicates a trajectory (or sequence) through time. The distribution Q may be any arbitrary distribution but takes on an interesting interpretation when the free energy is minimised. A rearrangement of the equation above gives

$$F = D_{\text{KL}}\left[Q(\tilde{s},\pi)\|P(\tilde{s},\pi|\tilde{o})\right] - \ln P(\tilde{o})$$

This shows that the difference between surprise and the free energy is the KL-divergence between Q and the posterior distribution. This is minimal when the two distributions are the same, indicating that free energy minimisation entails finding Q that approximates the posterior probability. Under mean-field assumptions (Feynman 1998), we rewrite the variational free energy:

$$\begin{aligned} F &= E_Q[F(\pi)] + D_{\text{KL}}[Q(\pi)\|P(\pi)] \\ F(\pi) &= -E_Q\left[\ln P(\tilde{o},\tilde{s}|\pi) - \ln Q(\tilde{s}|\pi)\right] \end{aligned}$$

This allows us to separate out perceptual inference (optimisation of $Q(\tilde{s}|\pi)$) from planning (optimisation of $Q(\pi)$). Planning requires us to define a prior probability distribution over possible action sequences. A self-consistent prior for a free energy minimising agent is that the most probable policies are those that minimise expected free energy ($G(\pi)$) (Friston et al. 2017a):

$$
\begin{aligned}
P(\pi) &= \sigma(-G(\pi)) \\
G(\pi) &= \sum_{\tau>t} G(\pi,\tau) \\
G(\pi,\tau) &= -E_{\tilde{Q}}[\ln P(o_\tau, s_\tau|\pi) - \ln Q(s_\tau|\pi)] \\
\tilde{Q}(o_\tau, s_\tau|\pi) &= P(o_\tau|s_\tau)Q(s_\tau|\pi)
\end{aligned}
$$

Under these self-evidencing prior beliefs, the probability of each policy is defined by applying a softmax function (σ) to the negative expected free energy of policies or sequences of actions that are currently entertained (e.g. 'where to look next'). To gain some intuition for the consequences of minimising expected free energy (i.e. expected surprise), we can express the expected free energy in the following form:

$$
G(\pi,\tau) = -E_{\tilde{Q}}\left[\underbrace{\ln P(s_\tau|o_\tau,\pi) - \ln Q(s_\tau|\pi)}_{\text{Epistemic value}} + \underbrace{\ln P(o_\tau)}_{\text{Extrinsic value}}\right]
$$

The first pair of terms on the right shows that the expected free energy will be smaller if the (expected) difference between the posterior belief before and after the next observation is large. In other words, a policy will be more likely if its sensory consequences bring about a larger change in beliefs. This epistemic value is also known as salience, information gain, Bayesian surprise or epistemic affordance. The final term says that policies are more probable if they fulfil prior beliefs (preferences) about sensations that will be sampled under that policy. Together, these terms ensure explorative, information seeking (epistemic) and exploitative, goal seeking (pragmatic) behaviour (Friston et al. 2015).

It remains for us to specify the form of the generative model (P). In this chapter, we appeal to a generative model that takes the form of a Markov decision process, as depicted in Fig. 1. This is a discrete state-space model that incorporates the variables considered above, namely, states of the world and the action sequences that change those states. In a Markov decision process, hidden states evolve through time according to a probability transition matrix (**B**). At each time, a state depends upon only on the state at the previous time and upon the policy pursued. States give rise to observations as determined by a likelihood distribution (**A**). The probability of initial states is given by the distribution **D**, while preferences are defined by **C**. In the next section, we will look at the interplay between perceptual inference (i.e. optimising beliefs about states of the world) and planning (i.e. optimising beliefs about policies from which actions are selected). This inference calls for a form of working memory that could be mediated by persistent neuronal firing rates (Funahashi et al. 1989; Friston et al. 2017b; Parr and Friston 2017d). We then turn to beliefs about the probability distributions and an elemental form of

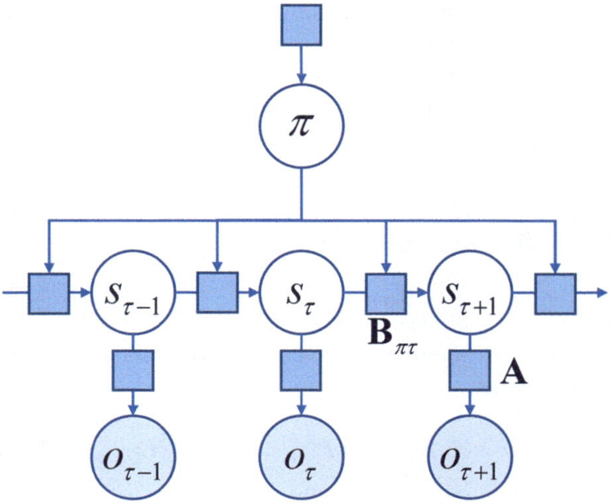

$$P(\tilde{o},\tilde{s},\pi) = P(s_1)P(\pi)\prod_\tau P(o_\tau \mid s_\tau)P(s_{\tau+1} \mid s_\tau,\pi)$$

$$P(o_\tau \mid s_\tau) = Cat(\mathbf{A})$$

$$P(s_{\tau+1} \mid s_\tau,\pi) = Cat(\mathbf{B}_{\pi\tau})$$

$$P(o_\tau) = Cat(\mathbf{C})$$

$$P(s_1) = Cat(\mathbf{D})$$

$$P(\pi) = \sigma(-\mathbf{G})$$

$$Q(\tilde{s},\pi) = Q(\pi)\prod_\tau Q(\tilde{s}_\tau \mid \pi)$$

Fig. 1 Markov decision process. This graphic shows a Markov decision process in factor graph form (Loeliger et al. 2007). Blue squares indicate probability distributions (factors of the generative model). In the lower panel, these factors are expressed in terms of probability matrices. 'Cat' denotes a categorical distribution. The mean-field factorisation of the approximate distribution, Q, is shown. This figure has been adapted from Parr and Friston (2018)

short-term memory that could be synaptically mediated (Hempel et al. 2000; Wang et al. 2006; Mongillo et al. 2008; Parr and Friston 2017b).

4 The Neurobiology of Active Inference

While the mathematical formalism presented above might seem a little abstract, active inference is associated with a process theory (Friston et al. 2017a) that interprets these quantities in terms of their biological substrates. For example, the approximate posterior beliefs (Q) can be thought of as neuronal (firing rate) representations. The conditional probability distributions (**A**, **B**) then represent the synaptic connections between different populations of neurons. Quantities like the expected free energy (**G**) are computed from the above variables to evaluate competing plans of action – this evaluation could take place in cortico-basal ganglia-cortical loops. There have been many papers linking the computational anatomy to functional anatomy in the human brain. These range from treatments of cortical hierarchies to associative plasticity. In the next sections, we will focus on precision (the confidence in a given synapse) – that may be mediated by ascending neuromodulatory influences – and learning, which involves plastic modulation of synapses through long-term potentiation or depression.

5 Inference and Salience

Perceptual inference is the process of optimising beliefs about states (under a given policy). By minimising free energy, we find that inference or belief updating can be expressed as

$$Q(s_\tau|\pi) = \sigma\big(E_{Q(s_{\tau-1}|\pi)}[\ln P(s_\tau|s_{\tau-1},\pi)] + E_{Q(s_{\tau+1}|\pi)}\big[\ln P\big(s_{\tau+1}|s_\tau,\pi\big)\big] + \ln P\big(o_\tau|s_\tau\big)\big)$$

The first term on the right entails a form of memory, because beliefs about a state in the past are used to inform beliefs about the present (via the transition probabilities). However, the capacity for the past to inform the present depends upon the degree of uncertainty associated with transitions. Intuitively, if the environment changes randomly from one time to the next, the past is of no use in inferring the present. Conversely, in a deterministic environment, the present can be reliably inferred using beliefs about the past. We have previously described a way to parameterise transition probabilities so that we can manipulate the precision (ω) of these transitions (Parr and Friston 2017c). When $\omega = 0$, there is no relationship between past and present. As $\omega \to \infty$, transitions tend towards being deterministic. Another way to think about this is that ω scales with the negative entropy of transition probabilities, also known as volatility (Parr and Friston 2017d). Neurobiologically, this suggests that this parameter may play the role of a synaptic gain – controlling the degree to which neurons representing the past can influence those representing the present. In turn, this implicates neuromodulatory transmitters, such as noradrenaline, that have been associated with signalling volatility (Dayan and Yu 2006; Marshall et al. 2016).

This has important consequences for visual search and the selection of saccadic targets that minimise expected free energy. If we treat eye position as a hidden state (with transitions dependent on policies or saccadic sequences), this hidden state determines which other hidden states are informed by foveal visual data (Mirza et al. 2016). Figure 2 shows four objects (squares of various shades). Performing a saccade to a location means foveal vision will provide precise information about the object at that location. If the shade of each square can change, with a certain probability, this has important consequences for our belief about the shades of squares not currently foveated. Specifically, if ω is high, the confidence in $Q(s_\tau| \pi)$ will be relatively high, and the expected free energy of a saccade to that location will be high. In other words, actively sampling other squares would have a low epistemic affordance. This would make for a poor perceptual experiment, as it would do little

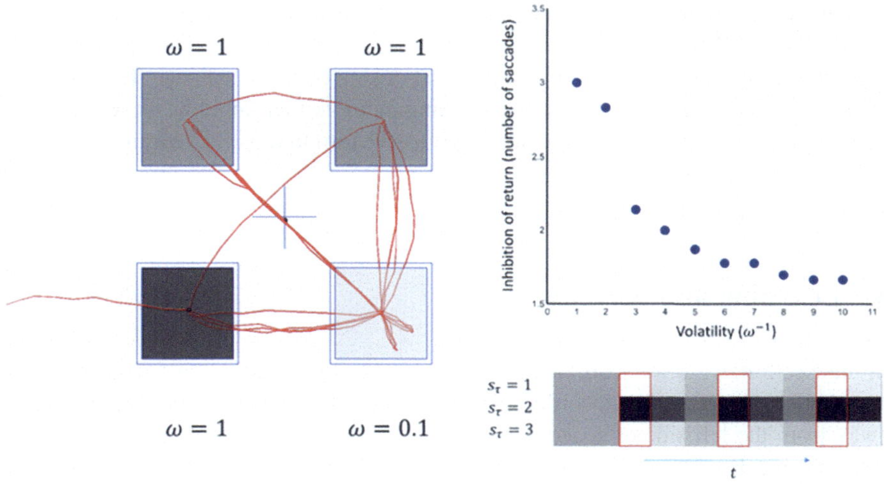

Fig. 2 Persistent activity and inhibition of return. *On the left,* we show the simulated eye-tracking data (over 19 saccades), when each hidden case is associated with a differing level of volatility (ω^{-1}). *The right upper plot* shows the influence of the volatility of beliefs on the simulated inhibition of return (quantified by the average number of saccades between fixations) for one of the stimulus locations. *The right lower plot* shows the simulated neuronal encoding over time for three units, each representing a different hypothesis about the identity (1, 2 or 3) of one of the hidden states. Darker shades indicate greater firing rates. The red boxes indicate the times during which the location corresponding to this hidden state is foveated. This figure has been reproduced from Parr and Friston (2017c)

to change beliefs. If low, the confidence in $Q(s_\tau | \pi)$ will decrease with each time step, making it an increasingly plausible (i.e. salient) saccadic target.

The simulated eye-tracking trace (Parr and Friston 2017c) in Fig. 2 shows that those locations associated with precise (relatively deterministic) transition probabilities are less frequently the targets of saccades compared to the more volatile lower right square. This formulation of perisaccadic working memory, and its role in directing the eyes towards salient locations, reproduces the phenomenon of 'inhibition of return' (Posner et al. 1985; Klein 2000) and shows how a Bayes optimal agent would modulate this at various levels of transition precisions. The lower right plot shows how expectations that a square possesses one of three alternative shades become more dispersed (as uncertainty accumulates) over time until a saccade is made to that location. This can be interpreted as a 'raster plot' showing the firing rates of three neurons (each row) over time. Each of these neurons encodes a probability that the square has one of three shades over time. This is exactly the profile of neuronal responses one sees in delayed saccade experiments used to elicit the neuronal correlates of working memory: see, for example, Kojima and Goldman-Rakic (1982). In short, simply inferring the best thing to do next mandates the encoding of trajectories of (hidden) states of the world that necessarily requires a neuronal encoding of beliefs about the past (and future). Crucially, this neuronal (working memory) encoding transcends the current moment, equipping the perceptual inference with a mnemonic aspect that allows for both prediction and postdiction. It is this encoding we suggest as a canonical form of working memory.

In this section, we have examined the relationship between salience (the potential to resolve uncertainty) and the sort of working memory that might be represented by persistently active neurons. In the next section, we turn to novelty, learning and their link with synaptic memory.

6 Learning and Novelty

In this section, we draw from observations concerning visual neglect (Halligan and Marshall 1998). This is a disorder of active vision that biases saccadic exploration away from the left side of space (Husain et al. 2001; Fruhmann Berger et al. 2008; Karnath and Rorden 2012). A common pen-and-paper test – used to assess this deficit – is the line cancellation task (illustrated in Fig. 3) (Albert 1973; Fullerton et al. 1986; Ferber and Karnath 2001). Patients are asked to cancel, by crossing out, all of the lines on a sheet of paper. Typically, neglect patients will cross out those on the right but miss those on the left. Similar effects have been found in oculomotor variants of this task but have demonstrated remarkably few re-cancellations in healthy participants (Mannan et al. 2005), even when there is no explicit (visual) cue to record that a target has been cancelled. The middle and right panels of Fig. 3 make the point that maintaining a memory of previously visited locations would

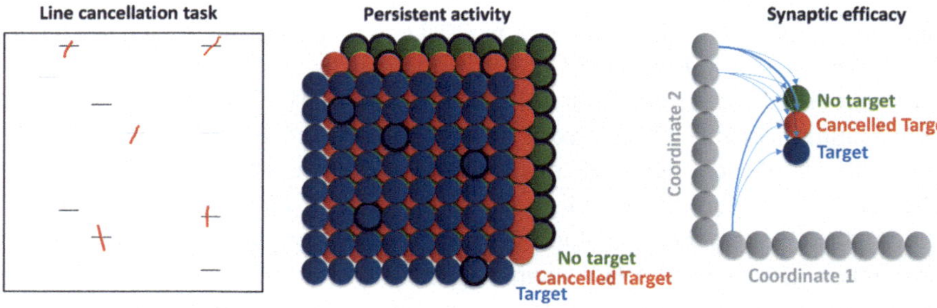

Fig. 3 Persistent activity or synaptic efficacy? *On the left,* an example of a line cancellation task is shown. The subject is presented with a sheet of paper with a set of horizontal lines and is asked to cancel (red marks) each of these lines. *The middle panel* shows the ensemble of 192 neurons that would be required to represent the subject's beliefs about where the lines are, and whether they have cancelled them, if the memory of previously visited locations were stored as persistent activity in a neuronal population. The currently active neurons are represented by a black outline. *The panel on the right* shows a much more efficient way to represent this information, in terms of a mapping from a representation of space to representations of each of the possible observations that could be made on visiting a particular location. Clearly it is more efficient to make use of synaptic efficacy when storing transient, high-dimensional memories. In short, synaptic efficacy represents probabilistic mappings (i.e. 'if I were to look there, I would see that') as opposed to beliefs about the current state of the world (i.e. 'I am looking there' or 'seeing that') encoded by synaptic activity. This figure has been adapted from Parr and Friston (2017a)

require many more neurons, if stored as persistent or delay-period activity, than if this information were stored through short-term plastic changes (Parr and Friston 2017a).

Under the process theory (Friston et al. 2017a) associated with active inference, we can think of synaptic plasticity in terms of changes in beliefs about conditional probabilities (Friston et al. 2016). This is because synapses mediate the influence of beliefs represented in one population of neurons on another. For these to change, we supplement our generative model with beliefs about the parameters of the mapping from 'where I am looking' to 'what I see'. When we do this for the parameters of the likelihood mapping, the expected free energy becomes

$$\begin{aligned} G(\pi,\tau) = \; & -\underbrace{E_{\tilde{Q}}[\ln P(s_\tau|,A,o_\tau,\pi) - \ln Q(s_\tau|\pi)]}_{\text{Salience}} \\ & -\underbrace{E_{\tilde{Q}}[\ln P(A|o_\tau,\pi) - \ln Q(A)]}_{\text{Novelty}} \\ & -\underbrace{E_{\tilde{Q}}[\ln P(o_\tau)]}_{\text{Extrinsic value}} \end{aligned}$$

Here, we have decomposed the epistemic value into the salience term used in the previous section, and a novelty term, that relates to the expected change in beliefs about the parameters of the likelihood distribution. Figure 4 shows a simulation of an oculomotor cancellation task (Parr and Friston 2017b). This incorporates beliefs about what would be seen conditioned upon the eye position. These beliefs are optimised by accumulation of Dirichlet parameters (Beal 2003; Blei et al. 2003; Friston et al. 2016) – this closely resembles activity-dependent plasticity (Hebb 1949; Brown et al. 2009), as the element of the likelihood matrix (**A**) representing a state-outcome mapping is increased whenever the two occur simultaneously. This means that, on fixating a location, uncertainty is resolved about the visual data obtained by saccades to that location, and its novelty is decreased (Schwartenbeck et al. 2013).

Although phenomenologically, this sort of memory is very different from the working memory of the previous section, they both emerge from the minimisation expected free energy or expected surprise (i.e. entropy or uncertainty). When resolving uncertainty about states of affairs in the world, we sample salient information. However, when resolving uncertainty about the contingencies, the corresponding epistemic affordance becomes novelty, i.e. the opportunity to resolve uncertainty about 'what would happen if I did that?' If updating beliefs about states of the world and parameters – that underwrite probabilistic contingencies – corresponds to perceptual inference and learning, respectively, then novelty is to learning as salience is to inference.

Figure 4 also shows the consequences of lesioning the generative model (Parr and Friston 2017b). Increasing the Dirichlet parameters for the left side of space means that (despite representing the same likelihood distribution) the capacity for the

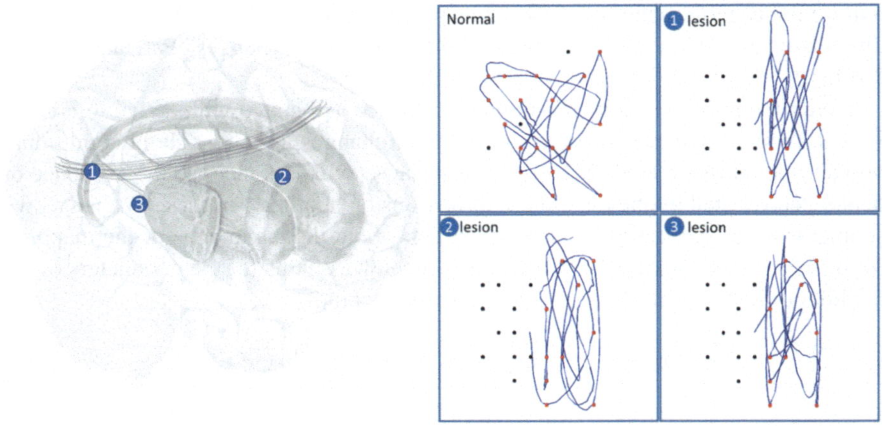

Fig. 4 The computational anatomy of visual neglect. *On the left,* three lesions implicated in visual neglect are highlighted here. (1) Disconnection of the second branch of the right superior longitudinal fasciculus [a white matter tract that connects dorsal frontal with ventral parietal regions (Makris et al. 2004; Bartolomeo et al. 2012)]; (2) Unilateral lesion to the right putamen (Karnath et al. 2002); (3) Unilateral lesion to the right pulvinar (a thalamic nucleus). This has been reproduced from Parr et al. (2018). *On the right,* each panel shows the simulated eye-tracking data (blue) during 20 saccades. In all cases, the target array was the same. The upper left panel shows the performance of the model with no simulated lesions. The upper right panel shows the results when the Dirichlet parameters associated with the likelihood (i.e. hidden state to outcome) mapping were increased for the left hemifield, corresponding to a functional disconnection of the second branch of the right superior longitudinal fasciculus. The lower left panel shows performance when there is a biasing of policy selection, simulating a lesion of the putamen. The lower right panel represents a lesion of the prior beliefs about proprioceptive outcomes, which relates to a deficit in the inputs to the dorsal parietal cortex, likely from the pulvinar. Notably, the lesioned scan paths look very similar to one another. By eye, it is almost impossible to infer which path was generated by which lesion. Despite this, it is possible to disambiguate between the cortical disconnection (1) and the two subcortical lesions (2 and 3) using Bayesian model comparison (i.e. fitting each lesioned model to synthetic data and comparing the evidence for each model) – for details, please see Parr and Friston (2017b). The scan paths here have been reproduced from Parr and Friston (2017b)

beliefs about the likelihood to change is diminished. Cutting the connection between two regions would have this effect, as no plastic changes can occur following disconnection. This means there is little novelty to resolve on the left, biasing saccades to the right. This is consistent with the idea that visual neglect is a 'disconnection syndrome' (Geschwind 1965; Catani and ffytche 2005; Bartolomeo et al. 2007; He et al. 2007). We additionally show the effect of including a preference (**C**) for proprioceptive outcomes on the right of space and of directly biasing the prior beliefs over policies in favour of rightward saccades. The image on the left of Fig. 4 shows common neuroanatomical sites for lesions that induce neglect (Doricchi and Tomaiuolo 2003; Thiebaut de Schotten et al. 2005; Karnath and Rorden 2012) and relates these to the computational lesions shown on the right. In brief, this account suggests that the superior longitudinal fasciculus, that connects frontal to temporoparietal regions, might be the substrate of the likelihood distribution representing the visual consequences of each fixation. Plastic changes in this tract would then underwrite the learning described above, and its dysfunction would compromise the resolution of novelty following a given fixation. The dorsal pulvinar (Ungerleider and Christensen 1979), that projects to parietal cortex (Weller et al. 2002; Behrens et al. 2003), may signal prior beliefs (preferences) about the 'where' data that the brain might seek out. Damage could bias these towards the left. The evaluation of competing saccadic policies is likely to implicate cortico-basal ganglia circuits (Schiller et al. 1980, 1987; Hikosaka and Wurtz 1985), and disruption of these could manifest as a bias towards rightward saccades. In this – and the previous section – we have shown that disruption of any part of the expected free energy (salience, novelty or extrinsic value) can have drastic consequences for active vision.

7 Conclusion

In this chapter, we have reviewed the basic principles of active inference and have illustrated the importance of holding beliefs about hidden states and parameters for planning and action. In brief, uncertainty about hidden states renders those actions that resolve uncertainty salient. Uncertainty about parameters presents the opportunity for resolving uncertainty by selecting actions that entail novelty. Interestingly, each of these beliefs equips actively inferring creatures with a form of memory. Optimisation of beliefs about states relies in part upon the propagation of beliefs about the past to the present, suggesting a form of working memory mediated by persistent activity in neuronal populations representing relevant beliefs. However, it is more efficient, for higher-dimensional memories, to represent context-sensitive states of the sampled world as relationships between variables, e.g. the mapping from 'where I look' to 'what I see'. This implicates changes in the connections between neuronal populations and short-term plasticity. We have illustrated the consequences of the requisite belief updating – for active vision – and the repertoire of computational lesions that could lead to visual neglect.

Acknowledgements TP is supported by the Rosetrees Trust (Award Number 173346). KJF is a Wellcome Principal Research Fellow (Ref: 088130/Z/09/Z).

Disclosure Statement The authors have no disclosures or conflict of interest.

References

Albert ML (1973) A simple test of visual neglect. Neurology 23(6):658
Andreopoulos A, Tsotsos J (2013) A computational learning theory of active object recognition under uncertainty. Int J Comput Vis 101(1):95–142

Bartolomeo P, Thiebaut de Schotten M, Doricchi F (2007) Left unilateral neglect as a disconnection syndrome. Cereb Cortex 17(11):2479–2490

Bartolomeo P, Thiebaut de Schotten M, Chica AB (2012) Brain networks of visuospatial attention and their disruption in visual neglect. Front Hum Neurosci 6:110

Beal MJ (2003) Variational algorithms for approximate Bayesian inference. University of London, London, UK

Behrens TEJ, Johansen-Berg H, Woolrich MW, Smith SM, Wheeler-Kingshott CAM, Boulby PA, Barker GJ, Sillery EL, Sheehan K, Ciccarelli O, Thompson AJ, Brady JM, Matthews PM (2003) Non-invasive mapping of connections between human thalamus and cortex using diffusion imaging. Nat Neurosci 6(7):750–757

Blei DM, Ng AY, Jordan MI (2003) Latent dirichlet allocation. J Mach Learn Res 3:993–1022

Brown TH, Zhao Y, Leung V (2009) Hebbian plasticity A2 – squire, Larry R. Encyclopedia of neuroscience. Academic Press, Oxford, pp 1049–1056

Catani M, ffytche DH (2005) The rises and falls of disconnection syndromes. Brain 128(10): 2224–2239

Dayan P, Yu AJ (2006) Phasic norepinephrine: a neural interrupt signal for unexpected events. Netw Comput Neural Syst 17(4):335–350

Doricchi F, Tomaiuolo F (2003) The anatomy of neglect without hemianopia: a key role for parietal-frontal disconnection? Neuroreport 14(17):2239–2243

Ferber S, Karnath H-O (2001) How to assess spatial neglect-line bisection or cancellation tasks? J Clin Exp Neuropsychol 23(5):599–607

Feynman RP (1998) Statistical mechanics: a set of lectures. Avalon Publishing, New York

Friston KJ, Daunizeau J, Kilner J, Kiebel SJ (2010) Action and behavior: a free-energy formulation. Biol Cybern 102(3):227–260

Friston K, Adams RA, Perrinet L, Breakspear M (2012) Perceptions as hypotheses: saccades as experiments. Front Psychol 3:151

Friston K, Rigoli F, Ognibene D, Mathys C, Fitzgerald T, Pezzulo G (2015) Active inference and epistemic value. Cognit Neurosci 6(4):187–214

Friston K, FitzGerald T, Rigoli F, Schwartenbeck P, O'Doherty J, Pezzulo G (2016) Active inference and learning. Neurosci Biobehav Rev 68:862–879

Friston K, FitzGerald T, Rigoli F, Schwartenbeck P, Pezzulo G (2017a) Active inference: a process theory. Neural Comput 29(1):1–49

Friston KJ, Rosch R, Parr T, Price C, Bowman H (2017b) Deep temporal models and active inference. Neurosci Biobehav Rev 77:388–402

Fruhmann Berger M, Johannsen L, Karnath H-O (2008) Time course of eye and head deviation in spatial neglect. Neuropsychology 22(6):697–702

Fullerton KJ, McSherry D, Stout RW (1986) Albert's test: a neglected test of perceptual neglect. Lancet 327(8478):430–432

Funahashi S, Bruce CJ, Goldman-Rakic PS (1989) Mnemonic coding of visual space in the monkey's dorsolateral prefrontal cortex. J Neurophysiol 61(2):331

Geschwind N (1965) Disconnexion syndromes in animals and Man1. Brain 88(2):237–237

Gregory RL (1980) Perceptions as hypotheses. Phil Trans Roy Soc Lond B Biol Sci 290(1038):181

Halligan PW, Marshall JC (1998) Neglect of awareness. Conscious Cogn 7(3):356–380

He BJ, Snyder AZ, Vincent JL, Epstein A, Shulman GL, Corbetta M (2007) Breakdown of functional connectivity in frontoparietal networks underlies behavioral deficits in spatial neglect. Neuron 53(6):905–918

Hebb DO (1949) The first stage of perception: growth of the assembly. The organization of behavior. Wiley, New York, pp 60–78

Hempel CM, Hartman KH, Wang XJ, Turrigiano GG, Nelson SB (2000) Multiple forms of short-term plasticity at excitatory synapses in rat medial prefrontal cortex. J Neurophysiol 83(5):3031

Hikosaka O, Wurtz RH (1985) Modification of saccadic eye movements by GABA-related substances. I. Effect of muscimol and bicuculline in monkey superior colliculus. J Neurophysiol 53(1):266

Hohwy J (2016) The self-evidencing brain. Noûs 50(2):259–285
Husain M, Mannan S, Hodgson T, Wojciulik E, Driver J, Kennard C (2001) Impaired spatial working memory across saccades contributes to abnormal search in parietal neglect. Brain 124(5):941–952
Karnath H-O, Rorden C (2012) The anatomy of spatial neglect. Neuropsychologia 50(6):1010–1017
Karnath HO, Himmelbach M, Rorden C (2002) The subcortical anatomy of human spatial neglect: putamen, caudate nucleus and pulvinar. Brain 125(2):350–360
Klein RM (2000) Inhibition of return. Trends Cogn Sci 4(4):138–147
Kojima S, Goldman-Rakic PS (1982) Delay-related activity of prefrontal neurons in rhesus monkeys performing delayed response. Brain Res 248(1):43–49
Loeliger HA, Dauwels J, Hu J, Korl S, Ping L, Kschischang FR (2007) The factor graph approach to model-based signal processing. Proc IEEE 95(6):1295–1322
Makris N, Kennedy DN, McInerney S, Sorensen AG, Wang R, Caviness JVS, Pandya DN (2004) Segmentation of subcomponents within the superior longitudinal fascicle in humans: a quantitative, in vivo, DT-MRI study. Cereb Cortex 15(6):854–869
Mannan SK, Mort DJ, Hodgson TL, Driver J, Kennard C, Husain M (2005) Revisiting previously searched locations in visual neglect: role of right parietal and frontal lesions in misjudging old locations as new. J Cogn Neurosci 17(2):340–354
Marshall L, Mathys C, Ruge D, de Berker AO, Dayan P, Stephan KE, Bestmann S (2016) Pharmacological fingerprints of contextual uncertainty. PLoS Biol 14(11):e1002575
Mirza MB, Adams RA, Mathys CD, Friston KJ (2016) Scene construction, visual foraging, and active inference. Front Comput Neurosci 10:56
Mongillo G, Barak O, Tsodyks M (2008) Synaptic theory of working memory. Science 319(5869): 1543–1546
Ognibene D, Baldassarre G (2014) Ecological active vision: four bio-inspired principles to integrate bottom-up and adaptive top-down attention tested with a simple camera-arm robot. IEEE transactions on autonomous mental development 7(1):3–25
Parr T, Friston KJ (2017a) The active construction of the visual world. Neuropsychologia 104: 92–101
Parr T, Friston KJ (2017b) The computational anatomy of visual neglect. Cereb Cortex 1–14
Parr T, Friston KJ (2017c) Uncertainty, epistemics and active inference. J R Soc Interface 14(136)
Parr T, Friston KJ (2017d) Working memory, attention, and salience in active inference. Sci Rep 7(1):14678
Parr T, Friston KJ (2018) The discrete and continuous brain: from decisions to movement—and back again. Neural Comput (Early Access):1–29
Parr T, Rees G, Friston KJ (2018) Computational neuropsychology and Bayesian inference. Front Hum Neurosci 12(61)
Posner MI, Rafal RD, Choate LS, Vaughan J (1985) Inhibition of return: neural basis and function. Cogn Neuropsychol 2(3):211–228
Rudrauf D, Bennequin D, Granic I, Landini G, Friston K, Williford K (2017) A mathematical model of embodied consciousness. J Theor Biol 428:106–131
Schiller PH, True SD, Conway JL (1980) Deficits in eye movements following frontal eye-field and superior colliculus ablations. J Neurophysiol 44(6):1175
Schiller PH, Sandell JH, Maunsell JH (1987) The effect of frontal eye field and superior colliculus lesions on saccadic latencies in the rhesus monkey. J Neurophysiol 57(4):1033
Schwartenbeck P, FitzGerald T, Dolan RJ, Friston K (2013) Exploration, novelty, surprise, and free energy minimization. Front Psychol 4:710
Thiebaut de Schotten M, Urbanski M, Duffau H (2005) Direct evidence for a parietal-frontal pathway subserving spatial awareness in humans. Science 309
Ungerleider LG, Christensen CA (1979) Pulvinar lesions in monkeys produce abnormal scanning of a complex visual array. Neuropsychologia 17(5):493–501
Wang Y, Markram H, Goodman PH, Berger TK, Ma J, Goldman-Rakic PS (2006) Heterogeneity in the pyramidal network of the medial prefrontal cortex. Nat Neurosci 9(4):534–542

Weller RE, Steele GE, Kaas JH (2002) Pulvinar and other subcortical connections of dorsolateral visual cortex in monkeys. J Comp Neurol 450(3):215–240

Wurtz RH, McAlonan K, Cavanaugh J, Berman RA (2011) Thalamic pathways for active vision. Trends Cogn Sci 5(4):177–184

Zimmermann E, Lappe M (2016) Visual space constructed by saccade motor maps. Front Hum Neurosci 10(225)

Prefrontal Contributions to Attention and Working Memory

Zahra Bahmani, Kelsey Clark, Yaser Merrikhi, Adrienne Mueller, Warren Pettine, M. Isabel Vanegas, Tirin Moore, and Behrad Noudoost

Contents

Abstract The processes of attention and working memory are conspicuously interlinked, suggesting that they may involve overlapping neural mechanisms. Working memory (WM) is the ability to maintain information in the absence of sensory input. Attention is the process by which a specific target is selected for further processing, and neural resources directed toward that target. The content of WM can be used to direct attention, and attention can in turn determine which

Z. Bahmani
School of Cognitive Sciences, Institute for Research in Fundamental Sciences (IPM), Tehran, Iran

K. Clark, M. Isabel Vanegas, and B. Noudoost (✉)
Department of Ophthalmology and Visual Sciences, University of Utah, Salt Lake City, UT, USA
e-mail: behrad.noudoost@utah.edu

Y. Merrikhi
Department of Physiology & Pharmacology, The Brain and Mind Institute, University of Western Ontario, London, ON, Canada

Robarts Research Institute, University of Western Ontario, London, ON, Canada

A. Mueller and T. Moore
Department of Neurobiology, Stanford University, Stanford, CA, USA

Howard Hughes Medical Institute, Stanford University, Stanford, CA, USA

W. Pettine
Center for Neural Science, New York University, New York, NY, USA

Curr Topics Behav Neurosci (2019) 41: 129–154
DOI 10.1007/7854_2018_74
Published Online: 10 February 2019

information is encoded into WM. Here we discuss the similarities between attention and WM and the role prefrontal cortex (PFC) plays in each. First, at the theoretical level, we describe how attention and WM can both rely on models based on attractor states. Then we review the evidence for an overlap between the areas involved in both functions, especially the frontal eye field (FEF) portion of the prefrontal cortex. We also discuss similarities between the neural changes in visual areas observed during attention and WM. At the cellular level, we review the literature on the role of prefrontal DA in both attention and WM at the behavioral and neural levels. Finally, we summarize the anatomical evidence for an overlap between prefrontal mechanisms involved in attention and WM. Altogether, a summary of pharmacological, electrophysiological, behavioral, and anatomical evidence for a contribution of the FEF part of prefrontal cortex to attention and WM is provided.

Keywords Attention · Dopamine · Working memory

1 Models of Attention and WM

Numerous computational models of WM and attention exist and exploit diverse strategies to recreate the key features of neural activity during these states. One type of computational model often used for both WM and attention is conceptually defined by neural population "attractor states." WM models primarily focus on the maintenance and evolution of these attractor states, while attention models have characterized how attractor states can be influenced by top-down activity.

Neurons can fire action potentials at specific rates and patterns, contingent on the stimulus. Given that neurons have limited range in firing rates and limited selectivity to stimuli, there is a finite set of firing "states" that neurons and populations of neurons can occupy. Factors such as recurrent inhibition and excitation cause some of these firing rate states to be more stable than others. Those relatively stable states "attract" the firing rate activity of individual neurons and populations, creating "attractor states." When a stimulus is present, these attractor states can be thought of as neural representations of its qualities. In periods when the stimulus is no longer present, these attractor states represent a memory maintained in the system.

In WM networks, attractor states have been simulated in neural populations using stable firing rates (Compte et al. 2000), time-varying firing rates (Druckmann and Chklovskii 2012), and even "activity-silent" mechanisms that maintain representations synaptically (Goldman 2009). Neural populations can only support a finite number of attractor states at any one moment. These states can thus be thought of as competing with one another for representation (either in discrete slots or for shared resources), with some "winning," and thus maintained through the delay in WM (Wang 2008).

Various models of attention can be conceived as ways to direct the competition of attractor states or align their transmission from one brain area to another. The sources of such signals are often thought to arise from prefrontal cortex and be directed to other brain areas, such as visual cortex. At the most basic, models of attention selectively increase the activity of an attractor state that represents an "attended" stimulus. That particular state then has a competitive bias relative to others (Ardid et al. 2007). One can also model more complicated experimental findings, such as normalization effects of attention, through a variety of mathematical frameworks (Carandini and Heeger 2012), including network models (Kraynyukova and Tchumatchenko 2018). The reduction in trial-to-trial variability seen with attention has been modeled as a reduction in excursion between various attractor states from one trial to the next, again caused by biased competition (Deco and Hugues 2012). This increased consistency caused by attention can also be applied to population oscillations, which has been used to model the communication of attractor content from one brain area to another (Deco and Kringelbach 2016).

Instantiations of these network models using spiking units or rate units allow one to capture patterns of response, the variability of that response from trial to trial, and more general oscillations in currents. Spiking networks additionally allow one to characterize spike timing and other discrete properties. These and other modeling tools are proving incredibly powerful for elucidating the underlying mechanisms of WM and attention.

2 FEF's Role in Attention and WM

Psychophysical and electrophysiological studies indicate that the FEF is heavily involved in both covert visual attention and WM. The first evidence suggesting the FEF's key role in high-level brain cognitive functions comes from its anatomical properties. Through massive reciprocal connections, this area is interconnected with many visual cortical and subcortical brain areas (Anderson et al. 2011; Markov et al. 2014b; Schall et al. 1995; Stanton et al. 1995). Some of these connected brain areas have been suggested as sources of visual attention, including adjacent prefrontal areas (Stanton et al. 1993), lateral intraparietal cortex (LIP) (Anderson et al. 2011; Stanton et al. 1995), and the superior colliculus (SC) (Sommer and Wurtz 2000).

Originally, the FEF was identified as an area involved in making saccadic eye movements (Robinson and Fuchs 1969). Electrical stimulation of a FEF site with microampere-magnitude currents (microstimulation) results in a fixed vector saccade to a specific part of space which is considered as the response field (RF) of that site (Bruce and Goldberg 1985). Neurons in the FEF show diverse responses including any combination of visual, memory-related, or saccade-related activity (Bruce and Goldberg 1985; Lawrence et al. 2005). During the last couple of decades, several studies have examined the role of the FEF in covert attention and WM. Psychophysical studies revealing a tight link between attention and saccadic eye movements (Deubel and Schneider 1996) initially suggested the FEF as a source of visual attention. In the first study showing the causal role of the FEF in the control of visual attention, it was shown that the microstimulation of FEF sites, with the currents less than what is needed to produce a saccadic eye movement (subthreshold microstimulation), improves the animal's performance in detecting a contrast change for stimuli presented within the area of space represented by the stimulated FEF site (Moore and Fallah 2001, 2004). Thus, stimulating the FEF results in the behavioral benefits of visual attention. Another study showed the link between FEF activation and the neuronal signatures of visual attention: subthreshold microstimulation of FEF is accompanied by an increase in firing rate and selectivity of V4 visual neurons with RFs overlapping with those of the stimulated FEF site (Moore and Armstrong 2003). Neurons in FEF also encode the locus of visual attention (Armstrong et al. 2009). Interestingly, the same neurons in the FEF that carry memory signals during the delay period are more likely to reflect target selection during the attention portion of a task (Armstrong et al. 2009), which will be discussed further in the context of identifying the FEF signal sent to visual areas. FEF inactivation profoundly impairs behavioral performance in both spatial WM (Noudoost et al. 2014) and attention-dependent tasks such as covert visual search and visual discrimination tasks (Wardak et al. 2006). In sum, the FEF portion of PFC appears to play a crucial role in both attention and WM.

3 Signatures of Attention and How WM-Induced Changes Resemble Attention

Similarities between the neural signatures of attention and WM maintenance within sensory areas further support the theory that these two processes have overlapping mechanisms within prefrontal cortex.

The neural signatures of attention at the level of individual neurons include an increase in response magnitude (Green and Swets 1966), reduced neuronal response latency (Galashan et al. 2013; Sundberg et al. 2012; Lee et al. 2005), a shift of RFs toward the locus of attention and shrinking RFs at the attended location (Anton-Erxleben et al. 2009; Connor et al. 1996, 1997; Kusunoki and Goldberg 2003; Womelsdorf et al. 2006; Suzuki and Cavanagh 1997; Anton-Erxleben & Carrasco 2013), reduced burstiness (Anderson et al. 2013), and reduced variability of visual responses (Mitchell et al. 2007). Each of these signatures can contribute to increasing the signal-to-noise ratio and thus lead to better discrimination and perception (Cohen and Maunsell 2009; Ling et al. 2009; Mitchell et al. 2009) or a faster reaction (Albrecht 1995; Bell et al. 2006; Oram et al. 2002; Raiguel et al. 1999). Attention also affects the correlated noise of simultaneously recorded pairs of neurons (Cohen and Kohn 2011; Cohen and Maunsell 2009; Ruff and Cohen 2014). The shared trial-to-trial fluctuations in firing rates in response to the same stimulus are often called the noise correlation. Attention can reduce the noise correlation in V4; interestingly, this reduction in noise correlation enhances the signal-to-noise ratio of the pooled neural activity more than the changes in firing rates do (Mitchell et al. 2009). A study by Cohen and Maunsell reported that over 80% of the attentional-induced enhancement in V4 population sensitivity was due to the reduced noise correlations between the firing rates of neurons (Cohen and Maunsell 2009).

WM plays an essential role in our daily life, determining how we interact with the world based on our current goals (D'Esposito and Postle 2015). WM enhances the processing of information related to the items held in WM, in part by altering the processing of sensory information (Awh and Jonides 2001; Postle 2005). Finding the neural basis by which WM engages sensory areas and enhances sensory representations in these areas is crucial to understanding the mechanisms underlying goal-directed behavior. Our recent study revealed that sensory areas receive a WM-rich signal from the FEF part of the prefrontal cortex during a memory-guided saccade (MGS) task (Fig. 1a) (Merrikhi et al. 2017). This evidence, together with studies

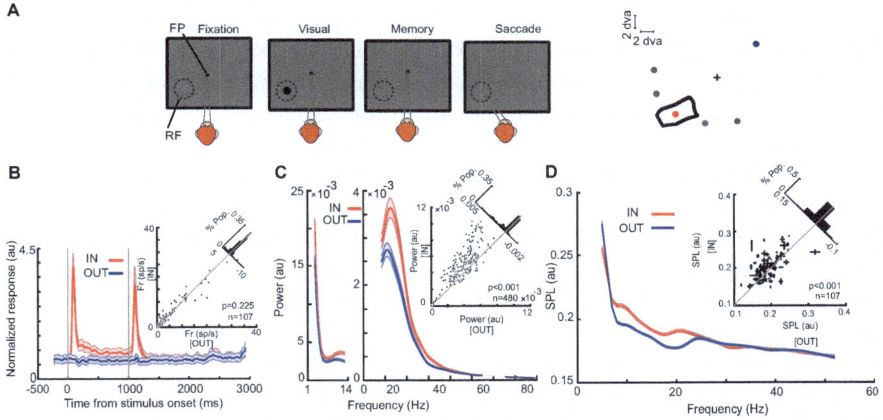

Fig. 1 Changes in MT oscillatory power and spike timing during WM. (**a**) Schematic of the MGS task. The monkey fixates on a central fixation point (FP), and a cue stimulus appears in one of six positions arranged around the neuron's RF location (right). The cue stimulus disappears, and the monkey maintains fixation throughout a blank delay period. Following the disappearance of the fixation point, the monkey saccades to the remembered location to receive a reward. (**b**) Firing rate of MT neurons does not change based on WM location. The bottom plot shows the normalized firing rate of 107 MT neurons across the course of the MGS task, when the memorized location is inside (IN, red) and outside (OUT, blue) of the neurons' RFs. The upper-right scatter plot shows raw firing rates during the last 500 ms of the memory period, and diagonal histogram shows the distribution of firing rate changes. (**c**) αβ LFP power reflects WM location. The average LFP power spectrum during the memory period across frequencies ($n = 480$ LFP recordings), for memory IN (red) and memory OUT (blue) condition. The scatter plot shows αβ power (8–25 Hz) during memory IN vs. OUT. The diagonal histogram shows the distribution of differences in αβ power for all LFPs. (**d**) αβ SPL reflects WM location. The SPL for memory IN (red) and memory OUT (blue), across frequencies for all pairs of neurons and simultaneously recorded LFPs ($n = 1,605$ neuron-LFP pairs). Inset scatter plot shows the SPL at αβ for memory IN compared to memory OUT, with the SPL values for multiple simultaneously recorded LFPs averaged for each neuron ($n = 107$ neurons). Shading and error bars show standard error. Adapted from Bahmani et al. (2018)

demonstrating the FEF's role in attention (reviewed in Clark et al. 2014), supports the idea that one important purpose of persistent signals sent from the FEF is to change neural processing within sensory areas (Gazzaley and Nobre 2012; Postle 2006).

Surprisingly, despite the WM-rich signal sent to extrastriate visual areas, there is no change in the firing rate of these neurons during the memory period of WM tasks (Fig. 1b) (Bahmani et al. 2018; Lee et al. 2005; Mendoza-Halliday et al. 2014; Zaksas and Pasternak 2006). The behavioral impact of the content of WM on perception in psychological studies (Awh and Jonides 2001), despite the lack of firing rate modulation, suggests that the WM signal sent to visual areas may be exerting a subthreshold effect on neurons in these areas, which in turn renders them more sensitive to incoming visual signals. Oscillatory power can provide a signature of such subthreshold effects. Indeed, the content of WM reflected is reflected in the αβ LFP power in area MT (Fig. 1c) (Bahmani et al. 2018). Moreover, this signature was correlated with memory performance: monkeys had more precise and accurate saccadic responses in trials with higher values of αβ power (Bahmani et al. 2018). This change also affected the temporal pattern of firing in MT neurons: spikes were locked to the phase of αβ ongoing oscillations during the memory period (Fig. 1d). Furthermore, in the presence of a bottom-up sensory signal, evoked visual responses were enhanced (Fig. 2a). WM improved the ability of neurons in visual areas, including V4 and MT, to distinguish between stimuli presented near the memorized location by shifting and expanding their RFs toward the location held in WM (Fig. 2b) (Merrikhi et al. 2017). WM was also able to enhance the discriminability of stimuli based on oscillatory patterns of spikes in the αβ frequency range (Fig. 2c). In fact, when visual information was presented during the delay period of the MGS task, the gain and discriminability of spiking activity increased in a αβ-phase-dependent manner: the enhancement of gain and discriminability was more for the spikes acquired at the preferred phase of ongoing αβ oscillations (Bahmani et al. 2018). Quantitatively, changes in spike timing had a greater effect on information encoding during memory than changes in firing rate (Bahmani et al. 2018). Altogether, these findings suggest that changes in spike timing in visual areas, driven by a WM signal sent from the prefrontal cortex, could form the basis for enhanced sensory representations and the accompanying benefits for visual perception.

4 Prefrontal Dopamine, a Common Modulator of Attention and WM

Dopamine signaling within PFC has been shown to play a role both in WM activity and behavior, and in generating attention-like modulations of visual signals elsewhere in the brain.

Iontophoretic application of dopaminergic drugs can increase persistent activity within PFC during a spatial WM task (Williams and Goldman-Rakic 1995). This

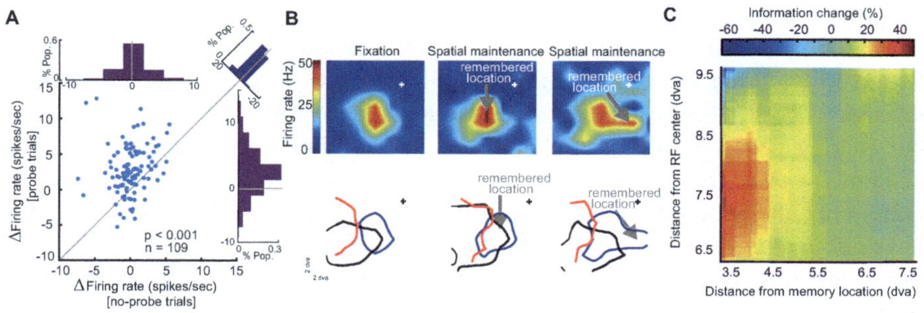

Fig. 2 Changes in MT visual sensitivity during WM. (**a**) During the memory period, visually evoked activity increased, but delay activity in the absence of probes was unchanged. A revised version of MGS task with probe presentation was used; visual probes appeared on 91% of trials, during both the fixation and delay periods of the MGS task. Scatter plot of WM-induced changes in the visually evoked spiking activity (probe trials) against WM-induced changes in the delay period activity (no-probe trials). Top histogram indicates the WM-induced change in delay period activity; right histogram indicates the WM-induced change in visually evoked activity. Diagonal histogram illustrates the difference between the effects of WM on firing rates in the presence and absence of visual probes. (**b**) RFs shift toward the WM location. Heat map showing the RF of an example MT neuron during fixation (at cross); RF of the same neuron measured while the monkey remembered a location inside of the RF, indicated by the arrow; and RF of the same neuron while the monkey remembered a location to the right of the fixation RF. Lower plots show the RF outlines of three simultaneously recorded MT neurons during fixation (left) and the delay period when the monkey remembered different locations (right). The blue outline is the RF of the neuron shown in upper plot. (**c**) Visual information encoded in spike phases increases near the memory location. The increase in mutual information during WM depends on the distance between the probes and the RF center or memory location. Color scale shows the change in mutual information (memory – fixation) between the spikes' phases (αβ) and probe location for pairs of probes. The change is plotted as a function of the probes' distance from the RF center (*y*-axis) and distance from the memory location (*x*-axis). The geometric mean of the two probe positions was used to calculate distances. Adapted from Bahmani et al. (2018) and Merrikhi et al. (2017)

dopamine-modulated change in persistent activity displays an "inverted-U" dose dependency, with the greatest persistent activity at an intermediate level of dopamine signaling (Vijayraghavan et al. 2007). Shifts in the baseline dopaminergic tone within PFC may account for changes in WM observed due to stress or aging (Arnsten et al. 1994, 1995; Arnsten 2000; Gamo and Arnsten 2011). Further work has shown that dopamine alters the probability of glutamate release within PFC (Gao et al. 2001) and increases the coincident firing of pyramidal PFC neurons, consistent with an increase in the synaptic reliability of their inputs (Castner and Williams 2007). More recent work suggests that the effect of prefrontal dopamine on excitatory and inhibitory neurons may differ (Jacob et al. 2013), reducing the activity of inhibitory neurons while increasing the activity and reliability of excitatory neurons.

Different classes of dopamine receptors within PFC have different effects on PFC activity and behavior. Dopamine receptors are generally divided into two classes: D1-like receptors (D1Rs) and D2-like receptors (D2Rs) (Missale et al. 1998; Seamans and Yang 2004); D1Rs are expressed in both superficial and deep cortical layers, but D2R expression is restricted to the deep layers (Lidow et al. 1991).

(Expression of dopamine receptors within PFC will be discussed in greater detail in a later section.) While D1R manipulation was found to enhance persistent activity within PFC during a spatial WM task (Williams and Goldman-Rakic 1995), as discussed above, D2R manipulation instead modulates motor-related activity, without altering persistent activity (Wang et al. 2004). Similarly, during a rule-based oculomotor task, D1Rs affected PFC neuronal tuning for rule and memory, while D2Rs only altered motor signals (Vijayraghavan et al. 2016). D1Rs and D2Rs also have different effects on oscillations within PFC, with power in the beta and gamma bands sensitive to D1R and D2R signaling, respectively (Ott et al. 2018); both receptor types altered alpha and theta power. Behaviorally, both D1R and D2R manipulation can bias the animal toward saccading to the retinotopic location corresponding to the infusion site; however, they have differential effects on the effect of experienced reward on subsequent choices (Soltani et al. 2013). It should also be noted that those effects which are similar for the two receptor types are usually obtained using a D1R *antagonist* and a D2R *agonist* (Ott et al. 2018; Soltani et al. 2013). Thus, dopamine signaling via D1Rs vs. D2Rs produces distinct and contrasting effects on prefrontal neurophysiology and behavior.

The effects of dopamine signaling within PFC are not limited to the activity of prefrontal neurons. Noudoost and colleagues have shown that pharmacologically manipulating dopamine signaling within the FEF can alter sensory responses in extrastriate visual cortex (Noudoost and Moore 2011b). Local pharmacological infusions into the FEF produced localized biases in saccade target selection in a free-choice task, causing an increased tendency to choose the portion of space represented by neurons near the infusion site (which can be estimated using electrical stimulation, see Fig. 3a). Infusing a D1R antagonist (SCH23390) into the FEF enhances visual responses at a retinotopically corresponding location of the V4 representation. This enhanced visual response included greater peak firing rates, greater reliability, and greater feature selectivity (Fig. 3b, c). The effects observed in V4 following FEF D1R manipulation resemble the reported signatures of spatial attention (reviewed in Clark et al. 2015; Noudoost et al. 2010; Squire et al. 2013). A D2R agonist, in contrast, produced a bias in saccadic target selection but no change in visual responses. Inactivating FEF with a GABA agonist (muscimol) resulted in a decrease in the selectivity of V4 visual responses, consistent with a net excitatory effect of a D1R antagonist within PFC. In combination with the previously discussed iontophoresis experiments showing that D1R but not D2R signaling enhances persistent activity within PFC, these results are consistent with the hypothesis that persistent activity in FEF is a key mediator of the attentional modulation of extrastriate visual responses.

5 DA Imbalance and Attention-WM Impairments

The coincident impairment of WM and attention in many neurological disorders associated with disruptions in dopamine signaling also suggests that both processes rely on this pathway for normal function. Here we review findings on the effect of

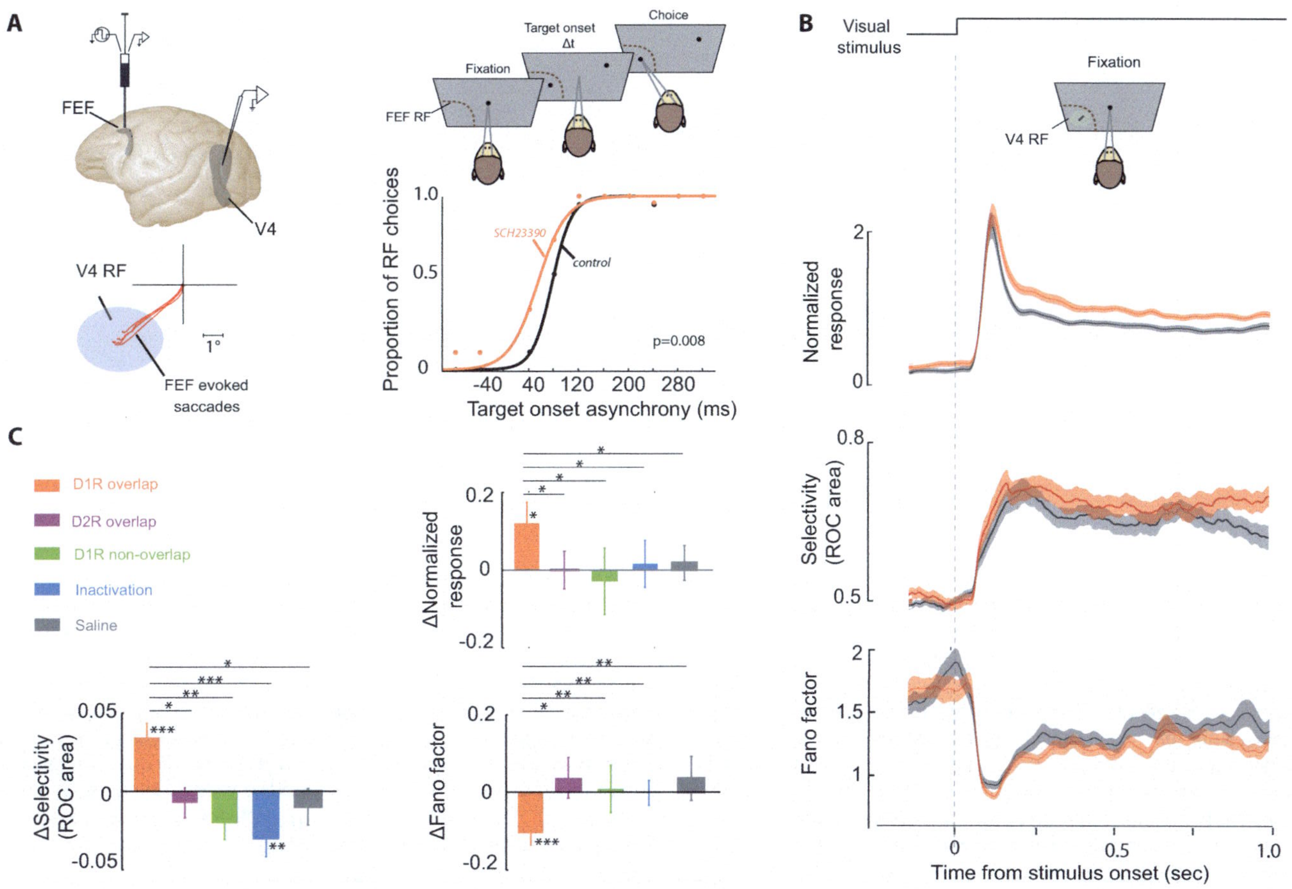

Fig. 3 PFC D1R manipulation enhances visual responses in V4. (**a**) A micro-injectrode (Noudoost and Moore 2011a) was used to deliver a small volume of D1R antagonist into the FEF. Electrical microstimulation of the FEF prior to pharmacological infusion allowed estimation of the affected area of space based on the

Fig. 3 (continued) endpoints of electrically evoked saccades (red traces). D1R antagonist infusion biased the animal's saccades toward the RF location in a two target free-choice saccade task (right). (**b**) Visual responses were recorded from V4 neurons during a passive fixation task, both before and after FEF D1R manipulation (gray and red traces, respectively). Results are shown for an example V4 neuron. Following FEF manipulation, normalized responses were greater (top), orientation selectivity increased (middle), and variability decreased (bottom). (**c**) Summary of effects of FEF D1R manipulation for the population of V4 neurons ($n = 37$). In the "overlap" condition, V4 RFs corresponded with the endpoints of electrically evoked saccades. Bar graphs to the right show the change in V4 orientation selectivity, normalized response, and variability (Fano Factor) after FEF D1R manipulation compared to baseline (orange). V4 visual response magnitude and selectivity increased, while cross-trial variability decreased, following FEF D1R manipulation. Infusing GABA agonist muscimol into the FEF reduced the selectivity of V4 responses, without altering overall firing rate or variability (blue). No changes in V4 activity, selectivity, or variability were observed when the D1R manipulation occurred at an FEF site not overlapping with the V4 RFs (green) or when a D2R agonist (magenta) or saline (gray) was infused at an overlapping FEF site. In all cases D1R effects were significantly different from all other conditions. Single, double, and triple asterisks denote significance at $p < 0.05$, $p < 0.01$, and $p < 0.001$, respectively. Adapted from Noudoost and Moore (2011b)

dopamine-related disorders on attention and WM task performance, focusing on Parkinson's disease (PD).

Visual processing and cognitive functions are considerably altered in populations with dopamine dysfunction such as PD, schizophrenia, and attention deficit hyperactivity disorder (ADHD) (Barkley 1997; Dakin et al. 2005; Gurvich et al. 2007; Karatekin and Asarnow 1998; Seymour et al. 2013). WM and tasks requiring top-down control are remarkably dependent on the integrity of dopamine function in prefrontal cortex (Cools et al. 2002; Miller et al. 1996; Moustafa et al. 2008; Williams and Goldman-Rakic 1995). In PD patients, dopamine dysfunction leads to disruption of frontal cortical networks and explains performance deficits (i.e., slower reaction time and higher error rates) often found in cognitive tasks that involve central planning, WM, motor learning, conflict resolution, and prior information about stimulus uncertainty (Brown et al. 1993; Brown and Marsden 1988; Ghilardi et al. 2003; Herz et al. 2016; Jahanshahi and Frith 1998; Perugini et al. 2016; Stern et al. 1983). Furthermore, visuo-motor abnormalities and visual sensory disturbances in PD have been attributed to dopamine deficiency in striatal circuits of the basal ganglia and the retina (Jankovic 2008; Patel et al. 2014; Rodnitzky 1998). Contrast sensitivity tests in PD patients show a significantly elevated threshold for stimulus detection compared to healthy controls (Bodis-Wollner 1990; Bodis-Wollner et al. 1987; Kupersmith et al. 1982).

The challenge of isolating "attentional deficits" from perceptual and executive control deficits in PD was originally addressed by increasing levels of complexity in choice reaction time tasks, in which subjects were instructed to respond according to stimulus features (i.e., color, shape or tone, or the combination of all three) (Cooper et al. 1994). Longer choice reaction time in PD patients suggested impaired central processing. Moreover, PD patients being slower in tests of WM, such as the London tower task, suggested an increased thinking time rather than deficits in the conception of logical steps toward a solution, given that strategies were comparable with control subjects (Morris et al. 1988). This was also proposed to be an issue with attention switching (Morris et al. 1988). The fact that PD patients exhibited WM deficits and difficulty with internal deployment of "top-down" mechanisms raised the question of whether exogenously driven attention was equally impaired. In a simple cued and non-cued Stroop task (i.e., word and ink cues), patients performed better when a cue relevant to the stimulus attribute was presented before each trial (Brown and Marsden 1988). Further studies confirmed that behavioral performance in PD is highly task dependent, with longer reaction times and more errors when tasks rely on internal processing and relatively normal reaction times when external cues are provided (Jahanshahi et al. 1995; Siegert et al. 2002). This benefit of external cues on task performance offers further insight on the ways that sensory stimuli can facilitate movement initiation, hypothetically explained by the pathophysiology of PD (Praamstra et al. 1998).

Several studies have aimed to establish a relationship between task performance and dopaminergic therapy or its withdrawal, using tests that are sensitive to frontal lobe function (Gurvich et al. 2007; Lange et al. 1992, 1995; Moustafa et al. 2008). Treatment with levodopa, a dopamine precursor, helps alleviate the motor symptoms of PD (Cotzias et al. 1969; Yahr et al. 1969) and has been associated with increased blood flow in PFC (Roshan Cools et al. 2002). Withdrawal of levodopa caused performance impairments in tests such as spatial memory and the tower of London (Lange et al. 1992). Furthermore, levodopa withdrawal could either impair or enhance performance on cognitive tasks, depending on the basal levels of dopamine and the nature of the task, suggesting a more complex link (Cools and D'Esposito 2011). An inverted U-shaped action of dopamine could explain the various effects of dopamine signaling on cognitive control (Cools and D'Esposito 2011; Cools et al. 2001, 2010). These observations support the inverted U-shaped findings in animal studies, although on a behavioral rather than neural level (Vijayraghavan et al. 2007).

Relevant studies in PD, schizophrenia, and ADHD are described with more detail in Table 1.

6 Anatomical Basis of Prefrontal Dopamine's Contribution to Attention and WM

Anatomical evidence, including patterns of receptor expression and the specific properties of neurons projecting from the PFC, also suggests a unique and important role for dopamine in the prefrontal networks that modulate sensory processing.

FEF neurons exhibit a mixture of visual, motor, and memory-related activity in varying proportions (Bruce and Goldberg 1985; Lawrence et al. 2005; Sommer and Wurtz 2000, 2001; Umeno and Goldberg 2001). FEF neurons also project to multiple cortical and sub-cortical areas (Leichnetz 1982; Markov et al. 2014a; Schnyder et al. 1985; Sommer and Wurtz 2000; Stanton et al. 1995). In order to determine which kinds of FEF activity are sent via these projections, projecting neurons must be identified, e.g., via antidromic stimulation in the target area (Ferraina et al. 2002; Sommer and Wurtz 2000), and their response characteristics assessed. For example, a combination of visual, motor, and delay activity is sent from the FEF to the SC (Sommer and Wurtz 2000, 2001). Recently, the content of FEF's projection to extrastriate visual cortex was examined using antidromic stimulation (Merrikhi et al. 2017), revealing that memory-related, delay period activity is the defining feature of FEF neurons projecting to extrastriate visual cortex. This memory activity was significantly stronger in the FEF projection to V4 than in the FEF population as a whole (Fig. 4). Motor activity, in contrast, was present in a much smaller fraction of V4-projecting FEF neurons than in the FEF population. Since modulation of FEF activity has been shown to modulate V4 activity in ways that mimic the effects of selective attention (Moore and Armstrong 2003; Noudoost and Moore 2011b), and we know now that the direct projection to V4 contains primarily memory-related activity (Merrikhi et al. 2017), this suggests that neurons exhibiting delay-period activity could be a source of attentional modulation in extrastriate cortex. In fact, the same neurons in FEF that display persistent

Table 1 Studies of attention and WM performance in PD, schizophrenia, and ADHD

Author	Task	Disease, treatments	Performance change
Mathis et al. (2011)	Attentional blink	Schizophrenia	Patients show an exaggerated attentional blink effect compared to healthy controls, with significantly lower detection of the second target. Results suggest deficits in both visual processing and general attentional deficits
Javitt et al. (1995)	WM, auditory oddball task	Schizophrenia	Impaired mismatch negativity, increased reaction time and decreased number of correct detections
Stablein et al. (2016)	WM, visual motion task	Schizophrenia	Decreased task accuracy, attributed to visual deficits and impaired attention in patients
Clementz et al. (2008)	Target identification sustained attention	Schizophrenia	Abnormal target identification in patients, attributed to visual deficits in schizophrenia
Fuller et al. (2006)	Visual search tasks, four types	Schizophrenia	Rates of search were slower in patients, especially when the task required precise attentional control. Results suggest primary impairment of attention in schizophrenia
Slagter et al. (2016)	Attentional blink	Parkinson's disease on and off dopaminergic therapy (levodopa and agonists)	Patients exhibited poorer performance than healthy controls An optimum level of dopamine for cognitive function is suggested
Cools and D'Esposito (2011)	Review, tower of London, WM task, attentional set-shifting	Parkinson's disease on and off medication	PD patients had deficits on the Tower of London planning task, WM and attentional set-shifting tests
Rodriguez-Oroz et al. (2009)	Review of tasks in which shifting of attention is impaired in PD	Parkinson's disease	Issues when switching attention and double task demands

(continued)

Table 1 (continued)

Author	Task	Disease, treatments	Performance change
Bulens et al. (1987) and Hutton et al. (1993)	Contrast sensitivity test	Parkinson's disease on and off levodopa treatment	Loss of contrast sensitivity in patients. Levodopa improves contrast sensitivity function, close to normal levels
Lange et al. (1992)	Tower of London and spatial WM task	Parkinson's disease; on and off L-dopa medication	L-dopa withdrawal dramatically impaired performance in the tower of London test, with times slower than baseline condition (on L-dopa). Increase of errors in the spatial WM task during L-dopa withdrawal
Moustafa et al. (2008)	WM task, continuous performance AX-CPT	Parkinson's disease; on and off dopaminergic medications (L-dopa and agonists)	PD patients showed deficits for ignoring distractors and attentional shifting
Kempton et al. (1999)	Cambridge Neuropsychological Test Automated Battery (CANTAB)	ADHD; stimulant methylphenidate, D-amphetamine	Medicated ADHD children did not show impairment in most of the executive function tasks, except the spatial recognition memory task. These results suggested that stimulant medication improved executive function performance
Karatekin and Asarnow (1998)	Dot test of visuospatial WM and digit span subtest of the Wechsler intelligence scale (digit recall)	Childhood onset schizophrenia and ADHD	Normal children were able to recall more digits than children with schizophrenia and ADHD. Both schizophrenic and ADHD children had greater distance error on the dot test

memory-related activity are more likely to show attentional modulation (Armstrong et al. 2009) – suggesting that the same memory-selective FEF neurons which project to visual areas drive both the modulations of these areas during WM (see Fig. 3) and also attentional modulation of incoming visual signals. Consistent with this is the fact that D1Rs in PFC selectively modulate persistent memory-related activity there (Williams and Goldman-Rakic 1995) and also modulate activity in visual areas (Noudoost and Moore 2011b). In sum, dopamine's modulation of persistent prefrontal activity (Williams and Goldman-Rakic 1995), which we know to be strongly

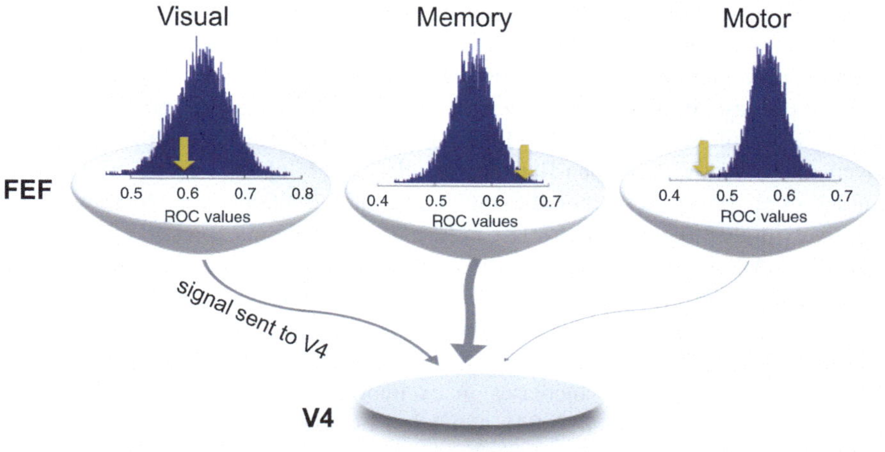

Fig. 4 Comparison of visual, memory, and motor activity in V4-projecting FEF neurons and the FEF population as a whole. Visual, memory, and motor selectivity were assessed using a memory-guided saccade task in which the cue appeared inside or opposite the RF of the FEF neurons being recorded (see Fig. 1a). Histograms show the distribution of average visual, memory, and motor selectivity for 1,000 ensembles of 15 FEF neurons chosen at random from the population ($n = 307$ non-projecting FEF neurons). Yellow arrow shows the mean selectivity for the V4-projecting FEF neurons ($n = 15$). Selectivity was measured based on the ROC value for trials in which the cue appeared inside vs. outside the FEF RF (during the visual, delay, or motor epochs of the task). Memory selectivity was significantly stronger in the V4-projecting FEF population, and motor selectivity was significantly weaker, compared to the non-projecting FEF population. Modified from (Merrikhi et al. 2017)

represented in FEF's projections to visual areas (Merrikhi et al. 2017), drives attention-like modulations of visual cortex (Noudoost and Moore 2011b).

Examinations of patterns of dopamine receptor expression suggest a specialized role for dopamine in prefrontal processing. Dopamine receptors are enriched in the prefrontal cortex compared to more caudal regions in both macaque (Lidow et al. 1991) and rodent (Gaspar et al. 1995). Dopamine terminals converge onto pyramidal neurons and a class of fast-spiking interneurons (Sesack et al. 1995; Verney et al. 1990). However, the contacts are more often made onto pyramidal neurons than interneurons (Goldman-Rakic et al. 1989; Sesack et al. 1995; Smiley and Goldman-Rakic 1993; Smiley et al. 1992). Goldman-Rakic and colleagues (Bergson et al. 1995; Smiley and Goldman-Rakic 1993) found that D1Rs were most prominent on pyramidal neurons in macaque dlPFC. Mueller et al. (2018) examined the expression of D5Rs (which are part of the D1 family of receptors) on different classes of neurons in the macaque PFC and found that D5Rs are more prevalent on pyramidal neurons than inhibitory interneurons and are especially prevalent on anatomically defined putative long-range projection neurons. These findings suggest that dopamine will exert a strong influence on the majority of pyramidal neurons through the D1 family of dopamine receptors.

The expression of dopamine receptors also varies across cortical layers. Previous in situ results have shown that within the macaque PFC, mRNA encoding all dopamine receptor subtypes was expressed most strongly in layer V; however, this was not the case for all cortical regions (Lidow 1998). Others (Bergson et al. 1995; Smiley and Goldman-Rakic 1993) found that D1R staining was most prominent in layers II–III and V. Goldman-Rakic and colleagues (Goldman-Rakic et al. 1990; Smiley et al. 1994) used autoradiography and immunofluorescence to show comparatively higher expression of D1Rs in layers I–II, medium expression in layers V and VI, and lower expression in layers IIIb and IV of dlPFC. They found D2R expression was strongest in layer V. Collectively these results suggest dopamine is likely to influence pyramidal neurons through D1 and D5 dopamine receptors in the macaque PFC, particularly in layer II–III and V. Because D1R activation potentiates NMDA receptor-mediated increases in excitability (Gonzalez-Islas and Hablitz 2003; Tseng and O'Donnell 2004), activation of D1Rs on recurrently connected pyramidal neurons could directly influence the increase in sustained activity in attention and WM.

Some interneuron classes in primate dlPFC differentially express D1Rs in their processes (Glausier et al. 2009). However, only very few studies have examined the expression of D1Rs on different neuronal populations in the macaque PFC. Muly et al. (1998) examined the distribution of D1Rs across different interneuron cell types across PFC layers. They found no variation in expression of D1R+/calretinin+ neurons but did for D1R+/calbindin+ and D1R+/parvalbumin+ neurons: there were fewer D1R+/parvalbumin+ neurons in superficial layers (I and II) than other layers, and D1R+/calbindin+ expression peaked in layer III (Muly et al. 1998). This suggests that the expression of D1Rs may also be laminarly specific to facilitate differential processing of sustained signals (WM- or attention-related) or motor signals.

There are also many differences in dopamine neuroanatomy between different species (Fig. 5). In contrast to dopamine receptor expression in the macaque, in the mouse PFC, it was shown that D1R+ neurons are mainly in deep layers and D2R+ neurons are mainly in superficial neurons (Wei et al. 2018). In the rat several studies show D1Rs and D2Rs are both mainly found in deeper layers (Berger et al. 1988, 1991; Davidoff and Benes 1998; Descarries et al. 1987; Gaspar et al. 1995; van Eden et al. 1987). Again in contrast to the macaque, immunological studies in rat mPFC showed that the density of D1R+ neurons was almost one third lower in layers II–III than layers V–VI and also that there was a higher density of D2R+ neurons than D1R + neurons in layers II–III (Vincent et al. 1993). Further in situ of different classes of neurons in the rat mPFC found that D1R expression was lower on pyramidal neurons compared to inhibitory neurons (11–21% vs 25–52%) (Santana and Artigas 2017) – again a stark contrast to dopamine receptor expression in the macaque. Combined, these very divergent results suggest a different role of neuromodulators in PFC-related cognitive functions across different species.

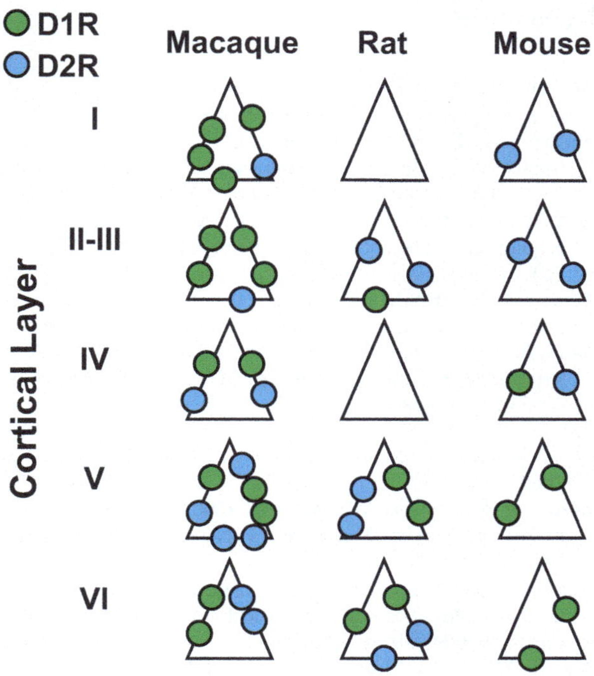

Fig. 5 Distribution of D1Rs and D2Rs across cortical layers in different species. D1Rs (green) and D2Rs (blue) are more abundant in macaque than rodent species. In the macaque, D1R expression tends to decrease with cortical depth, while D2R expression increases. The opposite is true in both rodent species

7 Conclusion

Numerous lines of evidence indicate the involvement of the PFC, and specifically the FEF, in both attention and WM. The FEF shows both memory- and attention-related activity, largely in the same neurons, and its activity is tied to performance on both attention and WM tasks. The FEF sends memory-related activity to visual areas, and the modulation of visual responses observed during spatial WM resembles that caused by covert spatial attention. Dopamine signaling and persistent activity within PFC may be the key mechanisms linking these two processes. Patients with altered dopamine signaling, such as PD, show disruptions of WM performance and tasks involving top-down control. Dopamine receptors display specific expression patterns within primate PFC, and manipulating PFC dopamine signaling both selectively modulates memory-related activity within PFC and reproduces the signatures of attention within visual cortex. In sum, dopamine-modulated activity in prefrontal cortex appears to be a critical player in both the behavioral and neural signatures of attention and WM.

References

Albrecht DG (1995) Visual cortex neurons in monkey and cat: effect of contrast on the spatial and temporal phase transfer functions. Vis Neurosci 12(6):1191–1210

Anderson JC, Kennedy H, Martin KAC (2011) Pathways of attention: synaptic relationships of frontal eye field to V4, lateral intraparietal cortex, and area 46 in macaque monkey. J Neurosci 31(30):10872–10881. https://doi.org/10.1523/JNEUROSCI.0622-11.2011

Anderson EB, Mitchell JF, Reynolds JH (2013) Attention-dependent reductions in burstiness and action-potential height in macaque area V4. Nat Neurosci 16(8):1125–1131. https://doi.org/10.1038/nn.3463

Anton-Erxleben K, Carrasco M (2013) Attentional enhancement of spatial resolution: linking behavioural and neurophysiological evidence. Nat Rev Neurosci 14(3):188–200

Anton-Erxleben K, Stephan VM, Treue S (2009) Attention reshapes center-surround receptive field structure in macaque cortical area MT. Cereb Cortex 19(10):2466–2478. https://doi.org/10.1093/cercor/bhp002

Ardid S, Wang X-J, Compte A (2007) An integrated microcircuit model of attentional processing in the neocortex. J Neurosci 27(32):8486–8495. https://doi.org/10.1523/JNEUROSCI.1145-07.2007

Armstrong KM, Chang MH, Moore T (2009) Selection and maintenance of spatial information by frontal eye field neurons. J Neurosci 29(50):15621–15629. https://doi.org/10.1523/JNEUROSCI.4465-09.2009

Arnsten AF (2000) Stress impairs prefrontal cortical function in rats and monkeys: role of dopamine D1 and norepinephrine alpha-1 receptor mechanisms. Prog Brain Res 126:183–192. https://doi.org/10.1016/S0079-6123(00)26014-7

Arnsten AF, Cai JX, Murphy BL, Goldman-Rakic PS (1994) Dopamine D1 receptor mechanisms in the cognitive performance of young adult and aged monkeys. Psychopharmacology 116(2):143–151

Arnsten AF, Cai JX, Steere JC, Goldman-Rakic PS (1995) Dopamine D2 receptor mechanisms contribute to age-related cognitive decline: the effects of quinpirole on memory and motor performance in monkeys. J Neurosci 15(5 Pt 1):3429–3439

Awh E, Jonides J (2001) Overlapping mechanisms of attention and spatial working memory. Trends Cogn Sci 5(3):119–126

Bahmani Z, Daliri MR, Merrikhi Y, Clark K, Noudoost B (2018) Working memory enhances cortical representations via spatially specific coordination of spike times. Neuron 97(4):967–979.e6. https://doi.org/10.1016/j.neuron.2018.01.012

Barkley RA (1997) Behavioral inhibition, sustained attention, and executive functions: constructing a unifying theory of ADHD. Psychol Bull 121(1):65–94

Bell AH, Meredith MA, Van Opstal AJ, Munoz DP (2006) Stimulus intensity modifies saccadic reaction time and visual response latency in the superior colliculus. Exp Brain Res 174(1):53–59. https://doi.org/10.1007/s00221-006-0420-z

Berger B, Trottier S, Verney C, Gaspar P, Alvarez C (1988) Regional and laminar distribution of the dopamine and serotonin innervation in the macaque cerebral cortex: a radioautographic study. J Comp Neurol 273(1):99–119. https://doi.org/10.1002/cne.902730109

Berger B, Gaspar P, Verney C (1991) Dopaminergic innervation of the cerebral cortex: unexpected differences between rodents and primates. Trends Neurosci 14(1):21–27

Bergson C, Mrzljak L, Smiley JF, Pappy M, Levenson R, Goldman-Rakic PS (1995) Regional, cellular, and subcellular variations in the distribution of D1 and D5 dopamine receptors in primate brain. J Neurosci 15(12):7821–7836

Bodis-Wollner I (1990) The visual system in Parkinson's disease. Res Publ Assoc Res Nerv Ment Dis 67:297–316

Bodis-Wollner I, Marx MS, Mitra S, Bobak P, Mylin L, Yahr M (1987) Visual dysfunction in Parkinson's disease – loss in spatiotemporal contrast sensitivity. Brain 110:1675–1698. https://doi.org/10.1093/brain/110.6.1675

Brown RG, Marsden CD (1988) Internal versus external cues and the control of attention in Parkinsons-disease. Brain 111:323–345. https://doi.org/10.1093/brain/111.2.323
Brown RG, Jahanshahi M, Marsden D (1993) Response choice in Parkinsons-disease – the effects of uncertainty and stimulus-response compatibility. Brain 116:869–885. https://doi.org/10.1093/brain/116.4.869
Bruce CJ, Goldberg ME (1985) Primate frontal eye fields. I. Single neurons discharging before saccades. J Neurophysiol 53(3):603–635
Bulens C, Meerwaldt JD, Vanderwildt GJ, Vandeursen JBP (1987) Effect of levodopa treatment on contrast sensitivity in Parkinsons-disease. Ann Neurol 22(3):365–369. https://doi.org/10.1002/ana.410220313
Carandini M, Heeger DJ (2012) Normalization as a canonical neural computation. Nat Rev Neurosci 13(1):51–62. https://doi.org/10.1038/nrn3136
Castner S a, Williams GV (2007) Tuning the engine of cognition: a focus on NMDA/D1 receptor interactions in prefrontal cortex. Brain Cogn 63(2):94–122. https://doi.org/10.1016/j.bandc.2006.11.002
Clark KL, Noudoost B, Schafer RJ, Moore T (2014) Neuronal mechanisms of attentional control: frontal cortex. In: Kastner S, Nobre AC (eds) Handbook of attention. Oxford University Press, Oxford
Clark K, Squire RF, Merrikhi Y, Noudoost B (2015) Visual attention: linking prefrontal sources to neuronal and behavioral correlates. Prog Neurobiol 132:59–80. https://doi.org/10.1016/j.pneurobio.2015.06.006
Clementz BA, Wang J, Keil A (2008) Normal electrocortical facilitation but abnormal target identification during visual sustained attention in schizophrenia. J Neurosci 28 (50):13411–13418. https://doi.org/10.1523/jneurosci.4095-08.2008
Cohen MR, Kohn A (2011) Measuring and interpreting neuronal correlations. Nat Neurosci 14 (7):811–819. https://doi.org/10.1038/nn.2842
Cohen MR, Maunsell JH (2009) Attention improves performance primarily by reducing interneuronal correlations. Nat Neurosci 12(12):1594–1600. https://doi.org/10.1038/nn.2439
Compte A, Brunel N, Goldman-Rakic PS, Wang X-J (2000) Synaptic mechanisms and network dynamics underlying spatial working memory in a cortical network model. Cereb Cortex 10 (9):910–923
Connor CE, Gallant JL, Preddie DC, Van Essen DC (1996) Responses in area V4 depend on the spatial relationship between stimulus and attention. J Neurophysiol 75(3):1306–1308
Connor CE, Preddie DC, Gallant JL, Van Essen DC (1997) Spatial attention effects in macaque area V4. J Neurosci 17(9):3201–3214
Cools R, D'Esposito M (2011) Inverted-U-shaped dopamine actions on human working memory and cognitive control. Biol Psychiatry 69(12):E113–E125. https://doi.org/10.1016/j.biopsych.2011.03.028
Cools R, Barker RA, Sahakian BJ, Robbins TW (2001) Enhanced or impaired cognitive function in Parkinson's disease as a function of dopaminergic medication and task demands. Cereb Cortex 11(12):1136–1143
Cools R, Stefanova E, Barker RA, Robbins TW, Owen AM (2002) Dopaminergic modulation of high-level cognition in Parkinson's disease: the role of the prefrontal cortex revealed by PET. Brain 125(Pt 3):584–594
Cools R, Miyakawa A, Sheridan M, D'Esposito M (2010) Enhanced frontal function in Parkinson's disease. Brain 133(Pt 1):225–233. https://doi.org/10.1093/brain/awp301
Cooper JA, Sagar HJ, Tidswell P, Jordan N (1994) Slowed central processing in simple and go no-go reaction-time tasks in Parkinson's disease. Brain 117:517–529. https://doi.org/10.1093/brain/117.3.517
Cotzias GC, Papavasiliou PS, Gellene R (1969) Modification of Parkinsonism – chronic treatment with L-dopa. N Engl J Med 280(7):337–345. https://doi.org/10.1056/NEJM196902132800701
D'Esposito M, Postle BR (2015) The cognitive neuroscience of working memory. Annu Rev Psychol 66(1):115–142. https://doi.org/10.1146/annurev-psych-010814-015031

Dakin S, Carlin P, Hemsley D (2005) Weak suppression of visual context in chronic schizophrenia. Curr Biol 15(20):R822–R824. https://doi.org/10.1016/j.cub.2005.10.015

Davidoff SA, Benes FM (1998) High-resolution scatchard analysis shows D1 receptor binding on pyramidal and nonpyramidal neurons. Synapse 28(1):83–90. https://doi.org/10.1002/(SICI)1098-2396(199801)28:1<83::AID-SYN10>3.0.CO;2-Z

Deco G, Hugues E (2012) Neural network mechanisms underlying stimulus driven variability reduction. PLoS Comput Biol 8(3):e1002395. https://doi.org/10.1371/journal.pcbi.1002395

Deco G, Kringelbach ML (2016) Metastability and coherence: extending the communication through coherence hypothesis using a whole-brain computational perspective. Trends Neurosci 39(3):125–135

Descarries L, Lemay B, Doucet G, Berger B (1987) Regional and laminar density of the dopamine innervation in adult rat cerebral cortex. Neuroscience 21(3):807–824

Deubel H, Schneider WX (1996) Saccade target selection and object recognition: evidence for a common attentional mechanism. Vis Res 36(12):1827–1837

Druckmann S, Chklovskii DB (2012) Neuronal circuits underlying persistent representations despite time varying activity. Curr Biol 22(22):2095–2103

Ferraina S, Paré M, Wurtz RH (2002) Comparison of cortico-cortical and cortico-collicular signals for the generation of saccadic eye movements. J Neurophysiol 87(2):845–858. https://doi.org/10.1152/jn.00317.2001

Fuller RL, Luck SJ, Braun EL, Robinson BM, McMahon RP, Gold JM (2006) Impaired control of visual attention in schizophrenia. J Abnorm Psychol 115(2):266–275. https://doi.org/10.1037/0021-843x.115.2.266

Galashan FO, Saßen HC, Kreiter AK, Wegener D (2013) Monkey area MT latencies to speed changes depend on attention and correlate with behavioral reaction times. Neuron 78(4):740–750. https://doi.org/10.1016/j.neuron.2013.03.014

Gamo NJ, Arnsten AF (2011) Molecular modulation of prefrontal cortex: rational development of treatments for psychiatric disorders. Behav Neurosci 125(3):282–296. https://doi.org/10.1037/a0023165

Gao W-J, Krimer LS, Goldman-Rakic PS (2001) Presynaptic regulation of recurrent excitation by D1 receptors in prefrontal circuits. Proc Natl Acad Sci 98(1):295–300. https://doi.org/10.1073/pnas.011524298

Gaspar P, Bloch B, Le Moine C (1995) D1 and D2 receptor gene expression in the rat frontal cortex: cellular localization in different classes of efferent neurons. Eur J Neurosci 7(5):1050–1063

Gazzaley A, Nobre AC (2012) Top-down modulation: bridging selective attention and working memory. Trends Cogn Sci 16(2):129–135. https://doi.org/10.1016/j.tics.2011.11.014

Ghilardi MF, Eidelberg D, Silvestri G, Ghez C (2003) The differential effect of PD and normal aging on early explicit sequence learning. Neurology 60:1313–1319. https://doi.org/10.1212/01.wnl.0000059545.69089

Glausier JR, Khan ZU, Muly EC (2009) Dopamine D1 and D5 receptors are localized to discrete populations of interneurons in primate prefrontal cortex. Cereb Cortex 19(8):1820–1834. https://doi.org/10.1093/cercor/bhn212

Goldman MS (2009) Memory without feedback in a neural network. Neuron 61(4):621–634

Goldman-Rakic PS, Leranth C, Williams SM, Mons N, Geffard M (1989) Dopamine synaptic complex with pyramidal neurons in primate cerebral cortex. Proc Natl Acad Sci 86(22):9015–9019

Goldman-Rakic PS, Lidow MS, Gallager DW (1990) Overlap of dopaminergic, adrenergic, and serotoninergic receptors and complementarity of their subtypes in primate prefrontal cortex. J Neurosci 10(7):2125–2138

Gonzalez-Islas C, Hablitz JJ (2003) Dopamine enhances EPSCs in layer II-III pyramidal neurons in rat prefrontal cortex. J Neurosci 23(3):867–875

Green D, Swets DJ (1966) Signal detection theory and psychophysics. Wiley, New York

Gurvich C, Georgiou-Karistianis N, Fitzgerald PB, Millist L, White OB (2007) Inhibitory control and spatial working memory in Parkinson's disease. Mov Disord 22(10):1444–1450. https://doi.org/10.1002/mds.21510

Herz DM, Bogacz R, Brown P (2016) Neuroscience: impaired decision-making in Parkinson's disease. Curr Biol 26(14):R671–R673. https://doi.org/10.1016/j.cub.2016.05.075

Hutton JT, Morris JL, Elias JW (1993) Levodopa improves spatial contrast sensitivity in Parkinsons-disease. Arch Neurol 50(7):721–724. https://doi.org/10.1001/archneur.1993.00540070041012

Jacob SN, Ott T, Nieder A (2013) Dopamine regulates two classes of primate prefrontal neurons that represent sensory signals. J Neurosci 33(34):13724–13734. https://doi.org/10.1523/JNEUROSCI.0210-13.2013

Jahanshahi M, Frith CD (1998) Willed action and its impairments. Cogn Neuropsychol 15 (6–8):483–533. https://doi.org/10.1080/026432998381005

Jahanshahi M, Jenkins H, Brown RG, Marsden CD, Passingham RE, Brooks DJ (1995) Self-initiated versus externally triggered movements. 1. An investigation using measurement of regional cerebral blood-flow with pet and movement-related potentials in normal and Parkinson's disease subjects. Brain 118:913–933. https://doi.org/10.1093/brain/118.4.913

Jankovic J (2008) Parkinson's disease: clinical features and diagnosis. J Neurol Neurosurg Psychiat 79(4):368–376. https://doi.org/10.1136/jnnp.2007.131045

Javitt DC, Doneshka P, Grochowski S, Ritter W (1995) Impaired mismatch negativity generation reflects widespread dysfunction of working-memory in schizophrenia. Arch Gen Psychiatry 52 (7):550–558

Karatekin C, Asarnow RF (1998) Working memory in childhood-onset schizophrenia and attention-deficit/hyperactivity disorder. Psychiatry Res 80(2):165–176. https://doi.org/10.1016/s0165-1781(98)00061-4

Kempton S, Vance A, Maruff P, Luk E, Costin J, Pantelis C (1999) Executive function and attention deficit hyperactivity disorder: stimulant medication and better executive function performance in children. Psychol Med 29(3):527–538. https://doi.org/10.1017/s0033291799008338

Kraynyukova N, Tchumatchenko T (2018) Stabilized supralinear network can give rise to bistable, oscillatory, and persistent activity. Proc Natl Acad Sci 115(13):3464–3469. https://doi.org/10.1073/pnas.1700080115

Kupersmith MJ, Shakin E, Siegel IM, Lieberman A (1982) Visual system abnormalities in patients with Parkinson's disease. Arch Neurol 39(5):284–286

Kusunoki M, Goldberg ME (2003) The time course of perisaccadic receptive field shifts in the lateral intraparietal area of the monkey. J Neurophysiol 89(3):1519–1527. https://doi.org/10.1152/jn.00519.2002

Lange KW, Robbins TW, Marsden CD, James M, Owen AM, Paul GM (1992) L-dopa withdrawal in Parkinson's disease selectively impairs cognitive performance in tests sensitive to frontal lobe dysfunction. Psychopharmacology 107(2–3):394–404

Lange KW, Paul GM, Naumann M, Gsell W (1995) Dopaminergic effects on cognitive performance in patients with Parkinson's disease. J Neural Trans Suppl 46:423–432

Lawrence BM, White RL, Snyder LH (2005) Delay-period activity in visual, visuomovement, and movement neurons in the frontal eye field. J Neurophysiol 94(2):1498–1508. https://doi.org/10.1152/jn.00214.2005

Lee H, Simpson GV, Logothetis NK, Rainer G (2005) Phase locking of single neuron activity to theta oscillations during working memory in monkey extrastriate visual cortex. Neuron 45 (1):147–156. https://doi.org/10.1016/j.neuron.2004.12.025

Leichnetz GR (1982) Connections between the frontal eye field and pretectum in the monkey: an anterograde/retrograde study using HRP GEL and TMB neurohistochemistry. J Comp Neurol 207(4):394–402. https://doi.org/10.1002/cne.902070410

Lidow MS (1998) Nonhuman primate model of the effect of prenatal cocaine exposure on cerebral cortical development. Ann N Y Acad Sci 846:182–193

Lidow MS, Goldman-Rakic PS, Gallager DW, Rakic P (1991) Distribution of dopaminergic receptors in the primate cerebral cortex: quantitative autoradiographic analysis using [3H] raclopride, [3H]spiperone and [3H]SCH23390. Neuroscience 40(3):657–671

Ling S, Liu T, Carrasco M (2009) How spatial and feature-based attention affect the gain and tuning of population responses. Vis Res 49(10):1194–1204. https://doi.org/10.1016/j.visres.2008.05.025

Markov NT, Ercsey-Ravasz MM, Ribeiro Gomes AR, Lamy C, Magrou L, Vezoli J et al (2014a) A weighted and directed interareal connectivity matrix for macaque cerebral cortex. Cereb Cortex 24(1):17–36. https://doi.org/10.1093/cercor/bhs270

Markov NT, Vezoli J, Chameau P, Falchier A, Quilodran R, Huissoud C et al (2014b) Anatomy of hierarchy: feedforward and feedback pathways in macaque visual cortex. J Comp Neurol 522 (1):225–259. https://doi.org/10.1002/cne.23458

Mathis KI, Wynn JK, Breitmeyer B, Nuechterlein KH, Green MF (2011) The attentional blink in schizophrenia: isolating the perception/attention interface. J Psychiatr Res 45(10):1346–1351. https://doi.org/10.1016/j.jpsychires.2011.04.002

Mendoza-Halliday D, Torres S, Martinez-Trujillo JC (2014) Sharp emergence of feature-selective sustained activity along the dorsal visual pathway. Nat Neurosci 17(9):1255–1262. https://doi.org/10.1038/nn.3785

Merrikhi Y, Clark KK, Albarran E, Parsa M, Zirnsak M, Moore T, Noudoost B (2017) Spatial working memory alters the efficacy of input to visual cortex. Nat Commun 8:15041. https://doi.org/10.1038/ncomms15041

Miller EK, Erickson C a, Desimone R (1996) Neural mechanisms of visual working memory in prefrontal cortex of the macaque. J Neurosci 16(16):5154–5167

Missale C, Nash SR, Robinson SW, Jaber M, Caron MG (1998) Dopamine receptors: from structure to function. Physiol Rev 78(1):189–225

Mitchell JF, Sundberg KA, Reynolds JH (2007) Differential attention-dependent response modulation across cell classes in macaque visual area V4. Neuron 55(1):131–141. https://doi.org/10.1016/j.neuron.2007.06.018

Mitchell JF, Sundberg KA, Reynolds JH (2009) Spatial attention decorrelates intrinsic activity fluctuations in macaque area V4. Neuron 63(6):879–888. https://doi.org/10.1016/j.neuron.2009.09.013

Moore T, Armstrong KM (2003) Selective gating of visual signals by microstimulation of frontal cortex. Nature 421:370–373. https://doi.org/10.1038/nature01285.

Moore T, Fallah M (2001) Control of eye movements and spatial attention. Proc Natl Acad Sci 98 (3):1273–1276. https://doi.org/10.1073/pnas.021549498

Moore T, Fallah M (2004) Microstimulation of the frontal eye field and its effects on covert spatial attention. J Neurophysiol 91(1):152–162. https://doi.org/10.1152/jn.00741.2002

Morris RG, Downes JJ, Sahakian BJ, Evenden JL, Heald A, Robbins TW (1988) Planning and spatial working memory in Parkinson's disease. J Neurol Neurosurg Psychiatry 51(6):757–766. https://doi.org/10.1136/jnnp.51.6.757

Moustafa AA, Sherman SJ, Frank MJ (2008) A dopaminergic basis for working memory, learning and attentional shifting in Parkinsonism. Neuropsychologia 46(13):3144–3156. https://doi.org/10.1016/j.neuropsychologia.2008.07.011

Mueller A, Shepard SB, Moore T (2018) Differential expression of dopamine D5 receptors across neuronal subtypes in macaque frontal eye field. Front Neural Circuits 12:12. https://doi.org/10.3389/fncir.2018.00012

Muly EC, Szigeti K, Goldman-Rakic PS (1998) D1 receptor in interneurons of macaque prefrontal cortex: distribution and subcellular localization. J Neurosci 18(24):10553–10565

Noudoost B, Moore T (2011a) A reliable microinjectrode system for use in behaving monkeys. J Neurosci Methods 194(2):218–223. https://doi.org/10.1016/j.jneumeth.2010.10.009

Noudoost B, Moore T (2011b) Control of visual cortical signals by prefrontal dopamine. Nature 474 (7351):372–375. https://doi.org/10.1038/nature09995

Noudoost B, Chang MH, Steinmetz NA, Moore T (2010) Top-down control of visual attention. Curr Opin Neurobiol 20(2):183–190

Noudoost B, Clark KL, Moore T (2014) A distinct contribution of the frontal eye field to the visual representation of saccadic targets. J Neurosci 34(10):3687–3698. https://doi.org/10.1523/JNEUROSCI.3824-13.2014

Oram MW, Xiao D, Dritschel B, Payne KR (2002) The temporal resolution of neural codes: does response latency have a unique role? Philos Trans R Soc Lond Ser B Biol Sci 357 (1424):987–1001. https://doi.org/10.1098/rstb.2002.1113
Ott T, Westendorff S, Nieder A (2018) Dopamine receptors influence internally generated oscillations during rule processing in primate prefrontal cortex. J Cogn Neurosci 30:770–784. https://doi.org/10.1162/jocn_a_01248
Patel N, Jankovic J, Hallett M (2014) Sensory aspects of movement disorders. Lancet Neurol 13 (1):100–112
Perugini A, Ditterich J, Basso MA (2016) Patients with Parkinson's disease show impaired use of priors in conditions of sensory uncertainty. Curr Biol 26(14):1902–1910. https://doi.org/10.1016/j.cub.2016.05.039
Postle BR (2005) Delay-period activity in the prefrontal cortex: one function is sensory gating. J Cogn Neurosci 17(11):1679–1690. https://doi.org/10.1162/089892905774589208
Postle BR (2006) Working memory as an emergent property of the mind and brain. Neuroscience 139(1):23–38. https://doi.org/10.1016/j.neuroscience.2005.06.005
Praamstra P, Stegeman DF, Cools AR, Horstink M (1998) Reliance on external cues for movement initiation in Parkinson's disease – evidence from movement-related potentials. Brain 121:167–177. https://doi.org/10.1093/brain/121.1.167
Raiguel SE, Xiao DK, Marcar VL, Orban GA (1999) Response latency of macaque area MT/V5 neurons and its relationship to stimulus parameters. J Neurophysiol 82(4):1944–1956
Robinson DA, Fuchs AF (1969) Eye movements evoked by stimulation of frontal eye fields. J Neurophysiol 32(5):637–648
Rodnitzky RL (1998) Visual dysfunction in Parkinson's disease. Clin Neurosci 5(2):102–106
Rodriguez-Oroz MC, Jahanshahi M, Krack P, Litvan I, Macias R, Bezard E, Obeso JA (2009) Initial clinical manifestations of Parkinson's disease: features and pathophysiological mechanisms. Lancet Neurol 8(12):1128–1139. https://doi.org/10.1016/s1474-4422(09)70293-5
Ruff D, Cohen M (2014) Attention can either increase or decrease spike count correlations in visual cortex. Nat Neurosci 17(11):1591–1597
Santana N, Artigas F (2017) Laminar and cellular distribution of monoamine receptors in rat medial prefrontal cortex. Front Neuroanat 11:87. https://doi.org/10.3389/fnana.2017.00087
Schall JD, Morel A, King DJ, Bullier J (1995) Topography of visual cortex connections with frontal eye field in macaque: convergence and segregation of processing streams. J Neurosci 15 (6):4464–4487
Schnyder H, Reisine H, Hepp K, Henn V (1985) Frontal eye field projection to the paramedian pontine reticular formation traced with wheat germ agglutinin in the monkey. Brain Res 329 (1–2):151–160
Seamans JK, Yang CR (2004) The principal features and mechanisms of dopamine modulation in the prefrontal cortex. Prog Neurobiol 74(1):1–58. https://doi.org/10.1016/j.pneurobio.2004.05.006
Sesack SR, Snyder CL, Lewis DA (1995) Axon terminals immunolabeled for dopamine or tyrosine hydroxylase synapse on GABA-immunoreactive dendrites in rat and monkey cortex. J Comp Neurol 363(2):264–280. https://doi.org/10.1002/cne.903630208
Seymour K, Stein T, Sanders LLO, Guggenmos M, Theophil I, Sterzer P (2013) Altered contextual modulation of primary visual cortex responses in schizophrenia. Neuropsychopharmacology 38 (13):2607–2612. https://doi.org/10.1038/npp.2013.168
Siegert RJ, Harper DN, Cameron FB, Abernethy D (2002) Self-initiated versus externally cued reaction times in Parkinson's disease. J Clin Exp Neuropsychol 24(2):146–153. https://doi.org/10.1076/jcen.24.2.146.991
Slagter HA, van Wouwe NC, Kanoff K, Grasman R, Claassen DO, van den Wildenberg WPM, Wylie SA (2016) Dopamine and temporal attention: an attentional blink study in Parkinson's disease patients on and off medication. Neuropsychologia 91:407–414. https://doi.org/10.1016/j.neuropsychologia.2016.09.006

Smiley JF, Goldman-Rakic PS (1993) Heterogeneous targets of dopamine synapses in monkey prefrontal cortex demonstrated by serial section electron microscopy: a laminar analysis using the silver-enhanced diaminobenzidine sulfide (SEDS) immunolabeling technique. Cereb Cortex 3(3):223–238

Smiley JF, Williams SM, Szigeti K, Goldman-Rakic PS (1992) Light and electron microscopic characterization of dopamine-immunoreactive axons in human cerebral cortex. J Comp Neurol 321(3):325–335. https://doi.org/10.1002/cne.903210302

Smiley JF, Levey AI, Ciliax BJ, Goldman-Rakic PS (1994) D1 dopamine receptor immunoreactivity in human and monkey cerebral cortex: predominant and extrasynaptic localization in dendritic spines. Proc Natl Acad Sci 91(12):5720–5724

Soltani A, Noudoost B, Moore T (2013) Dissociable dopaminergic control of saccadic target selection and its implications for reward modulation. Proc Natl Acad Sci 110(9):3579–3584. https://doi.org/10.1073/pnas.1221236110

Sommer MA, Wurtz RH (2000) Composition and topographic organization of signals sent from the frontal eye field to the superior colliculus. J Neurophysiol 83(4):1979–2001

Sommer MA, Wurtz RH (2001) Frontal eye field sends delay activity related to movement, memory, and vision to the superior colliculus. J Neurophysiol 85(4):1673–1685

Squire RF, Noudoost B, Schafer RJ, Moore T (2013) Prefrontal contributions to visual selective attention. Annu Rev Neurosci 36(1):451–466. https://doi.org/10.1146/annurev-neuro-062111-150439

Stablein M, Sieprath L, Knochel C, Landertinger A, Schmied C, Ghinea D et al (2016) Impaired working memory for visual motion direction in schizophrenia: absence of recency effects and association with psychopathology. Neuropsychology 30(6):653–663. https://doi.org/10.1037/neu0000267

Stanton GB, Bruce CJ, Goldberg ME (1993) Topography of projections to the frontal lobe from the macaque frontal eye fields. J Comp Neurol 330(2):286–301. https://doi.org/10.1002/cne.903300209

Stanton GB, Bruce CJ, Goldberg ME (1995) Topography of projections to posterior cortical areas from the macaque frontal eye fields. J Comp Neurol 353(2):291–305. https://doi.org/10.1002/cne.903530210

Stern Y, Mayeux R, Rosen J, Ilson J (1983) Perceptual motor dysfunction in Parkinson's disease: a deficit in sequential and predictive voluntary movement. J Neurol Neurosurg Psychiatry 46 (2):145–151. https://doi.org/10.1136/jnnp.46.2.145

Sundberg KA, Mitchell JF, Gawne TJ, Reynolds JH (2012) Attention influences single unit and local field potential response latencies in visual cortical area V4. J Neurosci 32 (45):16040–16050. https://doi.org/10.1523/JNEUROSCI.0489-12.2012

Suzuki S, Cavanagh P (1997) Focused attention distorts visual space: an attentional repulsion effect. J Exp Psychol Hum Percept Perform 23(2):443–463

Tseng KY, O'Donnell P (2004) Dopamine-glutamate interactions controlling prefrontal cortical pyramidal cell excitability involve multiple signaling mechanisms. J Neurosci 24 (22):5131–5139. https://doi.org/10.1523/JNEUROSCI.1021-04.2004

Umeno MM, Goldberg ME (2001) Spatial processing in the monkey frontal eye field. II. Memory responses. J Neurophysiol 86(5):2344–2352

van Eden CG, Hoorneman EM, Buijs RM, Matthijssen MA, Geffard M, Uylings HB (1987) Immunocytochemical localization of dopamine in the prefrontal cortex of the rat at the light and electron microscopical level. Neuroscience 22(3):849–862

Verney C, Alvarez C, Geffard M, Berger B (1990) Ultrastructural double-labelling study of dopamine terminals and GABA-containing neurons in rat anteromedial cerebral cortex. Eur J Neurosci 2(11):960–972

Vijayraghavan S, Wang M, Birnbaum SG, Williams GV, Arnsten AF (2007) Inverted-U dopamine D1 receptor actions on prefrontal neurons engaged in working memory. Nat Neurosci 10 (3):376–384. https://doi.org/10.1038/nn1846

Vijayraghavan S, Major AJ, Everling S (2016) Dopamine D1 and D2 receptors make dissociable contributions to dorsolateral prefrontal cortical regulation of rule-guided oculomotor behavior. Cell Rep 16(3):805–816. https://doi.org/10.1016/j.celrep.2016.06.031

Vincent SL, Khan Y, Benes FM (1993) Cellular distribution of dopamine D1 and D2 receptors in rat medial prefrontal cortex. J Neurosci 13(6):2551–2564

Wang X-J (2008) Decision making in recurrent neuronal circuits. Neuron 60(2):215–234. https://doi.org/10.1016/j.neuron.2008.09.034

Wang M, Vijayraghavan S, Goldman-Rakic PS (2004) Selective D2 receptor actions on the functional circuitry of working memory. Science 303(5659):853–856. https://doi.org/10.1126/science.1091162

Wardak C, Ibos G, Duhamel J-R, Olivier E (2006) Contribution of the monkey frontal eye field to covert visual attention. J Neurosci 26(16):4228–4235. https://doi.org/10.1523/JNEUROSCI.3336-05.2006

Wei X, Ma T, Cheng Y, Huang CCY, Wang X, Lu J, Wang J (2018) Dopamine D1 or D2 receptor-expressing neurons in the central nervous system. Addict Biol 23(2):569–584. https://doi.org/10.1111/adb.12512

Williams GV, Goldman-Rakic PS (1995) Modulation of memory fields by dopamine D1 receptors in prefrontal cortex. Nature 376(6541):572–575. https://doi.org/10.1038/376572a0

Womelsdorf T, Anton-Erxleben K, Pieper F, Treue S (2006) Dynamic shifts of visual receptive fields in cortical area MT by spatial attention. Nat Neurosci 9(9):1156–1160. https://doi.org/10.1038/nn1748

Yahr MD, Duvoisin RC, Schear MJ, Barrett RE, Hoehn MM (1969) Treatment of parkinsonism with levodopa. Arch Neurol 21(4):343–354. https://doi.org/10.1001/archneur.1969.00480160015001

Zaksas D, Pasternak T (2006) Directional signals in the prefrontal cortex and in area MT during a working memory for visual motion task. J Neurosci 26(45):11726–11742. https://doi.org/10.1523/JNEUROSCI.3420-06.2006

Functions of Memory Across Saccadic Eye Movements

David Aagten-Murphy and Paul M. Bays

Contents

Abstract Several times per second, humans make rapid eye movements called saccades which redirect their gaze to sample new regions of external space. Saccades present unique challenges to both perceptual and motor systems. During the movement, the visual input is smeared across the retina and severely degraded. Once completed, the projection of the world onto the retina has undergone a large-scale spatial transformation. The vector of this transformation, and the new orientation of

D. Aagten-Murphy (✉) and P. M. Bays
University of Cambridge, Cambridge, UK
e-mail: david.aagtenmurphy@gmail.com

Curr Topics Behav Neurosci (2019) 41: 155–184
DOI 10.1007/7854_2018_66
Published Online: 7 December 2018

the eye in the external world, is uncertain. Memory for the pre-saccadic visual input is thought to play a central role in compensating for the disruption caused by saccades. Here, we review evidence that memory contributes to (1) detecting and identifying changes in the world that occur during a saccade, (2) bridging the gap in input so that visual processing does not have to start anew, and (3) correcting saccade errors and recalibrating the oculomotor system to ensure accuracy of future saccades. We argue that visual working memory (VWM) is the most likely candidate system to underlie these behaviours and assess the consequences of VWM's strict resource limitations for transsaccadic processing. We conclude that a full understanding of these processes will require progress on broader unsolved problems in psychology and neuroscience, in particular how the brain solves the object correspondence problem, to what extent prior beliefs influence visual perception, and how disparate signals arriving with different delays are integrated.

Keywords Saccadic eye-movements · Transsaccadic processing · Visual working memory

1 Introduction

Light entering the human eye is focused by the crystalline lens onto the retina, a light-sensitive layer at the back of the eye. Within the retina, photoreceptor density falls off with distance from the fovea, the region corresponding to the centre of the visual image. This preferential distribution of neural resources to the centre of the visual field is reiterated in the subcortical and cortical visual areas that process and refine the retinal input. As a result, identification and discrimination of visual details is substantially better at the fovea than in the periphery. To bring different portions of the visual scene to project onto the fovea, six strong ocular muscles generate brief bursts of movement, termed saccades, that change the orientation of the eye in its socket. Natural viewing behaviour therefore consists primarily of a sequence of stable fixations directed to different locations in the environment, each lasting a few hundred milliseconds, separated by saccades.

Due to inherent limitations in the rate at which photoreceptors can respond to visual stimuli, integrating information over tens of milliseconds, any substantial movement of the eye smears the retinal image (Wurtz 2008) (Fig. 1a). To account for this, the saccadic system has evolved to move the eye at velocities of up to 500°/s, minimizing the period for which the nervous system does not receive a reliable visual input (Westheimer 1954; Wurtz 2008). Nonetheless, uptake of visual information from the external world is compromised for 30–50 ms with every saccade. Additionally, the rapid movement of the image across the retina would generate salient and distracting motion signals if not for a combination of passive and active suppression mechanisms (Wurtz 2008). The pre- and post-saccadic visual images may act as forward and backward masks, respectively, effectively hiding the intrasaccadic blur from perception (Castet et al. 2002). Active down-weighting of the magnocellular pathway has been reported to suppress visual information

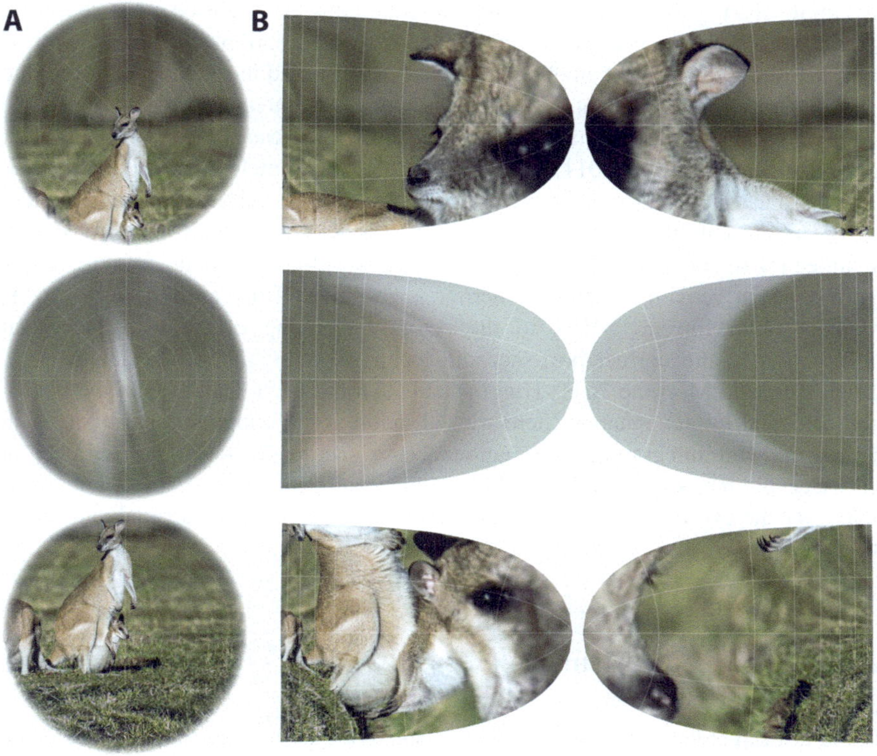

Fig. 1 Saccadic eye movements cause global changes to the retinal image and its cortical representation. (**a**) As gaze shifts from its pre-saccadic location (here centred on the face of the kangaroo; top panel), the rapid movement smears the retinal image (middle panel), before steady fixation is re-established at the new location (the joey in its mother's pouch; bottom panel). (**b**) Projection of the retinal image onto the cortical map (e.g. of V1). Note the substantial over-representation of the fovea. Comparison of visual information between fixations must take into account the large-scale reorganization of visual cortical representation due to the saccade

processing, particularly for motion (Burr et al. 1994). This suppression seems to begin even before saccade onset (Diamond et al. 2000; Bremmer et al. 2009), indicating that it is not merely a passive response arising from changes in input during an eye movement (see Ross et al. 2001 for a review). In combination, these suppression mechanisms ensure that disruptive visual signals associated with the eye movement are annulled, at the cost of slightly increasing the duration for which reliable visual information is unavailable.

With each saccade, representations of objects located in the periphery are suddenly shifted to the fovea and vice versa. This means that, at least in early visual areas where receptive fields are small, after the saccade the same object is represented by an entirely distinct neuronal population coding for a different location in the visual field (Fig. 1b). As such, decisions about the continuity of an object – or detecting whether the properties of an object have changed – cannot be determined by simple changes in firing rate of individual neurons, as might be possible during fixation. Furthermore, because resolution varies with eccentricity, the pre- and post-saccadic representations of an object will rarely correspond in spatial resolution. The challenge of understanding how such spatially distinct sources of information can be compared across a saccade has eluded researchers for decades (see Cavanagh et al. 2010; Burr and Morrone 2011; Tatler and Land 2011; Melcher and Morrone 2015; Higgins and Rayner 2015). In brief, explanations have varied from the problem being so difficult that nothing is stored across saccades, with stability simply assumed (O'Regan and Noë 2001), to dedicated world-based (spatiotopic) representations which remain unchanged across gaze shifts (Burr and Morrone 2011; Melcher and Morrone 2015), to the active remapping of pre-saccadic visual information to the correct post-saccadic location (Wurtz 2008; Cavanagh et al. 2010; Burr and Morrone 2011; Higgins and Rayner 2015). Rather than the neurophysiological implementation, here we will address the computations involved in comparing and integrating information across saccades. Our focus will be on transsaccadic memory – the internal representation of the pre-saccadic scene available after the eye movement – and the functions it serves in perceptual and motor systems.

2 Transsaccadic Memory

In the last three decades, substantial evidence about the properties of transsaccadic memory has accumulated from studies that ask observers to compare or directly report information from the pre-saccadic scene that is absent or changed in the post-saccadic input (Irwin et al. 1990; Irwin 1992, 1996; Irwin and Andrews 1996; Deubel and Schneider 1996; Hollingworth et al. 2008). While these results suggest an image-like snapshot of the whole visual scene is not stored across the saccade (Tatler and Land 2011; Higgins and Rayner 2015), detailed information about the colour, size, shape, and identity of select objects can be maintained (Henderson 1994; Henderson and Siefert 1999; Henderson and Hollingworth 1999, 2003; Melcher and Kowler 2001; Irwin and Zelinsky 2002; Tatler et al. 2003). In this respect, transsaccadic memory strongly resembles visual working memory (VWM), a limited store of visual information that survives masking inputs and decays over seconds (Baddeley and Hitch 1974; Luck and Vogel 1997).

Early research into transsaccadic memory, like VWM, assumed that the limits of representation could be captured by a single number, defining an upper limit on items stored (Miller 1956; Pashler 1988; Luck and Vogel 1997; Cowan 2001). Capacity estimates calculated on this basis have typically been similar whether recall was tested after an intervening saccade or at fixation with an intervening delay (Irwin 1992; Irwin and Andrews 1996; Gordon and Irwin 1998; Prime et al. 2007). Additionally, these results suggested that multiple features of the same object could be maintained across a saccade with little or no additional cost, mirroring results from VWM that suggested different feature dimensions are held in independent stores (Wheeler and Treisman 2002; Bays et al. 2011).

More recent conceptions of VWM recognize that the precision or resolution with which visual information is maintained can vary substantially as a function of memory load and the distribution of attention (Ma et al. 2014). Increasing the number of items in memory (Zhang and Luck 2008; Bays and Husain 2008; van den Berg et al. 2012), or enhancing the storage resolution of a specific item (Bays et al. 2011; Gorgoraptis et al. 2011; Melcher and Piazza 2011), conveys a cost to the precision of other remembered items. While the average precision of recall for a stimulus array changes little when a saccade is made to one of its elements, this finding hides a strong redistribution of memory resources towards the saccade target, with precision for that object increased at a cost to other items in the display (Bays and Husain 2008). When the information required for memorization is dissociated from the location of the saccade target, memory performance is considerably worse (Shao et al. 2010; Hanning et al. 2015; Tas et al. 2016; Ohl and Rolfs 2017; Jeyachandra et al. 2018), with task-irrelevant information unavoidably encoded. Demonstrating that this is not due to the saccade itself, no deficit was observed when subjects made eye movements to empty locations (Tas et al. 2016), indicating the effects originate from the cost of additionally encoding the saccade target. An exception is if the saccade is to an empty location matching the location of an item in memory, in which case changes in memory performance occur consistent with a redistribution of resources between the items in memory (Ohl and Rolfs 2017, 2018).

Memorization of the saccade target may be an unavoidable consequence of the obligatory shift of attention to the target location that precedes a saccade (Deubel and Schneider 1996; Peterson et al. 2004; Rolfs and Carrasco 2012) or may be an adaptive behaviour that ensures a reliable representation of the saccade target is available for post-saccadic comparison. In any case, this compulsory encoding occurs at the expense of some of the usual flexibility in distributing VWM resources (Gorgoraptis et al. 2011).

2.1 Other Memory Representations

In addition to VWM, researchers have explored whether other types of visual representation contribute to recall of a pre-saccadic scene. Visible persistence refers to the fact that a visual stimulus can appear to still be present very briefly after it is removed. During fixation, visible persistence means that rapidly interleaved sequential inputs can be perceived as a single image (perceptual fusion). However, for such a mechanism to operate across an eye movement would require persistence to be tied to spatial rather than retinal location. The possibility of fusion in spatial coordinates was investigated and discounted by Irwin and others (Irwin 1996), and while participants in one recent study reported spatiotopic fusion (Paeye et al. 2018), the very specific conditions required suggest its relevance to natural viewing is minimal (see Sect. 3.1 for further discussion).

Beyond the point at which an extinguished stimulus remains visible, there appears to be a further short-lived "informational persistence" (or iconic memory) – a very high capacity, pre-attentive memory of the visual image that is easily disrupted by masking (Sperling 1960; Gegenfurtner and Sperling 1993). Although iconic memory has commonly been assumed to be eliminated by eye movements, one study (Germeys et al. 2010) presented evidence that such a representation could survive and be spatially updated with the saccade. However, this transsaccadic informational persistence was masked by the presence of stimuli after the saccade. A role in typical vision, in which there are almost always post-saccadic stimuli to overwrite such representations, has not been established. However, we will return to this topic in the final section of our review.

In summary, while visible and non-visible high-capacity representations of a scene may survive long enough to contribute to transsaccadic memory, evidence for a role in natural viewing is sparse, and VWM-only accounts have generally proved sufficient to explain post-saccadic memory performance.

2.2 *Role of VWM Across Eye Movements*

In the following sections, we will consider evidence for three key functions that VWM has been proposed to fulfil with respect to saccadic eye movements. First, it could enable the detection of changes in the external world that occur during the saccade. Second, it could enable pre-saccadic information to supplement and enhance post-saccadic vision, so processing does not have to start anew with each new fixation. Third, it could support the correction of saccade errors and calibration of saccadic motor commands. Each of these roles will prove heavily interrelated, and common themes arising from their evaluation will be discussed in the final section.

3 Identifying Changes in the Environment

The world we live in is filled with dynamic stimuli capable of changing their appearance or position, often without warning. During fixation, such changes produce transient responses in the activity of visual neurons, supporting rapid detection and reorienting of attention to the changed stimulus (Boehnke and Munoz 2008). However, when a change occurs during a saccade, this transient response is unavailable, and the perceptual system must instead infer the occurrence of a change by comparing pre-saccadic visual information in memory to the current percept. Due to the substantial spatial reorganization caused by a saccade, this task is particularly challenging when attempting to detect spatial changes. In this section we will investigate our ability to infer the occurrence of changes in the external world across saccadic eye movements. As will become clear, this requires not only that we have a pre-saccadic memory representation of objects for comparison but also that we have an accurate estimate of the saccade vector and that we can determine how objects present after the saccade correspond to those seen before it.

3.1 Detection of Displacements

One of the most impressive demonstrations of the difficulty of inferring changes during an eye movement is saccadic suppression of displacement (SSD; Bridgeman et al. 1975) (Fig. 2a). Here, during execution of an eye movement to a target object, the location of the target can be shifted considerably backward or forward along the saccade path without observers' awareness (Bridgeman et al. 1975; Deubel et al. 1996; Niemeier et al. 2003). While the magnitude of displacement required for reliable detection varies across studies (Higgins and Rayner 2015) – from 10% (Li and Matin 1990) up to 30% (Bridgeman et al. 1975) or even 50% of the saccade amplitude (McConkie and Currie 1996) – within a single experiment the range of displacements tolerated scales linearly with saccade amplitude (Bridgeman et al. 1975; Li and Matin 1990; Deubel et al. 1996).

Early theories of visual stability had proposed a global subtraction or cancellation of the retinal shift based on the outgoing saccadic motor signal (efference copy)

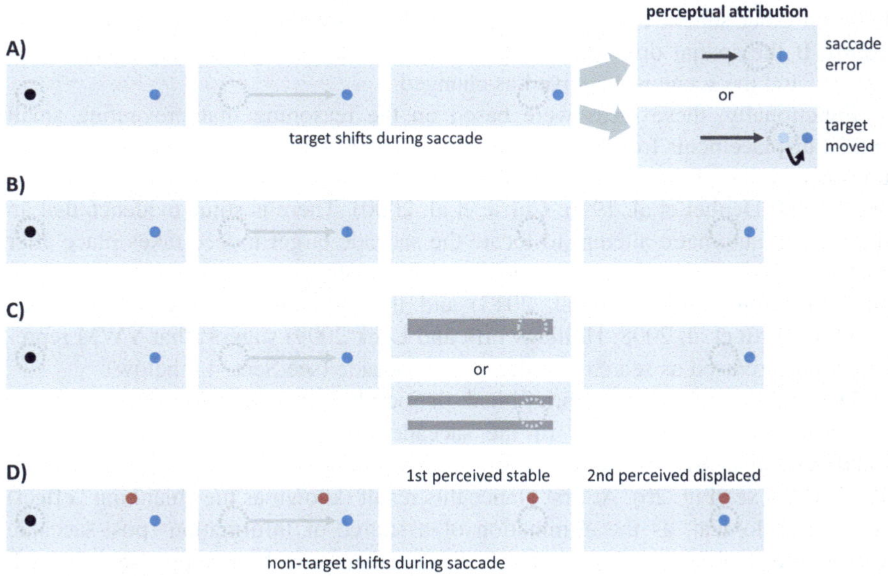

Fig. 2 Experimental methods for investigating saccadic suppression of displacement (SSD). (**a**) When the target of a saccade is displaced while the eyes are moving, the visual system can attribute the failure of gaze to land on the target to either error in the saccade (visual stability preserved) or movement of the stimulus (visual stability violated). Typical observers tolerate surprisingly large displacements without perceiving the target to have moved. (**b**) A brief delay or blank on saccade landing before reappearance of the target is sufficient to substantially reduce SSD, making the target movement readily detectible. (**c**) When the blank is replaced with an occluding object (top panel), the target movement again becomes hard to detect, but a similar object that would not have occluded the target (bottom panel) does not have this effect. (**d**) When multiple stimuli are present before the eye movement, the first post-saccadic stimulus to appear is assumed stable, regardless of whether it is the target or (as depicted) a different stimulus

(Bridgeman 2007; Higgins and Rayner 2015). In other words, the shift in the visual image across the retina would be exactly cancelled out by a contrary internal transformation based only on the intended saccade vector. Observations of SSD and related phenomena led researchers to instead suggest that a stable external world is the "null hypothesis" of the visual system (Mackay 1972; Bridgeman et al. 1994) and that post-saccadic visual input is critical in evaluating when this assumption is violated (Deubel et al. 1996; McConkie and Currie 1996; Currie et al. 2000). According to one version of this theory, there is a brief spatiotemporal "constancy window" in which the visual system searches for the saccade target within a few degrees of the post-saccadic landing position (Deubel et al. 1998; Deubel et al. 1998; Bridgeman 2007). If found, the post-saccadic input is deemed compatible with the assumption of a stable world, and previous spatial information is discarded or ignored. However, if the target is not found, then the assumption is violated, and eye proprioception, efference copy, and other "extraretinal" sources of information are consulted. This idea was conceptually similar to that of McConkie and Currie (1996), who emphasized the role of post-saccadic visual input and suggested a two-step search in which, after locating the saccade target, the relative distances between other salient objects in the surroundings are also examined (Currie et al. 2000). If the spatial discrepancies at either stage are too great, visual stability is broken, and the scene is perceived as changed.

Functionally, these ideas were based on the reasoning that preventing small spatial displacements from reaching awareness would be critical in enabling continuous perception of the world despite noise in sensory and motor signals (Bridgeman et al. 1994; Deubel et al. 1996; Currie et al. 2000). There is some evidence that an immediate, automatic attempt to locate the saccade target indeed takes place after each eye movement. The speed and accuracy of eye movements to matching singletons (Hollingworth et al. 2013) and the targeting of corrective saccades (Hollingworth et al. 2008; Hollingworth and Luck 2009) suggest that VWM representations are used as search templates post-saccade (see Sect. 4.1 below).

Deubel et al. introduced a brief blank period after the eye movement – delaying the post-saccadic appearance of the saccade target – and found a substantial improvement in the ability to identify displacements of the target (Deubel et al. 1996, 1998, see Fig. 2b). At first glance this result (known as the "blanking" effect) seems paradoxical, as the elimination of a source of information (post-saccadic vision) allows observers to judge target displacements much more precisely. Indeed, it was interpreted as a demonstration that the perceptual system *always* has access to high-quality information about the extent of the movement but routinely elects to ignore this information in favour of post-saccadic visual information (Deubel et al. 1996; Higgins and Rayner 2015). Only when forced, by the absence of the saccade target or other violations of the stability assumption, are saccade estimates consulted. Within this view, the perceptual system rigidly switches between two distinct states, either relying solely on post-saccadic visual input (assuming visual stability is maintained) or solely on the intended saccade vector (assuming visual stability is violated) to determine the location of post-saccadic gaze.

Niemeier et al. (2003) proposed an alternative to this heuristic account, suggesting that the inability to detect displacements across a saccade could actually represent the perceptual system making the best decision possible with noisy information. They used a Bayesian framework (Kersten et al. 2004) in which sensory inputs are used as evidence to update prior beliefs about the state of the world. Specifically, in their model, visual input and efference copy are optimally integrated with a prior belief that the visual scene is stable. When the saccade target is blanked, this discontinuity in the scene causes the prior to be weakened, making a displacement easier to detect. This model not only reproduced the patterns of spatial distortions and SSD typically observed in human observers but also made new predictions about the link between eye movement variability and magnitude of SSD that were experimentally verified (Niemeier et al. 2003, 2007; Wexler and Collins 2014). The Bayesian model captured both the increase in SSD with saccade magnitude (Li and Matin 1990) and the observation of greater sensitivity for detecting changes orthogonal than parallel to the saccade vector (van Opstal and van Gisbergen 1989). In both cases, the degree to which judgements are based on the prior assumption of stability is determined by the magnitude of uncertainty in saccadic landing position.

Subsequent findings have demonstrated that displacements of the saccade target which are orthogonal to the eye movement relieve SSD even when these shifts are irrelevant to the task (Wexler and Collins 2014). This further suggests that estimates of the spatial change induced by a saccade vector are critical in inferring the source of spatial inconsistencies across eye movements.

Atsma et al. (2016) recently expanded on this idea by incorporating causal inference into the integration process. They noted that the Niemeier model always integrates pre- and post-saccadic estimates of location – even for very large displacements – and argued that a causal judgement of whether the input is consistent with viewing a single stable object is critical to determine whether to integrate the two estimates. Rather than a heuristic decision to integrate or segregate inputs, they found the data were best described by a statistically optimal mixture of the two strategies. Finally, noting considerable variation in the strength of the "stable world" prior between participants, they suggested that experimental context and instructions might be idiosyncratically incorporated into individuals' beliefs that objects could change during a saccade, making some participants more sensitive to location changes than they would have been outside the laboratory.

3.2 Object Continuity

Spatiotemporal correspondence between objects represents one cue as to whether pre- and post-saccadic information relate to the same object or whether the post-saccadic stimulus represents an entirely new object (Cox et al. 2005; Li and DiCarlo 2008; Schneider 2013). Recent results indicate that changes to surface properties of objects also influence the magnitude of SSD, suggesting these properties are also used as cues to correspondence (Demeyer et al. 2010b; Tas et al. 2012; Poth et al. 2015; Poth and Schneider 2016). In one study (Tas et al. 2012) changes in surface features across a saccade decreased the spatial range of SSD – although not to the maximum detectability – suggesting they decreased the strength of the stability assumption but did not eradicate it. This indicates that, in addition to spatial information, information about surface features is automatically stored and maintained during an eye movement for comparison with post-saccadic input (Prime et al. 2007, 2011; Melcher and Colby 2008).

Relatedly, SSD has been found to be reduced for stimuli that are in motion before saccade onset (Gysen et al. 2002a). This could be interpreted as evidence that the prior for stability is down-weighted for these objects, suggesting that the "stable world" prior is in fact a "stable object" prior determined on an item-by-item basis.

Briefly replacing the saccade target with an elongated bar immediately after the saccade produces less reduction in SSD than a blank of the same duration (Deubel et al. 2002) (Fig. 2c; top panel). One hypothesis is that the bar presents less evidence for a break in object stability, because the stimulus could still be present but hidden by the bar. Supporting this interpretation, when the bar was split in two, so the saccade target should have been visible in the gap, the discrepancy from the blanking condition was eliminated (Deubel et al. 2002) (Fig. 2c; bottom panel).

Interruptions in object continuity benefit not only the detection of changes in spatial location but also detection of non-location changes such as spatial frequency (Weiß et al. 2015). This suggests that breaking continuity alters the way that new visual information is processed by our perceptual system. It has been suggested that a break in continuity causes post-saccadic object information to be represented separately to pre-saccadic information (Deubel et al. 1996; Tas et al. 2012; Schneider 2013), as opposed to the new information simply updating existing representations. This has important implications for VWM, given its very limited capacity: allocating memory resources separately to both pre- and post-saccadic objects will greatly diminish the fidelity of these representations, as set size has effectively doubled (Bays et al. 2009) – in a sequence of saccades the number of independent representations would quickly become unmanageable. Alternatively, if pre-saccadic representations are simply updated or overwritten with post-saccadic information, then subsequent comparison between pre- and post-saccadic information would no longer be possible, as only the combined representations persist (Schneider 2013). More research is needed to understand the circumstances under which visual information obtained in different fixations is segregated versus integrated.

3.3 Visual Landmarks

Although most studies focus on the role of the saccade target in determining the accuracy of saccadic eye movements, other objects in the pre-saccadic visual scene may also be remembered and influence our judgements (Deubel 2004). In contrast to the sparse visual displays used in most experiments, natural scenes typically contain an abundance of other objects. Remembered information about these objects – in particular their distance from the saccade target – may provide an additional, important source of information to aid localization and determine the magnitude of a saccade.

The ability to detect displacements in the location of saccade target and nontarget stimuli was investigated by Deubel et al. (1998), who independently blanked and/or displaced either the saccade target or the nontarget stimulus during the eye movement (Fig. 2d). They found that while the identity of the blanked stimulus (target or nontarget) had no effect, temporal order was critical. When the stimulus that remained visible after the saccade was displaced from its original location, and the blanked stimulus returned in its original location, observers typically incorrectly reported that the blanked stimulus had moved. This held true for displacements of the first stimulus of up to 50% of the saccade vector and for temporal asynchronies as short as 50 ms. When both stimuli were blanked, despite a temporal discontinuity occurring for both stimuli, the first stimulus to reappear was always perceived as stable. This suggests that the visual system prioritizes the first available visual feedback after a saccade. In general, post-saccadic objects which match the remembered pre-saccadic stimuli appear to be able to function as "anchor points" for aligning expectations about where other objects – even the saccade target – should be located after the saccade.

Ostendorf and Dolan (2015) examined how the proximity of nontarget objects to the target stimulus influenced the magnitude of SSD. They found that, as the distance to the landmark object increased, the precision of judgements of displacement decreased, suggesting that the reliability of visual distance estimates scaled inversely with eccentricity. Importantly, the presence of the landmark did not cause subjects to ignore the planned saccade vector but rather altered the weighting between visual and non-visual information, with the influence of nontargets depending on the reliability of the spatial evidence they provided. Particularly in more ecological settings, where visual landmarks are abundant, memory for the relative distances between objects is likely to have a dominant influence on discriminating transsaccadic displacements.

3.4 Conclusion

Saccades momentarily compromise our access to reliable visual information while also creating a dramatic spatial transformation of the retinal image. As a result, the signals that normally allow us to detect changes in the environment are largely eliminated. Inferring such changes therefore requires that information about the pre-saccadic scene be actively maintained and compared to post-saccadic vision. Our review of studies testing detection of transsaccadic changes – both spatial and nonspatial – leads us to conclude that the visual system, around the time of a saccade, engages in a dynamic process of integrating multiple sources of information to determine if and how visual objects have changed. Because the post-saccadic visual input provides the most reliable and precise evidence about object locations and features, it frequently outweighs other signals, such as a discrepancy between the retinal shift and planned saccade vector, that could indicate a change has taken place. This weighting of multiple sources of evidence is captured by Bayesian models of perception and stands in contrast to heuristic accounts that propose a stepwise decision process based primarily on the visual input.

4 Integrating Information Across Saccades

Visual exploration often involves a repeated process of identifying an object of interest in peripheral vision, followed by a saccade that brings it into foveal vision for detailed examination. While peripheral vision is relatively imprecise and often impaired by crowding (Pelli and Tillman 2008; Strasburger et al. 2011), it would be inefficient to discard information obtained from it entirely (Harrison and Bex 2014). In this section we investigate the ability to incorporate information from before the saccade into post-saccadic perception.

4.1 Transsaccadic Fusion

Early accounts of transsaccadic memory suggested that the visual system adds post-saccadic input to a high-resolution, image-like representation of the scene accumulated from previous fixations (see Higgins and Rayner 2015). This hypothesis, commonly termed the "integrative visual buffer" (McConkie and Rayner 1976) or "spatiotopic fusion" (Irwin 1992), would predict that observers could effortlessly combine partial pre- and post-saccadic images into a single percept, as if they had been presented simultaneously (Fig. 3a). While initial findings seemed to support this prediction (Wolf et al. 1980; Jonides et al. 1982), these results were subsequently found to be artifactual, due to the persistence of the CRT displays (Jonides et al. 1983; O'Regan and Lévy-Schoen 1983; Irwin et al. 1983; Rayner and Pollatsek 1983; Bridgeman and Mayer 1983), and the hypothesis of an integrative buffer was largely abandoned.

However, a recent study by Paeye et al. (2018) found surprising evidence for a form of transsaccadic fusion. Unlike the stimuli used in previous fusion experiments (e.g. the dot matrix stimuli of Irwin et al. 1983), they chose pre- and post-saccadic stimuli whose integration would not require precise spatial alignment. Specifically, they presented a single vertical bar before and a set of three horizontal bars after the saccade. They asked participants to report whether they saw neither stimuli, one stimulus, both in succession or both simultaneously. The stimuli were frequently reported as simultaneously occurring when the post-saccadic contrast was made very low. The authors suggested that, under normal circumstances, post-saccadic input acts as a backward mask of the pre-saccadic scene, preventing transsaccadic fusion.

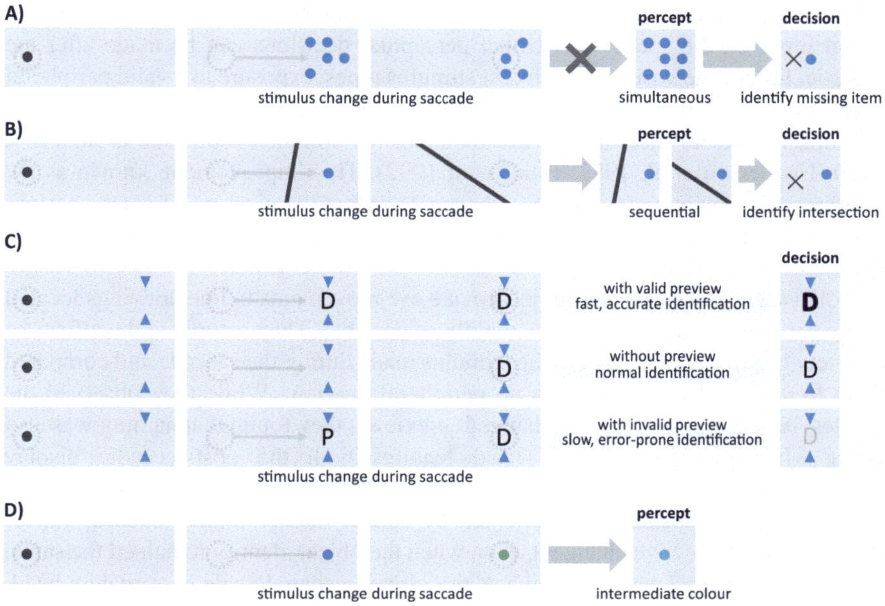

Fig. 3 Experimental methods for investigating the integration of information across saccades. (**a**) Partial images presented immediately before and after a saccade are in general not "fused" into a single perception of the full image. (**b**) While not experienced as a single image, different objects presented before and after saccade can be compared to make decisions which require both sources of information. (**c**) A valid peripheral "preview" of an object improves the speed and accuracy of its subsequent identification at the fovea, compared to a situation with no preview or a misleading, invalid preview. (**d**) When a small change is made to a surface feature (here, colour) of an object during a saccade, observers may report having seen a feature intermediate between the pre- and post-saccadic values

This implies that fusion would have little impact on everyday perception, where objects rarely change drastically in form or reduce substantially in contrast during a saccade.

4.2 *Transsaccadic Comparison and Preview Effects*

Despite the lack of support for explicit fusion, there is ample evidence that pre-saccadic information can be stored and processed alongside post-saccadic information. Such transsaccadic comparison has been shown for shapes and lines (Hayhoe et al. 1991; Prime et al. 2006), in studies where observers report information that is only accessible by comparison of the pre- and post-saccadic displays (Fig. 3b). It is important to note that stimuli are not merged into a single perception in these studies but rather that observers perceive and remember two presentations in sequence that they can relate together to make their decision.

Numerous studies have demonstrated an influence of pre-saccadic peripheral visual input on the speed with which perceptual decisions can be made after the saccade. For a wide variety of different stimulus types, exposure to a valid peripheral “preview” before an eye movement results in an approximately 100 ms reduction in the time it takes to discriminate details about the stimulus (Pollatsek et al. 1984, 1990; Henderson et al. 1987; Henderson 1992). This improvement, known as the transsaccadic preview effect (TPE), appears to reflect the memory of coarse peripheral information giving a “head-start” to object identification and perceptual decision-making (Fig. 3c).

Pollatsek et al. (1984) had subjects make eye movements to line drawings located in the periphery and name them as rapidly as possible. They examined the effects of a variety of different changes to the stimulus made during the saccade and compared them to a control condition with no peripheral preview. While they observed the greatest benefit for valid (i.e. unchanged) previews, they found that naming was also faster when previews shared visual features with the post-saccadic display (e.g. ball – tomato), compared to those that did not (ball – carrot). Additionally, the benefit was substantially weakened by spatial manipulations (such as mirror reflection or perspective changes), even when the object identity remained the same, but the benefit was unchanged when the object’s spatial scale was manipulated. These results suggested that the TPE reflects a specifically visual representation that is nonetheless, based on the tolerance for changes in scale, to some degree abstracted from the retinal image.

Subsequent work in which subjects made eye movements to a location in-between two objects – one of which disappeared during the saccade – indicated that a precise spatial correspondence was not critical (Pollatsek et al. 1990). Naming latency was largely unaffected when the remaining object switched locations, provided it matched one of the previewed stimuli. These researchers concluded that the TPE originates from priming of perceptual processes – such as the preactivation of long-term memories that can contribute to object identification, rather than a precise spatial integration of pre- and post-saccadic inputs. However, subsequent studies suggested that spatiotemporal alignment was important to observing the full preview benefit (Kahneman et al. 1992; Germeys et al. 2002). Indeed, Kahneman et al. (1992), using letter stimuli inside moving frames during fixation, found preview benefits only when the same letter appeared in the same frame that it had previously been shown in. Henderson et al. reported independent – and additive – spatially invariant and spatially dependent saccadic preview benefits. They observed that, compared to when new letters (Henderson and Anes 1994; Henderson 1994) or line drawings (Henderson and Siefert 2001) were used, repeating any stimulus from the pre-saccadic display sped up response times while repeating it at the same location boosted performance even more.

These results were based on tasks in which observers directed saccades to land equidistant between objects. Germeys et al. (2002) found that when observers directed gaze instead to one of the previewed objects, only a valid preview of the saccade target benefited performance, with no TPE when neighbouring objects were switched to the saccade target. Re-examining the findings of Henderson and

Anes (1994), they found that when saccade landing position was carefully controlled – eliminating trials where the eyes strayed to either object – the spatially independent benefit was eliminated, revealing that the object observers selected as the saccade target was critical. Finally, explicitly comparing saccade targets to surrounding flankers revealed a TPE for the saccade target regardless of its post-saccadic location, whereas flanker items required spatiotemporal continuity for any benefit to occur.

In summary, the TPE demonstrates that memory for visual attributes of pre-saccadic stimuli can facilitate post-saccadic visual processing, particularly when the stimulus is the saccade target. However, the extent to which the TPE represents a true integration of visual inputs, as distinct from priming of object and feature identification, has proven difficult to establish.

4.3 Transsaccadic Integration

While the hypothesis of transsaccadic fusion implies an image-like overlay of pre- and post-saccadic inputs, transsaccadic integration refers to an averaging of feature information extracted before and after the saccade (Fig. 3d). Because visual inputs are noisy, averaging could reduce error and uncertainty in perceptual estimates. In the lab, when pre- and post-saccadic features of an object are made to differ, transsaccadic integration is demonstrated by observers reporting an intermediate perception, falling between the two feature values.

Several lines of research have been suggestive of such a process. A number of studies have demonstrated that two static images seen before and after an eye movement may be perceived as a continuous motion or object transformation (Gysen et al. 2002b; Fracasso et al. 2010; Szinte and Cavanagh 2011). These results indicate that pre- and post-saccadic input can be combined to generate a continuous percept, and further that the representation maintained across the saccade includes shape as well as location information (Fracasso et al. 2010). The apparent motion results indicate that this process takes into account the shift in retinal location of objects caused by the saccade, although this compensation appears to be imperfect (Szinte and Cavanagh 2011).

Demeyer et al. (2010a) asked observers to make a judgement about the aspect ratio of an ellipse after making a saccade to it. When a small change was made to the aspect ratio during the eye movement, they observed that judgements were distributed around a point roughly mid-way between the two stimulus values. While the strongest evidence for integration was found in the condition where participants were reportedly not aware of the change, some integration occurred even when the two views were separated by a blank display, disrupting spatiotemporal continuity, and it took the presence of an interstimulus mask to fully eliminate the influence of the pre-saccadic stimulus. This finding was consistent with results of Wittenberg et al. (2008) who showed a small influence of stimuli presented during a previous fixation on judgements of briefly flashed colours, even though the pre- and post-saccadic colours were uncorrelated and separated by a blank, and observers were explicitly instructed to judge the colour of the flashed stimulus.

Oostwoud Wijdenes et al. (2015) demonstrated that the extent to which an integrated percept resembles either the pre- or the post-saccadic input depends on the relative reliability of each input. Examining integration of multiple coloured patches that shifted hue imperceptibly during a saccade, they found that manipulating the eccentricity of stimuli, or explicitly adding colour noise to one display, biased the integrated percept towards either the pre- or post-saccadic colour. They concluded that transsaccadic perception reflects a statistically optimal integration of visual inputs. A corollary of this result is that the information retained in transsaccadic memory does not consist only of point estimates of pre-saccadic feature values but also includes the reliability of those estimates.

Two subsequent studies confirmed using single-orientation stimuli that transsaccadic integration closely approximates the predictions of a maximum likelihood estimation model (Ganmor et al. 2015; Wolf and Schütz 2015). A similar result was obtained for numerosity judgements, using clouds of coloured dots (Hübner and Schütz 2017). Here it was observed that detectable changes in features of the stimuli (dot colour or individual dot locations) did not affect the degree of integration, even though they would be expected to break object continuity. The apparent discrepancy between these results and the effects of blanking described above could reflect the fact that the feature relevant to the task (numerosity) was not disrupted.

4.4 Conclusion

In situations where rapid decisions are required, the maintenance and use of pre-saccadic visual information could convey substantial benefits to an observer. When the post-saccadic information is congruent with the pre-saccadic memory, incorporating pre-saccadic information into perceptual judgements makes them both faster and more accurate. Indeed, the pre-saccadic input can directly influence the way that a post-saccadic stimulus is perceived, with the degree of influence dependent on the relative reliability of each view and the extent to which continuity of the relevant features can be established. In this way, the perceptual system makes the most of the information that it has available, improving both the efficiency and the continuity of our visual perception, despite the interruption of the saccade. Future work should investigate how transsaccadic integration operates in complex environments, where establishing object correspondence is often challenging, and the limited nature of transsaccadic memory is likely to impose an important constraint on what can be integrated.

5 Correcting Eye Movement Errors

Saccades, like every movement of the body, are subject to variability and inaccuracy. A number of mechanisms, operating on different timescales, exist to compensate or correct for saccade errors. When an eye movement fails to bring gaze precisely to the desired saccade target, it is often followed by a fast and automatic second eye movement, known as a corrective saccade, which typically brings gaze closer to the intended goal (Becker and Fuchs 1969). When the motor signal required to generate a particular eye movement changes – for example, as the result of fatigue or, on a longer timescale, advancing age (Warabi et al. 1984; Munoz et al. 1998; Hopp and Fuchs 2004; Dowiasch et al. 2015) – adaptation processes counter these effects to prevent systematic errors from developing (Herman et al. 2013). In this section we will look at how the storage of pre-saccadic visual information can contribute to immediate corrective saccades and longer-term recalibration of eye movements to minimize future errors.

5.1 Corrective Saccades

Debate about the role of both post-saccadic visual feedback and remembered pre-saccadic visual information in the initiation of corrective saccades has persisted for decades (Tian et al. 2013). Central to this discussion is the suggestion that the latency of some corrective saccades (which can be as short as 50 ms; Schut et al. 2017) is incompatible with a role for post-saccadic vision (Weber and Daroff 1972; Becker 1976). While some authors observed corrective saccades even when the saccade target was obscured (Barnes and Gresty 1973; Shebilske 1976; Ohl et al. 2013), others found that the absence of post-saccadic visual feedback eliminates their occurrence (Bonnetblanc and Baraduc 2007). Deubel et al. (1982) showed that introducing a brief blank after the saccade, followed by the reappearance of the saccade target, inhibited secondary saccades, as if delaying them until the return of visual feedback. Recent studies have attempted to draw a distinction between rapid (50–200 ms) corrective saccades that are pre-prepared based on the motor outflow and slower (200–300 ms) corrective saccades that are guided by visual feedback (Tian et al. 2013; Schut et al. 2017).

When available, post-saccadic visual information improves the accuracy of corrective saccades so that they compensate for about 70% of the error in the primary saccade (Prablanc et al. 1978; Tian et al. 2013), an increase from around 50% when only non-visual signals are available (Munuera et al. 2009; Tian et al. 2013). While many studies of corrective saccades have presented single saccade targets in darkness, in more naturalistic environments locating the saccade target in the post-saccadic scene is likely to require memory. Hollingworth et al. (2008) presented subjects with a circular array of coloured stimuli and exogenously cued one item as the saccade target. During the saccade, the array was rotated so that gaze landed in-between the cued target stimulus and another un-cued distractor. Participants consistently made corrective saccades to the item that had been cued rather than the distractor, even though both were equidistant from fixation and distinguishable only by reference to pre-saccadic appearance (Fig. 4a). This indicates that features of the saccade target (in this case, its colour) were stored in memory before the saccade and used post-saccade to reidentify the target and generate a corrective eye movement (see also Richard et al. 2008).

Hollingworth and Luck (2009) confirmed the role of VWM in this experiment by adding a concurrent task in which participants were instructed to remember the colour of a central stimulus in addition to making a saccade to a cued item in the periphery. As before, the array was rotated during the saccade so that gaze landed between the cued target and a distractor. When the distractor object matched the colour of the item the participants had been instructed to hold in memory, corrective saccades were slower and more frequently directed towards the distractor instead of the target, compared to when the distractor was an irrelevant colour.

One important caveat to these experiments is that participants were given an explicit instruction to fixate the object indicated by the cue and could not move on to the next trial until that goal had been accomplished. This is likely to have motivated participants to remember the target colour in order to complete the task quickly. In one experiment, Hollingworth et al. (2008) instructed participants to look away from the target (to an outer ring) if they perceived a movement of the array. They found that in many cases observers made a corrective saccade to the target before fixating the ring, suggesting they were unable to suppress the corrective eye movement. While this indicates that memory-guided corrective saccades can occur involuntarily, it does not clarify whether the act of memorizing the saccade target's features was itself automatic. Other lines of evidence do however support the idea that a saccade target is automatically assigned priority in memory (Bays and Husain 2008; Hanning et al. 2015; Tas et al. 2016; Ohl and Rolfs 2017).

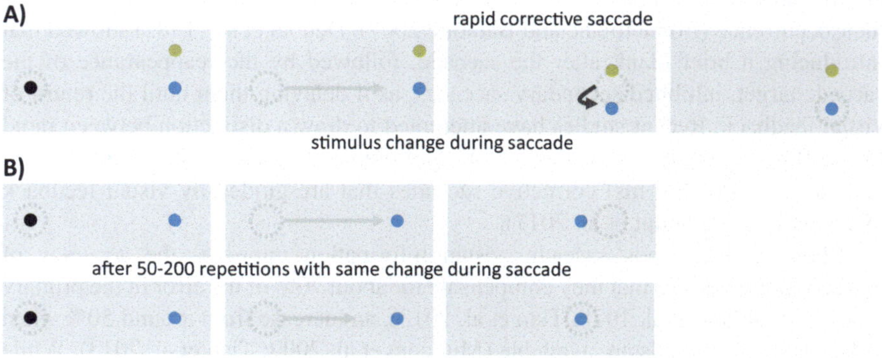

Fig. 4 Experimental methods for investigating correction and updating of saccades. (**a**) When a change is made to the display during an eye movement so the saccade lands at the midpoint between two stimuli, subsequent corrective saccades tend to be directed towards the stimulus that matches the surface features of the intended target. (**b**) When the saccade target is consistently displaced by the same distance over many trials, the saccade amplitude gradually adapts until gaze lands on the post-saccadic position of the target

5.2 Saccadic Adaptation

When saccade targets are systematically shifted during eye movements (Fig. 4b), so that the landing position consistently falls short of the target, then saccade amplitudes change progressively until gaze once again terminates close to the target on average (Hopp and Fuchs 2004; Pélisson et al. 2010; Prsa and Thier 2011). Subsequently eliminating the target shift will result in saccades overshooting its location, indicating that an adaptation of the saccade motor programme has taken place. Adaptation can be observed even when observers are unaware of the displacements of the target.

While it was initially proposed that corrective saccades provide the feedback required to adjust saccade amplitude (Albano and King 1989), the elegant demonstration of saccadic adaptation in the absence of corrective eye movements (Wallman and Fuchs 1998; Noto and Robinson 2001) demonstrated a role for visual feedback. Importantly, rather than the distance of the target from the fovea (Miller et al. 1981; Wallman and Fuchs 1998; Hopp and Fuchs 2004), saccadic adaptation is driven by a discrepancy between the predicted and actual location of the visual target in the post-saccadic image (Bahcall and Kowler 2000; Wong and Shelhamer 2011; Collins and Wallman 2012; Herman et al. 2013). For example, Collins and Wallman (2012) demonstrated that even when the trial-by-trial retinal error was identical across two experimental sessions, manipulating the predicted target position resulted in different magnitudes of saccadic adaptation.

These results suggest that memory for the pre-saccadic location of the saccade target is important for saccadic adaptation, but an influence of surface properties of the target initially proved difficult to establish. Bahcall and Kowler (2000) presented inward-shifting circular targets and outward-shifting square targets on alternating trials and found no evidence of adaptation. Similarly, after adaptation to a consistently displaced green cross, Deubel (1995) reported that the adaptation state persisted even when the target was changed to a red circle. In both cases, adaptation seems insensitive to visual features of the target. However, when the target is replaced during the saccade by two objects displaced in opposite directions, one visually matching the saccade target and the other distinct from it, saccadic adaptation is observed in the direction of the object that matches the pre-saccadic target (Madelain et al. 2010).

One explanation for the seeming discrepancy between studies is that saccadic adaptation is influenced by object continuity. When only a single object is present after the eye movement, there is no ambiguity as to the intended saccade target, and feature changes are ignored. In contrast, when a new object appears during the saccade, determining which object matches the intended target requires a comparison of surface features against memory of the pre-saccadic input.

When subjects are instructed to saccade to target objects that are presented overlaid on background images, consistent displacements of the background typically fail to induce saccadic adaptation (Ditterich et al. 2000; Robinson et al. 2000; Madelain et al. 2013). Only when observers are forced to rely on the background information – such as when the saccade target is blanked (Madelain et al. 2013) or when the target is itself defined as a specific location within a structured background (Deubel 1991; Ditterich et al. 2000) – do background displacements reliably induce adaptation. So, saccadic adaptation appears to be based on identification of the saccade target in the post-saccadic image where possible, but when this fails, comparisons to other pre-saccadic information may contribute (Herman et al. 2013). From the perspective of VWM, these findings are consistent with the evidence discussed above that the saccade target is compulsorily encoded into memory with high precision. When multiple objects are present, the saccade target will likely have the largest weight in assessing the accuracy of the eye movement because of its greater reliability. However, in its absence, other nearby objects or identifiable components of visual texture may be used.

5.3 Conclusion

The pre-saccadic content stored in VWM plays an important supporting role in correcting inaccurate eye movements, both through secondary saccades and adaptation of motor signals. While memory for the pre-saccadic location of the saccade target has the dominant influence, surface feature information and memory for nontarget objects both may be called upon when the location of the target in the post-saccadic scene is ambiguous. Future research could explore the dissociation between perception of transsaccadic change and saccadic adaptation, perhaps from the perspective of optimal decision where the cost of errors differs between the two mechanisms.

6 Discussion

By enabling the retention of visual information across the disruption of a saccade, VWM provides a means through which we can link pre- and post-saccadic information. This supports vital perceptual and oculomotor functions including detecting changes that occur during a saccade, permitting visual processing of stimuli to bridge a saccade, and correcting and compensating for errors in saccade endpoint. We note a number of consistent themes across the studies reviewed above that have investigated these different functions of transsaccadic memory.

The saccade target appears to have a unique status with respect to each of the functions we have considered. However, in every case we have also seen evidence that other elements of the visual scene are stored and compared across a saccade and can take on a dominant role when post-saccadic information about the saccade target is unavailable or compromised. While some theories have proposed that the saccade target is processed in a qualitatively different way from other items, we suggest that enhanced visual processing (Deubel and Schneider 1996) and prioritized encoding of the saccade target into VWM (Bays and Husain 2008) before the eye movement may be sufficient to account for its dominant influence on behaviour. According to this view, it is simply that the most precise and reliable evidence to survive the saccade is related to the target of the eye movement.

6.1 Optimality

One recurring theme is a shift away from heuristic models of processing (Gigerenzer and Brighton 2009) and towards models based on optimal inference (Knill and Pouget 2004). At one extreme, the process by which our perceptual system utilizes pre-saccadic representations has been depicted as a series of decisions unfolding in time after the saccade. In these models, the use of different sources of information hinges on the outcomes of preceding decisions, such that only subsets of the information present feed in to any decision. An example from the SSD literature would be the proposal that precise localisation information is always available after an eye movement but only consulted in certain circumstances, for example, if the saccade target cannot immediately be located. Another example is the (refuted) hypothesis that saccadic adaptation is driven solely by displacement of the target from the fovea. Both procedures would necessarily be suboptimal because some available, relevant information is not integrated into the decision-making process.

The contrasting hypothesis would be that all sources of information (from vision, memory, proprioception, and efference copy) are integrated in a statistically optimal manner (weighting each according to its reliability) to generate a single coherent and consistent set of estimates about the likely states of the observer and the world. These estimated states would determine our perception of changes in the environment, integrate pre- and post-saccadic features of object judged not to have changed, and drive corrective saccades and recalibration of oculomotor commands. According to Bayesian principles, these inferences would be based not only on the immediately available inputs but also on a lifetime of experience in how objects behave in everyday environments, expressed in the form of statistical priors.

While optimal inference models have been shown to capture aspects of transsaccadic processing better than heuristic models – for example, parsimoniously describing effects of saccade magnitude and direction on saccadic suppression of displacement (Niemeier et al. 2003) – the picture presented above is unlikely to be fully accurate. First, due to the complexity of the computations, in most cases exact Bayesian inference cannot feasibly be achieved by neural systems, meaning that an approximation of the optimal process is the most we can expect to find (Siegelmann and Holzman 2010; Penny 2012; Rich et al. 2015; Zhang et al. 2016). Second, there are experimental indications that different processes weigh information differently. For example, the information driving saccadic adaptation appears at least partially dissociable from the information contributing to awareness of the displacement of the saccade target (Souto et al. 2016). Also, limited feature integration appears to occur even between objects that observers judge changed or distinct (Wittenberg et al. 2008; Demeyer et al. 2010b). One possibility is that these dissociations arise from optimizing decisions under different cost functions, for example, our perceptual systems may be less averse to erroneously integrating two different objects than they are to failing to detect a change.

One area in need of more investigation is the role that previous experiences – prior information – have on how different sources of information are evaluated and used across a saccade. It is important to recognize that these priors may reflect not only our experience of the natural world but also the specific context described by a psychophysical experiment. During one trial of a typical study, participants briefly view a sparse, static visual image on a computer display before making a single saccade to an instructed location. While their judgements will undoubtedly be informed by experience of objects in the everyday world, it also seems reasonable to assume that participants are learning about the behaviour of these particular blobs of light as the experiment progresses. The extent to which mechanisms such as transsaccadic integration are malleable, and on what timescales, must be established for us to relate conclusions drawn from laboratory observations to natural vision.

6.2 Object Correspondence

Another clear theme to emerge from our review is the important role of object correspondence in determining how pre-saccadic visual information is used after the saccade. For each of the VWM functions considered, we have found evidence that changes to features of an object that are secondary to the task nonetheless modulate the influence of pre-saccadic memory. For example, the displacement of a target orthogonal to the saccade can relieve saccadic suppression of its shift in the direction of the saccade (Wexler and Collins 2014). Inferring which objects in one image correspond to objects in another is a challenging computational problem that remains unsolved in computer vision and machine learning. An object could move location, but otherwise remain the same object, and thus although seen at an unexpected location should still inherit remembered information about its content. Similarly, one feature of an object may unexpectedly change – such as a multicoloured LED lamp changing colour – without affecting other features such as shape and location. However, at other times these changes may correspond with the disappearance of one object and the appearance of an entirely new object, where inheritance of the previous properties would be undesirable. Beyond being simply present or absent, single objects can also be perceived to split into smaller objects, or objects can merge to form a single object.

The need to weigh up evidence for every possible such combination of changes to objects in a scene is part of the reason that full Bayesian inference is infeasible. Understanding what compromises the visual system makes to approximate optimal inference is an important avenue for future research.

6.3 *Memory Limitations*

As discussed in the Introduction, our ability to detect changes across an eye movement, or explicitly recall what we saw before our eyes moved, is typically found to reflect the limited capacity of VWM. Recent developments in our understanding of VWM have led to the idea that its limits are best expressed in terms of a limited resource, flexibly allocated to objects to determine the resolution of their storage (Ma et al. 2014). However, the implications of this new view of VWM for transsaccadic processing have received only preliminary investigation. For example, we know that the saccade target receives a disproportionate share of resources, but what factors influence the size of this share, and how are the remaining resources distributed between other elements of complex visual scenes?

A final consideration is that, while there is strong evidence that the shorter-lived forms of visual persistence contribute little or nothing to our explicit awareness of transsaccadic changes in real scenes, the possibility that they play a part in implicit processes, including transsaccadic integration and saccadic adaptation, has not been definitively ruled out. The critical test will be whether these processes are subject to the same strict resource limits as explicit change detection, indicating a reliance on VWM alone. These open questions, like many of those uncovered in this review, will only be answered by moving beyond single target displays to study more complex and naturalistic scenes.

Acknowledgement We thank Karl Gegenfurtner, Heiner Deubel, Martin Rolfs, Eckart Zimmerman, and Sebastian Schneegans for their helpful comments on a draft version of this manuscript. This work was supported by the Wellcome Trust (grant number 106926).

References

Albano JE, King WM (1989) Rapid adaptation of saccadic amplitude in humans and monkeys. Invest Ophthalmol Vis Sci 30:1883–1893

Atsma J, Maij F, Koppen M et al (2016) Causal inference for spatial constancy across saccades. PLoS Comput Biol 12:e1004766

Baddeley AD, Hitch G (1974) Working memory. Psychol Learn Motiv 8:47–89

Bahcall DO, Kowler E (2000) The control of saccadic adaptation: implications for the scanning of natural visual scenes. Vis Res 40:2779–2796

Barnes GR, Gresty MA (1973) Characteristics of eye movements to targets of short duration. Aerosp Med 44:1236–1240

Bays PM, Husain M (2008) Dynamic shifts of limited working memory resources in human vision. Science 321:851–854

Bays PM, Catalao RFG, Husain M (2009) The precision of visual working memory is set by allocation of a shared resource. J Vis 9(7):1–711

Bays PM, Wu EY, Husain M (2011) Storage and binding of object features in visual working memory. Neuropsychologia 49:1622–1631

Becker W (1976) Do correction saccades depend exclusively on retinal feedback? A note on the possible role of non-retinal feedback. Vis Res 16:425–427

Becker W, Fuchs AF (1969) Further properties of the human saccadic system: eye movements and correction saccades with and without visual fixation points. Vis Res 9:1247–1258
Boehnke SE, Munoz DP (2008) On the importance of the transient visual response in the superior colliculus. Curr Opin Neurobiol 18:544–551
Bonnetblanc F, Baraduc P (2007) Saccadic adaptation without retinal postsaccadic error. Neuroreport 18:1399–1402
Bremmer F, Kubischik M, Hoffmann K-P, Krekelberg B (2009) Neural dynamics of saccadic suppression. J Neurosci 29:12374–12383
Bridgeman B (2007) Efference copy and its limitations. Comput Biol Med 37:924–929
Bridgeman B, Mayer M (1983) Failure to integrate visual information from successive fixations. Bull Psychon Soc 21:285–286
Bridgeman B, Hendry D, Stark L (1975) Failure to detect displacement of the visual world during saccadic eye movements. Vis Res 15:719–722
Bridgeman B, Van Der Heijden CAH, Velichkovsky BM (1994) A theory of visual stability across saccadic eye movements. Behav Brain Sci 17:247–292
Burr DC, Morrone MC (2011) Spatiotopic coding and remapping in humans. Philos Trans R Soc Lond Ser B Biol Sci 366:504–515
Burr DC, Morrone MC, Ross J (1994) Selective suppression of the magnocellular visual pathway during saccadic eye movements. Nature 371:511–513
Castet E, Jeanjean S, Masson GS (2002) Motion perception of saccade-induced retinal translation. Proc Natl Acad Sci 99:15159–15163
Cavanagh P, Hunt AR, Afraz A, Rolfs M (2010) Visual stability based on remapping of attention pointers. Trends Cogn Sci 14:147–153
Collins T, Wallman J (2012) The relative importance of retinal error and prediction in saccadic adaptation. J Neurophysiol 107:3342–3348
Cowan N (2001) The magical number 4 in short-term memory: a reconsideration of mental storage capacity. Behav Brain Sci 24:87–114 Discussion 114–185
Cox DD, Meier P, Oertelt N, DiCarlo JJ (2005) "Breaking" position-invariant object recognition. Nat Neurosci 8:1145–1147
Currie CB, McConkie GW, Carlson-Radvansky LA, Irwin DE (2000) The role of the saccade target object in the perception of a visually stable world. Percept Psychophys 62:673–683
Demeyer M, de Graef P, Wagemans J, Verfaillie K (2010a) Parametric integration of visual form across saccades. Vis Res 50:1225–1234
Demeyer M, Graef PD, Wagemans J, Verfaillie K (2010b) Object form discontinuity facilitates displacement discrimination across saccades. J Vis 10:17–17
Deubel H (1991) Adaptive control of saccade metrics. In: Presbyopia research. Springer, Boston, pp 93–100
Deubel H (1995) Is saccadic adaptation context-specific? In: Findlay JM, Walker R, Kentridge RW (eds) Studies in visual information processing. North-Holland, Amsterdam, pp 177–187
Deubel H (2004) Localization of targets across saccades: role of landmark objects. Vis Cogn 11:173–202
Deubel H, Schneider WX (1996) Saccade target selection and object recognition: evidence for a common attentional mechanism. Vis Res 36:1827–1837
Deubel H, Wolf W, Hauske G (1982) Corrective saccades: effect of shifting the saccade goal. Vis Res 22:353–364
Deubel H, Schneider WX, Bridgeman B (1996) Postsaccadic target blanking prevents saccadic suppression of image displacement. Vis Res 36:985–996
Deubel H, Bridgeman B, Schneider WX (1998) Immediate post-saccadic information mediates space constancy. Vis Res 38:3147–3159
Deubel H, Schneider WX, Bridgeman B (2002) Transsaccadic memory of position and form. Prog Brain Res 140:165–180
Diamond MR, Ross J, Morrone MC (2000) Extraretinal control of saccadic suppression. J Neurosci 20:3449–3455

Ditterich J, Eggert T, Straube A (2000) The role of the attention focus in the visual information processing underlying saccadic adaptation. Vis Res 40:1125–1134

Dowiasch S, Marx S, Einhäuser W, Bremmer F (2015) Effects of aging on eye movements in the real world. Front Hum Neurosci 9:46

Fracasso A, Caramazza A, Melcher D (2010) Continuous perception of motion and shape across saccadic eye movements. J Vis 10:14

Ganmor E, Landy MS, Simoncelli EP (2015) Near-optimal integration of orientation information across saccades. J Vis 15:8

Gegenfurtner KR, Sperling G (1993) Information transfer in iconic memory experiments. J Exp Psychol Hum Percept Perform 19:845–866

Germeys F, de Graef P, Verfaillie K (2002) Transsaccadic perception of saccade target and flanker objects. J Exp Psychol Hum Percept Perform 28:868–883

Germeys F, Graef PD, Eccelpoel CV, Verfaillie K (2010) The visual analog: evidence for a preattentive representation across saccades. J Vis 10:9

Gigerenzer G, Brighton H (2009) Homo heuristicus: why biased minds make better inferences. Top Cogn Sci 1:107–143

Gordon RD, Irwin DE (1998) Eye movements, attention and trans-saccadic memory. Vis Cogn 5:127–155

Gorgoraptis N, Catalao RFG, Bays PM, Husain M (2011) Dynamic updating of working memory resources for visual objects. J Neurosci 31:8502–8511

Gysen V, de Graef P, Verfaillie K (2002a) Detection of intrasaccadic displacements and depth rotations of moving objects. Vis Res 42:379–391

Gysen V, Verfaillie K, de Graef P (2002b) Transsaccadic perception of translating objects: effects of landmark objects and visual field position. Vis Res 42:1785–1796

Hanning NM, Jonikaitis D, Deubel H, Szinte M (2015) Oculomotor selection underlies feature retention in visual working memory. J Neurophysiol 115:1071–1076

Harrison WJ, Bex PJ (2014) Integrating retinotopic features in spatiotopic coordinates. J Neurosci 34:7351–7360

Hayhoe M, Lachter J, Feldman J (1991) Integration of form across saccadic eye movements. Perception 20:393–402

Henderson JM (1992) Identifying objects across saccades: effects of extrafoveal preview and flanker object context. J Exp Psychol Learn Mem Cogn 18:521–530

Henderson JM (1994) Two representational systems in dynamic visual identification. J Exp Psychol Gen 123:410–426

Henderson JM, Anes MD (1994) Roles of object-file review and type priming in visual identification within and across eye fixations. J Exp Psychol Hum Percept Perform 20:826–839

Henderson JM, Hollingworth A (1999) The role of fixation position in detecting scene changes across saccades. Psychol Sci 10:438–443

Henderson JM, Hollingworth A (2003) Eye movements and visual memory: detecting changes to saccade targets in scenes. Percept Psychophys 65:58–71

Henderson JM, Siefert ABC (1999) The influence of enantiomorphic transformation on transsaccadic object integration. J Exp Psychol Hum Percept Perform 25:243–255

Henderson JM, Siefert ABC (2001) Types and tokens in transsaccadic object identification: effects of spatial position and left-right orientation. Psychon Bull Rev 8:753–760

Henderson JM, Pollatsek A, Rayner K (1987) Effects of foveal priming and extrafoveal preview on object identification. J Exp Psychol Hum Percept Perform 13:449–463

Herman JP, Blangero A, Madelain L et al (2013) Saccade adaptation as a model of flexible and general motor learning. Exp Eye Res 114:6–15

Higgins E, Rayner K (2015) Transsaccadic processing: stability, integration, and the potential role of remapping. Atten Percept Psychophys 77:3–27

Hollingworth A, Luck SJ (2009) The role of visual working memory in the control of gaze during visual search. Atten Percept Psychophys 71:936–949

Hollingworth A, Richard AM, Luck SJ (2008) Understanding the function of visual short-term memory: transsaccadic memory, object correspondence, and gaze correction. J Exp Psychol Gen 137:163–181
Hollingworth A, Matsukura M, Luck SJ (2013) Visual working memory modulates low-level saccade target selection: evidence from rapidly generated saccades in the global effect paradigm. J Vis 13:4
Hopp JJ, Fuchs AF (2004) The characteristics and neuronal substrate of saccadic eye movement plasticity. Prog Neurobiol 72:27–53
Hübner C, Schütz AC (2017) Numerosity estimation benefits from transsaccadic information integration. J Vis 17:12
Irwin DE (1992) Memory for position and identity across eye movements. J Exp Psychol Learn Mem Cogn 18:307–317
Irwin DE (1996) Integrating information across saccadic eye movements. Curr Dir Psychol Sci 5:94–100
Irwin DE, Andrews RV (1996) Integration and accumulation of information across saccadic eye movements. Atten Perform 16:122–155
Irwin DE, Zelinsky GJ (2002) Eye movements and scene perception: memory for things observed. Percept Psychophys 64:882–895
Irwin DE, Yantis S, Jonides J (1983) Evidence against visual integration across saccadic eye movements. Percept Psychophys 34:49–57
Irwin DE, Zacks JL, Brown JS (1990) Visual memory and the perception of a stable visual environment. Percept Psychophys 47:35–46
Jeyachandra J, Nam Y, Kim Y et al (2018) Transsaccadic memory of multiple spatially variant and invariant object features. J Vis 18:6
Jonides J, Irwin DE, Yantis S (1982) Integrating visual information from successive fixations. Science 215:192–194
Jonides J, Irwin DE, Yantis S (1983) Failure to integrate information from successive fixations. Science 222:188
Kahneman D, Treisman A, Gibbs BJ (1992) The reviewing of object files: object-specific integration of information. Cogn Psychol 24:175–219
Kersten D, Mamassian P, Yuille A (2004) Object perception as Bayesian inference. Annu Rev. Psychol 55:271–304
Knill DC, Pouget A (2004) The Bayesian brain: the role of uncertainty in neural coding and computation. Trends Neurosci 27:712–719
Li N, DiCarlo JJ (2008) Unsupervised natural experience rapidly alters invariant object representation in visual cortex. Science 321:1502–1507
Li W, Matin L (1990) The influence of saccade length on the saccadic suppression of displacement detection. Percept Psychophys 48:453–458
Luck SJ, Vogel EK (1997) The capacity of visual working memory for features and conjunctions. Nature 390:279–281
Ma WJ, Husain M, Bays PM (2014) Changing concepts of working memory. Nat Neurosci 17:347–356
Mackay DM (1972) Visual stability. Invest Ophthalmol Vis Sci 11:518–524
Madelain L, Harwood MR, Herman JP, Wallman J (2010) Saccade adaptation is unhampered by distractors. J Vis 10:29–29
Madelain L, Herman JP, Harwood MR (2013) Saccade adaptation goes for the goal. J Vis 13:9
McConkie GW, Currie CB (1996) Visual stability across saccades while viewing complex pictures. J Exp Psychol Hum Percept Perform 22:563–581
McConkie GW, Rayner K (1976) Asymmetry of the perceptual span in reading. Bull Psychon Soc 8:365–368
Melcher D, Colby CL (2008) Trans-saccadic perception. Trends Cogn Sci 12:466–473
Melcher D, Kowler E (2001) Visual scene memory and the guidance of saccadic eye movements. Vis Res 41:3597–3611

Melcher D, Morrone MC (2015) Nonretinotopic visual processing in the brain. Vis Neurosci 32: E017

Melcher D, Piazza M (2011) The role of attentional priority and saliency in determining capacity limits in enumeration and visual working memory. PLoS One 6:e29296

Miller GA (1956) The magical number seven, plus or minus two: some limits on our capacity for processing information. Psychol Rev 63:81–97

Miller JM, Anstis T, Templeton WB (1981) Saccadic plasticity: parametric adaptive control by retinal feedback. J Exp Psychol Hum Percept Perform 7:356–366

Munoz DP, Broughton JR, Goldring JE, Armstrong IT (1998) Age-related performance of human subjects on saccadic eye movement tasks. Exp Brain Res 121:391–400

Munuera J, Morel P, Duhamel J-R, Deneve S (2009) Optimal sensorimotor control in eye movement sequences. J Neurosci 29:3026–3035

Niemeier M, Crawford JD, Tweed DB (2003) Optimal transsaccadic integration explains distorted spatial perception. Nature 422:76–80

Niemeier M, Crawford JD, Tweed DB (2007) Optimal inference explains dimension-specific contractions of spatial perception. Exp Brain Res 179:313–323

Noto CT, Robinson FR (2001) Visual error is the stimulus for saccade gain adaptation. Cogn Brain Res 12:301–305

O'Regan JK, Lévy-Schoen A (1983) Integrating visual information from successive fixations: does trans-saccadic fusion exist? Vis Res 23:765–768

O'Regan JK, Noë A (2001) A sensorimotor account of vision and visual consciousness. Behav Brain Sci 24:939–973 Discussion 973–1031

Ohl S, Rolfs M (2017) Saccadic eye movements impose a natural bottleneck on visual short-term memory. J Exp Psychol Learn Mem Cogn 43:736–748

Ohl S, Rolfs M (2018) Saccadic selection of stabilized items in visuospatial working memory. Conscious Cogn 64:32–44

Ohl S, Brandt SA, Kliegl R (2013) The generation of secondary saccades without postsaccadic visual feedback. J Vis 13:11

Oostwoud Wijdenes L, Marshall L, Bays PM (2015) Evidence for optimal integration of visual feature representations across saccades. J Neurosci 35:10146–10153

Ostendorf F, Dolan RJ (2015) Integration of retinal and extraretinal information across eye movements. PLoS One 10:e0116810

Paeye C, Collins T, Cavanagh P, Herwig A (2018) Calibration of peripheral perception of shape with and without saccadic eye movements. Atten Percept Psychophys 80:723–737. https://doi.org/10.3758/s13414-017-1478-3

Pashler H (1988) Familiarity and visual change detection. Percept Psychophys 44:369–378

Pélisson D, Alahyane N, Panouillères M, Tilikete C (2010) Sensorimotor adaptation of saccadic eye movements. Neurosci Biobehav Rev 34:1103–1120

Pelli DG, Tillman KA (2008) The uncrowded window of object recognition. Nat Neurosci 11:1129–1135

Penny W (2012) Bayesian models of brain and behaviour. ISRN Biomath 785791:1–19. https://doi.org/10.5402/2012/785791

Peterson MS, Kramer AF, Irwin DE (2004) Covert shifts of attention precede involuntary eye movements. Percept Psychophys 66:398–405

Pollatsek A, Rayner K, Collins WE (1984) Integrating pictorial information across eye movements. J Exp Psychol Gen 113:426–442

Pollatsek A, Rayner K, Henderson JM (1990) Role of spatial location in integration of pictorial information across saccades. J Exp Psychol Hum Percept Perform 16:199–210

Poth CH, Schneider WX (2016) Breaking object correspondence across saccades impairs object recognition: the role of color and luminance. J Vis 16:1

Poth CH, Herwig A, Schneider WX (2015) Breaking object correspondence across saccadic eye movements deteriorates object recognition. Front Syst Neurosci 9:176

Prablanc C, Massé D, Echallier JF (1978) Error-correcting mechanisms in large saccades. Vis Res 18:557–560
Prime SL, Niemeier M, Crawford JD (2006) Transsaccadic integration of visual features in a line intersection task. Exp Brain Res 169:532–548
Prime SL, Tsotsos L, Keith GP, Crawford JD (2007) Visual memory capacity in transsaccadic integration. Exp Brain Res 180:609–628
Prime SL, Vesia M, Crawford JD (2011) Cortical mechanisms for trans-saccadic memory and integration of multiple object features. Philos Trans R Soc B 366:540–553
Prsa M, Thier P (2011) The role of the cerebellum in saccadic adaptation as a window into neural mechanisms of motor learning. Eur J Neurosci 33:2114–2128
Rayner K, Pollatsek A (1983) Is visual information integrated across saccades? Percept Psychophys 34:39–48
Rich D, Cazettes F, Wang Y et al (2015) Neural representation of probabilities for Bayesian inference. J Comput Neurosci 38:315–323
Richard AM, Luck SJ, Hollingworth A (2008) Establishing object correspondence across eye movements: flexible use of spatiotemporal and surface feature information. Cognition 109:66–88
Robinson F, Noto C, Watanabe S (2000) Effect of visual background on saccade adaptation in monkeys. Vis Res 40:2359–2367
Rolfs M, Carrasco M (2012) Rapid simultaneous enhancement of visual sensitivity and perceived contrast during saccade preparation. J Neurosci 32:13744–13752a
Ross J, Morrone MC, Goldberg ME, Burr DC (2001) Changes in visual perception at the time of saccades. Trends Neurosci 24:113–121
Schneider WX (2013) Selective visual processing across competition episodes: a theory of task-driven visual attention and working memory. Philos Trans R Soc Lond Ser B Biol Sci 368:20130060
Schut MJ, Fabius JH, der Stoep NV, der Stigchel SV (2017) Object files across eye movements: previous fixations affect the latencies of corrective saccades. Atten Percept Psychophys 79:138–153
Shao N, Li J, Shui R et al (2010) Saccades elicit obligatory allocation of visual working memory. Mem Cogn 38:629–640
Shebilske WL (1976) Extraretinal information in corrective saccades and inflow vs outflow theories of visual direction constancy. Vis Res 16:621–628
Siegelmann HT, Holzman LE (2010) Neuronal integration of dynamic sources: Bayesian learning and Bayesian inference. Chaos 20:037112
Souto D, Gegenfurtner KR, Schütz AC (2016) Saccade adaptation and visual uncertainty. Front Hum Neurosci 10:227
Sperling G (1960) The information available in brief visual presentations. Psychol Monogr Gen Appl 74:1–29
Strasburger H, Rentschler I, Jüttner M (2011) Peripheral vision and pattern recognition: a review. J Vis 11:13
Szinte M, Cavanagh P (2011) Spatiotopic apparent motion reveals local variations in space constancy. J Vis 11:4
Tas AC, Moore CM, Hollingworth A (2012) An object-mediated updating account of insensitivity to transsaccadic change. J Vis 12:18
Tas AC, Luck SJ, Hollingworth A (2016) The relationship between visual attention and visual working memory encoding: a dissociation between covert and overt orienting. J Exp Psychol Hum Percept Perform 42:1121–1138
Tatler BW, Land MF (2011) Vision and the representation of the surroundings in spatial memory. Philos Trans R Soc Lond Ser B Biol Sci 366:596–610
Tatler BW, Gilchrist ID, Rusted J (2003) The time course of abstract visual representation. Perception 32:579–592

Tian J, Ying HS, Zee DS (2013) Revisiting corrective saccades: role of visual feedback. Vis Res 89:54–64

van den Berg R, Shin H, Chou W-C et al (2012) Variability in encoding precision accounts for visual short-term memory limitations. Proc Natl Acad Sci 109:8780–8785

van Opstal AJ, van Gisbergen JAM (1989) Scatter in the metrics of saccades and properties of the collicular motor map. Vis Res 29:1183–1196

Wallman J, Fuchs AF (1998) Saccadic gain modification: visual error drives motor adaptation. J Neurophysiol 80:2405–2416

Warabi T, Kase M, Kato T (1984) Effect of aging on the accuracy of visually guided saccadic eye movement. Ann Neurol 16:449–454

Weber RB, Daroff RB (1972) Corrective movements following refixation saccades: type and control system analysis. Vis Res 12:467–475

Weiß K, Schneider WX, Herwig A (2015) A "blanking effect" for surface features: transsaccadic spatial-frequency discrimination is improved by postsaccadic blanking. Atten Percept Psychophys 77:1500–1506

Westheimer G (1954) Eye movement responses to a horizontally moving visual stimulus. AMA Arch Ophthalmol 52:932–941

Wexler M, Collins T (2014) Orthogonal steps relieve saccadic suppression. J Vis 14:13

Wheeler ME, Treisman AM (2002) Binding in short-term visual memory. J Exp Psychol Gen 131:48–64

Wittenberg M, Bremmer F, Wachtler T (2008) Perceptual evidence for saccadic updating of color stimuli. J Vis 8:9

Wolf C, Schütz AC (2015) Trans-saccadic integration of peripheral and foveal feature information is close to optimal. J Vis 15:1

Wolf W, Hauske G, Lupp U (1980) Interaction of pre- and postsaccadic patterns having the same coordinates in space. Vis Res 20:117–125

Wong AL, Shelhamer M (2011) Saccade adaptation improves in response to a gradually introduced stimulus perturbation. Neurosci Lett 500:207–211

Wurtz RH (2008) Neuronal mechanisms of visual stability. Vis Res 48:2070–2089

Zhang W, Luck SJ (2008) Discrete fixed-resolution representations in visual working memory. Nature 453:233

Zhang W, Chen A, Rasch MJ, Wu S (2016) Decentralized multisensory information integration in neural systems. J Neurosci 36:532–547

What Is Memory-Guided Attention? How Past Experiences Shape Selective Visuospatial Attention in the Present

Danlei Chen and J. Benjamin Hutchinson

Contents

Abstract What controls our attention? It is historically thought that there are two primary factors that determine selective attention: the perceptual salience of the stimuli and the goals based on the task at hand. However, this distinction doesn't neatly capture the varied ways our past experience can influence our ongoing mental processing. In this chapter, we aim to describe how past experience can be systematically characterized by different types of memory, and we outline experimental evidence suggesting how attention can then be guided by each of these different memory types. We highlight findings from human behavioral, neuroimaging, and neuropsychological work from the perspective of two related frameworks of human memory: the multiple memory systems (MMS) framework and the neural processing (NP) framework. The MMS framework underscores how memory can be separated based on consciousness (declarative and non-declarative memory), while the NP framework emphasizes different forms of memory as reflective of different brain processing modes (rapid encoding of flexible associations, slow encoding of rigid associations, and rapid encoding of single or unitized items). We describe how

D. Chen (✉)
Northeastern University, Boston, MA, USA
e-mail: d.chen@northeastern.edu

J. B. Hutchinson
University of Oregon, Eugene, OR, USA
e-mail: bhutch@uoregon.edu

Curr Topics Behav Neurosci (2019) 41: 185–212
DOI 10.1007/7854_2018_76
Published Online: 25 December 2018

memory defined by these frameworks can guide our attention, even when they do not directly relate to perceptual salience or the goals concerning the current task. We close by briefly discussing theoretical implications as well as some interesting avenues for future research.

Keywords Basal ganglia · Covert attention · Hippocampus · Long-term memory · Overt attention

1 Introduction

A core principle of selective visuospatial attention is that, at any given moment, our sensory receptors receive vastly more information than our brain can ultimately process. Past work has consistently suggested that which information ultimately gets attended to can be determined by factors both external to the observer, such as the perceptual salience of the information (exogenous or stimulus-driven attention), and those internal to the observer, such as the relevance of the information to an ongoing task (endogenous or goal-directed attention) (Carrasco 2011; Chun et al. 2011; Corbetta and Shulman 2002; Knudsen 2007). However, recent critiques of this dichotomy have highlighted instances where attention can be influenced by factors that are not quite captured by either the perceptual salience of the stimuli or the ongoing goals of the observer (Awh et al. 2012; Theeuwes 2018; Sisk et al. 2018). Here, we focus on one class of such influences in particular: memory (for related reviews, see Chun and Turk-Browne 2007; Dudukovic and Wagner 2006; Hutchinson and Turk-Browne 2012; Logan 2002; Võ and Wolfe 2015; Woodman and Chun 2006).

As will be reviewed below, there are many ways in which our past experience can shape our ongoing information processing. Indeed, even in our daily lives, there are countless examples of how our memory can guide attention and subsequently modify our behavior. For example, when you are at a noisy party where it is hard to hear what your friend next to you is talking about, your attention might be drawn to a background conversation that mentions your favorite childhood TV show. On the other hand, the repeated occurrence of the loud sound from the leaf blower made by your neighbor at the same time every morning will attract your attention less and less as you get used to it.

Generally, memory is critical for behavior as it can orient our attention toward information which has acquired relevance through experience (e.g., the so-called cocktail party effect (Moray 1959)). Just as importantly, memory can help us notice new things in our environment, allowing us to selectively process novel information in order to learn about the world (Fantz 1964; Tulving et al. 1996; Ranganath and Rainer 2003). Theoretically, the role of memory is important to consider in order to fully understand attention as it does not fit cleanly into the existing dichotomy of stimulus-driven and goal-directed attention. That is, memory is not fully stimulus-driven, since stimuli might exert attentional influence only as a result of prior experience, nor is it necessarily goal-directed, since we are able to encode (put into memory) and retrieve information about the world even if it is not directly relevant to the task at hand (e.g., Fan and Turk-Browne 2016; Hall et al. 2008; Kuhl et al. 2013; Neill et al. 1990; Turk-Browne et al. 2010).

Despite the potentially profound importance of memory for selective attention, there have been relatively few attempts to catalog how and when memory might influence attention. Here, we aim to provide context to the work described elsewhere in this book by giving an overview of how forms of memory beyond working memory might relate to visuospatial attention. Specifically, we will briefly summarize a representative cross-section of empirical findings concerning how memory shapes our attention. We consider these findings from the perspective of two related, non-mutually exclusive frameworks of human memory: a traditional multiple memory systems (MMS) framework and a more recently proposed neural processing (NP) framework. First, we introduce and present evidence for how memory defined in terms of the traditional MMS framework offers insight into how visuospatial attention can be modulated by past experience, highlighting the impact from both declarative and non-declarative forms of memory. Second, we discuss how recent neuroscientific definitions of memory might inform how the neural operations underlying visuospatial attention could be intertwined with those of memory. Finally, we conclude by highlighting theoretical implications and open questions awaiting future research.

2 Multiple Memory Systems and Their Role in Attention

2.1 Introduction to the Multiple Memory Systems (MMS) Framework

Over the past decades, memory research has deeply been shaped by the idea that memory is not a unitary or homogeneous entity but rather involves a confluence of multiple brain systems that mediate different yet interacting types of memory (Gabrieli 1998; Poldrack and Packard 2003; Squire et al. 1993). Scientists have sought to delineate these systems from one another and experimentally isolate their defining features using a wide range of behavioral, neuropsychological, neuroimaging, and computational modeling techniques.

The MMS framework took shape in the 1950s with the case study of patient H. M., who underwent a bilateral medial temporal lobectomy in an effort to cure his epilepsy. After his surgery, he was unable to form new memories about the events in his life (episodic memory) and also had trouble retrieving some memories of events which occurred before the surgery happened (Scoville and Milner 1957). Strikingly, other aspects of his memory, general intelligence, and other cognitive functions remained intact (Scoville and Milner 1957; Squire and Wixted 2011). For example, patient H.M. was able to perform a hand-eye coordination task that required motor

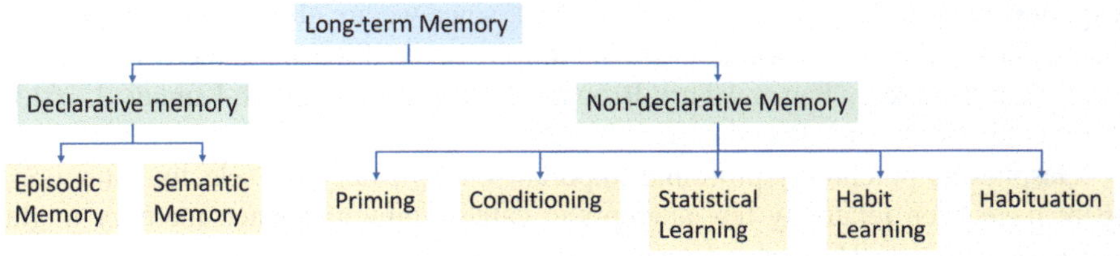

Fig. 1 Taxonomy of memory based on multiple memory systems. In the MMS framework, long-term memory generally can be divided into declarative and non-declarative types. Declarative memory consists of episodic memory and semantic memory, whereas non-declarative memory includes phenomena such as priming, conditioning, statistical learning, habit learning, and habituation

skill learning despite having no memory of actually performing the task (Milner 1962). These and related findings (Cohen and Squire 1980) strongly suggested that memory could not be considered a unitary entity and was instead composed of multiple memory systems. The work with patient H.M. laid the groundwork for MMS, and, based on these findings as well as many other neuropsychological studies, it was proposed that long-term memory could be broadly separated into at least two systems: declarative (sometimes referred to as explicit) and non-declarative (sometimes referred to as implicit) memory (Squire 1992) (Fig. 1).

Declarative Memory Declarative memory refers to knowledge based on past experience that is consciously available and can be further divided into two forms of long-term memory: semantic and episodic. Semantic memory refers to one's general factual knowledge about the world. For example, knowing who was the first president of the United States, how many degrees Fahrenheit is one degree Celsius, and what color combined with blue creates purple are all examples of semantic knowledge. Episodic memory, on the other hand, refers to one's specific memory about an event that most likely occurred only once. For example, remembering what you said to your friend this morning, which restaurant you went with your parents last week, and what type of cake you had on your last birthday are all likely to be episodic memories.

Non-declarative Memory In contrast, non-declarative memory refers to the knowledge that is not consciously accessible and yet can often automatically influence one's behavior. Non-declarative memory[1] includes, but is not limited to, priming, conditioning, statistical learning, habit learning, and habituation. Priming refers to how exposure to a stimulus can later facilitate a response to the same, or related, stimulus. For example, in a case of semantic priming, a flash of the word

[1]Note that we do not exhaustively cover the subcategories of non-declarative memories but rather focus on those that are most relevant to attentional guidance. Some studies further distinguish associative and nonassociative forms of memory within non-declarative memory. Here we do not stress this distinction as to provide more specific distinctions in terms of the relationship to attention literature.

"nurse" can help a person recognize the word "doctor" faster, when compared with the word "manager." Conditioning is a type of associative learning where responses can be elicited from the systematic pairing of stimuli with rewards or punishments. The most famous example, articulated by Ivan Pavlov, is that after pairing the presence of food with the sound of a metronome, his dog eventually began to salivate in response to the metronome without the food being present (Pavlov 1927). Statistical learning can broadly refer to the ability to learn the regularities present in the environment over time. For example, when we are young, we learn to group phonemes with higher statistical co-occurrence to effectively parse discrete words from what is in fact continuous auditory input (speech; Saffran et al. 1996). Habit learning can be acquired through repeated responding to certain stimuli so as to form context-response associations in memory. If, after you dropped your phone in the water several times, you spontaneously check your phone every time you hear the sound of running water, you have learned a habit. Lastly, habituation is an adaptive, nonassociative learning process, where a response to something decreases after repetition and is not driven by sensory adaptation or motor fatigue. For example, when there is the constant loud banging noise of construction right outside your office, you might initially find it distracting, but as you get accustomed to the noise over time, your attention is less likely to be captured by it.

Memory research has long benefited from this taxonomy of separable but interacting memory systems, and here we aim to highlight how such a viewpoint might help to better understand the manner in which memory can influence attention. We consider several key memory systems in turn.

2.2 *Studies of Attention and the Multiple Memory Systems Framework*

2.2.1 Episodic Memory

Just one or a few events, or episodes, in the past might be sufficient for guiding attention in the present. For example, you probably will not need to search for your desk effortfully on your second day at a new job, because your attention can be directed to the spot that you worked on your first day. Interactions between episodic memory and attention are important, as they might be critical in order to incorporate rapidly learned information into ongoing processing when that information is behaviorally relevant or useful for supporting further learning. Here, we briefly highlight a few lines of work which demonstrate the impact of episodic memory on attention.[2]

---

[2]Note that some of the studies discussed here contain more than a single encoding event, thus not meeting strict definitions of episodic memory which emphasize single exposure acquisition but do align with definitions which include multiple exposures (e.g., so-called "repisodic" memory (Neisser 1981)).

Selective visuospatial attention is intimately related to the location of our gaze (Corbetta and Shulman 2002; Hayhoe and Ballard 2005). It has been known for some time that eye movements can be influenced by a range of factors, including the current task or experimental context (e.g., Rothkopf et al. 2007; Yarbus 1967). Recent work has highlighted that episodic memory might also influence eye movements. For example, Ryan et al. (2007) showed that viewing behavior of previously studied or novel faces was different as early as the first fixation in the context of both the experiment and real-world exposure. The results further suggested that attention and representations stored in memory can be drawn, retrieved, and compared in an obligatory manner. Similarly, previewing of specific objects facilitates later search of that object embedded in a scene, such that studied objects were fixated sooner than unstudied ones (Chanon and Hopfinger 2008).

Additionally, numerous studies employing variants of common attention paradigms, such as visual search or change detection, have found evidence that past experience can facilitate subsequent performance. For example, Summerfield et al. (2006) had participants search for the location of targets embedded in scene images (e.g., an image of a key within an outdoor scene) repeatedly to learn the locations of the targets and then perform a visual search task for the target 1 or 2 days later. They found that the reaction time for finding targets was lower when they were in the same location as the prior session. Stokes et al. (2012), similarly, found facilitation of target search, when the target location within a specific scene was encoded 1 day earlier (Fig. 5a–c). Beyond visual search tasks, Rosen et al. (2014) found in a change blindness paradigm that participants were able to detect multiple changes after one or two exposures to a changing image. More broadly, there is evidence that a single, prior exposure to visual information, whether single objects/locations or entire scenes, can facilitate performance in tasks of attention (Castelhano and Henderson 2007; Hollingworth 2009). Taken together, the above findings highlight how only one or few experiences might be able to provide detailed information that can then be exploited in a manner which facilitates selective attention.

2.2.2 Semantic Memory

Our knowledge about the world, ranging from abstract concepts to everyday objects, contributes to the deployment of attention. Although attention is often studied in controlled laboratory conditions, it is not possible to truly remove the presence of this knowledge base. Thus, even in situations where the meaning of a word or image has no direct relevance for the task at hand, there still might be an impact on attention from semantic memory.

A hallmark of our semantic knowledge is that it specifies how different things are related to each other, and this general knowledge about associative relationships can help or hinder performance on attention tasks depending on the context (Becker and Rasmussen 2008; Davenport and Potter 2004; Neider and Zelinsky 2008; Torralba et al. 2006). For example, participants were able to identify a foreground object more

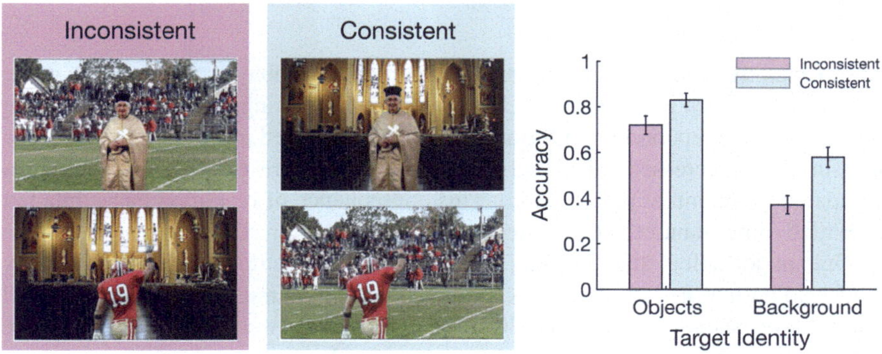

Fig. 2 Influence of semantic information on attentional guidance. In Davenport and Potter (2004), both foreground objects and the backgrounds were more likely to be reported accurately when they were consistent than when they were inconsistent

accurately when it was displayed in a semantically consistent setting (e.g., a football player in a football field) than when it was displayed in an inconsistent setting (e.g., a pope in a football field) with fewer eye movements (Davenport and Potter 2004; see Fig. 2). It has also been found that people direct their attention to the object that is semantically related to a key word held in working memory, such as attending to a motorbike helmet while maintaining the word motorbike (Moores et al. 2003; also see Seidl-Rathkopf et al. 2015). In one experiment which used a more natural viewing environment, participants searched through everyday objects on a table for a given target object adjacent to either a related object (e.g., a wallet when the credit card is the target) or an unrelated object (e.g., a marker when a pair of headphones are the target) (Mack and Eckstein 2011). Using data recorded from mobile eye-trackers, it was found that search time decreased in the condition where the target was located near a related object and participants were more likely to fixate at the related object than the rest of the nontarget objects available on the table.

Interestingly, semantic association is not inherently facilitative for attention in service of the task at hand. For example, visual search (e.g., for a bird) can be impaired in terms of accuracy and search time by the presence of a semantically related competitors (e.g., a feather) compared with when there was no such associate present in the condition when the target was absent (Belke et al. 2008; Moores et al. 2003). In another study, when participants were asked to search for a target traffic sign among several black-and-white signs and one colorful distractor sign, they were more likely to be distracted by a sign with related color of the target traffic sign (e.g., when the distractor was a red speed limit sign when the target was a red stop sign), compared to in the condition where the distractor was in an unrelated color (Olivers 2011). One study even found that saccade trajectories tended to deviate away from irrelevant, peripheral taboo words compared to neutral words (Weaver et al. 2011). Taken together, these results suggest that semantic knowledge permeates processing of visual information when everyday/real-world stimuli are being attended to.

2.2.3 Priming

Non-declarative forms of memory are also thought to be able to meaningfully guide attention. One such form of non-declarative memory, priming, generally refers to the observation that repetition of information leads to decreased processing time (e.g., as measured by decreased reaction times upon second presentation of a stimulus compared to the initial presentation). This consequence of repetition has subtle but potentially important effects on attention.

One major effect that priming might have on attention is that the fluency conferred on repeated information can influence the dynamics of attention in visual search. For example, Krueger (1970) found that people were able to find a single letter faster when embedded in real words compared to non-words and for common words compared to rare words, suggesting that repeated exposure might facilitate perceptual processing and influence the deployment of attention. Indeed this relationship is supported by the novelty bias effect (Reicher et al. 1976; Wolfe 2001) wherein participants are faster to find an unfamiliar target (e.g., inverted letter) among familiar distractors (e.g., non-inverted letters) than they are to find a familiar target among unfamiliar distractors (also compare to Flowers et al. 1981). Notably, as underscored by this particular set of examples, whether repeated information effectively attracts attention or effectively promotes attention to less fluently processed information varies depending on the exact experimental context.

Another aspect of priming that has been explored in numerous studies of visuospatial attention is the so-called "priming of pop-out" effect. Priming of pop-out refers to the phenomenon that during visual search involving a conspicuous pop-out object (e.g., a color singleton), intertrial repetition of the properties of the target and the distractor improves performance in both humans (Kristjnsson 2010; Maljkovic and Nakayama 1994, 1996) and monkeys (Bichot and Schall 1999) in an implicit manner (Maljkovic and Nakayama 2000). For example, when searching for a red singleton among green distractors, participants showed improved performance when the features of the target and distractors were repeated (Maljkovic and Nakayama 1994). This improvement is not observed when the target and distractor features are switched in-between trials; that is, switching the features between trials causes more erroneous selection (Becker and Rasmussen 2008; Leber et al. 2009; McPeek et al. 1999). Further work has suggested that this phenomenon cannot be accounted for by episodic memory retrieval (Becker 2008) and spans a range of perceptual features and experimental settings (e.g., Hillstrom 2000; Maljkovic and Nakayama 1994, 1996). Notably, the effect is observed even when the features of the target are irrelevant to the task at hand (Huang et al. 2004; Lee et al. 2009; Kristjnsson 2010) (also compare to Fecteau 2007). Priming of pop-out is an important consideration in tasks of attention and has figured into debates on, e.g., the limits of top-down influences on attentional capture (Folk and Remington 1998; Theeuwes and Burg 2011).

2.2.4 Conditioning

Conditioning, specifically classical conditioning, is another mnemonic phenomenon that might influence attention. Classical conditioning emphasizes the aspect of reinforcement in which a particular response known to be elicited from a given stimulus (unconditioned stimulus) can be associated with another arbitrary stimulus (conditioned stimulus) by, e.g., presenting the stimuli together in time. In the example of Pavlov's dog, salivation was ultimately elicited by the sound of a metronome (conditioned stimulus) because it had been consistently paired with the saliva-inducing presence of food (unconditioned stimulus) (Pavlov 1927). Although there is not much work in the attention literature which uses classical conditioning, here, we briefly highlight one study that assessed attention and novelty preference using a classical conditioning paradigm.

In this study, Snyder et al. (2008) instructed participants to view a series of polygon-shaped, solid color objects presented one at a time. Each polygon was immediately followed by either a positive image (e.g., a kitten), a negative image (e.g., a snake), or simply a grey square. Then each polygon was presented alongside another novel polygon, while participants' gaze location was recorded. For unconditioned neutral trials, looking times revealed a familiarity preference in the early time window followed by a shift of gaze toward the novel item. However, this lagged novelty preference was not found in the conditioned trials. Thus, although only a single study, the results are suggestive that learned or conditioned information can influence attentional dynamics (e.g., override a tendency to shift attention to novelty).

2.2.5 Statistical Learning

Statistical learning broadly refers to our ability to extract regularities from our environment based on the co-occurrence of multiple events, items, or features. Such a mechanism is thought to be critical for driving learning early in life (e.g., language acquisition; Saffran et al. 1996) as well as enabling specific predictions about future events (Chun and Turk-Browne 2008; Schapiro and Turk-Browne 2015).

Although very much an active area of research, there is evidence that statistical regularities of a series of events in and of itself might receive attentional priority over noisier, more randomly sequenced information (Yu and Zhao 2015; Zhao et al. 2013) and often in an implicit manner (e.g., Turk-Browne et al. 2005, 2008, 2010). For example, in one study (Zhao et al. 2013), participants were asked to view four streams of simultaneously changing symbols (Fig. 3). Critically, one stream contained repeated triplets of symbols, while other streams did not have any reliable sequential consistency. Periodically, a search array was presented with items at the locations of the streams, and participants had to perform a target discrimination task. Even though the statistical structure of the streams was not related to the target discrimination task, it was found that participants were faster to respond when the

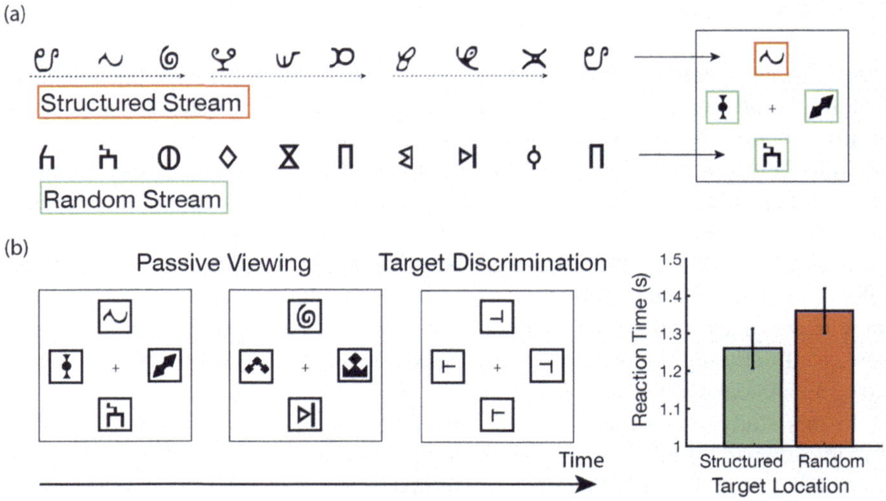

Fig. 3 Attentional guidance by statistical regularities. (**a**) In Zhao et al. (2013), four streams of symbols were shown at four fixed locations. One of the streams contained temporal regularities (the structured stream) in which triplets of symbols always appeared in the same order. The other three streams contained symbols without temporal regularities (each symbol was randomly chosen to be presented). (**b**) The four parallel streams of shapes were occasionally interrupted by a target discrimination task. Targets were discriminated faster when at the location of the structured stream than the random streams, despite participants reporting no explicit knowledge of the structure. Figure adapted with permission from Zhao et al. (2013)

target appeared at the location of the structured stream relative to the unstructured stream. Interestingly, this attentional bias toward statistical regularities was also found to extend to cases where regularities existed in other feature domains (e.g., color) or along entire feature dimensions (Zhao et al. 2013). It has also been found that this acquired bias is persistent and attention to the location can remain even when statistical regularities change location (Yu and Zhao 2015).

2.2.6 Habit Learning

Although they might take some time to acquire, habits are extremely powerful drivers of behavior. They can be instilled voluntarily such as when parents teach their children to wash their hands before eating but can also arise purely through repetition and persist beyond conscious initiation, such as how you might reach for your cell phone, even if you don't have it on you. Habit learning can be generally defined as the acquisition of the associated relationship between stimuli and responses across many repetitions.

In terms of visuospatial attention, habits can be experimentally induced by, e.g., increasing the probability of a target appearing at the same location (sometimes called "probability cuing"; Jiang 2018). For example, Shaw and Shaw (1977) asked participants to search for a target letter in one of the eight possible locations and found that recognition accuracy was higher at locations that were more likely than others to contain a target (also see, e.g., Miller 1988; Hoffmann and Kunde 1999). Interestingly, spatial probabilities governing target location can be considered an implicit attentional cue that differs from both typical explicit endogenous cues and exogenous cues (Geng and Behrmann 2005; Salovich et al. 2018). Much like the everyday habits described above, such search habits can persist even when no longer suitable for the task at hand. For example, even when the viewpoint changes and the target becomes equally likely to appear at any location, search habits based on initial probability cues can persist for hundreds of trials (Jiang and Swallow 2013). Notably, such search habits might also extend to not just learning target locations but learning about the properties of distractor items as well (also see, e.g., Goschy et al. 2014; Kelley and Yantis 2009).

2.2.7 Habituation

Habituation is a form of nonassociative implicit memory that allows people to tune out information after frequent exposure in order to attend to, e.g., more behaviorally relevant stimuli. There are many everyday examples of habituation such as when you learn to "tune out" distracting noises like the murmur of background chatter at the office.

Habituation of distraction can occur in the context of an attention paradigm and is subtly different from the case where locations frequently containing a target are prioritized (see last section). Habituation has been studied heavily in infants (e.g., Aslin 2007; Bornstein 1985), and, more generally, habituation of the orienting response in particular has been long known about (Sokolov 1963). Although there is comparatively less work in adults in studying visuospatial attention (e.g., in audition, Elliott and Cowan 2001), at least one recent line of work has explored the potential properties of habituation to distraction in the context of visuospatial attention (Turatto and Pascucci 2016; Turatto et al. 2018). Turatto et al. (2018) had participants perform a cued target discrimination task. On half of the trials, the onset of the target was immediately preceded by the sudden visual onset of a task-irrelevant distractor. Consistent with habituation, attentional capture by the distractor as measured by reaction time was diminished across the course of the session. Critically, participants were then divided into two groups. Half of the participants came back for a second session to perform the same task but without the distractors, serving as an extinction condition (i.e., the habituated orienting response to the distractor would be extinguished). The other half did not perform this second session. Then, in a final session, all of the participants performed the task with distractor. It was found that participants assigned to the extinction condition showed a spontaneous recovery of capture to the distractor (i.e., there was no habituation of the response any longer), while the non-extinction group did not show any sign of attentional capture, suggesting the habituated response remained. Interestingly, they found that the habituation effect persisted for many days, even after only roughly 45 min of training.

3 Neural Processing of Memory and Its Role in Attention

3.1 Introduction to the Neural Processing Framework

Although the above MMS framework has been extremely fruitful for guiding systematic research on human memory, there has been an emerging trend to consider memory not as a set of discrete systems defined by phenomenology but rather as a product of one or several basic neural operations or processing modes (Henke 2010). Here, we provide an overview of this related neural processing (NP) framework and highlight some neuroscientific findings suggesting how these basic neural characteristics might be involved in the dynamics of visuospatial attention.

Although the NP and MMS frameworks are far from mutually exclusive, they do differ in terms of how to think about what particular regions of the brain do in the service of memory. For example, there is notable divergence between how the two frameworks characterize the role of the key memory structure of the hippocampus. The MMS framework was heavily informed by the finding that damage to the medial temporal lobe, and the hippocampus in particular, led to the specific loss of the ability to form episodic memories. This causal relationship was effectively extended to equating the functionality of the hippocampus to declarative, long-term (episodic) memory. However, there is growing evidence suggesting the structure's involvement in rapid associative learning with or without conscious awareness and in both long- or short-term memory.

In terms of the role of the hippocampus in non-declarative memory, there is evidence that both declarative and non-declarative learning can be impaired by hippocampal damage. Numerous studies have suggested that healthy participants (Graf and Schacter 1985; Moscovitch et al. 1986) and patients without extensive hippocampal damage (Gabrieli et al. 1997; Goshen-Gottstein et al. 2000) display intact non-declarative memory for forming arbitrary associations (e.g., priming for semantically unrelated, studied word pairs). However, this ability is impaired in patients with hippocampal damage. Shimamura and Squire (1989) found that hippocampal amnesic patients, after they were presented with two unrelated words in the same sentences (e.g., A BELL was hanging over the baby's CRADLE), showed no evidence for associative priming from those two words, such that they did not complete the word displayed in the same context (e.g., BELL-CRA___) more than those in different context (e.g., APPLE-CRA___), while healthy participants who performed the same task did. Further, Hannula et al. (2007) found that after viewing a scene image, healthy participants showed disproportionately greater looking time toward face images associated with the scenes, despite not being able to explicitly identify the correct associated face. However, this implicit measure of associative memory was not seen in amnesic patients with hippocampal damage (also see Degonda et al. 2005; Hannula and Ranganath 2009; Henke et al. 2003a, b).

In terms of the role of the hippocampus in non-long-term memory, there is evidence for hippocampal involvement in short-term and working memory tasks and even some perceptual tasks (Ranganath and Blumenfeld 2005). For example, several studies have observed activity in the hippocampus and its neighboring medial temporal structures during performance of short-term memory tasks in both primate (Davachi and Goldman-Rakic 2001) and humans (Hannula and Ranganath 2008; Ranganath et al. 2005; Ranganath and D'Esposito 2001). Lesion studies have also found that patients with hippocampal damage display impaired visual working memory in particular settings (Olson et al. 2006a, b). Even in a perceptual discrimination task, there is evidence of hippocampal involvement under certain conditions. Aly et al. (2013) found that patients with hippocampal damage had difficulty detecting changes between similar pairs of images which differed slightly in terms of how their components related to each other (e.g., by contracting or "pinching" together one image compared to the other) (c.f., e.g., Urgolites et al. 2018) (Fig. 4).

Together, the above evidence is suggestive that the hippocampus is not exclusively linked to long-term, declarative memories but rather supports rapid encoding of flexible associations with or without conscious awareness. Indeed, when considering other brain regions, the NP framework suggests that instead of linking

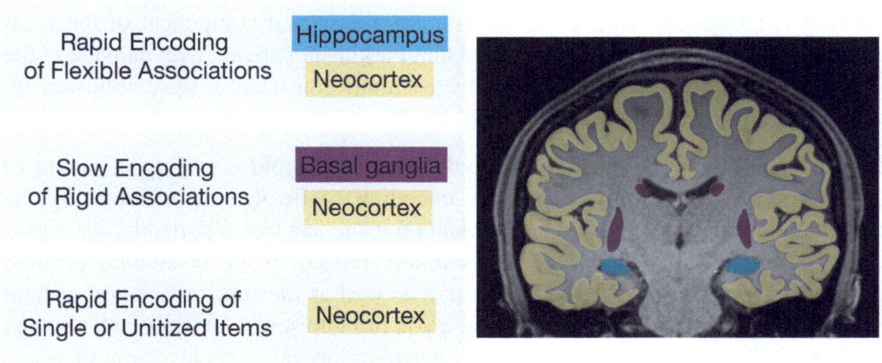

Fig. 4 A simplified depiction of the neural-based processing framework for memory. This framework does not distinguish memory based on consciousness per se but rather on different modes that can generate qualitatively distinct memories. Here we focus on three such modes: rapid encoding of flexible associations, slow encoding of rigid associations, and rapid encoding of single or unitized items. Figure adapted with permission from Henke (2010)

particular brain regions with particular memory systems per se, it instead might be preferable to consider the brain's contribution to memory in terms of its supported component operations or processing modes. Specifically, Henke (2010) distinguishes several dimensions along which memory differs and highlights the various neural regions thought to support such differences. That is, memories can be rapidly created (what type of cake was at your last birthday party) or slowly formed (what goes on at birthday parties in general). They can contain associations of a number of arbitrary items or features (remembering the name that goes with a particular person) or instead be focused on a single item (recognizing that person when you pass them in the hallway). They can be flexible and compositional (recalling either the setting of a play or the actors in it depending on your goal) or rigid and unitized (recognizing the exact costume of one character). Here, following the emphasis of the NP framework, we briefly cover three key types of memory in terms of these underlying dimensions:

Rapid Encoding of Flexible Associations The rapid encoding of flexible associations refers to remembering based on a single trial that involves many different features of an event. This processing mode depends on *the hippocampus* and the neocortex. In general, this processing mode supports the encoding of rich, multisensory events and complex spatial information after only one or few exposures. It is thought that memory might first be enabled by synaptic changes in the hippocampus and then eventually "consolidated" into the cortex through thalamically mediated communication between the hippocampus and neocortex.

Slow Encoding of Rigid Associations The slow encoding of rigid associations requires learning from a large number of trials and is thought to engage *the basal ganglia*, cerebellum, and cortex. The formation of semantic memory, habit learning, categorical learning, and procedural learning can be included in this category of memory (Foerde and Shohamy 2011). For example, stimulus-response association learning is thought to depend on the striatum, the largest component of the basal ganglia. Knowlton et al. (1996) reported that Parkinson patients with damage in the striatum were impaired at learning which stimuli consistently were followed by which outcomes without any explicit feedback.

Rapid Encoding of Single or Unitized Items The rapid encoding of single or unitized items often complements the encoding of flexible associations by the hippocampus but critically is not dependent on it and can be supported by *the cortex* alone. This mode of processing is potentially related to the behavioral priming phenomena we described in the last section as well as the subjective sense of faint familiarity in certain situations (e.g., when you run into someone that you know that you've met before, but you cannot recall where or any other details about them).

Below, we summarize a collection of empirical findings which highlight how these modes and key neural structures are involved in various manipulations of attention.

3.2 *Studies of Attention and the Neural Processing Framework*

3.2.1 The Hippocampus and the Rapid Encoding of Flexible Associations in Attention

As perhaps the most studied brain structure in memory, the hippocampus is thought to support associative relationships and thus plays an important role in, but not restricted to, the formation and retrieval of episodic memories where multiple aspects of an event need to be associated together (e.g., Eichenbaum et al. 2007; Moscovitch et al. 2016; Scoville and Milner 1957). In the context of memory-guided attention, for example, Summerfield et al. (2006) reported hippocampal involvement when visual search could be facilitated by remembering how a particular object was associated with a particular scene (i.e., its location) in the past. Specifically, on day one, participants were presented with scene images, some of which contained a target object at a fixed location. The next day, participants performed a visual search task where they were instructed to locate target objects embedded in scene images. Critically, the scene images were the same from the day before, and the location of the target was sometimes cued perceptually (by a white dotted frame) or via memory (the scene had contained a target object in the last session). Both perception and memory cues were always valid: they predicted the subsequent target location. As indicated by comparing reaction times to a neutral (non-cued) condition, the participants were able to orient their attention efficiently to the target when validly cued by either memory or perception. And, interestingly, performance benefited more from memory cues than perception cues. Consistent with the behavioral results, the authors also found hippocampal activation in valid vs neutral trials in memory-cued trials relative to those in perception-cued trials. Moreover, the magnitude of hippocampal activation was positively correlated with the magnitude of reaction time difference produced by the validity effect more so in memory-cued trials than in perception-cued trials, suggesting a role for hippocampally mediated memory in guiding attention in a similar manner as a perceptual cue might. Stokes et al. (2012) (Fig. 5) further found that hippocampal activity is specifically triggered by the presentation of the memory cue instead of the target (but see, e.g., Rosen et al. 2016), suggesting the important role of the hippocampus during memory-guided attention (Aly and Turk-Browne 2017).

Similar to how memory for a target object embedded in a particular scene might serve to cue attention through hippocampal involvement, it is possible that memory for targets embedded in basic visual search arrays, such as in contextual cueing paradigms, might also rely on the hippocampus. The contextual cueing effect refers to instances where visual search is facilitated when entire search arrays are repeated (see Fig. 6a, b for details) (Chun and Jiang 1998). Critically, this effect is typically not accompanied by explicit awareness of the repetition (Chun and Jiang 1999, 2003; Chun and Turk-Browne 2008; Chun 2000). Contextual cueing effects can be retained in memory immediately (e.g., Chun and Jiang 1998) and persist up to a

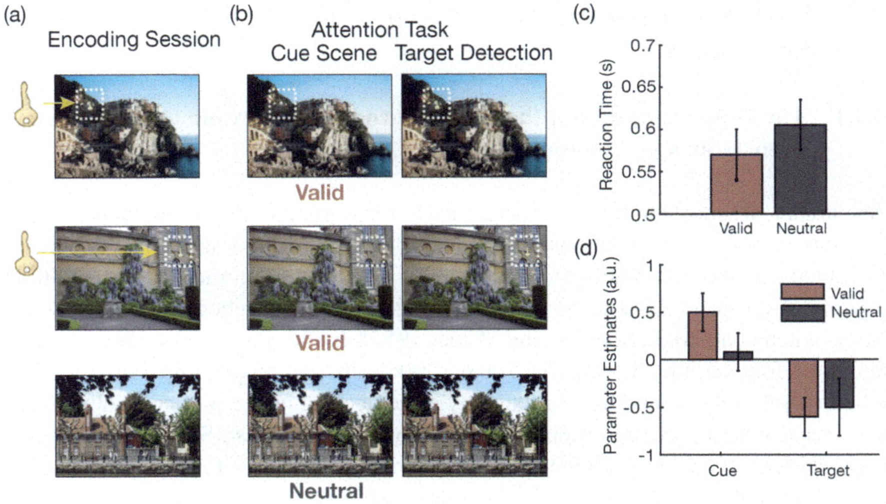

Fig. 5 Visual search and the hippocampus. (**a**) Participants were asked to search for an object (e.g., key) embedded in scenes on some trials (valid) and not others (neutral). (**b**) The following day, the same scenes were first presented as cues without targets, and then the participants searched for the target which would appear, on 50% of the trials, at the learned location. (**c**) Detection of the target was faster after valid memory-based cues than after neutral cues. (**d**) The hippocampus showed higher activation for memory-cued trials than for neutral trials during the cue period and not the target period suggesting hippocampal involvement in guiding subsequent search and not the appearance of the familiar target. Figure adapted with permission from Stokes et al. (2012)

week after acquisition (Chun and Jiang 2003). In terms of hippocampal involvement in contextual cueing, it was first reported by Chun and Phelps (1999) that patients with medial temporal lobe damage showed impaired contextual cueing, despite having intact perceptual learning. Using fMRI, Greene et al. (2007) reported that healthy adults show a relationship between hippocampal activity and repeated versus novel arrays without explicit recognition for the repeated arrays. The involvement of the hippocampus in contextual cueing has been noted in different experimental paradigms (e.g., as measured by eye movements (Neider and Zelinsky 2008) and using real-world scenes (Brockmole and Henderson 2006)).

3.2.2 Basal Ganglia and the Slow Encoding of Rigid Associations in Attention

The basal ganglia, and in particular the striatum, is crucial for creating rigid associations between, e.g., stimulus and response in memory (Packard et al. 1989), and its function is largely separable from that of the hippocampus (McDonald and White 1994). In contrast with the sizable literature regarding hippocampal involvement in attention, less is known on how striatal learning mechanisms might interact with attention. One interesting comparison between these two systems, however, was provided by Goldfarb et al. (2016) who found dissociable contributions of hippocampal and striatal memory using fMRI during a visual search task. In a modification of the classic contextual cueing task, participants were asked to locate a rotated T-shaped target among L-shaped distractors. Unbeknownst to the participants, some trials were designed to recruit the striatum by containing probabilistic stimulus-response (SR) associations such that the color of the presented items probabilistically (80% of the time) predicted both the quadrant of the target location and the correct response (the direction of the target). Other trials were designed to recruit the hippocampus through a more traditional contextual cueing (CC) approach (arrays were repeated even though the correct responses varied). SR and CC trials were randomly interleaved with trials that did not provide mnemonic cues, allowing a comparison of different forms of memory-guided attention in the same task. The behavioral results showed general learning as reflected by decreased reaction times with greater improvement of performance on CC and SR trials versus those with no mnemonic component (Fig. 6c). CC and SR cued trials did not differ in reaction times, suggesting a similar enhancement of attention, and participants did not demonstrate any explicit awareness of either manipulation. Also, participants were not faster at locating the target when the color of the items did not match the learned SR color cue, indicating the rigid association between the target and the SR cue. The authors found that hippocampal activity predicted subsequent attention benefits indexed by reaction times on CC trials (interestingly, lower hippocampal activation related to faster subsequent RTs), while striatal activity did not, and the striatal activity predicted attention on subsequent SR trials, while the hippocampus did not (with more striatal activity related to faster subsequent RTs; Fig. 6d). Also, reflective of the difference in the putative learning time courses of the hippocampus (rapid) and basal ganglia (slow), CC-guided attention correlated more with hippocampal activity early on, while SR-guided attention correlated more with striatal activity later in the task, suggesting the slow speed of the striatal encoding relative to the hippocampus. Although awaiting replication, these results indicate the separable involvement of the hippocampus and the striatum during in a single attention task.

Another common setting for rigid stimulus-response associations is in studies of reinforcement learning and reward. Reward and the reward history of stimuli are closely linked to basal ganglia function (Schultz 2016) and are thought to have an impact on attention. Indeed, Awh et al. (2012) have argued that the top-down and bottom-up dichotomy of attention is inadequate, in part, because it does not capture the influence of selection history, including reward, on visual selection. The subregions of the basal ganglia, including the substantia nigra (Sato and Hikosaka 2002), caudate nucleus (Lauwereyns et al. 2002; Watanabe et al. 2003), and striatum (Hollerman et al. 1998), have been found crucial in terms of expecting rewards, even in the context of a visual attention task. For example, Leong et al. (2017) had participants perform a task in which three multi-dimensional stimuli in each trial were probabilistically associated with different reward outcomes. Only one dimension was relevant for predicting reward in a given experimental round. Within that dimension, one feature was associated with a high probability of a reward, while other features in that dimension were associated with a low probability of reward.

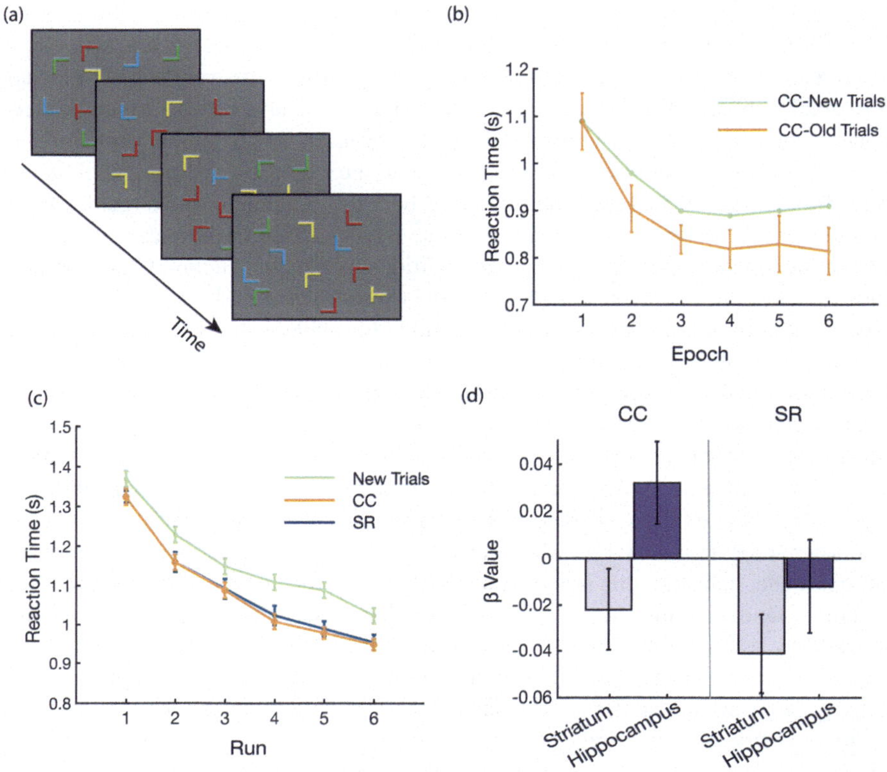

Fig. 6 Hippocampal and striatal involvement in attention. (**a**) A typical contextual cueing paradigm from Chun and Jiang (1998). Participants were asked to search for a rotated T-shaped target among L-shaped distractors. Critically, some of the trials contained configurations that were repeated multiple times throughout the experiment, while other trials contained new configurations. In the repeated trials, the targets appeared at consistent locations relative the distractors. (**b**) Search time of the target was faster in repeated configurations (CC – old trials) than in new configurations (CC – new trials), suggesting facilitated visual search. (**c**) Goldfarb et al. (2016) used a similar paradigm and found that repeated arrays using the standard contextual cueing (CC) manipulation or more specific stimulus-response (SR) repetition (see text for details) both led to reduced reaction times relative to novel configurations. (**d**) Model estimates of the relationship between trial-evoked BOLD responses and subsequent reaction times (indicative of facilitated attention). Specifically, only the hippocampus could significantly predict subsequent CC trial RTs, while only the striatum predicted subsequent SR trial RTs. Panel (**b**) adapted with permission from Chun and Jiang (1998) (also see Chun (2000)). Panels (**c**) and (**d**) adapted with permission from Goldfarb et al. (2016)

Without explicit cues, participants learned to choose the stimulus with the rewarding feature over time. Using a combination of computational modeling of learning, eye-tracking, and fMRI, it was found that allocation of attention on a trial-by-trial basis could be influenced by ongoing learning. In turn, the neural signals guiding learning ("prediction error") in the striatum were found to be influenced by attention, suggesting a bidirectional relationship between the allocation of attention and learning during decisionmaking. Further, given the tight relationship between the oculomotor system and selective attention (Corbetta and Shulman 2002), it is worth noting that several studies of nonhuman primates have shown that particular eye movements can be linked with neuronal activity in the basal ganglia when, e.g., saccades are associated with reward (Watanabe et al. 2003; Kawagoe et al. 1998). These results, in general, provide evidence that the basal ganglia might play an important role in reward-related attentional guidance, particularly in situations where the relationship between reward and attention is relatively fixed (e.g., to a particular location or feature dimension) and slowly acquired.

3.2.3 Cortical Dynamics and the Rapid Encoding of Single or Unitized Items in Attention

The cerebral cortex encompasses a vast and functionally diverse region of the brain. The cortex is closely linked to virtually all aspects of human psychology including executive function, decisionmaking, language, motor planning, and sensory perception. As noted above, the cortex plays at least some role in all three types of processing modes. Although we will not attempt to catalog all ways in which the cortex might facilitate interactions of memory and attention, a shortlist of key dynamics has been already noted within the context of the biased competition model of selective attention. This model suggests that, at any given time, objects in the visual field compete for processing capacity (Desimone and Duncan 1995; Desimone 1996) and that the competition can be biased by, e.g., what information is most relevant for the task at hand. Notably, Desimone (1996) outlined several ways in which past experience alters cortical activity in ways which might influence attention: decreased response after repetition (attenuation), increased response after repetition (enhancement), persistent response after stimulus offset (delay activity), tuning (as in perceptual learning), and association (e.g., potentially as mediated by the hippocampus). Rather than assess all the ways in which the cortex might be involved in supporting the relationship between memory and attention, here, we highlight recent work on the first of these memory-related dynamics, repetition attenuation or a decreased response upon repetition of the same stimulus (Grill-Spector et al. 2006), and highlight how it might fit into competitive dynamics underlying attention.

One study has looked at such repetition attenuation and its potential role in guiding attention. In two fMRI experiments by Hutchinson et al. (2015), participants viewed a series of images which were organized in a triplet trial structure. That is, in each trial, the first two events contained the same image repeated twice in a row, but then the third event contained either that same image presented at the same time as a novel image or instead two novel images appearing at the same time (see examples in Fig. 7a). In line with the predictions of the biased competition framework, it was hypothesized that when old and new information was presented concurrently in the third event, an attenuated response to the repeated image should promote processing of the novel image which did not have an attenuated response. Consistent with this hypothesis, it was found that subsequent memory, a measure tightly related to

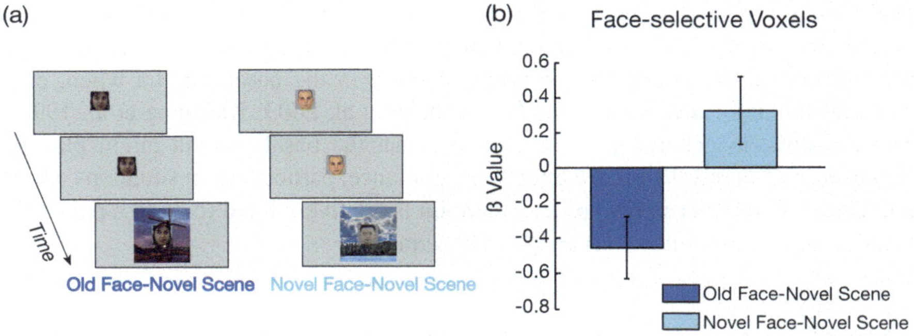

Fig. 7 Repetition attenuation and competitive processing. (**a**) Two example trials from Hutchinson et al. (2015). In the critical Old Face-Novel Scene condition, a face image was repeated twice in a row and followed by the same face embedded in a novel scene. In the control condition (Novel Face-Novel Scene), the third event contained two novel images. (**b**) Using logistic regression, it was found that there was a significant negative relationship between BOLD activity in face-selective voxels and subsequent scene memory. That is, the more the response for the repeated face was attenuated, the more likely the novel, competing scene would be processed (encoded into memory). Figure adapted with permission from Hutchinson et al. (2015)

selective attention, was greater for novel images presented with the repeated images than with other novel images, suggestive of attenuation influencing processing. Critically, a neural measure of repetition could be calculated in regions of the brain which responded preferentially to the repeated stimuli. Relating this neural measurement of activity to the repeated item to the processing of the concurrently presented novel item (as indexed by subsequent memory) was then done across trials for each participant (Fig. 7b). Across experiments, it was found that there was a reliable relationship wherein the greater the amount of attenuation for the repeated item, the more likely it was that the concurrently presented novel item would be later remembered. Although a single study, these results reflect how cortical dynamics might influence attentional processing in situations where both old and new information is present.

4 Conclusion

In this chapter, we have attempted to provide a concise overview of the ways in which different forms of memory might systematically influence selective visuospatial attention. Attention has been historically studied in the context of sensory systems: how attentional deployment is influenced by the current perceptual environment and how attention, in turn, impacts the sensory processing of the perceptual input. Here, we highlighted work which instead speaks to the influence of our internal mnemonic environment on attention. We have provided some evidence suggesting memory can meaningfully guide attention and, at the same time, does not fit cleanly into the prevailing dichotomy of stimulus-driven or goal-directed attention. Memory often acts on perceptual information independent of the attentional task at hand and is defined by prior experience, not purely perceptual attributes. We highlighted two related taxonomies of human memory, the multiple memory systems framework and the neural processing framework, in an attempt to provide a broader context of the work to date on memory-guided attention, as well as provide potential guiding principles for research moving forward. The classic view of memory systems, MMS, has influenced the field of memory research and serves as useful way to catalog different empirical, particularly behavioral, findings in the attention literature. Although MMS has been extremely fruitful for exploring the different manner in which memory interacts with other cognitive domains including attention, the neural processing framework offers insight into how particular neural mechanisms of memory might be integrated with what we know about how attention is enabled by the brain.

Beyond the departure from the traditional stimulus-goal dichotomy described above, considering the role of memory has other consequences for theories of attention. For example, historic debates over early versus late attentional selection (Broadbent 1958; Pashler 1998) might be revisited by recasting various experimental factors (e.g., learned importance (Moray 1959)) in terms of their putative memory system, when possible. Also, as noted in Sect. 3, the biased competition model offers specific insight as to how various neural signatures of memory might influence competitive processing (Desimone 1996). Other neural consequences of remembering, such as cortical reinstatement (e.g., elevated activity in the visual cortex when you vividly recollect visual details; Wheeler et al. 2000), might be evaluated through the lens of competitive processing. Also of potential interest will be understanding how neural representations in structures such as the hippocampus might themselves be influenced by attention (Aly and Turk-Browne 2016a, b). More broadly, a major open theoretical question at the heart of memory-guided attention is the degree to which past experience is compartmentalized from ongoing processing (Waugh and Norman 1965; Glanzer and Cunitz 1966; Cowan 1988) versus fundamentally intertwined (e.g., Rao and Ballard 1999; Barrett and Simmons 2015; Clark 2013; Panichello et al. 2013).

Future empirical work on attention might also benefit from incorporating, or accounting for, specific elements from the field of memory research. For example, decisions concerning optimal trial sequence, condition frequencies, stimuli choice, and trial-to-trial featural overlap can be informed by many of the studies described above. Also, studies which seek to understand how past experience might influence a particular measure of attention can vary item history in a manner that is explicitly related to a specific form of memory as outlined by either the MMS or NP frameworks. Critically, as the understanding of how memory guides attention is still maturing, research of patient groups with specific damage to key memory structures or systems is vital. Patient studies contribute greatly in informing us of the necessity of a given brain region for various mnemonic and attentional processes (e.g., Chun and Phelps 1999; Hannula et al. 2007; Schapiro et al. 2014; Shimamura and Squire 1989). And new, targeted studies exploring how such groups perform on tasks of selective attention involving memory will provide key insight as to the range and specific attributes of memory-guided attention.

In conclusion, we highlight how memory can and does meaningfully guide attention and suggest that different types of memory, whether indexed by the MMS or NP framework, can be systematically explored in experimental settings. It is not hard to find in our daily lives instances where previous experience can direct our attention and a single episode in the present might alter your attention for years in the future. Although studied in isolation for decades, the interactions between memory and attention guide much of our conscious experience, determine what we learn about our world, and, compared to either faculty alone, largely await future research.

Acknowledgements We gratefully acknowledge Thomas M. Biba and Sarah DuBrow's comments on an earlier version of this manuscript. We also thank the authors who provided their permission for use of their figures.

References

Aly M, Turk-Browne NB (2016a) Attention promotes episodic encoding by stabilizing hippocampal representations. Proc Natl Acad Sci 113(4):E420–E429

Aly M, Turk-Browne NB (2016b) Attention stabilizes representations in the human hippocampus. Cereb Cortex 26(2):783–796

Aly M, Turk-Browne NB (2017) How hippocampal memory shapes, and is shaped by, attention. In: The hippocampus from cells to systems. Springer, Cham, pp 369–403

Aly M, Ranganath C, Yonelinas A (2013) Detecting changes in scenes: the hippocampus is critical for strength-based perception. Neuron 78(6):1127–1137

Aslin RN (2007) What's in a look? Dev Sci 10(1):48–53

Awh E, Belopolsky AV, Theeuwes J (2012) Top-down versus bottom-up attentional control: a failed theoretical dichotomy. Trends Cogn Sci 16(8):437–443

Barrett LF, Simmons WK (2015) Interoceptive predictions in the brain. Nat Rev Neurosci 16 (7):419–429

Becker SI (2008) The mechanism of priming: episodic retrieval or priming of pop-out? Acta Psychol (Amst) 127(2):324–339

Becker MW, Rasmussen IP (2008) Guidance of attention to objects and locations by long-term memory of natural scenes. J Exp Psychol Learn Mem Cogn 34(6):1325–1338

Belke E, Humphreys GW, Watson DG, Meyer AS, Telling AL (2008) Top-down effects of semantic knowledge in visual search are modulated by cognitive but not perceptual load. Percept Psychophys 70(8):1444–1458

Bichot NP, Schall JD (1999) Saccade target selection in macaque during feature and conjunction visual search. Vis Neurosci 16(1):81–89

Bornstein MH (1985) Habituation of attention as a measure of visual information processing in human infants: summary, systematization, and synthesis. In: Measurement of audition and vision in the first year of postnatal life: a methodological overview. Ablex, Westport, CT, pp 253–300

Broadbent DE (1958) Perception and communication. Pergamon Press, Elmsford, NY

Brockmole JR, Henderson JM (2006) Recognition and attention guidance during contextual cueing in real-world scenes: evidence from eye movements. Q J Exp Psychol 59(7):1177–1187

Carrasco M (2011) Visual attention: the past 25 years. Vision Res 51(13):1484–1525

Castelhano MS, Henderson JM (2007) Initial scene representations facilitate eye movement guidance in visual search. J Exp Psychol Hum Percept Perform 33(4):753–763
Chanon VW, Hopfinger JB (2008) Memory's grip on attention: the influence of item memory on the allocation of attention. Vis Cogn 16(2–3):325–340
Chun MM (2000) Contextual cueing of visual attention. Trends Cogn Sci 4(5):170–178
Chun MM, Jiang Y (1998) Contextual cueing: implicit learning and memory of visual context guides spatial attention. Cogn Psychol 36(1):28–71
Chun MM, Jiang Y (1999) Top-down attentional guidance based on implicit learning of visual covariation. Psychol Sci 10(4):360–365
Chun MM, Jiang Y (2003) Implicit, long-term spatial contextual memory. J Exp Psychol Learn Mem Cogn 29(2):224–234
Chun MM, Phelps EA (1999) Memory deficits for implicit contextual information in amnesic subjects with hippocampal damage. Nat Neurosci 2(9):844–847
Chun MM, Turk-Browne NB (2007) Interactions between attention and memory. Curr Opin Neurobiol 17(2):177–184
Chun MM, Turk-Browne NB (2008) Associative learning mechanisms in vision. In: Visual memory. Oxford University Press, New York, pp 209–245
Chun MM, Golomb JD, Turk-Browne NB (2011) A taxonomy of external and internal attention. Annu Rev Psychol 62(1):73–101
Clark A (2013) Whatever next? Predictive brains, situated agents, and the future of cognitive science. Behav Brain Sci 36(3):181–204
Cohen N, Squire L (1980) Preserved learning and retention of pattern-analyzing skill in amnesia: dissociation of knowing how and knowing that. Science 210(4466):207–210
Corbetta M, Shulman GL (2002) Control of goal-directed and stimulus-driven attention in the brain. Nat Rev Neurosci 3(3):201–215
Cowan N (1988) Evolving conceptions of memory storage, selective attention, and their mutual constraints within the human information-processing system. Psychol Bull 104(2):163–191
Davachi L, Goldman-Rakic PS (2001) Primate rhinal cortex participates in both visual recognition and working memory tasks: functional mapping with 2-DG. J Neurophysiol 85(6):2590–2601
Davenport JL, Potter MC (2004) Scene consistency in object and background perception. Psychol Sci 15(8):559–564
Degonda N, Mondadori CR, Bosshardt S, Schmidt CF, Boesiger P, Nitsch RM, Hock C, Henke K (2005) Implicit associative learning engages the hippocampus and interacts with explicit associative learning. Neuron 46(3):505–520
Desimone R (1996) Neural mechanisms for visual memory and their role in attention. Proc Natl Acad Sci 93(24):13494–13499
Desimone R, Duncan J (1995) Neural mechanisms of selective visual attention. Annu Rev Neurosci 18(1):193–222
Dudukovic NM, Wagner AD (2006) Attending to remember and remembering to attend. Neuron 49 (6):784–787
Eichenbaum H, Yonelinas AP, Ranganath C (2007) The medial temporal lobe and recognition memory. Annu Rev Neurosci 30(1):123–152
Elliott EM, Cowan N (2001) Habituation to auditory distractors in a cross-modal, colorword interference task. J Exp Psychol Learn Mem Cogn 27(3):654–667
Fan JE, Turk-Browne NB (2016) Incidental biasing of attention from visual long-term memory. J Exp Psychol Learn Mem Cogn 42(6):970–977
Fantz RL (1964) Visual experience in Infants: decreased attention to familiar patterns relative to novel ones. Science 146(3644):668–670
Fecteau JH (2007) Priming of pop-out depends upon the current goals of observers. J Vis 7(6):1–1
Flowers JH, Polansky ML, Kerl S (1981) Familiarity, redundancy, and the spatial control of visual attention. J Exp Psychol Hum Percept Perform 7(1):157–166
Foerde K, Shohamy D (2011) The role of the basal ganglia in learning and memory: insight from Parkinsons disease. Neurobiol Learn Mem 96(4):624–636

Folk CL, Remington R (1998) Selectivity in distraction by irrelevant featural singletons: evidence for two forms of attentional capture. J Exp Psychol Hum Percept Perform 24(3):847–858
Gabrieli JDE (1998) Cognitive neuroscience of human memory. Annu Rev Psychol 49(1):87–115
Gabrieli JD, Keane MM, Zarella MM, Poldrack RA (1997) Preservation of implicit memory for new associations in global amnesia. Psychol Sci 8(4):326–329
Geng JJ, Behrmann M (2005) Spatial probability as an attentional cue in visual search. Percept Psychophys 67(7):1252–1268
Glanzer M, Cunitz AR (1966) Two storage mechanisms in free recall. J Verbal Learn Verbal Behav 5(4):351–360
Goldfarb EV, Chun MM, Phelps EA (2016) Memory-guided attention: independent contributions of the hippocampus and striatum. Neuron 89(2):317–324
Goschy H, Bakos S, Mller HJ, Zehetleitner M (2014) Probability cueing of distractor locations: both intertrial facilitation and statistical learning mediate interference reduction. Front Psychol 5:1195
Goshen-Gottstein Y, Moscovitch M, Melo B (2000) Intact implicit memory for newly formed verbal associations in amnesic patients following single study trials. Neuropsychology 14 (4):570–578
Graf P, Schacter DL (1985) Implicit and explicit memory for new associations in normal and amnesic subjects. J Exp Psychol Learn Mem Cogn 11(3):501–518
Greene AJ, Gross WL, Elsinger CL, Rao SM (2007) Hippocampal differentiation without recognition: an fMRI analysis of the contextual cueing task. Learn Mem 14(8):548–553
Grill-Spector K, Henson R, Martin A (2006) Repetition and the brain: neural models of stimulus-specific effects. Trends Cogn Sci 10(1):14–23
Hall N, Gjedde A, Kupers R (2008) Neural mechanisms of voluntary and involuntary recall: a PET study. Behav Brain Res 186(2):261–272
Hannula DE, Ranganath C (2008) Medial temporal lobe activity predicts successful relational memory binding. J Neurosci 28(1):116–124
Hannula DE, Ranganath C (2009) The eyes have It: hippocampal activity predicts expression of memory in eye movements. Neuron 63(5):592–599
Hannula DE, Ryan JD, Tranel D, Cohen NJ (2007) Rapid onset relational memory effects are evident in eye movement behavior, but not in hippocampal amnesia. J Cogn Neurosci 19 (10):1690–1705
Hayhoe M, Ballard D (2005) Eye movements in natural behavior. Trends Cogn Sci 9(4):188–194
Henke K (2010) A model for memory systems based on processing modes rather than consciousness. Nat Rev Neurosci 11(7):523–532
Henke K, Mondadori CRA, Treyer V, Nitsch RM, Buck A, Hock C (2003a) Nonconscious formation and reactivation of semantic associations by way of the medial temporal lobe. Neuropsychologia 41(8):863–876
Henke K, Treyer V, Nagy ET, Kneifel S, Drsteler M, Nitsch RM, Buck A (2003b) Active hippocampus during nonconscious memories. Conscious Cogn 12(1):31–48
Hillstrom AP (2000) Repetition effects in visual search. Percept Psychophys 62(4):800–817
Hoffmann J, Kunde W (1999) Location-specific target expectancies in visual search. J Exp Psychol Hum Percept Perform 25(4):1127–1141
Hollerman JR, Tremblay L, Schultz W (1998) Influence of reward expectation on behavior-related neuronal activity in primate striatum. J Neurophysiol 80(2):947–963
Hollingworth A (2009) Two forms of scene memory guide visual search: memory for scene context and memory for the binding of target object to scene location. Vis Cogn 17(1–2):273–291
Huang L, Holcombe AO, Pashler H (2004) Repetition priming in visual search: episodic retrieval, not feature priming. Mem Cogn 32(1):12–20
Hutchinson JB, Turk-Browne NB (2012) Memory-guided attention: control from multiple memory systems. Trends Cogn Sci 16(12):576–579
Hutchinson JB, Pak SS, Turk-Browne NB (2015) Biased competition during long-term memory formation. J Cogn Neurosci 28(1):187–197

Jiang YV (2018) Habitual versus goal-driven attention. Cortex 102:107–120
Jiang YV, Swallow KM (2013) Spatial reference frame of incidentally learned attention. Cognition 126(3):378–390
Kawagoe R, Takikawa Y, Hikosaka O (1998) Expectation of reward modulates cognitive signals in the basal ganglia. Nat Neurosci 1(5):411–416
Kelley TA, Yantis S (2009) Learning to attend: effects of practice on information selection. J Vis 9 (7):16–16
Knowlton BJ, Mangels JA, Squire LR (1996) A neostriatal habit learning system in humans. Science 273(5280):1399–1402
Knudsen EI (2007) Fundamental components of attention. Annu Rev Neurosci 30(1):57–78
Kristjnsson CG (2010) Where perception meets memory: a review of repetition priming in visual search tasks. Atten Percept Psychophys 72(1):5–18
Krueger LE (1970) Search time in a redundant visual display. J Exp Psychol 83(3, Pt. 1):391–399
Kuhl BA, Johnson MK, Chun MM (2013) Dissociable neural mechanisms for goal-directed versus incidental memory reactivation. J Neurosci 33(41):16099–16109
Lauwereyns J, Watanabe K, Coe B, Hikosaka O (2002) A neural correlate of response bias in monkey caudate nucleus. Nature 418(6896):413–417
Leber AB, Ji K, Gabari Y (2009) Long-term, abstract learning of attentional set. J Exp Psychol Hum Percept Perform 35(5):1385–1397
Lee H, Mozer MC, Vecera SP (2009) Mechanisms of priming of pop-out: stored representations or feature-gain modulations? Atten Percept Psychophys 71(5):1059–1071
Leong YC, Radulescu A, Daniel R, DeWoskin V, Niv Y (2017) Dynamic interaction between reinforcement learning and attention in multidimensional environments. Neuron 93(2):451–463
Logan GD (2002) An instance theory of attention and memory. Psychol Rev 109(2):376–400
Mack SC, Eckstein MP (2011) Object co-occurrence serves as a contextual cue to guide and facilitate visual search in a natural viewing environment. J Vis 11(9):9–9
Maljkovic V, Nakayama K (1994) Priming of pop-out: I. Role of features. Mem Cognit 22 (6):657–672
Maljkovic V, Nakayama K (1996) Priming of pop-out: II. The role of position. Percept Psychophys 58(7):977–991
Maljkovic V, Nakayama K (2000) Priming of popout: III. A short-term implicit memory system beneficial for rapid target selection. Vis Cogn 7(5):571–595
McDonald RJ, White NM (1994) Parallel information processing in the water maze: evidence for independent memory systems involving dorsal striatum and hippocampus. Behav Neural Biol 61(3):260–270
McPeek RM, Maljkovic V, Nakayama K (1999) Saccades require focal attention and are facilitated by a short-term memory system. Vision Res 39(8):1555–1566
Miller J (1988) Components of the location probability effect in visual search tasks. J Exp Psychol Hum Percept Perform 14(3):453–471
Milner B (1962) Memory disturbance after bilateral hippocampal lesions. In: Cognitive processes and the brain. Van Nostrand, Princeton, NJ, pp 97–111
Moores E, Laiti L, Chelazzi L (2003) Associative knowledge controls deployment of visual selective attention. Nat Neurosci 6(2):182–189
Moray N (1959) Attention in dichotic listening: affective cues and the influence of instructions. Q J Exp Psychol 11(1):56–60
Moscovitch M, Winocur G, McLachlan D (1986) Memory as assessed by recognition and reading time in normal and memory-impaired people with Alzheimer's disease and other neurological disorders. J Exp Psychol Gen 115(4):331–347
Moscovitch M, Cabeza R, Winocur G, Nadel L (2016) Episodic memory and beyond: the hippocampus and neocortex in transformation. Annu Rev Psychol 67(1):105–134
Neider MB, Zelinsky GJ (2008) Exploring set size effects in scenes: identifying the objects of search. Vis Cogn 16(1):1–10

Neill WT, Beck JL, Bottalico KS, Molloy RD (1990) Effects of intentional versus incidental learning on explicit and implicit tests of memory. J Exp Psychol Learn Mem Cogn 16(3):457
Neisser U (1981) John Dean's memory: a case study. Cognition 9(1):1–22
Olivers CNL (2011) Long-term visual associations affect attentional guidance. Acta Psychol (Amst) 137(2):243–247
Olson IR, Moore KS, Stark M, Chatterjee A (2006a) Visual working memory is impaired when the medial temporal lobe Is damaged. J Cogn Neurosci 18(7):1087–1097
Olson IR, Page K, Moore KS, Chatterjee A, Verfaellie M (2006b) Working memory for conjunctions relies on the medial temporal lobe. J Neurosci 26(17):4596–4601
Packard MG, Hirsh R, White NM (1989) Differential effects of fornix and caudate nucleus lesions on two radial maze tasks: evidence for multiple memory systems. J Neurosci 9(5):1465–1472
Panichello MF, Cheung OS, Bar M (2013) Predictive feedback and conscious visual experience. Front Psychol 3
Pashler HE (1998) The psychology of attention. MIT Press, Cambridge, MA
Pavlov PI (1927) Conditioned reflexes: an investigation of the physiological activity of the cerebral cortex. Oxford University Press, London
Poldrack RA, Packard MG (2003) Competition among multiple memory systems: converging evidence from animal and human brain studies. Neuropsychologia 41(3):245–251
Ranganath C, Blumenfeld RS (2005) Doubts about double dissociations between short- and long-term memory. Trends Cogn Sci 9(8):374–380
Ranganath C, D'Esposito M (2001) Medial temporal lobe activity associated with active maintenance of novel information. Neuron 31(5):865–873
Ranganath C, Rainer G (2003) Cognitive neuroscience: neural mechanisms for detecting and remembering novel events. Nat Rev Neurosci 4(3):193–202
Ranganath C, Cohen MX, Brozinsky CJ (2005) Working memory maintenance contributes to long-term memory formation: neural and behavioral evidence. J Cogn Neurosci 17(7):994–1010
Rao RPN, Ballard DH (1999) Predictive coding in the visual cortex: a functional interpretation of some extra-classical receptive-field effects. Nat Neurosci 2(1):79–87
Reicher GM, Snyder CR, Richards JT (1976) Familiarity of background characters in visual scanning. J Exp Psychol Hum Percept Perform 2(4):522–530
Rosen ML, Stern CE, Somers DC (2014) Long-term memory guidance of visuospatial attention in a change-detection paradigm. Front Psychol 5
Rosen ML, Stern CE, Michalka SW, Devaney KJ, Somers DC (2016) Cognitive control network contributions to memory-guided visual attention. Cereb Cortex 26(5):2059–2073
Rothkopf CA, Ballard DH, Hayhoe MM (2007) Task and context determine where you look. J Vis 7 (14):16–16
Ryan JD, Hannula DE, Cohen NJ (2007) The obligatory effects of memory on eye movements. Memory 15(5):508–525
Saffran JR, Aslin RN, Newport EL (1996) Statistical learning by 8-month-old infants. Science 274 (5294):1926–1928
Salovich NA, Remington RW, Jiang YV (2018) Acquisition of habitual visual attention and transfer to related tasks. Psychon Bull Rev 25(3):1052–1058
Sato M, Hikosaka O (2002) Role of primate substantia nigra pars reticulata in reward-oriented saccadic eye movement. J Neurosci 22(6):2363–2373
Schapiro A, Turk-Browne N (2015) Statistical learning. In: Brain mapping: an encyclopedic reference. Elsevier, Amsterdam, pp 501–506
Schapiro AC, Gregory E, Landau B, McCloskey M, Turk-Browne NB (2014) The necessity of the medial temporal lobe for statistical learning. J Cogn Neurosci 26(8):1736–1747
Schultz W (2016) Reward functions of the basal ganglia. J Neural Transm 123(7):679–693
Scoville WB, Milner B (1957) Loss of recent memory after bilateral hippocampal lesions. J Neurol Neurosurg Psychiatry 20(1):11–21
Seidl-Rathkopf KN, Turk-Browne NB, Kastner S (2015) Automatic guidance of attention during real-world visual search. Atten Percept Psychophys 77(6):1881–1895

Shaw ML, Shaw P (1977) Optimal allocation of cognitive resources to spatial locations. J Exp Psychol Hum Percept Perform 3(2):201–211

Shimamura AP, Squire LR (1989) Impaired priming of new associations in amnesia. J Exp Psychol Learn Mem Cogn 15(4):721–728

Sisk CA, Remington RW, Jiang YV (2018) The risks of downplaying top-down control. J Cognit 1 (1):23

Snyder KA, Blank MP, Marsolek CJ (2008) What form of memory underlies novelty preferences? Psychon Bull Rev 15(2):315–321

Sokolov EN (1963) Higher nervous functions: the orienting reflex. Annu Rev Physiol 25 (1):545–580

Squire LR (1992) Declarative and nondeclarative memory: multiple brain systems supporting learning and memory. J Cogn Neurosci 4(3):232–243

Squire LR, Wixted JT (2011) The cognitive neuroscience of human memory since H.M. Annu Rev Neurosci 34(1):259–288

Squire LR, Knowlton B, Musen G (1993) The structure and organization of memory. Annu Rev Psychol 44(1):453–495

Stokes MG, Atherton K, Patai EZ, Nobre AC (2012) Long-term memory prepares neural activity for perception. Proc Natl Acad Sci 109(6):E360–E367

Summerfield JJ, Lepsien J, Gitelman DR, Mesulam MM, Nobre AC (2006) Orienting attention based on long-term memory experience. Neuron 49(6):905–916

Theeuwes J (2018) Visual selection: usually fast and automatic; seldom slow and volitional. J Cognit 1(1):21

Theeuwes J, Burg EVD (2011) On the limits of top-down control of visual selection. Atten Percept Psychophys 73(7):2092

Torralba A, Oliva A, Castelhano MS, Henderson JM (2006) Contextual guidance of eye movements and attention in real-world scenes: the role of global features in object search. Psychol Rev 113 (4):766–786

Tulving E, Markowitsch HJ, Craik FIM, Habib R, Houle S (1996) Novelty and familiarity activations in PET studies of memory encoding and retrieval. Cereb Cortex 6(1):71–79

Turatto M, Pascucci D (2016) Short-term and long-term plasticity in the visual-attention system: evidence from habituation of attentional capture. Neurobiol Learn Mem 130:159–169

Turatto M, Bonetti F, Pascucci D (2018) Filtering visual onsets via habituation: a context-specific long-term memory of irrelevant stimuli. Psychon Bull Rev 25(3):1028–1034

Turk-browne NB, Jung JA, Scholl BJ (2005) The automaticity of visual statistical learning. J Exp Psychol Gen (134):552–564

Turk-Browne NB, Scholl BJ, Chun MM, Johnson MK (2008) Neural evidence of statistical learning: efficient detection of visual regularities without awareness. J Cogn Neurosci 21 (10):1934–1945

Turk-Browne NB, Scholl BJ, Johnson MK, Chun MM (2010) Implicit perceptual anticipation triggered by statistical learning. J Neurosci 30(33):11177–11187

Urgolites ZJ, Levy DA, Hopkins RO, Squire LR (2018) Spared perception of object geometry and object components after hippocampal damage. Learn Mem 25(7):330–334

Võ MLH, Wolfe JM (2015) The role of memory for visual search in scenes. Ann N Y Acad Sci 1339(1):72–81

Watanabe K, Lauwereyns J, Hikosaka O (2003) Neural correlates of rewarded and unrewarded eye movements in the primate caudate nucleus. J Neurosci 23(31):10052–10057

Waugh NC, Norman DA (1965) Primary memory. Psychol Rev 72(2):89–104

Weaver MD, Lauwereyns J, Theeuwes J (2011) The effect of semantic information on saccade trajectory deviations. Vision Res 51(10):1124–1128

Wheeler ME, Petersen SE, Buckner RL (2000) Memory's echo: vivid remembering reactivates sensory-specific cortex. Proc Natl Acad Sci 97(20):11125–11129

Wolfe JM (2001) Asymmetries in visual search: an introduction. Percept Psychophys 63 (3):381–389

Woodman GF, Chun MM (2006) The role of working memory and long-term memory in visual search. Vis Cogn 14(4–8):808–830

Yarbus AL (1967) Eye movements during perception of complex objects. In: Eye movements and vision. Springer, Boston, MA, pp 171–211

Yu RQ, Zhao J (2015) The persistence of the attentional bias to regularities in a changing environment. Atten Percept Psychophys 77(7):2217–2228

Zhao J, Al-Aidroos N, Turk-Browne NB (2013) Attention is spontaneously biased toward regularities. Psychol Sci 24(5):667–677

Superstitious Perception: Comparing Perceptual Prediction by Humans and Neural Networks

Patrick Laflamme and James T. Enns

Contents

Abstract Recent developments in convolutional neural networks (CNNs) have introduced new ways to model the complex processes of human vision. To date, the comparison of human vision and CNNs has focused on internal representations (i.e., receptive fields), with behavioral comparisons left largely unexplored. Here, we probe the influence of cognitive strategy on the similarity between CNN output and human behavior. We gave study participants a superstitious perception task (i.e., we asked them to detect an assigned target in white noise) while asking them to engage in either an active or passive attentional strategy. Previous research has shown that an active attentional strategy tends to engage central executive functions, whereas a passive strategy allows perceptual processes to unfold with limited central control. The results showed that the pattern of human responses in the superstitious perception task depended significantly on task strategy. Specifically, detecting targets

P. Laflamme and J. T. Enns (✉)
Department of Psychology, University of British Columbia, Vancouver, BC, Canada
e-mail: jenns@psych.ubc.ca

Curr Topics Behav Neurosci (2019) 41: 213–244
DOI 10.1007/7854_2018_65
Published Online: 1 May 2019

superstitiously (i.e., false alarms) was correlated with evidence of a target's presence in the passive condition, but not in the active condition.

Human data were compared to the performance of a CNN performing the same task, with the decision criterion of the CNN set to match the false alarm rates observed in the two strategy conditions of the human participants. CNN responses resembled those of human participants in the passive condition more closely than those in the active condition. This observation suggests that the CNN does a better job of mimicking human behavior when central executive functions are not engaged than when they are engaged. This, in turn, has important implications for what human participants are doing in the superstitious perception task. Namely, it implies that superstitious perception may have two important ingredients that are somewhat dissociable. First, there is the ability to detect weak signals in noise that correspond to the target image. This appears to be what participants are doing under passive strategy conditions; they allow externally generated signals to dominate their perceptual experience. Second, there is the ability to ignore the noise in favor of basing responses solely on internally generated signals. This seems to correspond more closely to what participants are doing under active strategy conditions, when attention is controlled by representations in memory. This research emphasizes the importance of modeling the full range of human responsiveness in even a simple noisy detection task.

Keywords Cognitive strategies · Computational neural networks · Deep learning · Executive functions · Superstitious perception

1 Introduction

In this chapter we describe a set of experiments designed to specifically compare the behavior of humans and convolutional neural nets on a superstitious perception task. In doing so, we are forced to confront the riddle facing every contributor to this volume on attention and working memory: "To what extent are perceptual experiences driven by the visual signal and to what extent are they influenced by long term memories?" We think the superstitious perception task is an ideal and daily relevant microcosm in which to pose these questions. We also hastily admit that the answers that emerge are as much a surprise to us as they may be for the reader.

1.1 Superstitious Perception Revealed Through Classification Images

A time-honored approach to studying the complex workings of the human visual system is to focus on the instances in which it performs unexpectedly. For example, studies of visual-geometric illusions have helped establish the human bias to interpret two-dimensional images as representing three-dimensional scenes (Haffenden et al. 2001; Leibowitz et al. 1969; Massaro and Anderson 1971; Rock and Kaufman 1962; Shepard 1981).

Apophenia is another instance of unexpected behavior. First coined by Klaus Conrad, apophenia is the experience of seeing patterns in random or meaningless data (Fyfe et al. 2008; Shermer 2008). Common examples include seeing animals in clouds and seeing an image of Jesus in a piece of toast (Liu et al. 2014), which are likely an outcome of the general human tendency to detect and identify structure in spatial and temporal patterns (Yu and Zhao 2015; Zhao et al. 2013; Zhao and Yu 2016).

One practical problem that arises when studying apophenia concerns how to quantify the extent to which the source image resembles (or contains evidence consistent with) the percept that the observer claims to see. At one extreme, the percept may be generated entirely from within the visual system and therefore bears no resemblance at all to the percept. On the other hand, the visual system may be detecting evidence in some portions of the image that is objectively consistent with the apophenic percept, despite the ground truth that the image was generated entirely independent of such a percept. The difficulty therefore lies in trying to quantify the extent of any such resemblance, when it is present.

One technique that has been developed to investigate apophenia in humans at the behavioral level is the superstitious perception task (Gosselin and Schyns 2003). This technique offers a way to measure of the apophenic percept through a classification image. The classification image is then used to measure the extent to which any source image resembles the apophenic percept. The technique involves asking people to identify geometric shapes in random noise that looks like static on an old TV. On each trial, participants are asked to identify a specific shape (e.g., the letter "s") within a noisy image that is presented on a computer screen. They do this for several thousand trials of randomly generated images. Participants report having little confidence in their judgments when doing this task (Gosselin and Schyns 2003). Yet, when all of the images identified as having a target are added up and averaged, and then subtracted from an average of all of the images that had not been identified, a representation of the target becomes visible. The image that results from this process is called a classification image (CI), and Gosselin and Schyns claim that the CI is an image of the internal representation of the target that participants were detecting in the white-noise images.

The stated theoretical goal of the superstitious perception task (Gosselin and Schyns 2003) is to measure the internal representation that participants are using when they falsely detect target signals in noise. The rationale for this method is built upon the same logic that underpins the use of reverse correlation to estimate the receptive field properties of individual neurons (see Ringach and Shapley 2004 for a review). Reverse correlation was introduced by researchers who stimulated single neurons in the primary visual cortex of cats with random white visual noise input to the retina. They then recorded each of neuron's spikes in response to the stimulation (Jones and Palmer 1987) and modeled a neuron's receptive field by computing the average stimulus pattern that causes a neuron to fire. These spike-triggered averages were interpreted as templates for the neuron's receptive field because they represented the stimulus that was most likely to cause the neuron to fire.

1.2 The Importance of External Prediction

Critics of this interpretation cautioned that before accepting this interpretation of spike-triggered averages, it was important to describe how well the receptive field predicted the neuron's activity (DeAngelis et al. 1993). This was done by estimating a neuron's receptive field using reverse correlation. Once the receptive field was estimated, one could predict the neuron's firing rate in response to the individual stimuli used to estimate the receptive field. This is a test of internal reliability, and it is illustrated in Fig. 1c. Internal prediction was thus a measure of how well the estimated receptive field predicted the neuron's response to the individual stimuli

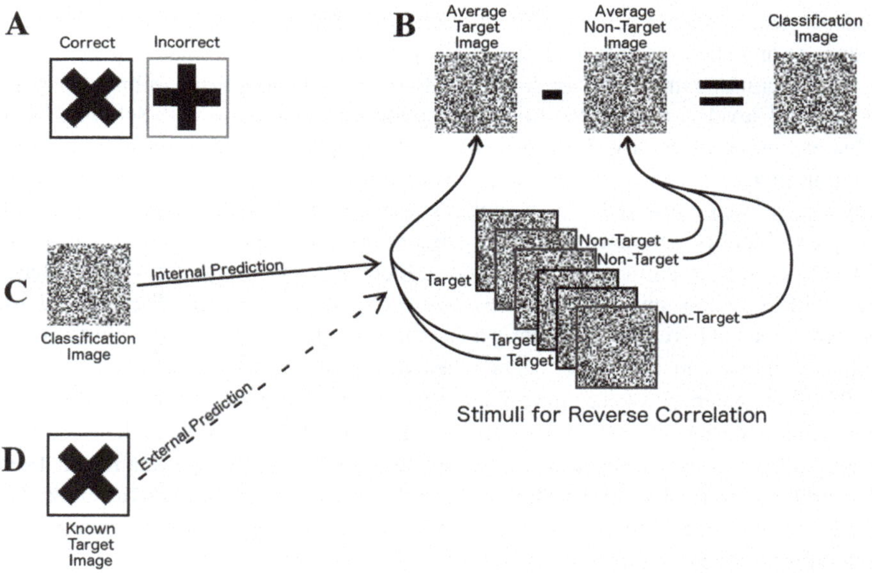

Fig. 1 A summary of the study design, procedures, and rationale. (**a**) Human participants were assigned to one of two possible targets in Experiment 1. A CNN was assigned to the x target in Experiment 2. (**b**) Reverse correlation is the process of generating a classification image by averaging all stimuli that generated a false alarm and subtracting the average of all stimuli that did not generate a false alarm. (**c**) Internal reliability is the process of using a CI to predict participants' responses to the white-noise images that were also used to generate the CI. Note that this process is referred to as backward reliability in the neural spike train literature, but here we call it internal reliability. (**d**) External reliability is computed the same as internal reliability but differs in using the participant's assigned target to predict responses to white-noise images. Note that this process is similar to what is called forward prediction in the neural spike train literature and serves to give a less biased estimation of the goodness of fit between target images and responses

that had contributed to the estimation. If the neural response of individual stimuli was not sufficiently predicted by the receptive field, one could conclude that the calculated receptive field of the neuron was not adequately descriptive of the neuron's function.

DeAngelis et al. (1993) performed internal predictions on the receptive fields of early visual neurons in cat and was able to show a high degree of accuracy, thus validating the procedure for further use. Reverse correlation soon became the standard for interpreting neuron spike trains and reconstructing representations of the receptive fields of visual and auditory neurons (Jones and Palmer 1987; Ringach et al. 1997; Theunissen et al. 2001). However, Ringach et al. (1997) suggested another important step to the validation procedure. These authors argued that one should also use CIs to make predictions of a neuron's response to novel stimuli, based on the receptive field suggested by reverse correlation. The critical difference between internal and external prediction is whether the stimuli used to generate the receptive field are the same as those used to predict responses (internal prediction) or whether the receptive field is correlated with novel stimuli (external prediction). When, as in Fig. 1c, the same stimuli are used for creation of the CI and for its evaluation of predictive power, the measurement procedure is vulnerable to capitalization on chance (or measurement error). This approach therefore suffers from the well-known dangers of overfitting data (e.g., Burnham and Anderson 2002). When novel stimuli are tested instead, these dangers are avoided. Alternatively, if the receptive field is known, one can get an external prediction estimate using the known receptive field, shown in Fig. 1d. Stated in the most neutral way possible, internal prediction and external prediction can be considered to be liberal and conservative bounds of the predictive power of a given receptive field.

In order to apply these concepts to the superstitious perception technique, several important differences must be noted. First, behaving participants do not respond with neural spikes, as neurons do, but rather with responses that can be coded as false alarms or correct rejections, in a signal detection sense. If the response is a false alarm, it suggests that the image may be target-like in some respects. The activation of a target template by the participant is analogous to the activation of a neuron in the single-cell work. On the other hand, if the response is a correct rejection, the implication is that the image is not sufficiently target-like to activate the target template. Note that each response to an image is in a sense even more information than we get from a neuron, which typically only responds in a positive way by increasing its activity above some baseline level. Gosselin and Schyns (2003) put this extra information to good use in their application of reverse correlation to psychophysical data, by creating an average-target image and an average-nontarget image, as shown in Fig. 1b.

Second, participants' responses do not have a temporal component analogous to time-stamped spikes in a volley of neural firing. Instead, a participant's response depends solely on the current image. In the neural realm, a prediction is calculated based on an aggregate of spikes over time, in the units of firing rate (Hz) (Dyan and Abbott 2001). In the superstitious perception realm, we can instead interpret the aggregate prediction as a likelihood of false alarm for a given trial. For this reason, we chose to aggregate trials into bins of comparable target similarity when estimating the extent of explained variance in responses. Since there are no guidelines for bin size, a range of bin sizes was used in the experiments we describe to ensure that our conclusions did not depend on bin size. The statistical logic behind the reverse correlation technique, however, remains the same as that in the neural spiking domain (Gosselin et al. 2001).

A third difference is that participants in behavioral experiments are less predictable than neurons when presented with ambiguous stimuli (Rieth et al. 2011). While neurons are generally thought to have a fixed pattern that they respond to, participants have a wide array of patterns that they can detect in parallel. As a result, presenting white noise to participants is likely to yield few to no superstitious perceptions. Instead, participants require a more elaborate ruse to report these superstitious perceptions. Rieth et al. (2011) achieved this by having two consecutive practice phases: (1) participants were shown targets that were simple to identify in noise; they called this the easy practice condition; and then (2) participants were asked to identify targets that were very heavily overlaid with noise, making them hard to identify – they called this the hard practice condition. The goal here was to encourage participants to see superstitiously. Notably, for our purposes, this also aids in stressing to participants that the targets they will be shown are identical on all trials, differing only in the difficulty associated with detecting them. This design consideration helped us strengthen our experimental design over previous studies (Gosselin and Schyns 2003).

1.3 Four Goals of the Present Study

1.3.1 External Prediction

Our first goal was to apply the concept of external prediction to behavioral data. The application of reverse correlation in the neural domain helped to establish that external reliability was essential for determining the template of a putative receptive field. Surprisingly, this practice has not been adopted by researchers of superstitious perception. To do so, one should attempt to predict participants' false alarm rates in response to novel white-noise images (external reliability) in addition to using the classification images used in establishing internal reliability. Without this crucial step, one does not know the degree to which the CI over-fit the data from which it was generated. Unfortunately, this has not been done, likely because of the daunting number of trials required to do so. In this work, we take another approach by correlating noise-free target images with the participant's response.

We therefore achieved our first goal by making comparisons not only between a participant's CI and their responses to noise images (internal prediction) but also between the specific target image assigned to participants and the noise images (external prediction). This design allowed us to test whether the CIs distinguished between a given participant's assigned target and other potential targets (assigned to other participants). If the CI did not distinguish between similar targets assigned to different participants, we would have to conclude that the individual differences observed by Gosselin and Schyns (2003) in the CIs of the participants' target "S" reflected sampling noise rather than individual differences in participants' search templates.

1.3.2 Does Task Strategy Matter?

The second goal of the present study was to examine the role of central executive functioning in the superstitious perception task. Executive function is an umbrella term for a wide array of cognitive control functions, including inhibition, interference control, working memory, and cognitive flexibility. Together, these functions are thought to enable higher-level cognitive functions such as decision-making and reasoning (Diamond 2013). A consideration of these functions in the superstitious task is a natural one, given that the task requires a decision on every trial – whether or not to report that the image appears to contain a target. At the same time, the degree to which executive functions are recruited to the task could impact the type of information that is used to reach a decision. For instance, if participants respond to an image based on an internal representation that differs from the specific details of the target image they have been assigned, then the external reliability of their responses may differ markedly from the internal reliability of the same responses.

Studies in related visual tasks have had considerable success modulating the contributions of executive functioning through instructions (Jacoby and Brooks 1984; Marcel 1983; Smilek et al. 2006; Van Selst and Merikle 1993; Whittlesea and Brooks 1994). In short, giving participants *passive* strategy instructions encourages them to use sensory information to perform simple pattern matching, thereby minimizing the role of cognitive control. *Active* strategy instructions, on the other hand, encourage participants to be consciously aware of the bases of their decisions and to exert control over the particular sensory pathways that are relevant in coming to a decision. In the realm of object categorization, participants were observed to be more accurate during an item classification task over a small set of features when adopting a feature-focused, analytical strategy (active instructions) when compared to taking a more holistic, nonanalytical strategy (passive instructions) (Jacoby and Brooks 1984; Whittlesea and Brooks 1994). The opposite was true when the classification task involved a large number of features. Even when percepts never reached conscious awareness, the effects of unconscious stimuli on task outcome were stronger when participants let the answer passively "pop" into mind, as opposed to trying to actively determine whether a stimulus was present in a preceding trial (Van Selst and Merikle 1993). In the field of visual search, participants' search efficiency was modulated by instructions they are given before the task. Participants were faster and more accurate in their search when told to let the target of their search passively "pop" into their minds as opposed to when they were told to actively search for the target of interest (Smilek et al. 2006). While the literature makes it clear that task strategy influences task outcome, specifically through the active engagement or inhibition of executive control, the optimal strategy for any given task appears to vary, with analytical, active strategies being the better strategy in those tasks that require executive control, but not in others tasks where executive control puts the observer at a disadvantage.

The theoretical proposal of Smilek et al. (2006) is that task outcome differences that occur in response to the adoption of passive versus active instructions rely on the recruitment of executive functions. From their perspective, adopting a passive strategy in a visual search task recruits executive functions to a lesser degree than adopting an active strategy (Smilek et al. 2006). It is also consistent with the previous findings (Jacoby and Brooks 1984; Marcel 1983; Van Selst and Merikle 1993; Whittlesea and Brooks 1994), where the optimal strategy is consistent for a given task but varies between different tasks. This is because tasks differ with respect to the requirement of various cognitive functions to effectively complete the task in question. For example, a memory task might benefit more from an increase in recruitment of memory functions, while an executive functions task may see less of an improvement from the same change in recruitment. In this way, identifying the optimal strategy for a given task can give important insights into the cognitive functions upon which the task depends.

Given that the superstitious perception task is intended to measure internal representations (Gosselin and Schyns 2003), identifying the optimal strategy for the task may give insight into the functions from which those internal representations are derived. If the task depends primarily on the visual sensory system, then the images selected as containing a target during the task will reflect the images that more closely match the template of the target of interest. If, however, the task is also dependent on executive functions, this would suggest that the images selected are chosen due to internally represented perceptual hypotheses that have been overlaid onto the images. Moreover, these perceptual hypotheses may be without objective support in the noisy image signals. In an effort to disambiguate these possibilities, we compared the results obtained with active and passive task strategies on the outcomes of the superstitious perception task.

In a passive task strategy, participants were asked to simply let the target "pop" into their mind when viewing the trial images. This strategy is meant to minimize the influence of top-down effects on the final decision about a targets' presence in each trial image. In an active task strategy, participants were asked to actively search for evidence of the targets' presence in each trial image. This strategy is designed to increase reliance on participants' executive functions during the task. If a passive task strategy is more effective than an active task strategy at creating reliable classification images, the internal representations are likely being derived directly from the visual sensory system. However, if the active task strategy is most effective for creating reliable classification images, then executive functions are likely necessary for the manifestation of these internal representations.

1.3.3 Humans and Machines

A third goal of this study was to compare humans and convolutional neural networks (CNNs) on the same superstitious perception task. Recent advances in computing technology have expanded the capabilities of computational models used for simulating neural connections in the human brain. The advent of highly parallelized processing means that we can increase the number of simultaneously activated nodes (units) corresponding to neurons in the human visual system. From these new technologies, a particular set of models try to mimic that the human visual system translates points of light on the retina into representations of pictorial scenes. These models, known as CNNs, have shown neural representations of stimuli that are remarkably similar to the human visual system (Cichy et al. 2016; Khaligh-Razavi and Kriegeskorte 2014; Yamins et al. 2014). However, we are not aware of any attempts to test these models by using their outputs to predict human behavior that goes beyond testing the accuracy of image categorization.

The superstitious perception task goes beyond most previous tasks studied in CNNs in that it forces a decision under conditions of great uncertainty. We know in advance that humans "go beyond the information given" at almost every turn in tasks of perception. But is this behavior intrinsic to a dynamic, hierarchically organized, visual system? By comparing the outcomes of a superstitious perception task for humans and CNNs, we hope to come to a better understanding of how these two systems are at present both similar and different.

The hierarchical nature of CNNs mimics parts of our current understanding of the architecture of the human visual system. Neurons at higher levels of the processing hierarchy have receptive fields that are larger relative to the visual input and select for more complex visual features relative to neurons at lower levels of the processing hierarchy (Dyan and Abbott 2001; Yamins et al. 2014). These CNNs are a class of neural network models that search for patterns in incoming data using "filters." These "filters" label subsections of data that match a specific pattern and pass that information on to higher processing layers, which search for specific patterns among the activations of lower layers. In order to distinguish the biological neurons of the human brain from the simulated neurons of the CNNs, the simulated neurons will be referred to as nodes. In the first layer of the network, a set of filters are applied to the input image, to detect patterns of interest (e.g., Gabor filters of various orientations). One node's activity represents the degree to which the subsection of the input image matches one of the filters. When taken together, the nodes in the first layer of a CNN represent a set of spatial maps, one for each filter, indicating the locations in the image where patterns in the input image match the pattern in a filter. The activations of the nodes in the first layer are then used as the input data for the second layer, where the same procedure is followed. In essence, the first layer of the network searches for specific patterns in the image; the second layer searches for specific second-order patterns in the collection of patterns identified by the first layer; the third layer searches for specific patterns in the collection of patterns identified by the second layer, and so on.

One well-studied CNN implementation is that of Krizhevsky et al. (2012), commonly known as AlexNet. AlexNet consists of eight layers of artificial neurons. In the first five layers (layers 1–5), the receptive fields are smaller than the width of the previous layer, meaning that the spatial locations are preserved from one layer to the next. The next two layers (layers 6 and 7) have filters that cover the entirety of the previous layer, meaning that each node has a unique filter – each node is sensitive to patterns in the entirety of the image. The final layer is another layer where the filters are the size of the previous layer except that it is used as the "classification layer." In the classification layer, each node corresponds to a class that the neural network is trying to identify in the images. The activation of nodes in this layer can be interpreted as the networks' confidence in the presence of that object in the input image. This procedure is similar to the human visual system in that the receptive fields of neurons in the visual system increase in size relative to the visual input as one increases from low-level areas to higher level areas (Dyan and Abbott 2001).

The representations of nodes may be similar in the human visual system and in CNNs. In the first layer of the CNN, the nodes are directly referencing the image itself and tend to be sensitive to contrast and oriented lines. As a result, the receptive fields of those neurons can be directly interpreted. These low-level receptive fields appear remarkable similar to those found in V1 (Krizhevsky et al. 2012). However, once you proceed beyond the first layer of the network, it becomes increasingly difficult to interpret the nodes' representations in the same way. These upper layers tend to represent more abstract concepts, such as "dog snout" or "clock face" instead of lines or edges, which cannot easily be visualized comprehensively.

Nonetheless, researchers have developed techniques to compare activity in the macaque visual cortex and in the high-order layers of a CNN in response to the same set of stimuli (Khaligh-Razavi and Kriegeskorte 2014; Yamins et al. 2014). These comparisons have concluded that there is remarkable similarity in the ways that the two systems represent visual stimuli. Further work has suggested that the similarities observed at the high-order layers extend also to the lower-level areas (Cichy et al. 2016).

At a behavioral level, we see similar trends when comparing human vision and CNNs. Recent CNN models have been able to achieve accuracy in view-variant classification tasks that are indistinguishable from human performance on the same task (Kheradpisheh et al. 2016). Taken together, these observations suggest they are reasonable models of human vision (Yamins and DiCarlo 2016).

In summary, the third goal of this study is to compare the behavior of a CNN to that of the human visual system during the superstitious perception task. We aimed to evaluate one CNN model of human vision by testing its tendency to perceive superstitiously, like humans. This approach has a number of advantages over previous comparisons between CNNs and humans. Most notably, it allows for the isolation of the cognitive functions being simulated by the CNN. We expected that our CNN would approximate human performance in the passive condition due to previous work observing CNN-primate visual system similarities primarily in the visual cortices (Cichy et al. 2016; Khaligh-Razavi and Kriegeskorte 2014; Yamins et al. 2014; Yamins and DiCarlo 2016). But what would occur in the active condition, where central control functions are also likely contributing to the decision?

1.3.4 Machines with Strategies?

The final goal of this study was to explore the effect of task strategy on superstitious perception in CNNs. One challenge for this goal was to create a CNN that perceives either in an "active" mode or a "passive" mode. Our approach here was to match the rate false alarm rate of humans in the same mode with the false alarm rates of CNNs. Any behavioral differences in CNN performing under these two false alarm rates should help to identify which aspects of human vision the CNN is modeling. If the CNN mirrors human behavior in both conditions, it would suggest that variations in false alarm rates alone are responsible for the differences attributed to instructional strategies in humans. However, if the CNN resembles the passive instruction responses of humans, but not the active instructions, it would be consistent with the hypothesis that the CNN models human sensory responses but not central executive contributions to the task. Finally, if the CNN also resembles the active instruction condition, it would suggest that aspects of both the visual system and executive functions are being modeled in the CNN.

2 Experiment 1: Superstitious Humans

This experiment addressed our first two goals by (1) comparing the internal and external reliability of human responses in the superstitious perception task and by (2) manipulating participant's cognitive strategy via instructions in order to study the role of executive functioning in superstitious perception.

Participants were assigned to one of two targets in the superstitious perception task. We examined participants' responses on each trial using both the participants' CIs (measure of internal reliability) and each participant's assigned target (measure of external reliability). The magnitude of the difference between the predictive power of the CI versus that of the target image will be taken as a measure of the validity of the CI. Specifically, a CI that has much higher predictive power (internal prediction) than the original target image (external prediction) is an indication that the CI is capitalizing largely on sampling noise and is not a good estimate of the participant's search template.

One half the participants were asked to perform the superstitious perception task with an active strategy, in an effort to engage executive functions in the task (Smilek et al. 2006). The other half were asked to perform the superstitious perception task with a passive strategy, in an effort to reduce executive functioning. If the internal representations that are measured in the standard superstitious perception task are influenced by executive functions, then the CI-image correlations should be higher in the active than in the passive instruction conditions.

2.1 Methods

2.1.1 Participants

Fourteen participants (nine female) between the ages of 18 and 66 (mean age = 26.57, SD = 12.02) were recruited from the paid study participant pool at the University of British Columbia. All participants self-reported normal or corrected-to-normal vision and were tested individually.

2.1.2 Stimuli and Procedures

Participants were tested twice in a 2-week period in order to collect 4,000 experimental trials in total. Each 1 h session consisted of a practice phase (involving hard-to-see targets in order to acquaint participants with the procedures) and an experimental phase (in which there were no targets to generate the CI for each participant). The practice phase was composed of two conditions, following Rieth et al. (2011): (1) an easy condition, intended to allow for practice identifying the target for which they were searching, and (2) a hard condition intended to encourage participants by giving them the impression that targets were present but difficult to identify. The task of the participant was the same in all phases: participants viewed images in the center of the screen, and following each one, they were asked to indicate whether a target image was present or absent, using different key presses.

The target images in the practice phase were overlaid by white pixel noise. Nontarget images were generated by overlaying pixel noise on a solid gray square with the same average luminance as the target image. Target image contrast, as measured with weber fractions, differed between the easy (weber fraction of 0.66) and hard (weber fraction of 0.11) practice conditions by 0.55 units. Over these contrast-reduced stimuli, we added random noise, sampled from a normal distribution with a mean of 0, and a standard deviation of 0.2. Notably, the average luminance of each trial image stayed constant throughout the experiment, with minor variations due to random sampling of the noise. During the experimental phase of the experiment, all stimuli were generated by overlaying white pixel noise atop the solid gray square, whose luminance matched that of the average luminance in the target.

Participants were randomly assigned to search for either a target "x" or "+," shown in Fig. 1a. All participants completed the task on an iMac late 2009 model, and the task was constructed using Matlab 2010a and PsychToolBox v11 (Brainard 1997; Kleiner et al. 2007; Pelli 1997). Images presented to participants occupied about 2° of visual angle and were 50 × 50 pixels in resolution.

Participants were also randomly assigned to one of two instruction groups: an *active* instructions group and a *passive* instructions group. The instructions for each group were modified slightly from Smilek et al. (2006) to match the superstitious perception task.

In the *active* group, participants were told:

> The best strategy for this task, and the one that we want you to use in this study, is to be as active as possible and to "search" for the target in the image as you look at the screen. The idea is to deliberately direct your attention to determine your response. Sometimes people find it difficult or strange to "direct their attention" – but we would like you to try your best. Try to respond as quickly and accurately as you can while using this strategy. Remember, it is very critical for this experiment that you actively search for the target in the image. If you cannot find any reason to suspect that the target is present, respond that the target is not present.

In the *passive* group, participants were told:

> The best strategy for this task, and the one that we want you to use in this study, is to be as receptive as possible and see if the target "pops" into your mind as you look at the image. The idea is to let the display and your intuition determine your response. Sometimes people find it difficult or strange to tune into their "gut feelings" – but we would like you to try your best. Try to respond as quickly and accurately as you can while using this strategy. Remember, it is very critical for this experiment that you allow the target to just "pop" into your mind. If this does not happen, respond that the target is not present.

2.1.3 Data Analysis

Measures of sensitivity (d') and decision bias ($\ln\beta$) in the practice phase indexed the degree to which participants understood the task and were following instructions (data from one participant was omitted because d' was near zero). Analysis of the practice phase was performed using a mixed effects ANOVA, with instruction group as a between-subject factor with two levels (active and passive) and target difficulty as a within-subject factor with two levels (easy and hard). One analysis compared sensitivity (d') across factors, and a second analysis compared decision bias, $\ln\beta$.

Classification images were generated for each participant based on their responses in the test phase. This was done by adding together all of the trial images in which a target was identified and then subtracting all of the images in which a target was not identified, as shown in Fig. 1b (Gosselin and Schyns 2003). A Gaussian filter was then used to eliminate spatial frequencies higher than three cycles per image, in order to eliminate random noise (Gosselin and Schyns 2003).

The CIs were assessed with regard to their similarity to the target in two ways. A first assessment involved examining the image-wise correlation (IWC) between each participant's CI and each of the two possible targets. The predicted target for each participant was the target with which their CI correlated most strongly. A second assessment of image-target similarity involved *internal prediction* across participants, based on the likelihood of participants as a group selecting a given trial image as containing a target. This was done by computing image-wise correlations between the CI and the given trial image, as shown in Fig. 1c.

The IWCs were generated for every single image that was presented to participants. Then the trials were sorted from least to greatest with respect to their IWCs and further split into 400 bins in order to estimate the likelihood of an image with a given IWC being selected as containing a target. The number of bins was selected due to its allowance for enough trials in each bin for the resulting proportion to be considered continuous. To ensure that bin size had no effect on the reliability of any resulting relationships, the same procedures were also conducted using 50 and 1,000 bins. The proportion of images identified as having a target was then calculated for each bin, along with the mean IWC value. The relationship between these two variables was then observed and interpreted. This procedure, predicting participants' responses using the CI will be referred to as internal prediction.

An assessment of image-target similarity involved *external prediction* across participants, based on the likelihood of participants selected a given trial image in the test phase as containing a target. An IWC was computed for each trial image with respect to the target. As with the internal predictions, this was done by computing image-wise correlations between the CI and trial images that were not involved in forming the Cis (hence the use of nontarget trials in the practice phase trials), as shown in Fig. 1d. The same binning procedures were also followed.

Both internal and external prediction measures yielded a single number for each trial image, interpreted as the degree of similarity between the trial image and the target the participants were assigned to identify.

2.2 *Results*

2.2.1 Practice Phase

Figure 2a shows that easy targets were more visible to participants than hard targets, as assessed by mean d', $F(1,22) = 16.90$, $p < 0.001$, but that there was no influence of instructions on d', $F(1,22) = 0.58$, $p = 0.46$, nor was there an instruction x difficulty interaction, $F(1,22) = 1.52$, $p = 0.23$.

Figure 2b shows that there was a trend for easy targets to result in more liberal responding than hard targets, but none of the effects in this analysis were statistically significant: main effect of instructions on decision bias, $F(1,22) = 0.68$, $p = 0.42$; main effect of target difficulty, $F(1,22) = 2.23$, $p = 0.15$; and interaction between instructions and target difficulty, $F(1,22) = 0.05$, $p = 0.82$.

2.2.2 Test Phase

Figure 3 shows that the active instructions resulted in more false alarms (reports of superstitious perception) than the passive instructions, $t(9.74) = 2.17$, $p = 0.056$. Not shown in the figure is that participants in both instruction conditions also responded about equally quickly (active mean RT = 1,438 ms, SD = 942 ms; passive mean RT = 855 ms, SD = 364 ms, $t(9.54) = 1.60$, $p = 0.14$).

Inspection of Fig. 4 suggests that there were more distinctive visible patterns in the CIs of the passive condition than in the active condition. But we did not rely on visual inspection to assess the quality of these CIs. To test how well the CIs were

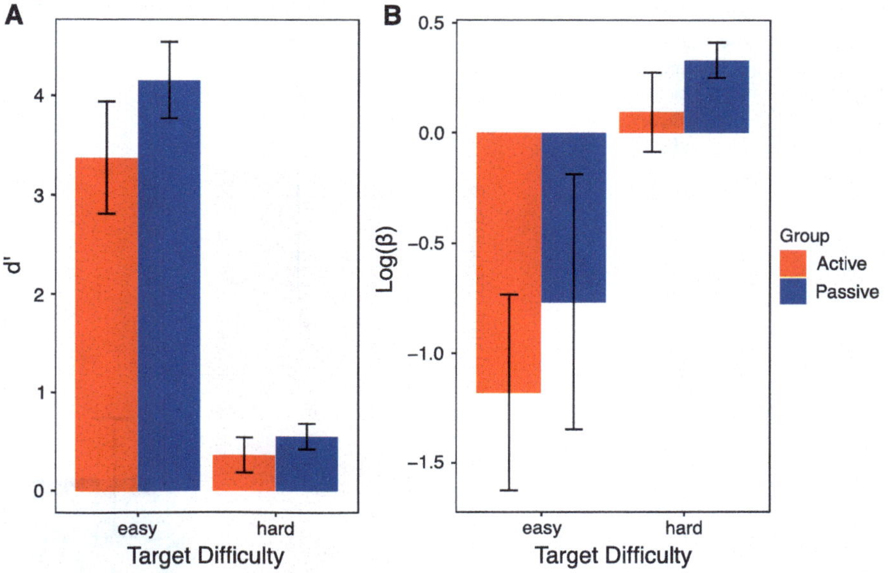

Fig. 2 (**a**) Mean sensitivity of participants' responses to the visible target, as indexed by d-prime. (**b**) Mean decision bias in participants' responses to the visible target, as indexed by log beta. Error bars represent the standard error of the mean

able to discriminate between their respective target and other potential targets, an IWC was calculated between each participant's CI and the two potential targets in order to estimate the degree to which the CI can discriminate the target for which it was generated (x or +). Discriminability ratings were averaged across assigned targets for each condition to estimate how well the overall condition was able to generate discriminating CIs. In both conditions, discriminability was very low. In the active condition, CIs had an average similarity score of $r = 0.003$ with their respective targets, a similarity score of $r = -0.001$ for the high similarity target, and $r = -0.002$ for the low similarity score. In the passive condition, CIs had an average similarity score of $r = 0.045$ for their respective targets, a score of $r = 0.040$ for high similarity targets, and $r = 0.027$ for low similarity targets. The CIs from the passive condition, while having a low degree of similarity with any targets, were able to discriminate more effectively between the possible targets than were CIs from the active condition.

2.2.3 Internal Prediction

An IWC was calculated between each participant's CI and each of their trial images. Larger, positive IWCs indicate higher similarity between the participant's CI and a trial image. Negative IWCs indicate lower similarity between the participant's CI

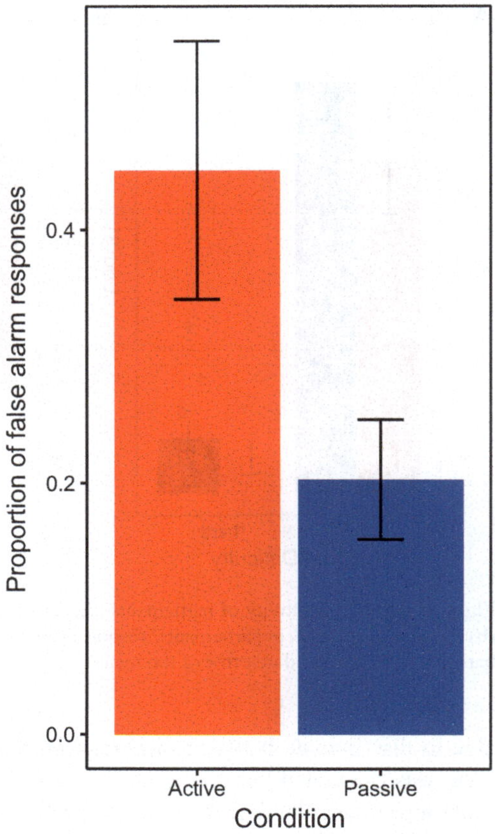

Fig. 3 The mean proportion of false alarms in the test phase for humans in Experiment 1, where there was no visible target. Error bars represent the standard error of the mean

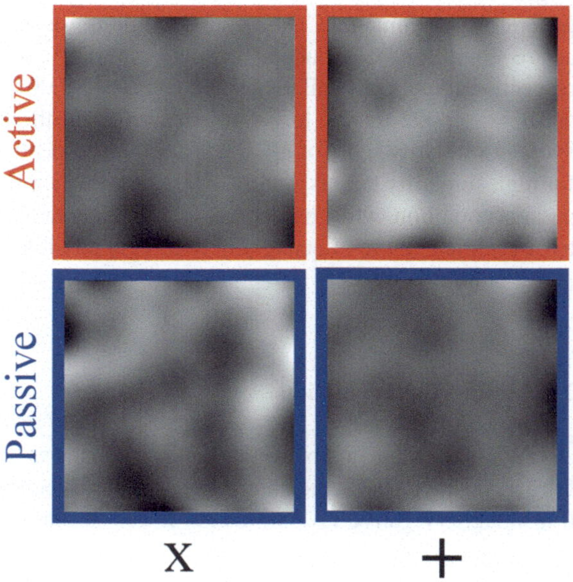

Fig. 4 Classification images generated for each target from participant's responses in the two strategy conditions of Experiment 1

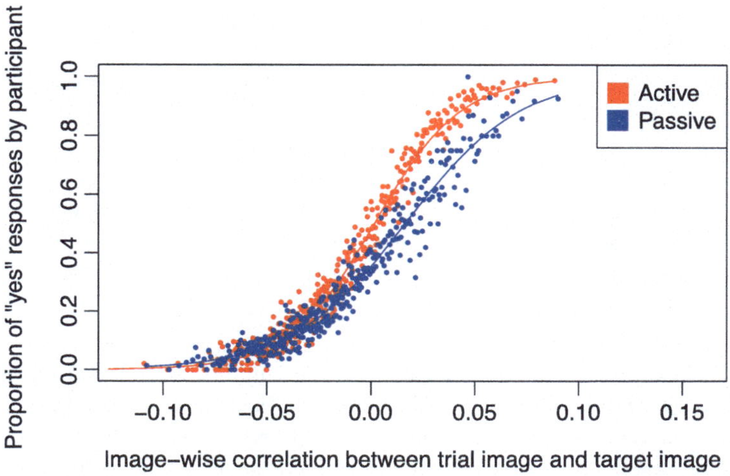

Fig. 5 A scatterplot of the relationship between the IWC of the CI and each trial image and the proportion of images in each bin selected as containing a target in Experiment 1

and the trial image. These IWCs were used to predict the likelihood of a participant identifying a target in a trial image.

Figure 5 shows the sigmoidal relationship between the IWC for a participants' CI and each trial image and the proportion of false alarms for images with that IWC value, grouped into 400 bins. To quantify the strength of these relationships, a sigmoidal transform was performed to create a linear relationship, before a correlation was computed. The sigmoidal relationship is very strong in both cases, $r(397) = 0.99$, $p < 0.001$, in the active condition, and $r(397) = 0.98$, $p < 0.001$ in the passive condition. The two conditions diverge, however, with respect to the shape of the sigmoidal relationship. The slope of the relationship is steeper in the active condition than in the passive condition. This was verified by fitting a sigmoid curve to each condition, showing that the active condition data, slope = 48.7 (SE = 0.7), did indeed have a steeper slope than the passive condition, slope = 36.7 (SE = 0.6). A test of significance of the difference between two independently observed slopes was conducted as suggested by Cohen et al. (2003), $z = 13.22$, $p < 0.001$. However, the relationship between bin size and the strength of the relationship observed in Experiment 1 was still present in both conditions, with a very strong linear relationship between the proportion of false alarms in each bin and sigmoid-transformed IWC values in both the passive and active conditions. These values are shown in Table 1.

These tests of internal prediction were also conducted with bins of size 50 and 1,000 to examine whether they were altered by bin size. The observed slope for the two conditions remained the same for all three bins, suggesting that bin size does not affect these values. In addition, the relationship between the proportion of false alarms and the sigmoid-transformed mean IWC for each bin decreased slightly as the

Table 1 Internal prediction: image-wise correlations between participants' classification images and their responses to white-noise images

Condition	Number of bins	Observed correlation	Statistical significance
Active	50	$r(48) = 0.998$	$p < 0.001$
	400	$r(398) = 0.99$	$p < 0.001$
	1,000	$r(998) = 0.97$	$p < 0.001$
Passive	50	$r(48) = 0.997$	$p < 0.001$
	400	$r(398) = 0.98$	$p < 0.001$
	1,000	$r(998) = 0.94$	$p < 0.001$

Note that the results depend very little on bin size

number of bins increased. The decrease in the strength of the relationship is likely due to less aggregation of the data as the number of bins increases, and thus a weaker signal-to-noise ratio is obtained. Nonetheless, the reliability of these relationships remained relatively constant, likely due to an increase in degrees of freedom as the bin number increases, which strongly suggests that the relationship is present. These results are shown in Table 1.

2.2.4 External Prediction

A second set of IWCs were calculated between the target assigned to each participant and their respective trial images. As was the case for the IWCs involving the CIs, these can be interpreted as the similarity between a given trial image and the target assigned to the participant. As in Experiment 1, the IWC scores can be used to predict the likelihood with which participants will identify a target in a given trial image.

Figure 6 shows the relationship between the IWC with participants' target images and the proportion of false alarms in each of 400 equally sized bins of IWC observations. In the passive condition, there was a moderate relationship between the observed IWC and the proportion of false alarms, $r(397) = 0.25$, $p < 0.001$. However, the active condition shows no relationship between the IWC with participants' target images and the proportion of false alarms, $r(397) = 0.01$, $p = 0.87$. These relationships were significantly different from one another, $z = 3.45$, $p < 0.001$. To verify that this difference was not influenced by the number of bins, the two correlations were recalculated for 50 and 1,000 bins. The results were similar, showing the same trends as observed in the previous experiments, but with reliability remaining consistent. A summary of these results can be found in Table 2.

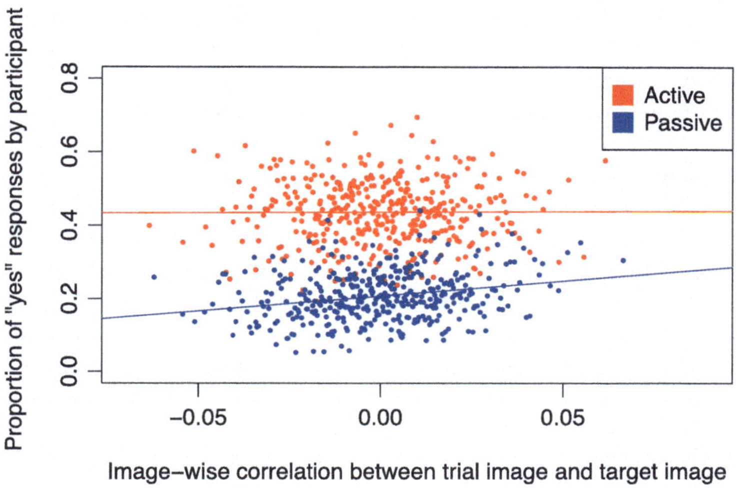

Fig. 6 A scatterplot of the relationship between the likelihood of an image being selected as a target and the IWC between that image and the target image in Experiment 1

Table 2 External prediction: image-wise correlations between participants' assigned targets and their responses to white-noise images

Condition	Number of bins	Observed correlation	Statistical significance
Active	50	$r(48) = 0.04$	$p = 0.76$
	400	$r(398) = 0.01$	$p = 0.87$
	1,000	$r(998) = 0.00$	$p = 0.95$
Passive	50	$r(48) = 0.57$	$p < 0.001$
	400	$r(398) = 0.25$	$p < 0.001$
	1,000	$r(998) = 0.15$	$p < 0.001$

Note that the results depend very little on bin size

3 Experiment 2: Superstitious Machines

Experiment 2 was designed to address our third goal: to test for the presence of superstitious perceptions within a neural network without using human responses for training. In the past, comparisons between CNNs and humans have been made after training a CNN on human responses and then measuring the similarity in responses between the trained CNN and human responses (Cichy et al. 2016; Khaligh-Razavi and Kriegeskorte 2014; Kheradpisheh et al. 2016; Peterson et al. 2017; Yamins et al. 2014; Yamins and DiCarlo 2016). A criticism of this approach stems from the observation that neural networks can learn to represent any mathematical function (Hornik et al. 1989).

Here we first trained a CNN to identify the presence or absence of a target "X" within noise images, making the task very similar to the one engaged by our study participants in the practice phase of Experiment 1. If the neural network displays a pattern of results for CIs and their success in internal and external predictions that are similar to humans, it would be evidence that the underlying processes in the two systems may be similar.

Our fourth goal was addressed in this experiment by attempting to model the role of cognitive strategy in CNNs by varying response bias. This was done by matching the response bias of the CNN to that of humans in the two instructions condition. A liberal responding condition and a conservative responding condition were taken as proxies for the active and passive instructions conditions in Experiment 1. Previous work suggests that CNN models do a better job of modeling that activity of the human cortical visual areas, than they do of modeling the neural activity to the areas responsible for central executive functions (Cichy et al. 2016). Consistent with this, we expected CNN responses in the conservative responding condition to be most similar to humans who are not actively engaging their executive functions on the task (i.e., the "passive" condition Experiment 1).

3.1 Methods

3.1.1 Model Description

The CNN model used in this experiment was a modified, pre-trained form of AlexNet (Krizhevsky et al. 2012), a CNN which attained state-of-the-art image classification accuracy for the ImageNet benchmark in 2012. This model was chosen due to its relative popularity in comparing CNNs to the human visual system (Khaligh-Razavi and Kriegeskorte 2014; Yamins et al. 2014). With the additional information given by the aforementioned comparisons, AlexNet was the model with the highest documented similarity with the human visual system, making it the ideal model to use when comparing the ability of neural networks and humans to perceive superstitiously.

Since the model was not originally trained to identify an "x" in noise, the network needed to be modified in order to perform such a task. To do so, the weights of the nodes in the final two fully connected layers of AlexNet were randomly initiated, and a new binary classifier replaced the original classification layer of the network. These new layers were trained to perform the novel detection task. The decision to retrain the final two fully connected layers was made due to these layers being the point in the network hierarchy in which spatial information is lost. All previous layers of the network were convolutional layers, meaning that spatial information in those previous layers is preserved from one layer to the next (Krizhevsky et al. 2012). Notably, however, the weights in the five convolutional layers were not modified from their trained state, as provided by its creators (Krizhevsky et al. 2012).

Training of the three randomly initiated layers of the network was performed using stimuli generated in the same fashion as the easy training images used in Experiment 1. These stimuli were generated either by reducing the contrast slightly between the target and the image background (target present trials) or by creating a consistent gray image, luminance matched to the target (target absent trials). All training images were then overlayed with Gaussian random noise. Unlike previous chapters, the network was trained exclusively on images containing an “x” as the target image, learning to distinguish between images that contained an “x” hidden in noise from images that contained pure noise. The network was trained on 100,000 training images (50% of them containing a target) for 19 epochs, meaning that the network was shown each training image 19 times during training.

The number of epochs for training was chosen in a very specific manner. After each training epoch, the network was testing on novel “easy training” images and on “hard training” images from Experiment 1. From the network’s responses, a sensitivity and bias score was calculated. Training continued until the sensitivity and bias at the end of the epoch were approximately matched in sensitivity and bias to the participants observed in Experiment 1.

3.1.2 Stimuli and Procedures

White-noise trial images were generated in the same way as in Experiment 1: each image was generated pixel by pixel, randomly sampling a luminance from a Gaussian distribution centered at 0.5, which a standard deviation of 0.1 for each of the 2,500 pixels. As an analogue to the superstitious perception task completed by humans, the network was fed 16,000 50 $\times$ 50 px noise images centered in a 227 $\times$ 227 px gray image exactly as done for the human participants. The size of the overall image, gray border included, was set to match the 227 $\times$ 227 px input the AlexNet architecture was built to accept (Krizhevsky et al. 2012). The network’s reported confidence in the presence of a target was recorded for each of the 16,000 images resulting in a 16,000 response vector.

Once the network confidence was recorded for each trial image, decision criteria were set at two different points. In a liberal condition, the decision criterion was set such that the false alarm rate during the test phase was matched with the mean false alarm rate observed in the active condition of Experiment 1. In a conservative condition, the decision criterion was set such that the false alarm during the test phase was matched with the mean false alarm rate observed in the passive condition of Experiment 1. Once again, due to the complete lack of any targets among the images presented to the network in the test phase, any positive identification of the presence of a target was a false alarm.

3.1.3 Data Analysis

Data analysis procedures were similar to those performed in Experiment 1 but modified slightly to accommodate conservative and liberal criterion settings. All analyses were performed on the two groups separately, and then the results statistically compared.

Analysis of the practice phase was performed using mixed effects ANOVA, with criterion setting as a between-subject factor with two levels (liberal and conservative) and target difficulty as a within-subject factor with two levels (easy and hard). One analysis compared sensitivity (d') across factors, and a second analysis compared decision bias, lnβ.

The analysis of the test phase was identical to that of Experiment 1, with the liberal and conservative criterion groups being compared with t-tests of means and a z-test to compare the two discrimination slopes. A Fischer's Z-test compared the strength of the relationships between target image IWCs and proportion of false alarms for the two groups.

3.2 *Results*

3.2.1 Practice Phase: Training and Validation Stimuli

Figure 7 shows a summary of the neural network's performance on trials generated in the same way as the easy and hard trials in Experiment 1. For this phase of the experiment, the network ran without human intervention. If the network was more than 50% confident that a target was present, it responded "target present." Figure 7a shows a large difference in the network's sensitivity (d') to targets in the easy ($d' = 6.34$) and hard target ($d' = 0.57$) difficulty sessions. Figure 7b shows a large

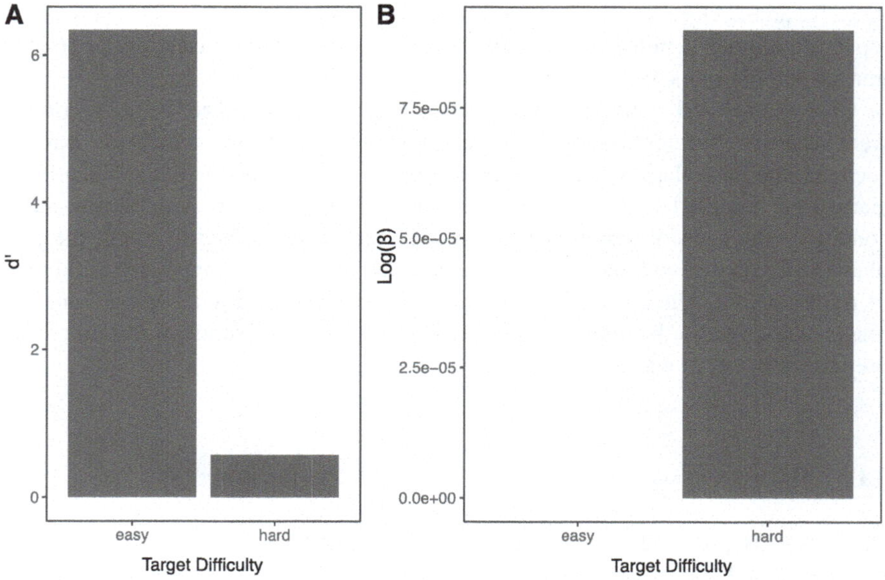

Fig. 7 The mean proportion of false alarms in the test phase for CNNs in Experiment 2, where there was no visible target. Error bars represent the standard error of the mean

difference between the neural network and human participants. Namely, the neural network showed little to no bias in its decisions in both the easy ($\ln\beta = 0$) and hard ($\ln\beta = 0$) target difficulty sessions, unlike in the human data.

3.2.2 Test Phase

Figure 8 shows the distribution of confidence ratings made by the neural network for the presence of a target image in the white-noise test trials. These ratings could range from zero to one, with zero denoting a high degree of confidence that no target is present and a one denoting a high degree of confidence that a target is present. The standard approach to making decisions based on the neural network output is to consider a target to be present if the confidence is above 0.5. Based on this criterion, the network did not perceive superstitiously at all, with its confidence ratings ranging from 0.0006 to 0.0117. However, by accepting a lower decision boundary, we can assess the remaining few superstitious responses made by the network.

We mimicked the observed proportions of false alarms in the active instruction conditions of Experiment 1 by using a 55.3 percentile point in the CNN data to match the 44.7% false alarm rate in human participants. In the passive condition, the boundary was set at 79.7 percentile point to mimic the 20.3% false alarm rate observed in humans. These decision boundaries are depicted in Fig. 8, with a red vertical bar depicting the liberal condition and a blue vertical bar the conservative condition. If the network's confidence was above the decision boundary in a given condition, the trial was labeled as a false alarm.

Figure 9 shows the CIs calculated from the CNN responses in the liberal and conservative condition. Notably, the two CIs are very similar in form. The image-wise correlation between the two CIs was $r = 0.64$. An IWC was calculated between each CI and the target image in order to evaluate their similarity to the target. The CI generated from the conservative responses correlated with the target with a strength

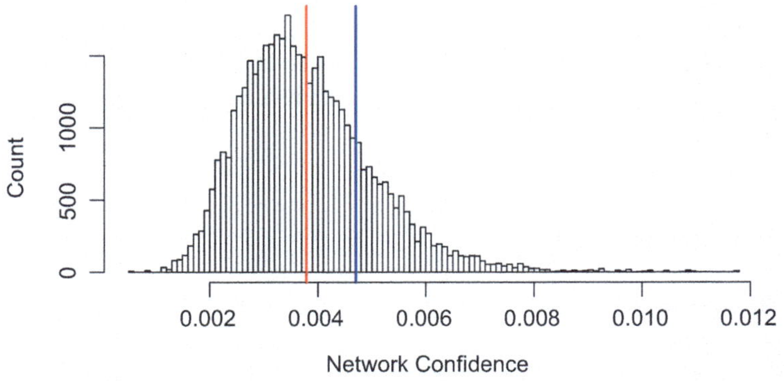

Fig. 8 A histogram of the confidence of the neural network for the presence of a target image in a white-noise trial images. The red vertical line denotes the decision cutoff for the liberal condition. The blue vertical line denotes the decision cutoff for the conservative condition

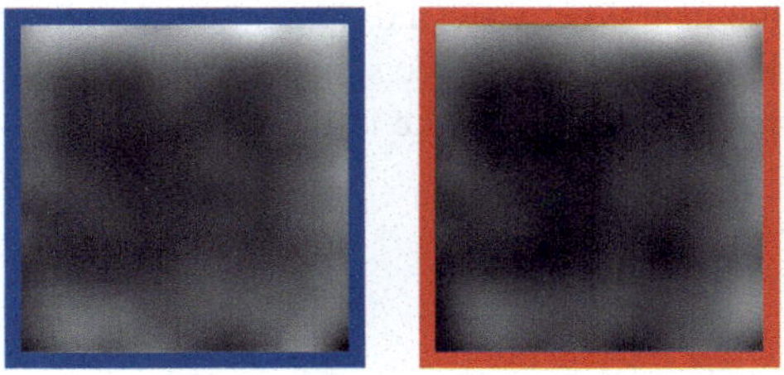

Fig. 9 Classification images generated for each target from CNN's responses in the two decision bias conditions of Experiment 2

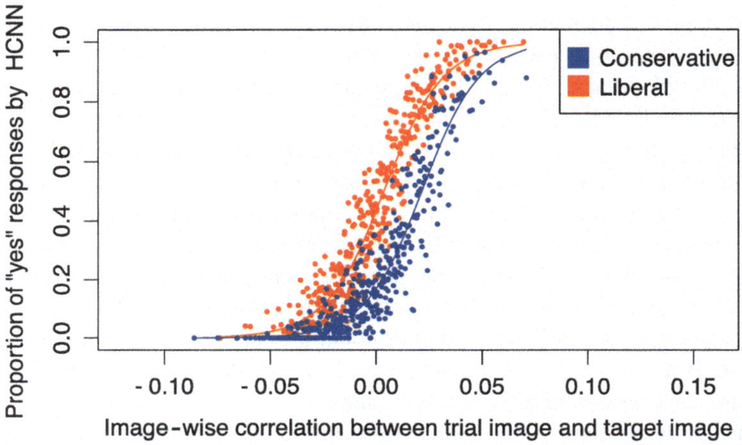

Fig. 10 A scatterplot of the relationship between the IWC for the CNN's CI and each trial image and the proportion of images in each bin selected as containing a target in Experiment 2

of $r = 0.55$, while the CI generated from the liberal responses correlated with the target with a strength of $r = 0.54$. To test the ability of the generated CIs to discriminate what the target of the network from a low similarity target, the IWCs were calculated between the conservative and liberal CIs and the "+" target. The IWCs were both markedly smaller than for the correct target, $r = 0.38$ and $r = 0.33$, respectively.

3.2.3 Internal Prediction

Figure 10 shows a sigmoidal relationship between the IWC generated for each trial image and the neural network's CIs and the proportion of false alarms. The figure shows the relationship for both the conservative and liberal conditions. The proportion of false alarms was generated by binning responses into equal groups based on similar IWC scores between the CI and each trial image. A sigmoid function was fit to each set of data, resulting in a slope and center constant for both the conservative

Table 3 Internal prediction: image-wise correlations between CNN classification images and their responses to white-noise images

Condition	Number of bins	Observed correlation	Statistical significance
Liberal	50	$r(48) = 0.99$	$p < 0.001$
	400	$r(398) = 0.96$	$p < 0.001$
	1,000	$r(998) = 0.90$	$p < 0.001$
Conservative	50	$r(48) = 0.99$	$p < 0.001$
	400	$r(398) = 0.94$	$p < 0.001$
	1,000	$r(998) = 0.87$	$p < 0.001$

Note that the results depend very little on bin size

and liberal conditions. In the liberal condition, there was a relatively steep slope of 71.7 (SE = 1.9), and a center constant of 0.004 (SE = 0.0003). In the conservative condition, there was also a relatively steep slope of 74.2 (SE = 2.2), and a center constant of 0.022 (SE = 0.004). The two slope coefficients were not significantly different from one another, $z = 0.84$, $p = 0.40$. Furthermore, in order to test the goodness of fit of the two models, we calculated the correlation between the sigmoid-transformed IWC scores and proportion of false alarms per bin. The goodness of fit was very high in both the liberal condition, $r(398) = 0.96$, $p < 0.001$, and in the conservative condition, $r(398) = 0.993$, $p < 0.001$. A range of bin sizes were calculated in order to estimate the effect of the number of bins on the magnitude of the observed goodness of fit. We performed the same analysis on 50, 400, and 1,000 bins generated from 16,000 trial images fed to the network; the results are summarized in Table 3.

3.2.4 External Prediction

Figure 11 shows a linear relationship between IWC scores for each trial image with the target "x" image and the proportion of false alarm responses by the neural network at both the conservative and liberal decision points. The proportion of false alarm scores was generated by taking the proportion of false alarms from equal bins of responses whose trial images had similar IWC scores with the target. Figure 11 shows the relationship as observed when the data are grouped into 400 equally sized bins. In the liberal condition, there was a relatively strong linear relationship between IWC scores and proportion of false alarms, $r(398) = 0.58$, $p < 0.001$. In the conservative condition, there was also a relatively strong relationship between IWC scores and the proportion of false alarms, $r(398) = 0.60$, $p < 0.001$. Most notably, there was no significant difference in the strength of the relationship between the conservative and liberal conditions, $z = 0.43$, $p = 0.67$. In order to verify that the number of bins had no effect on our conclusions, we tested the

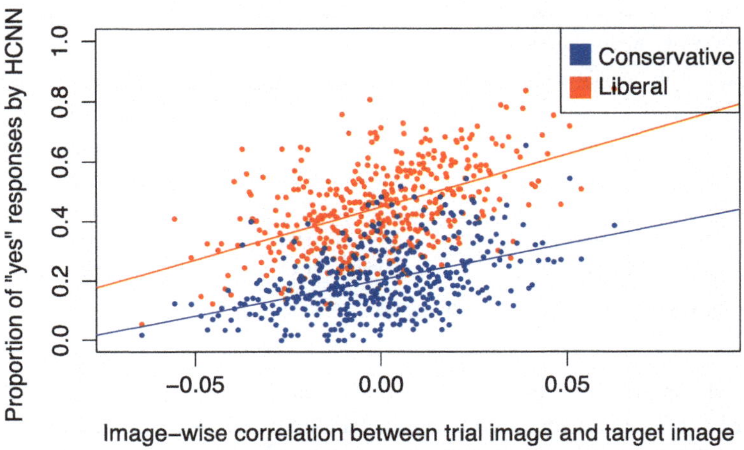

Fig. 11 A scatterplot of the relationship between the IWC for the CNN's target image and the proportion of images in each bin identified as containing a target in Experiment 2

Table 4 External prediction: image-wise correlations between the CNN's assigned target and their responses to white-noise images

Condition	Number of bins	Observed correlation	Statistical significance
Liberal	50	$r(48) = 0.87$	$p < 0.001$
	400	$r(398) = 0.58$	$p < 0.001$
	1,000	$r(998) = 0.42$	$p < 0.001$
Conservative	50	$r(48) = 0.90$	$p < 0.001$
	400	$r(398) = 0.60$	$p < 0.001$
	1,000	$r(998) = 0.41$	$p < 0.001$

Note that the results depend very little on bin size

relationship at bin sizes of 50, 400, and 1,000. A summary of the results for the different bin sizes are found in Table 4.[1]

[1]In order to ensure the stability of the CNN model, all model weights and biases were truncated to four decimal places as opposed to the unlimited number of significant digits allowed under normal CNN procedures. All analyses covered in this chapter were performed anew using the model with truncated weights. The changes had no influence on the various analyses but did noticeably reduce the quality of the generated CIs. The shape of the target was still present in both the liberal and conservative conditions. Nonetheless, the CIs lacked the clarity observed in the CIs generated from the model without truncated weights. These observations demonstrate the sensitivity of the CNN model to restricting the significant digits of the model weights. However, the observations do not change the overall conclusions of this work.

4 General Discussion

The first goal of our study was to compare the internal and external reliability of human responses in a superstitious perception task. Internal reliability was measured by the correlation between the classification image generated from participants "detect" versus "not detect" responses and the same white-noise images that were used to generate the classification images. The magnitude of this correlation is vulnerable to data over-fitting since there is no independent measure of the participant's search template. External reliability was measured by the correlation between the specific images assigned to participants as targets and the white-noise images classified as "detect" versus "not detect" by the participant. Any difference between the internal and external prediction measures can be used to assess the degree to which a CI generated in the superstitious perception task effectively captures an internal representation of an object in human visual perception. Our main finding in Experiment 1 was that participant's responses had much stronger internal than external reliability, implying that the classification images reflect sampling noise much more than they reflect individual differences in participants' search templates.

The second goal was to examine the role of cognitive strategy on response reliability in the superstitious perception task. We did this in Experiment 1 by randomly assigning participants to either *passive* or *active* task instructions in the superstitious perception task, so named because they are intended to manipulate the involvement of central executive functioning in the task. The results showed that the CIs generated by participants in the passive condition had stronger external reliability than those generated by participants in the active condition. Greater external reliability in the *passive* condition suggests that recruitment of executive functions to the superstitious perception task *hinders* the generation of a CI that resembles the target.

A third goal was to compare humans and CNNs on the superstitious perception task, while a fourth goal was to compare the strategy differences of humans in Experiment 1 with the responses of CNNs making decisions under either conservative or liberal response tendencies. We achieved both of these goals in Experiment 2, where we compared the responses of humans and CNNs on the same superstitious perception task. The results showed some remarkable similarities in the responses but also some important differences. In particular, the response patterns of CNNs were most similar to human responses when the CNN's were directed to make decisions with a conservative response bias (i.e., making relatively fewer false alarms). Moreover, under both conservative (few false alarms) and liberal decision-making tendencies (more false alarms), the responses of CNN resembled those of the humans in Experiment 1 most closely when humans were instructed to use a passive strategy. This finding implies that CNNs, at least as currently constituted, do not capture the contributions of executive functioning in the superstitious perception task.

4.1 Implications for the Study of Superstitious Perception in Humans

The classification images averaged over all participants, as shown in Fig. 4, showed only slight visual resemblances to each of their targets. Yet, despite there being such a small visible (qualitative) difference between the two CIs, when the CI's of each participant were compared quantitatively to the noise images, the passive instructions yielded significantly greater quantitative resemblance than the active condition. This implies that the CI comes to resemble the internal search target of the participant when they are using a passive rather than an active strategy. Our interpretation of this difference is that the passive instructions discourage the engagement of central executive functions in the decision-making of the task.

An analysis of the discriminative ability of the CIs in the two conditions further indicates that the passive instructions yielded CIs that were better able to discriminate between the two targets. This supports the notion that the superstitious perception task is more reliably linked to the objective signals of resemblance between the noise images and the target when participants are engaged in a passive strategy, that is, a cognitive strategy intended to minimize the involvement of central executive brain functions.

Figure 6 shows a very large difference in the predictive accuracy of the classification images between strategy conditions when the external reliability of participant's responses was assessed. Recall that these analyses did not depend on the formation of the CIs. In the active condition, the small and nonsignificant correlation between target-image similarity and the false alarm rate suggests that the participants' target images did not form the basis of participants' responses. In the passive condition, we observed a moderate, significant correlation between target-image similarity and the false alarm rate. The presence of this correlation in the passive condition suggests that participants' assigned target images formed the basis of participants' responses. This difference in outcome between the active and passive conditions indicates that participants' ability to detect coincidental signals in random noise images is modulated by the recruitment of executive functions. These results are in line with previous research in the visual categorization literature showing that judgements based on overall shape similarity are performed most effectively when executive functions are least likely to be recruited for the task (Jacoby and Brooks 1984).

A critical finding of the strategy manipulation in Experiment 1 was that the false alarm rate was significantly greater for participants instructed to be active rather than passive. One explanation might be that participants in the active condition calibrated their responses more sensitively to task instructions, which were that 50% of the images contained targets. However, we favor the interpretation that the active instructions also encouraged participants to recruit executive functions for the task, and in doing so, used memorial (non-sensory) information to guide their responses more so than participants given passive instructions. This interpretation is bolstered by the findings that responses in the active condition had lower external reliability (lower resemblance to the assigned target) than in the passive condition. A simple increase in the false alarm rate is unlikely to lead to that same result on its own.

There are other possible reasons why the active condition may have had lower external reliability than the passive condition. For example, in keeping with the instructions, participants in the active condition may have focused on smaller subsets of the trial image, detecting small portions that appear particularly target-like while disregarding the rest of the image. Since the image-wise correlations we computed considered the image as a whole, these local correlations may have been invisible to the procedure. The responses of passively instructed participants, in contrast to this, showed significantly more sensitivity to overall shape similarity: their false alarms rates increased in direct proportion to the similarity between target and noise images.

Experiment 2, where we tested a CNN on the same task, added to our understanding of superstitious perception in humans in several ways. First, a simple change in the response tendency to false alarm in the CNN does not account for the differences attributable to task strategy in humans. The responses of the CNN, under both conservative and liberal response biasing modes, more closely resembled human participants in the passive strategy condition than the active one. We interpret this finding to suggest that human participants following passive instructions respond based on weak, yet detectable, evidence of a target resemblance in a white-noise image. Participants following active instructions, on the other hand, while reporting more false alarms, seem to be reporting evidence favoring a target without any objective signal for its existence in the white noise. This raises the intriguing possibility for future work that there are two distinct classes of superstitious perception. One form, the kind exhibited by passively instructed participants, is the detection of weak but objective evidence favoring a target image. Another form, the kind exhibited by actively instructed participants, involves the imposition of internally generated signals (presumably based on long term memories) on the sensory registers in the visual cortex. Finding techniques to distinguish these two possible variants of superstitious perception will be an important challenge for future theory development in this area.

4.2 Implications for Superstitious Perceptions by Machine

The CIs generated from CNN responses, as shown in Fig. 9, resembled the target shapes better than the CIs generated from human responses. This was true both qualitatively and quantitatively. There is a distinctive x-like form visible in both conditions for the CNN, and these CIs each have high IWC scores with the target image and lower IWC scores with the nontarget image (the +). So, by every standard that has been applied to human responses in the past, this CNN was responding as superstitiously as humans and likely even more so. The only caveat to this claim concerns the baseline rate of false alarms in the CNN, which never reached the high human levels we observed in Experiment 1. Yet, when we examined the CNN responses, using decision criteria that matched those of the human participants, the resulting Cis were as strong, if not stronger evidence of superstitious perception than in humans.

There was a moderate-to-strong relationship in the CNN responses between the trial-target IWC and the proportion of false alarms for both conditions, as shown in Fig. 11. Such strong correlations suggest that CNN responses were driven by the similarity of the trial images to the target more so than for humans. In keeping with our interpretation of the human data in the previous section, it also suggests that CNN responses were based on the kind of superstitious perception that is based on the detection of weak or trace signals of the assigned target in the white-noise images. Notably, we found no evidence that CNN responses were based on the kind of superstitious perception that derives from imposing memorial (non-sensory) influences on the sensory registers of the visual cortex. Evidence consistent with that interpretation would be a reduction in the correlation between target image and noise images as the false alarm rate was increased; something we did not find. Taking the human and CNN data together, it suggests that CNNs at present are only able to model the weak trace detection form of superstitious perception.

Acknowledgments This research was supported by a Natural Sciences and Engineering Research Council of Canada grant (No. RGPIN-2013-36796) to J.T.E. Portions of this research were conducted as part of P. L's MA thesis at the University of British Columbia (Aug 2017) entitled "Superstitious perception in humans and convolutional neural networks."

References

Brainard DH (1997) The psychophysics toolbox. Spat Vis 10(4):433–436

Burnham KP, Anderson D (2002) Model selection and multimodel inference: a practical information-theoretic approach. Springer, New York

Cichy RM, Khosla A, Pantazis D, Torralba A, Oliva A (2016) Deep neural networks predict hierarchical spatio-temporal cortical dynamics of human visual object recognition. Sci Rep 6(1):27755

Cohen J, Cohen P, West SG, Aiken LS (2003) Applied multiple regression/correlation analysis for the behavioral sciences, 3rd edn. L. Erlbaum Associates, Mahwah

DeAngelis GC, Ohzawa I, Freeman RD (1993) Spatiotemporal organization of simple-cell receptive fields in the cat's striate cortex. II. Linearity of temporal and spatial summation. J Neurophysiol 69(4):1118–1135

Diamond A (2013) Executive functions. Annu Rev Psychol 64(1):135–168

Dyan P, Abbott L (2001) Theoretical neuroscience: computational modeling of neural systems. Massachusetts Institute of Technology Press, Cambridge

Fyfe S, Williams C, Mason OJ, Pickup GJ (2008) Apophenia, theory of mind and schizotypy: perceiving meaning and intentionality in randomness. Cortex 44(10):1316–1325

Gosselin F, Bonnar L, Paul LK, Schyns PG (2001) "Superstitious" perceptions to depict pure internal object representations. J Vis 1:46

Gosselin F, Schyns PG (2003) Superstitious perceptions reveal properties of internal representations. Psychol Sci 14(5):505–509

Haffenden AM, Schiff KC, Goodale MA (2001) The dissociation between perception and action in the Ebbinghaus illusion: nonillusory effects of pictorial cues on grasp. Curr Biol 11(3):177–181

Hornik K, Stinchcombe M, White H (1989) Multilayer feedforward networks are universal approximators. Neural Netw 2(5):359–366

Jacoby L, Brooks L (1984) Nonanalytic cognition: memory, perception, and concept learning. Psychol Learn Motiv 18:1–47
Jones JP, Palmer LA (1987) The two-dimensional spatial structure of simple receptive fields in cat striate cortex. J Neurophysiol 58(6):1187–1211
Khaligh-Razavi SM, Kriegeskorte N (2014) Deep supervised, but not unsupervised, models may explain IT cortical representation. PLoS Comput Biol 10(11):e1003915
Kheradpisheh SR, Ghodrati M, Ganjtabesh M, Masquelier T (2016) Humans and deep networks largely agree on which kinds of variation make object recognition harder. Front Comput Neurosci 10:92
Kleiner M, Brainard D, Pelli D (2007) What's new in psychtoolbox-3? Perception 36(14):1–16
Krizhevsky A, Sutskever I, Hinton G (2012) ImageNet classification with deep convolutional neural networks. Adv Neural Inf Proces Syst 25
Leibowitz H, Brislin R, Perlmutter L, Hennessy R (1969) Ponzo perspective illusion as a manifestation of space perception. Science 166(3909):1174–1176
Liu J, Li J, Feng L, Li L, Tian J, Lee K (2014) Seeing Jesus in toast: neural and behavioral correlates of face pareidolia. Cortex 53:60–77
Marcel AJ (1983) Conscious and unconscious perception: experiments on visual masking and word recognition. Cogn Psychol 15(2):197–237
Massaro DW, Anderson NH (1971) Judgmental model of the Ebbinghaus illusion. J Exp Psychol 89(1):147–151
Pelli DG (1997) The VideoToolbox software for visual psychophysics: transforming numbers into movies. Spat Vis 10(4):437–442
Peterson JC, Abbott JT, Griffiths TL (2017) Adapting deep network features to capture psychological representations. In: Proceedings of the 38th conference of the cognitive science society
Rieth CA, Lee K, Lui J, Tian J, Huber DE (2011) Faces in the mist: illusory face and letter detection. i-Perception 2(5):458–476
Ringach D, Sapiro G, Shapley R (1997) A subspace reverse correlation technique for the study of visual neurons. Vis Res 37(17):2455–2464
Ringach D, Shapley R (2004) Reverse correlation in neurophysiology. Cogn Sci 28(2):147–166
Rock I, Kaufman L (1962) The moon illusion. Sci Am 207(1):120–131
Shepard RN (1981) Psychological complementarity. In: Perceptual organization. Routledge, Abingdon
Shermer M (2008) Patternicity: finding meaningful patterns in meaningless noise. Sci Am 299:74–75
Smilek D, Enns JT, Eastwood JD, Merikle PM (2006) Relax! Cognitive strategy influences visual search. Vis Cogn 14(4–8):543–564
Theunissen FE, David SV, Singh NC, Hsu A, Vinje WE, Gallant JL (2001) Estimating spatio-temporal receptive fields of auditory and visual neurons from their responses to natural stimuli. Network 12(3):289–316
Van Selst M, Merikle PM (1993) Perception below the objective threshold? Conscious Cogn 2(3):194–203
Whittlesea B, Brooks L (1994) After the learning is over: factors controlling the selective application of general and particular knowledge. J Exp Psychol Learn Mem Cogn 20(2):259
Yamins DLK, DiCarlo JJ (2016) Using goal-driven deep learning models to understand sensory cortex. Nat Neurosci 19(3):356–365
Yamins DLK, Hong H, Cadieu CF, Solomon EA, Seibert D, DiCarlo JJ (2014) Performance-optimized hierarchical models predict neural responses in higher visual cortex. Proc Natl Acad Sci U S A 111(23):8619–8624
Yu RQ, Zhao J (2015) The persistence of the attentional bias to regularities in a changing environment. Atten Percept Psychophys 77(7):2217–2228
Zhao J, Al-Aidroos N, Turk-Browne NB (2013) Attention is spontaneously biased toward regularities. Psychol Sci 24(5):667–677
Zhao J, Yu RQ (2016) Statistical regularities reduce perceived numerosity. Cognition 146:217–222

Dynamic Protention: The Architecture of Real-Time Cognition for Future Events

Mark A. Elliott and Liam Coleman

Contents

Abstract For over 30 years now, a body of physiological evidence has been acquired which indicates that cognitive operations coordinate via the phase synchronization of neuronal firing. While usually ascribed to "binding," i.e., the putting together of basic perceptual features to form more complex perceptual units, this ascription is not without critics, who identify phase synchronization as a function of sensorimotor coordination. From the perspective of an experimental paradigm used to measure the effects of stimulus synchronization, we discuss what is "bound" and attempt a reconciliation between perceptual and sensorimotor accounts of oscillatory synchronization. Our evidence identifies a role for synchronization in protentive coding, this is to say, coding in anticipation of a future event, and hence describes the architecture of real-time cognition for future events.

Keywords Generalized phase angle hypothesis · Perception · Priming · Protention · Return phase hypothesis · Synchronization

---

M. A. Elliott (✉) and L. Coleman
School of Psychology, National University of Ireland Galway, Galway, Ireland
e-mail: mark.elliott@nuigalway.ie

Curr Topics Behav Neurosci (2019) 41: 245–254
DOI 10.1007/7854_2019_94
Published Online: 1 May 2019

1 Do Synchronized Assemblies Bind? A Physiological Debate

The synchronization of oscillatory neural activity at frequencies in the range 30–70 Hz has been argued to relate to the organization of visual events, in particular the "binding together" of visual features to form larger and more complex perceptual objects (the physiological "Binding Hypothesis" summarized by Gray 1999; also Singer 1999). This point of view, while supported by numerous studies, from several laboratories is not without its share of speculation and critic. Related to the current topic, a speculation and consequent disagreement revolved around the link made by Herrmann et al. (1999) between stimulus-"evoked" gamma activity [i.e., frequencies recorded in the electroencephalogram (EEG) at between 30 and 70 Hz] and visual grouping, suggesting that at least the onset of the gamma response is time locked to a stimulus event (see also Herrmann and Mecklinger 2001; Herrmann and Bosch 2001; Tallon-Baudry 2009; Martinovic and Busch 2011 for reviews). Why this finding is controversial lies in the distinction between an evoked EEG response and an induced response. In the case of the former, the response is found to be reliably time locked to stimulus, and this is what Herrmann and colleagues reported. In the case of an induced response, there is no reliable temporal relationship between stimulus events and the phase of the EEG response. Proponents of the physiological binding hypothesis did not find evidence to identify a particular temporal relationship between stimulus and the neuronal response to the stimulus at gamma-band frequencies. This leads not only to the idea that the phase of oscillatory gamma activity is unrelated to the timing of stimulus events but also that it makes sense it is not, as this would avoid the gamma phase being reset by changes in stimulus event structure. Changes in the gamma phase are considered equivalent to changes in perceptual structure, and so, if this idea holds, the independence of the oscillation from stimulus events helps preserve perceptual structure (see Tallon-Baudry and Bertrand 1999 as well as Pantev 1995; Fries et al. 2007, for reviews).

A recent and similarly relevant focus has involved discussion of the relationship between small eye movements (microsaccades) and induced cortical gamma activity. It has been shown that recordings of cortical gamma-band activity, occurring in the time window between 200 and 350 ms after stimulus onset, pick up a response matching the frequency of muscular movements in the eye (Yuval-Greenberg et al. 2008, 2009; Melloni et al. 2009b; Bosman et al. 2009; Melloni et al. 2009a). This has led to debate on the idea that, while induced gamma activity is related to perception, it is not dedicated to binding visual features. As a rejoinder to this debate, there are two EEG studies that report: first, that the morphology of the induced gamma-band activity, which is neither transiently evoked nor evoked by electromyogenic activity linked to eye movements, is sensitive to object recognition in the 200–350 ms post stimulus-onset time window (Hassler et al. 2011). However second, and as reported by Hassler et al. (2013), microsaccade rates were modulated by object familiarity in a time window from 100 to 300 ms after stimulus onset, while later object-coding processes that involve induced gamma-band activity seem independent of microsaccadic activity. The early response in this second study is interpreted generally in terms of the deployment of early (but unknown) mechanisms in visual perception.

In the following, we will describe the largely neglected experimental case for dynamic binding and how the evidence that forms the body of this case informs on this debate. In fact, we will go as far as to say that techniques such as those described in the following sections and the results of the derivative studies offer one of the only ways in which frequency responses and the complex interactions between frequencies relate to the processes involved in perception.

2 The Experimental Case for Binding

One question posed by debate on the physiological and electrophysiological literature concerns the role of an early stimulus-evoked oscillatory response in perception. In the experimental psychology literature, there are a class of paradigms that have employed stimulus synchronizations presented below detection thresholds to prime or bring about the Gestalt organization of the synchronized stimulus elements (Elliott and Müller 1998; Usher and Donnelly 1998). These studies have a priori addressed critics that challenge the necessity and sufficiency of some stimulus synchronization paradigms (Farid 2002; Elliott et al. 2006b) and acknowledge a now large body of evidence indicating that elements of a visual scene are often bound ahead of attentional deployment and are very unlikely to be coded in the same mechanisms as those mediating direct conscious experience of the Gestalt (e.g., Duncan and Humphreys 1989; Rensink and Enns 1995; Driver et al. 2001).

Usher and Donnelly showed that synchronization of all elements at a given orientation at a frequency in the gamma band significantly biases subsequent orientation judgments even though observers cannot reliably report the synchronized orientation. In Elliott and Müller's paradigm (Fig. 1), embedding a figurally relevant grouping as one phase of a multiphase premask that flickered at 40 Hz leads to faster detection reaction times (RTs) to a subsequently presented target grouping, without prior attentional deployment to the location of the grouping in the premask. Observers could not detect the presence of the premask grouping, and because presentation of this stimulus does not cue target presentation, it is referred to as a prime.

Prima facie, the results reported by Elliott and Müller (1998) make two contributions to the aforementioned debates: first, binding may be driven by stimulus activity. In fact, two lines of evidence support this directly: (1) Binding for form perception may be assumed on the basis of the results presented by Shi and Elliott (2007), who found that the magnitude of priming differed for different geometric forms and was correlated positively and significantly with measures of subjective complexity. (2) Binding, as the term refers to neural mechanisms responsible for coding relations between object features, may be assumed on the basis of the findings presented by Elliott et al. (2006a). They found that when the efficiency of

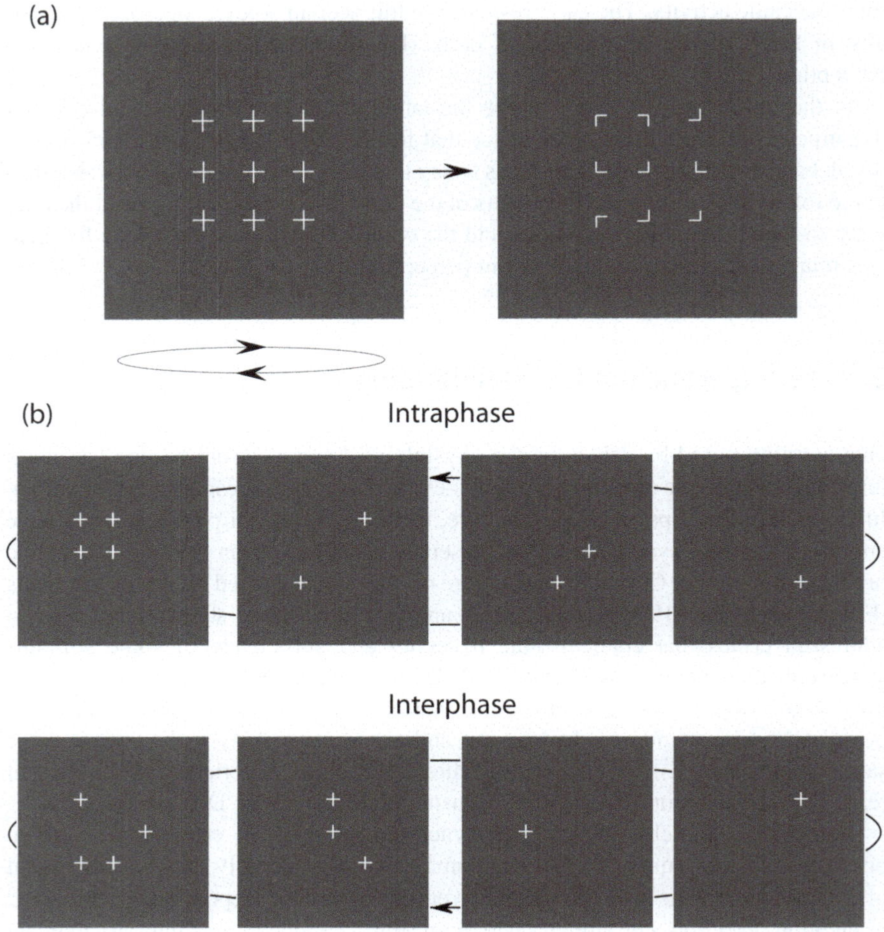

Fig. 1 The premask-matrix presentation paradigm: in (**a**) upon termination the oscillating premask matrix was immediately followed by presentation of a target display comprising 90°-corner junctions to which observers had to make a speeded target (i.e., Kanizsa square) present or absent response. In (**b**) are shown example sequences of four image frames that comprise the premask matrix. In the intraphase conditions (upper panels), one premask frame consists of four crosses in square arrangement; see the upper far left panel. In the intraphase condition (lower panels), these four crosses are distributed across frames

GABAergic interneurons, responsible for modulating synchronized activity, are influenced by drug administration, priming effects are substantially different to control performance. However, these effects were improved only when the premask crosses flickered around static lines that terminated adjacent to the unspecified continuance between the premask crosses. This shows clearly that the flickering premask had effects at the stage of processing during which the continuance between the crosses was coded. Interestingly, and while this process might be expected to occur during very early visual processing, Elliott and Müller (1998) had already found the same priming effects across durations in the range 300–4,800 ms; priming would not appear specific to any particular processing time.

3 Protention or Memory? A Review of the Frequency-Specific Effects of Priming

The second contribution concerns the precision of timing: the effects described above were specific to premask frames presented at 40 Hz. This has two implications: first, one is obliged to consider the effects as "evoked" by stimulus timing, using the moniker adopted by the physiology literature and second, it is not sufficient to consider just any old frequency as enough to promote binding: Elliott and Müller (1998) had examined 25, 33, 40, 50, 66, and 100 Hz and only found effects at 40 Hz.

The 40 Hz specificity of the original effect was discussed by Elliott and Müller (2004) with the contributing methods and data then described in full by Elliott (2014). Elliott and Müller (2004) referred to a set of experiments in which premask matrices were presented at frequencies in singe Hertz steps over the range 30–50 Hz. These experiments extended upon the original finding reported by Elliott and Müller (1998) in that they showed priming effects not to be confined to 40 Hz. Instead, priming was found when premask matrices flickered at 33 Hz, 39–40 Hz, and 46–47 Hz. Described in terms of a "generalized phase angle hypothesis" (GPAH), Elliott and Müller (2004) observed that priming occurs for primes presented within premask rhythms that would all be in phase alignment at regular 148 ms intervals, implicating modulation of the premask-presentation rhythm by a slower (EEG theta) rhythm of approximately 6.75 Hz with which they would share a common phase angle (see Fig. 2). The GPAH predicts that, for a given priming frequency f and corresponding period duration $\tau = 1/f$, facilitation reoccurs at every time point

$$J(\tau) = (n\tau + ½) \cdot \tau - T$$

where $n\tau$ is a frequency-specific integer multiplier and T denotes a constant quantal time delay. The term + ½ accounts for the observation that, for $f = 40$ Hz, maximal

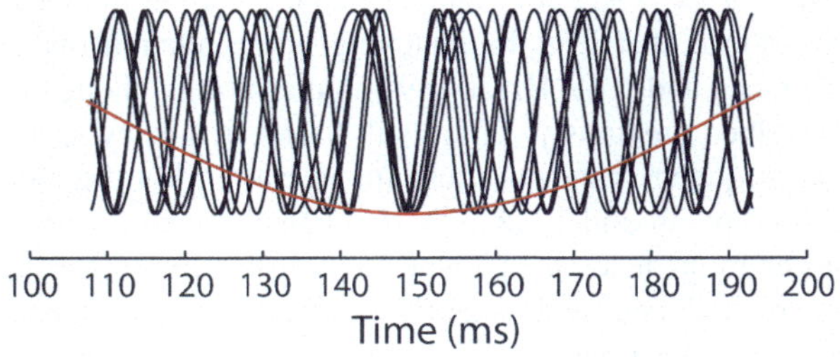

Fig. 2 The GPAH model of oscillatory priming in which the priming frequencies (33 Hz, 39–40 Hz, and 46–47 Hz – black functions) align phases at 148 ms and are thus also phase aligned with a stimulus-evoked rhythm of 6.75 Hz (red function)

facilitation occurs at phase angles of 180° relative to the rhythm of premask-matrix presentation (Elliott and Müller 2000).

Prime generation given regularly ordered stimulus frequencies in phase with a slower, presumably endogenous rhythm are only one part of the story. An analysis of the same dataset, published earlier by Kompass and Elliott (2001), reported priming to vary in magnitude (or priming was or was not present) for frequencies according to the time of target presentation expressed in terms of the phase of the premask-matrix presentation frequency (referred to in terms of a "return phase hypothesis" or RPH for oscillatory priming). In fact priming was maximal for targets presented at a time slightly ahead of the phase at which the priming stimulus *would have been presented* if a target matrix had not interrupted the premask-matrix presentation. This indicates the prime to be a cognitive response that can develop *in advance* of the priming stimulus, most likely as a function of the rhythmic nature of premask-matrix presentation.

The GPAH and RPH are not mutually exclusive: Elliott (2014) described the outcome of a broader examination of the effects of variation in priming stimulus presentation frequency as well as target presentation times. In this study, the GPAH was generally reproduced insofar as for all frequencies in the range 27–67 Hz, which would, if presented concurrently, align in phase at 149 ms or, alternatively, at a matching phase with a hypothetical 6.69 Hz rhythm. The RPH was also reproduced in that the maximum differences between intra- and interphase premask RTs (Fig. 1) occur for targets presented at times that peak midway through the frame 4 phase and so are slightly ahead in phase relative to frame 1. In this respect, the data presented by Elliott (2014) is overall consistent with the RPH described by Kompass and Elliott (2001) and indicates that a protentive prime comes to develop in advance of priming stimulus presentation. Protention refers here to coding that is carried out in anticipation of a future event. Both RPH and GPAH describe the pattern of priming over frequency when targets are presented with 0 ms interstimulus interval after 600 m ms of premask-matrix presentation, which is also very close to the fourth multiple of 149 ms. Elliott (2014) then tested the RPH, using similar frequencies but varying premask-matrix presentation time. In this he examined the hypotheses that if the target is presented at a time inconsistent with the regular phase alignment of frequencies as described by the GPAH, priming effects would be reduced, and perhaps the RPH would no longer be found.

In a first follow-up experiment, frequencies in the range 28–51 Hz and premask-matrix presentation time of 700 ms. This presentation time does not harmonically relate to the time equivalent to the GPAH rhythm and so allows examination of the hypothesis that it is the matching phase relation that exists between priming frequencies and 6.69 Hz at 149 ms that brings about the protentive priming effects described previously. Elliott still found frequency-specific priming and within similar frequency bands to those found earlier, but these bands appear to be rightward shifted by around 1 Hz in the frequency domain, with priming occurring at slightly faster frequencies than found previously. Importantly, protentive priming was no longer evident, with slightly reduced priming effects maximal precisely phase aligned with target presentation. A second follow-up tested the RPH directly by using a small set of frequencies (35–40 Hz) and by setting premask-matrix presentation times at the return phase for each frequency (which is also the return phase of 6.69 Hz), as well as at other phases for which no priming effects would be expected. Accordingly, this experiment also questioned the frequency-specificity of oscillatory priming. Consistent with expectations, this experiment found that if premask-matrix presentation times vary and are not in phase with the slow rhythm of 6.69 Hz, prime maxima will be aligned with or follow the frame phase at which target presentation aligns with frame 1 or priming stimulus presentation. In other words and as expected, priming is not protentive when target presentation does not align with the common phase of priming stimulus and slow rhythm. When this alignment occurs, priming was found to be protentive with priming found at a slightly different set of frequencies than found in any previous experiment, but not found at 40 Hz. This experiment indicates that it is return phase and not frequency alone that determines which premask-matrix presentation frequencies prime target detection.

4 Putting the Evidence Together: The What and When of Primes and Brain

4.1 The Prime

Previous speculations on what this form of oscillatory priming brings about were confined to the idea that the prime synchronized across basic feature coding mechanisms (Elliott and Müller 1998; Elliott et al. 2006a, b), contributing to a low-level Gestalt coding (Shi and Elliott 2007) that was neither sensitive to motion signals (Elliott and Müller 2001), while the priming stimulus, of itself, did not encourage deployment of focal attentional mechanisms (Elliott and Müller 1998). Although this may be a matter of debate based upon the study carried out by Bauer et al. (2009), certainly and unlike much of the thinking related to neural synchronization, binding does not occur post-attentional selection.

In terms of its temporal characteristics, oscillatory priming has been shown to exhibit a temporal modulation that matches the frequency of premask-matrix presentation (Elliott and Müller 2000). This is likely attributable to local $GABA_A$ interneuron mechanisms which are known to modulate the firing pattern of excitatory neurons (Elliott et al. 2000, 2006a, b).

Considering the data reported by Elliott (2014), the picture develops to include the idea of a pre-activation of neurons coding the priming stimulus as a function of the interaction between one or more phases of the premask-presentation rhythm and an inferred but not directly observed rhythm of around 6.69 Hz.

This picture is however complicated by the absence of a corresponding EEG response matching the 40 Hz rhythm of premask presentation (Elliott et al. 2001), but the presence nevertheless of a 33–34 Hz response (a rhythm predicted by the GPAH to encourage prime formation; see Elliott et al. 2003). Given that prime formation is very likely to occur preattentively and is based upon a stimulus that is not detected (Elliott and Müller 1998; Shi and Elliott 2007), and while it is not possible to refer prime generation as strictly "protentive" after Husserl's (1928) definition (which directly concerns an "experienced" future state), it seems that the prime can, under some circumstances, represent the temporal advancement (or pre-activation) of visual cognition.

4.2 The Brain

Oscillatory priming doesn't correspond with the idea of post-attentional binding and is more consistent with an early rather than a later binding process, although the correlation between variation in the magnitude of priming with estimates of subjective complexity, as reported by Shi and Elliott (2007), does suggest that the outcomes of early processes are not easily differentiable for those that must occur later. Importantly, oscillatory priming cannot be identified with the idea that perceptual binding is mediated by an induced gamma response. There is no question that oscillatory priming relates to binding, while the study reported by Elliott et al. (2006a, b) even identifies an important neurophysiological locus in the GABAergic interneuron system, so psychophysiological evidence supports the relevance of the oscillatory priming paradigm to measures of binding reported in the physiology literature. However, a fundamental aspect of the paradigm is the regular temporal relationship between stimulus and response. Consequently, the outcomes of the paradigm would be more consistent with the effects of evoked gamma. But what could this effect actually be?

Recall the findings of Hassler et al. (2011, 2013): in these there is a late, induced gamma response in the 200–350 ms post stimulus-onset time window that is sensitive to object recognition. In addition, there is an effect relating the frequency of microsaccades, which varied with object familiarity in a time window from 100 to 300 ms after stimulus onset. The effects of oscillatory priming are inconsistent with the former, but are not inconsistent with the latter response. One possible inconsistency concerns the absence of an effect of repetition priming in the early time window; however, Hassler et al. did not test very fast frequency repetitions of the sort used in the oscillatory priming paradigm, to test for the effects of these in aligning the phase of the gamma-band response to that of the stimulus: in other words, examining whether under some circumstances the effects of an early induced gamma-band response can be the same as the effects of an evoked response.

Thus, it is possible that the early, object-related response measured in the EEG by Hassler et al. (2013) is the same mechanism as that bringing about oscillatory priming, but what is this mechanism doing? Under conditions in which there is a phase relationship between the low (EEG theta, also recorded by Hassler et al. 2013) and stimulus gamma rhythms, the prime starts to form ahead of priming stimulus presentation, and perhaps this is a successful strategy to guide motor as well as other cognitive systems to respond appropriately. Considering the situation in which a rugby ball is suddenly passed to you during a match, in order to respond appropriately, you need to a priori compute the position in space at a certain time to be able to deploy yourself to successfully catch the ball. However, you will also need to quickly identify the event as involving the ball and articulate your response to be able to catch the irregularly shaped ball. Not only does this process require precognition – or a form of protentive processing – but also this processing needs to have taken into account shape and perhaps also form for the response to be successful. This early, protentive process is sensitive to the identity of that which is predicted to occur.

Oscillatory priming may thus be a form of evoked gamma response, which encodes the shape of the prime using mechanisms otherwise engaged in our "pre-cognitive" interaction with the form-related structures of the environment. The mechanisms mediating this interaction are essentially concerned with the formulation of correct responses to immediate future events (an idea not inconsistent with the Deutsch and Deutsch's 1963 late attentional model). However, this mechanism is also sensitive to shape-based attributes of events; it is likely to be sensitive to the spatial reference system of the body and may also be capable of formulating a response based upon more complex, form-based attributes. Using the oscillatory priming paradigm, not only have we information on the stimulus characteristics to which priming is sensitive; we also have very precise information on the dynamics of the mechanism, information that could prove extremely useful for the disambiguation of electrophysiological and other physiological data related to early perceptual processing.

References

Bauer F, Cheadle SW, Parton A, Müller HJ, Usher M (2009) Gamma flicker triggers attentional selection without awareness. Proc Natl Acad Sci 106(5):1666–1671

Bosman CA, Womelsdorf T, Desimone R, Fries P (2009) A microsaccadic rhythm modulates gamma-band synchronization and behavior. J Neurosci 29:9471–9480

Deutsch JA, Deutsch D (1963) Attention: some theoretical considerations. Psychol Rev 70:80–90

Driver J, Davis G, Russell C, Turatto M, Freeman E (2001) Segmentation, attention and phenomenal visual objects. Cognition 80:61–95

Duncan J, Humphreys GW (1989) Visual search and stimulus similarity. Psychol Rev 96:433–458

Elliott MA (2014) Atemporal equilibria: pro- and retroactive coding in the dynamics of cognitive microstructures. Front Psychol 5:990. https://doi.org/10.3389/fpsyg.2014.00990

Elliott MA, Müller HJ (1998) Synchronous information presented in 40-Hz flicker enhances visual feature binding. Psychol Sci 9:277–283

Elliott MA, Müller HJ (2000) Evidence for a 40-Hz oscillatory short-term visual memory revealed by human reaction time measurements. J Exp Psychol Learn Mem Cogn 26:703–718

Elliott MA, Müller HJ (2001) Effects of stimulus synchrony on mechanisms of perceptual organization. Vis Cognit 8:655–677

Elliott MA, Müller HJ (2004) Synchronization and stimulus timing: implications for temporal models of visual information processing. In: Kaernbach C, Schröger E, Müller H (eds) Psychophysics beyond sensation. Lawrence Erlbaum and Associates, Mahwah, pp 137–156

Elliott MA, Becker C, Boucart M, Müller HJ (2000) Enhanced GABAA inhibition enhances synchrony coding in human perception. Neuroreport 11:3403–3407

Elliott MA, Herrmann CS, Mecklinger A, Müller HJ (2001) The loci of 40-Hz visual-object priming mechanisms: a combined electroencephalographic and reaction-time study. Int J Psychophysiol 38(3):211–225
Elliott MA, Conci M, Müller HJ (2003) Prefrontal cortex maintains visual information in very short-term oscillatory persistence. Behav Brain Sci 26:733–734
Elliott MA, Giersch A, Seifert D (2006a) Some facilitatory effects of lorazepam on dynamic visual binding. Psychopharmacology (Berl) 184:229–238
Elliott MA, Shi Z, Kelly SD (2006b) A moment to reflect upon perceptual synchrony. J Cogn Neurosci 18:1880–1883
Farid H (2002) Temporal synchrony in perceptual grouping: a critique. Trends Cogn Sci 6:284–288
Fries P, Nikolić D, Singer W (2007) The gamma cycle. Trends Neurosci 30:309–316
Gray CM (1999) The temporal correlation hypothesis of visual feature integration: still alive and well. Neuron 24:31–47
Hassler U, Trujillo-Barreto N, Gruber T (2011) Induced gamma band responses in human EEG after the control of miniature saccadic artifacts. Neuroimage 57:1411–1421
Hassler U, Friese U, Martens U, Trujillo-Barreto N, Gruber T (2013) Repetition priming effects dissociate between miniature eye movements and induced gamma-band responses in the human electroencephalogram. Eur J Neurosci 38(3):2425–2433
Herrmann CS, Bosch V (2001) Gestalt perception modulates early visual processing. Neuroreport 12:901–904
Herrmann CS, Mecklinger A (2001) Gamma activity in human EEG is related to highspeed memory comparisons during object selective attention. Vis Cognit 8:593–608
Herrmann C, Mecklinger A, Pfeifer E (1999) Gamma responses and ERPs in a visual classification task. Clin Neurophysiol 110:636–642
Husserl E (1928) Zur Phänomenologie des inneren Zeitbewusstseins. Niemeyer, Halle a. S.
Kompass R, Elliott MA (2001) Modeling as part of perception: a hypothesis on the function of neural oscillations. In: Sommerfeld E, Kompass R, Lachmann T (eds) Fechner Day 2001. Proceedings of the seventeenth annual meeting of the international society of psychophysics. Pabst Science Publishers, Lengerich, pp 130–135
Martinovic J, Busch N (2011) High frequency oscillations as a correlate of visual perception. Int J Psychophysiol 79:32–38
Melloni L, Schwiedrzik CM, Rodriguez E, Singer W (2009a) (Micro)Saccades, corollary activity and cortical oscillations. Trends Cogn Sci 13:239–245
Melloni L, Schwiedrzik CM, Wibral M, Rodriguez E, Singer W (2009b) Response to: Yuval-Greenberg et al., "Transient Induced Gamma-Band Response in EEG as a Manifestation of Miniature Saccades." Neuron 58:429–441. Neuron 62:8–10
Pantev C (1995) Evoked and induced gamma-band activity of the human cortex. Brain Topogr 7:321–330
Rensink RA, Enns JT (1995) Preemption effects in visual search: evidence for low-level grouping. Psychol Rev 102:101
Shi Z, Elliott MA (2007) Oscillatory priming and form complexity. Percept Psychophys 69:193–208
Singer W (1999) Neuronal synchrony: a versatile code for the definition of relations? Neuron 24:49–65
Tallon-Baudry C (2009) The roles of gamma-band oscillatory synchrony in human visual cognition. Front Biosci 14:321–332
Tallon-Baudry C, Bertrand O (1999) Oscillatory gamma activity in humans and its role in object representation. Trends Cogn Sci 3:151–162
Usher M, Donnelly N (1998) Visual synchrony affects binding and segmentation processes in perception. Nature 394:179–182
Yuval-Greenberg S, Tomer O, Keren AS, Nelken I, Deouell LY (2008) Transient induced gamma-band response in EEG as a manifestation of miniature saccades. Neuron 58:429–441
Yuval-Greenberg S, Keren AS, Tomer O, Nelken I, Deouell LY (2009) Response to Letter: Melloni et al., "Transient Induced Gamma-Band Response in EEG as a Manifestation of Miniature Saccades". Neuron 58:429–441. Neuron 62:10–12

The Relationship Between Spatial Attention and Eye Movements

Amelia R. Hunt, Josephine Reuther, Matthew D. Hilchey, and Raymond M. Klein

Contents

Abstract The nature of the relationship between spatial attention and eye movements has been the subject of intense debate for more than 40 years. Two ideas have dominated this debate. First is the idea that spatial attention shares common neural mechanisms with eye movement programming, characterizing attention as an eye movement that has been prepared but not executed. Second, based on the observation that attention shifts to saccade targets, several theories have proposed that saccade programming necessarily recruits attentional resources. In this chapter, we review the evidence for each of these ideas and discuss some of the limitations and challenges in confirming their predictions. Although they are clearly dependent under some circumstances, dissociations between spatial attention and eye movements, and clear differences in their basic functions, point to the existence of two interconnected, but separate, systems.

A. R. Hunt (✉) and J. Reuther
University of Aberdeen, Aberdeen, UK
e-mail: a.hunt@abdn.ac.uk

M. D. Hilchey
University of Toronto, Toronto, ON, Canada

R. M. Klein
Dalhousie University, Halifax, NS, Canada

Curr Topics Behav Neurosci (2019) 41: 255–278
DOI 10.1007/7854_2019_95
Published Online: 30 April 2019

Keywords Attention · Eye movements · Premotor theory · Spatial orienting

The focus of the eyes is often assumed to also be where attention is focused. Eye movements bring new regions of the visual field to the fovea, making more detailed visual analysis of that region possible. It therefore makes sense that attention would generally accompany the direction of gaze when it shifts. However, spatial attention can facilitate faster and deeper processing of information at selected locations, even in the absence of overt gaze shifts. In Posner's classic Bartlett Lecture (1980), based on the state of the field nearly 40 years ago, he concluded that eye movements and attention have a "close functional but no intrinsic physiological relationship" (p. 13). A by-product of the intense discussion over the ensuing decades has been important insights about spatial attention, but we will argue here that the conclusion that can be drawn remains broadly the same.

In this chapter, we will discuss three hypotheses about the nature of the relationship between attention and eye movements (see Fig. 1). The first hypothesis comes from early proposals (one from Klein 1980 and one from Rizzolati et al. 1987) suggesting that spatial attention and eye movement preparation are essentially the same process. The second hypothesis is that shared central attentional selection processes are involved in both visual identification and motor selection (e.g., Schneider 1995; Deubel and Schneider 1996). A third plausible hypothesis is that eye movements and attention are independent. Independence is often cast in the role of the null hypothesis against which theories about dependence can be tested. But the many dissociations between the two systems evident in the existing literature depict a pair of systems that often shifts together, and relies on similar information, but can diverge when required. After providing some definitions and background, we will review the evidence for these three separate classes of ideas about the relationship between covert attention and eye movements, by answering three questions:

1. Does spatial attention depend on the eye movement system?
2. Does eye movement execution depend on covert attention?
3. Are eye movements and attention independent?

1 Definitions and Background

Eye movements involve selection. The peripheral retina prioritizes sensitivity over acuity, and the central retina prioritizes acuity over sensitivity. This division of labor is mediated by eye movements, which allow fast and frequent repositioning of visual input over different parts of the retina. Reorienting the eyes to bring particular regions of space into the higher-acuity central retina (known as a saccade) is often interpreted as a direct reflection of the priorities of the viewer, revealing their expectations about which locations are most profitable in terms of information gain. As we will discuss later, this may not be the principle driving all saccades, but is certainly true for some of them. These particular saccades can be thought of as

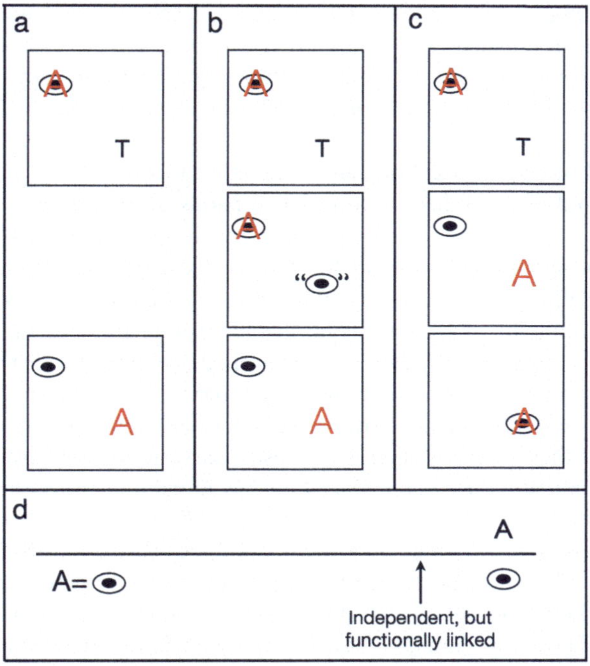

Fig. 1 The spatial relations that have been observed and hypothesized to exist between attention (denoted as a red "A") and gaze direction are illustrated in panels (**a**)–(**c**). In each of these panels, T denotes a location in space that is the target for attention (**a**, **b**) or a saccade (**c**). (**a**) As first demonstrated by von Helmholtz (1866), it is possible for an observer to separate where the eyes are gazing from the region of space that is attended. (**b**) Illustrated here is the hypothesis (c.f., Klein 1980; Rizzolati et al. 1987) that preparing an eye movement, but not executing it (denoted as an eye in quotes), might be the mechanism by which attention could be voluntarily separated from gaze direction. (**c**) Illustrated here is the proposal that a shift of attention to the location to be fixated is a prerequisite for a shift of gaze. (**d**) Redrawn from Posner (1980), the horizontal line represents a continuum of possible relations between attention and the eye movement system, from complete identity on the left to complete independence on the right. The view presented by Posner, and with which we agree, is indicated by the upward pointing arrow

the final outcome of a selection process, where one location wins out over the others and is consequently fixated.

Selection is also an essential function of attention. The information coming in through the senses on a moment-to-moment basis is more than the brain is able to process, so it is important to have a mechanism for selecting relevant information. Attention is defined as the process that achieves this selection (e.g., Broadbent 1957). The typical behavioral marker of attention is more detailed and efficient processing of perceptual events following a cue intended to direct attention toward these events. Experiments on attention tend to use visuospatial cues for expediency, but it is important to remember that attention can equally be allocated to nonspatial visual features, to other modalities, and to mental imagery or memories. That noted, our understanding of the relation between attention and eye movements has, quite naturally, been dominated by studies of visuospatial attention. Throughout this review, we will use the term "spatial attention" to refer to this spatial selectivity, which gives rise to faster responses and higher sensitivity to particular locations over others.

The control of spatial attention can be thought of as reflexive ("bottom-up") or voluntary ("top-down"). When prototypical versions of the Posner cueing task are used, reflexive control is explored by peripherally presented visual cues that provide no information about the upcoming location of the target. In contrast, voluntary control is typically explored using centrally presented cues that are spatially informative and is thought of as a strategic and deliberate form of orienting. Using versions of this paradigm, studies have identified fundamental differences in the properties of spatial attention summoned by reflexive versus voluntary cues. For example, in an early distinction between voluntary and reflexive attention, Jonides (1981) showed that reflexive attention is faster and less affected by memory load than voluntary attention. Research since then has found numerous other differences (e.g., Posner and Cohen 1984; Briand and Klein 1987; Lu and Dosher 2000; Taylor and Klein 1998; for a review, see Klein 2009). Corbetta and Shulman (2002) suggested that different neural networks implement these different kinds of control of attention. They propose that the dorsofrontal parietal attention network, which involves dorsal posterior parietal areas, especially the intraparietal sulcus (IPS), and an area around the frontal eye field (FEF) in the frontal cortex, is responsible for the implementation of goals and expectations about incoming sensory signals and anticipated motor responses to them. A second network, the right frontoparietal network, involves the right temporoparietal junction (TPJ) and the ventral frontal cortex (VFC) and is activated when an unexpected salient sensory event occurs. Corbetta and Shulman refer to this latter network as an independent "circuit breaker" that interrupts the voluntary attention activity in the dorsofrontal parietal network to direct attention to unexpected sensory events.

An analogous bifurcation of bottom-up and top-down control can be applied to eye movements, with eye movements to salient peripheral events thought of as "reflexive" and to task-relevant, informative locations as "voluntary." The superior colliculi (SC), on the dorsal surface of the midbrain, are central for the control of eye movements (e.g., Schiller 2013). A branch of the optic tract runs directly into the SC (Garey and Powell 1968), and the SC outputs to structures that mediate the circuitry of the eye muscles. The SC therefore represents a relatively simple interface between sensory input and motor output, usually associated with reflexive eye movements to basic visual stimuli. Upon closer inspection, things are much more complicated: the SC is a multilayered and multifunctional structure that receives and sends both sensory and motor input to and from a wide range of cortical and subcortical brain areas that also play important roles in eye movement control (e.g., White and Munoz 2017). The FEF can compensate when the SC is ablated, and vice versa (Schiller et al. 1987, 1980). The FEF and the SC were once thought to be the key structures in two independent pathways controlling eye movements, with the FEF being responsible for planning voluntary eye movements in response to internally generated goals and intentions and the SC for reflexive saccades to salient visual events (Schiller et al. 1980; Becker and Juergens 1979). More recent research, however, suggests that in the undamaged brain, the cortical "voluntary" pathway converges on the intermediate layers of the SC, where reflexive and voluntary inputs compete for control over eye movements on a common activation map, through lateral interactions (e.g., Findlay and Walker 1999; Godijn and Theeuwes 2002; Hunt et al. 2004; Kopesz 1995). For example, in the neural model developed by Trappenberg et al. (2001), "bottom-up" visual information reaches the SC quickly and affects groups of cells corresponding to a given location. Slower activation in the map occurs in response to probabilistic, strategic, or memory-based information and can modulate the activation caused by reflexive input.

In summary, spatial attention and eye movements have distinct voluntary and reflexive modes of orienting. The neural systems associated with spatial attention and eye movements are clearly distinct but interactive. There has been a sustained effort in the literature to understand the interplay between these two systems, both in the brain and also in behavior.

2 Question 1: Does Covert Attention Depend on Eye Movement Programming?

The functional similarity between eye movements and spatial attention has led to the suggestion that they share underlying neural mechanisms. Specifically, a covert shift of spatial attention has been suggested to be nothing more than a prepared but unexecuted (or suppressed) eye movement. This is known as the *oculomotor readiness hypothesis* (OMRH, Klein 1980) or *premotor theory* (Rizzolati et al. 1987; for a recent review, see Smith and Schenk 2012). Although both ideas have mostly been tested by examining the role of eye movements in spatial attention, premotor theory characterizes attention as a weak or subthreshold version of overt motor acts, a more general relationship than Klein (1980) proposed. A direct implication of this proposal is that an understanding of eye movement planning will also reveal the mechanisms of visuospatial attention. Given the intuitive appeal of "killing two birds with one stone," it is not surprising that this idea has catalyzed a large body of research examining the overlap in function, control, and neural networks involved in eye movements and attention. Several different lines of research have explored predictions of OMRH/premotor theory, which we review here. These include functional neuroanatomy and behavioral studies.

2.1 *Question 1a: Is There a Shared Functional Neuroanatomy Between Eye Movements and Attention?*

The specific question here, in light of the anatomical distinctions drawn in the previous section, is whether spatial attention depends on neural activity in the SC and/or FEF. This might seem like a straightforward question to answer, but it is complicated by the fact that eye movements and spatial attention are highly interconnected and both directed by spatial cues. Co-activation may occur because they are the same system or because they are different systems that happen to be influenced by visual events in very similar ways. Some neurophysiological studies have tried to resolve this issue by separating the focus of spatial attention from the location of a response. For example, Ignashchenkova et al. (2004) recorded from visual, motor, and visuomotor neurons within the intermediate layers of the SC, while eye movements were prepared and then executed on the basis of the identity of a peripheral target. Peripheral targets were preceded by peripheral cues, and these cues enhanced both visual acuity and the firing rate of visual neurons in response to the target. Motor neuron firing rates increased only when saccades were executed to their receptive fields. In contrast with visual neurons, visuomotor neurons sustained an elevated firing rate during the cue-target interval, and this discharge was unrelated to motor activity. In a second experiment, instead of peripheral cues, attention was directed to the target location by a central cue. Sustained firing of visuomotor neurons was not observed in response to these cues, despite the fact that visual acuity and the visual neurons' response to the target were still enhanced. The study therefore shows that visual cells in SC reflect the effects of spatial attention, but that spatial attention is independent of motor responses in the SC. On a similar note, Murthy et al. (2001) found visual neurons within the FEF that are responsible for visual selection of target locations, but are independent of saccade programming (see also Thompson et al. 2005; Juan et al. 2004). They suggest that saccade preparation and spatial attention are distinct neural processes that are both reflected within the FEF. The existence of these different classes of neurons within the FEF not only argue against premotor theory; they also mean it is difficult, if not impossible, to draw conclusions about the anatomical co-dependency of spatial attention and eye movement programming using techniques that have a low spatial resolution (including PET, fMRI, and studies based on patients with brain lesions).

The effects of microstimulation of FEF and SC pose interpretive problems as well. Müller et al. (2005) stimulated regions of the SC at a level well below the threshold that is required to induce a saccade and found increased motion sensitivity at locations corresponding to the stimulated regions of the SC, providing evidence that spatial attention is a subthreshold eye movement. However, the authors acknowledge that the increase in sensitivity resulting from SC stimulation may reflect antidromic activation of cortical areas that send inputs to the SC. In other words, stimulating the SC may send impulses back up axons leading to the SC from cortical areas involved in attention, thus activating these networks artificially. A similar criticism has been applied to similar microstimulation studies of the FEF

(Moore and Fallah 2001, 2004); it is difficult to know whether the electrical current to the stimulated region is responsible for the change in perceptual sensitivity when electrical signals spread easily through these highly interconnected regions.

There have been several attempts to understand the role of the SC in spatial attention using behavioral "markers" of collicular involvement. One line of research has examined the costs and benefits of attentional cues in the temporal (outer) and nasal (inner) hemifields (Berger and Henik 2000; Lambert et al. 2000; Rafal et al. 1989, 1991; Sapir et al. 2002; Shulman 1984; Simion et al. 1998; Zackon et al. 1997, 1999). This is based on the assertion that the temporal hemifields are overrepresented in the retinotectal visual pathway (from the retina to SC) relative to the primary visual pathway from the retina through the lateral geniculate nucleus (LGN) into V1. Accordingly, nasotemporal asymmetries in behavior could index retinotectal involvement. However, while this neurophysiological asymmetry occurs in cats (e.g., Sterling 1973), Williams et al. (1995) found that in monkeys, the ratio of temporal to nasal retinal cells projecting to the LGN was about the same as in the SC. Thus in primates, there may be nothing distinctive about the distribution of temporal and nasal projections from the retina to the SC relative to the geniculostriate pathway, calling into question the use of field differences as a valid marker of collicular involvement.

Two other putative behavioral markers for collicular involvement are similarly controversial. The first is color sensitivity. The S-cones of the retina do not project directly to the SC (e.g., Marrocco and Li 1977). Sumner et al. (2002) found that S-cone isolating spatial cues had no effect on the speed of eye movements to subsequent targets. The same cues did facilitate manual detection responses when the eyes remained fixed, suggesting covert attention was deployed to these cues independent of the oculomotor system. However, Hall and Colby (2014) have pointed out that S-cone information does reach the SC, just not directly from the retina. A second marker for collicular involvement is the fixation offset effect (FOE). The FOE is the speeding of eye movements that occurs when a fixated stimulus is removed just as a peripheral visual target appears, relative to when the fixated stimulus remains on (Saslow 1967). The FOE is thought to occur because of the mutually inhibitory relationship between the regions of the SC that project to the "move" and "fixate" centers of the oculomotor plant (Dorris et al. 1997). Kingstone and Klein (1993) showed that the effect of removing the fixation point is specific to the oculomotor system and does not influence manual responses. They also showed that the FOE does not interact with reflexive or voluntary covert orienting for either eye movement or manual responses. That said, there are several components at work in the FOE (Taylor et al. 1998), complicating a straightforward conclusion based on these findings.

It should be clear from the challenges noted above that the question of whether spatial attention and eye movement systems share an underlying anatomy is not straightforward to answer. Investigative tools with a course spatial resolution and those which rely on behavioral markers of anatomical pathways carry interpretive problems. At the single-unit level, motor activity can be dissociated from visual selectivity. Overall, the effects of attention are evident in neural responses in the SC and the FEF. This alone cannot be interpreted as a functional dependence of spatial attention on eye movements, but it does shed light on how and where they interact. We now turn to behavioral evidence for premotor theory.

2.2 *Question 1b: Does Eye Movement Preparation Influence Spatial Attention?*

The term "premotor theory" arose from a study by Rizzolati et al. (1987) showing that shifts of attention to visual cues share spatial anisotropies with eye movements. From their finding of increased reaction time costs when responding to targets that were not in the same visual quadrant as a spatial pre-cue, they inferred that attention was slower to cross the horizontal and vertical midlines than to move an equal distance that did not cross a midline. Based on this they made two claims: (a) reorienting the eyes across the midline involves activating different sets of muscles rather than simply modifying the activation of already-selected muscle sets and (b) even when the eyes are fixed, endogenous attention mirrors the behavior that is expected if the muscles driving the eyes were involved. The premotor theory of attention follows from this speculation: that attention is covertly shifted to a location when the saccadic program for moving the eyes to this location is ready to be executed. When a target appears in an unexpected location, the cost of reorienting attention represents the time needed to cancel one eye movement and prepare another. However, Reuter-Lorenz and Fendrich (1992) demonstrated that the meridian effect exists for eye movements directed by central cues, but not peripheral cues, suggesting this is not a stable property of the eye movement system. This same pattern is observed in spatial attention: meridian effects following central cues, but not peripheral cues. Further research from Hodgson et al. (1999) showed meridian effects for targets appearing 500 ms after the cue, but not at 100 ms (see also Hodgson and Müller 1999). Their results are consistent with a slower, more voluntary attention system that relies on internal representations of space in which the meridians have some special emphasis.

The OMRH had a similar inauspicious start; it was first advanced by Klein (1980), who promptly rejected it based on evidence described in the same paper. When directing eye movements and attention with endogenous cues, he showed that preparation of an eye movement to a given location was not accompanied by a corresponding shift of attention there. Likewise, a shift of spatial attention to a given location was not accompanied by the preparation of an eye movement there. A series of further studies made various improvements on Klein's original method and found the same result (Klein and Pontefract 1994; Hunt and Kingstone 2003a; Born et al. 2014). MacLean et al. (2015) found a similar independence when directing attention and the eyes with exogenous cues (see also Hunt and Kingstone 2003b; Klein 2004; MacInnes et al. 2015). Based on this large set of studies, there is strong and consistent evidence that spatial attention can be directed to particular locations (confirmed by better performance at these locations), independent of eye movement preparation.

A series of studies has assessed the prediction that restrictions in the ability to plan eye movements should also restrict the ability to allocate attentional resources. Craighero et al. (2004) compared effects of cue validity under monocular viewing conditions for straight gaze (no motor restriction) and maximally rotated gaze (40°) and found similar benefits of cue validity in the nasal and temporal visual field, but only for straight gaze. When the eye was rotated in a temporal direction, a cuing effect was only observed for the nasal, but not for the temporal visual field. Based on this, the authors claimed that this is "clear evidence of a strict dependence of attention on oculomotor processes in neurologically healthy subjects" (p. 332). However, this claim was not supported by an interaction between visual field and cueing validity. In two subsequent studies, this finding was not replicated (Smith et al. 2014, 2012). In these studies, effects of cues outside the oculomotor range were eliminated when the cues were reflexive, but remained for voluntary cues (but see Smith et al. 2010). Other research still suggests that targets presented beyond the oculomotor range, relative to within it, have effects on feature but not conjunction search, possibly in line with the idea that reflexive shifts of attention are most strongly linked to oculomotor preparation (Smith et al. 2014). However, other research demonstrates that distractors that should capture attention reflexively still have an impact on eye movement trajectories even when presented outside of the oculomotor range (Boon et al. 2017).

Using a similar logic, studies have examined patient AI, who as a result of congenital ophthalmoplegia lacks the ability to execute saccadic eye movements. Relative to a group of age-matched controls, she has a reduced cueing effect for reflexive but not for voluntary attention allocation (Smith et al. 2004; see also Gabay et al. 2010). However, this study neither draws direct statistical comparisons between AI and the control group, nor does it provide individual subject data, so it is unknown whether AI is outside the normal range with respect to reflexive attention orienting. Eye movements were not recorded, so it is also possible that control subjects are employing a strategy that is not available to AI (i.e., moving their eyes to the cue or the target). Mirroring this result, Khan et al. (2009) found a dissociation between spatial attention and eye movements in a patient with optic ataxia due to unilateral parietal damage. He had intact eye movements into the damaged field, but showed no evidence of elevated discrimination at this location, the measure typically used to demonstrate attention has shifted to the saccade target. A 2012 review by Smith and Schenk rejects a strong form of premotor theory on the basis of these and other results. They acknowledge that the collated evidence does leave room for a dependence of reflexive covert attention on the eye movement system but suggest an alternative account of the apparent link between reflexive attention and eye movements, whereby motor preparation biases neural competition in favor of to-be-fixated stimuli. In other words, locations to which the eyes are free to move may be prioritized for selection, but this is one input of many into the competition.

Further dissociations between eye movements and spatial attention have been documented. Saccades tend to curve away from salient visual distractors, even while spatial attention is captured by them (e.g., Godijn and Theeuwes 2004). Saccades also curve away from locations that are voluntarily attended; Sheliga et al. (1994) interpreted this as evidence for premotor theory, but the finding that spatial attention can be allocated in one direction while the eyes deviate in the opposite direction seems an unlikely consequence of perfectly overlapping mechanisms. A recent study by McCoy and Theeuwes (2018) trained participants to learn a relationship between particular eye movement targets and monetary rewards. Rewards speeded eye movements to those locations, but did not facilitate discrimination of targets in these locations relative to non-rewarded locations, suggesting eye movements and spatial attention can maintain separate spatial priorities. Taken together, these results all point to a system for programming and executing saccades that has different properties from the system responsible for allocating covert spatial attention.

Answer to Question 1 The weight of evidence suggests that spatial attention does not depend on eye movement programming. Although spatial attention and eye movement control processes share some common brain areas (especially FEF and SC), clear evidence has not yet demonstrated an obligatory relationship, whereby spatial attention requires saccade preparation. On the other hand, there are many dissociations between saccade preparation and spatial attention in the literature.

3 Question 2: Do Eye Movements Depend on Spatial Attention?

A second way the relationship between attention and eye movements has been described is that spatial attention determines the target of the next saccade. Several proposals have been made along this line, and they all share with premotor theory the prediction that eye movements and attention have an obligatory coupling. However, in this category of theories, attention is not dependent on eye movements. Rather, the dependency runs the other way, with the execution of eye movements depending on spatial attention. The initial basis for this idea is usually attributed to Deubel and Schneider (1996), who demonstrated elevated discrimination performance at the target of an upcoming eye movement (directed there by a central arrow cue). They presented this as evidence in favor of Schneider's (1995) visual attention model (VAM), in which a common attentional mechanism selects information both for object recognition and for action targeting. In a more recent iteration of this idea, Wischnewski et al. (2010) proposed an extension of Bundesen's (1990) theory of visual attention (TVA) that defines "proto-objects," which capture and combine static and dynamic low- and mid-level feature information (e.g., orientation, color, size, location, and direction) over areas that are defined by receptive field size, to form units that can be targeted by attention. These *proto-objects* are then subjected to bottom-up feature-dependent and/or top-down task-dependent prioritization, where the unit that receives the highest priority value (relative weight) becomes the target of the next saccade. Based on the nature of the proto-objects, the model can explain saccades driven by exogenous factors (e.g., a unique orientation or color, a sudden movement or onset), as well as saccades that are motivated intrinsically ("Where are my keys?") or extrinsically ("Follow the arrow cue!").

Although VAM is sometimes conflated with premotor theory, they are fundamentally different, both in terms of their theoretical implications and their predictions about how eye movements and attention should interact. Premotor theory is, at its core, an explanation of the mechanism of spatial attention, proposing attention is equivalent to eye movement preparation. In contrast, VAM describes attention's role in other cognitive processes. Premotor theory predicts that manipulations of eye movement-related processes, including preparation and priorities, should have measurable consequences for spatial attention. In the above section, we therefore reviewed studies that measured the effect of eye movement *preparation* on spatial attention, and based on these, we concluded these functions are independent. Crucially, VAM posits an obligatory relationship between eye movement *execution* and the orienting of spatial attention, as measured by facilitation of visual processing at the upcoming target of eye movements. Therefore the focus in this section is on the role of spatial attention in eye movement execution.

Several studies have demonstrated that attention tends to be allocated to a location immediately before a saccade is directed there. Sheperd et al. (1986) asked subjects to respond to the appearance of a peripheral target with a button press while moving their eyes in the direction indicated by a central arrow cue. Reaction times to detect the target were faster when it appeared in the location to which an eye movement was prepared, even when the appearance of the target preceded the saccade, suggesting that eye movement preparation can facilitate target detection at the to-be-fixated location. In a condition where the direction of the saccade and the direction of attention were pitted against each other (cue validity 20%), detection was still faster at the to-be-fixated location, suggesting that saccade preparation is a more powerful determinant of the locus of attention than the endogenous probability cue. Similar results were obtained by Hoffman and Subramanian (1995), Kowler et al. (1995), Peterson et al. (2004), and Born et al. (2014), all of whom found a facilitation of processing at the location to which a saccade was about to be executed.

Does attention to saccade targets reflect an *obligatory* relationship, wherein attention is a necessary precondition for saccadic eye movements? Examples of eye movements executed in a manner seemingly independent of attentional allocation are admittedly rare in the literature, Remington (1980) and Stelmach et al. (1997) being the only two studies using a typical attentional cueing approach to find evidence for saccade execution and attentional orienting directed to separate locations. More recently, however, trial history has been shown to play a critical role by Belopolsky and Theeuwes (Belopolsky and Theeuwes 2009, see also Belopolsky and Theeuwes 2012), who instructed eye movements to the left or right using a digit that could appear in a left or right location. The digit was preceded by a spatial cue that was valid 80% of the time. Eye movements were faster to valid cue locations, but only on trials that were preceded by trials where the saccade had been executed toward the digit. Whenever the digit was in one location and the saccade had to be executed to another location, the following trial showed a suppression, rather than an enhancement, of saccade preparation. This result suggests that a coupling of oculomotor preparation and covert attention can be selectively activated or suppressed, depending on recent experience.

A final rare instance of a dissociation in saccade execution and spatial attention comes from Van der Stigchel and de Vries (2015), who induced a consistent landing position error in saccades by presenting the saccade target together with a salient distractor. Although the eyes tended to land in between the target and the distractor (known as a *global effect* on saccade targeting), attentional benefits (elevated discrimination performance) were seen at the target and distractor locations, but not between them. This is evidence that at least some components of the saccade target selection process can be independent of spatial attention. This study is important, because it overcomes a limitation present in most studies addressing the relationship between spatial attention and eye movement execution. That is, before the saccade was executed, the saccade target was an empty space between two "relevant" locations, and participants were not instructed to look there, nor were their eyes drawn there by a salient visual signal. This is in contrast with typical studies examining the role of attention in eye movements, in which eye movements are directed to particular locations. In these studies, saccades are directed or guided using explicit instructions, spatial cues, and/or a limited set of visually distinct saccade targets in order to know where the saccades are most likely to go (so that attention can be measured in these locations prior to the saccade being executed). These studies typically find speeded responses or higher sensitivity for attentional probes presented in the locations to which a saccade is about to be executed relative to other locations. However, the cues that are used to guide saccades might also orient attention, thereby artificially introducing the association that is at the very core of these studies.

Saccades that are guided based on structures imposed by the experiment may engage attention in a way that may not generalize to the relationship between attention and eye movements that exists under more natural circumstances. The explicit selection of saccade targets in experimental settings may require attentional processes, but an important question that needs to be addressed is whether this is a valid reflection of how saccade target selection usually occurs.

3.1 Question 2a: Do Eye Movements Depend on Spatial Attention in the "Real World"?

A steadily increasing number of studies have documented eye movement behavior under less constrained conditions than the typical laboratory experiment. A now-classic study by Land et al. (1999) measured eye movements using a head-mounted eye tracker while participants made a cup of tea. About a third of eye movements could be directly linked to the task, and relevant objects were fixated immediately before being manipulated, nicely illustrating the information sampling function of eye movements. Given that these actions were highly automated (e.g., picking up a cup, turning on a tap), it is not straightforward to classify eye movements directly associated with components of these action sequences onto a voluntary versus reflexive dichotomy, and the authors explicitly note this awkward mapping of concepts based on laboratory research onto real-world behavior. On a similar note, Tatler et al. (2011) take aim at visual salience, a property describing stimuli that stand out from the background (e.g., because of a unique color, high contrast, motion). In the laboratory, high-salient events like sudden onsets have been shown to attract eye movements and attention in a reflexive manner (e.g., Theeuwes et al. 1998), but Tatler et al. (2011) point out that salience can account for very little variance in eye movement behavior in natural tasks. Indeed, eye movements are often driven to locations with no objects in them at all, in anticipation of an object arriving there (e.g., Land and McLeod 2000). In this volume, the chapter by Land (2018) makes the interesting point that saccadic eye movements evolved to keep the image stable on the retina when the head and body move. The optokinetic reflex rotates the eye against the direction of motion, and saccades are needed to rapidly recenter the eye in its orbit when the end of the range of eye motion is reached. These kinds of saccades are not driven by visual selection, but by biomechanical constraints. Recent reports that participants' awareness of their own eye movements is extremely impoverished (Clarke et al. 2017b; Võ et al. 2016) are also consistent with the notion that the majority of our eye movements are not as directed or deliberate as those generated in experimental settings in response to instructions.

Hayhoe and Ballard (2005, 2014) describe eye movements as guided by, and supporting, the sequence of component behaviors associated with achieving an agenda. They also characterize the eye movement selection process as driven by reward, with the currency of reward being uncertainty reduction. For example, while driving, fixations update current estimates of the state of various dynamic elements such as the speedometer, lane position, and distance of other cars. Fixations are driven to a particular information source when uncertainty exceeds a threshold; the threshold is influenced by the perceived value of the information. For example, when traveling at speed, the tolerance for uncertainty about the positions of other cars is lower than when traveling more slowly. The function of saccades in ongoing goal-directed actions in this and similar models is closely yoked to that of attention. Relative to laboratory tasks, where eye movements are directed with explicit signals, the ongoing task structure and environment direct the process of information sampling, gathering critical information just at the moment when it is needed. Attention and eye movements are almost certainly closely aligned in both of these situations.

Hayhoe and Ballard's model of eye movement control is broadly consistent with the *ideal observer* model in visual search (Najemnik and Geisler 2005). In this model, fixations are selected on the basis of expected information gain, given the limits of visual acuity and the locations that have already been inspected. The model matches human search in terms of the number of fixations needed to find a target, suggesting our fixation selection process maximizes information gain. However, fixation behavior in search also matches a stochastic model (Clarke et al. 2016), in which the sequence of fixations is a random walk, with each saccade selected from the population of saccade vectors produced from that region of the search array.[1] In other words, the same human search behavior has been well-described as both ideal and stochastic. The idea of saccade "heuristics," introduced by Morvan and Maloney (2012), could help to resolve this apparent contradiction: perceptual and motor constraints and biases in eye movement vectors that have evolved or developed to produce behavior that is, under most circumstances, sufficient to produce eye movement paths that are efficient in finding targets. If we randomly select from a restricted set of "useful" saccade vectors during search, the search path could approach a noisy optimal path without the need for the complex computations required under the ideal search model (in which estimating the information gained from each possible fixation is a prerequisite for selecting the best place to look).

Morvan and Maloney's heuristics proposal (2012) originated from a simple but elegant test of the idea that eye movements are driven by expected information gain. They asked participants to decide where to look to detect a target that could appear in one of two possible locations. When the two locations are close together, it makes sense to fixate between them, so the target can be reported regardless of where it appears. But when the two locations are too far apart to reliably detect the target from a central position, a better strategy is to fixate one of the two locations directly, ensuring the target is clearly visible 50% of the time. Participants did not conform to this optimal strategy; instead, they were just as likely to fixate the center when the targets were close together as when they were well outside the range at which they would be visible from center (see also Clarke and Hunt 2016). A similar conclusion was reached by Nowakowska et al. (2017) who created search arrays in which the target would be easy to detect in peripheral vision on one half of the background but required central vision to be detected on the other. Starting from center, eye movements to the easy side of the display can be considered superfluous, because they provide no new information. Nonetheless, nearly half of all eye movements were made to the easy side, and each of these saccades delayed finding the target by several hundred milliseconds. Even in uniformly easy displays, participants executed an average of seven eye movements, none of which were necessary to find the target; when shown the same search array with a 200 ms duration, too brief to execute any eye movements, participants were over 90% correct in determining whether the target was present or absent. Therefore, many eye movements appear to be "unguided," at least during visual search. Making a similar observation based on natural scenes, Tatler et al. (2011) note the existence of many "nonessential fixations," which lack the precision and purpose of fixations that can be directly linked to the task at hand.

The idea of saccade heuristics (Morvan and Maloney 2012) combined with a stochastic saccade selection process (Clarke et al. 2016) suggests that during natural

[1]Note that both these models describe fixation selection in search within a uniform texture; presumably other selection mechanisms come into play in more structured or meaningful scenes.

behavior, at least some eye movements are guided by a set of consistent biases, such as a bias to look at the middle of an image (Tatler 2007). Clarke et al. (2017a) formalized this idea in their saccadic flow model, which was developed and tested using 15 pre-existing databases of eye movements collected from a range of photographic images, tasks, and display durations. The flow model estimates the task- and image-independent likelihood that each location in the scene will be fixated next, given where the eyes are currently fixated. In essence, saccadic flow describes the background eye movement behavior upon which image- and task-relevant behavior takes place. The more general point here is that the drivers of fixation behavior are varied, and some are likely to engage attention, but others may not require it.

Answer to Question 2 Do eye movements depend on spatial attention? The answer is clearly yes, when those eye movements involve selection. Eye movements directed in response to cues or instructions in a laboratory task clearly involve selection, and attention has an obligatory relationship with these saccades. In "real-life" tasks, including search, a subset of eye movements complement chains of motor acts intended to accomplish a goal, but many are unrelated to the task and unguided by immediate visual details. We spend some of our time in states that require no eye movements; for example, imagine waiting for a kettle to boil and thinking about what to have for breakfast. While we do this, the eyes are still moving. Are even these eye movements dependent on spatial attention? It may be the case that they are, but this question has not yet been addressed. If saccade execution depends on spatial attention, *all* saccades would require attention to be executed, in addition to the highly controlled, instructed saccades that take place in a typical experiment.

4 Question 3: Are Eye Movements and Attention "Independent"?

Eye movements and attention clearly overlap in their functional roles: both can be modes of spatial selection, involved in accruing visual information from the environment, and they share many characteristics. Eye movements and attention also rely on shared information. However, shared information does not imply functional dependence, in the sense that the functioning of one depends *critically* upon the functioning of the other, or that directing one *necessarily* directs the other. As an analogy, consider the processes involved in visual perception of color and motion. Both depend on information represented in V1, and provide important information that can often come from a shared source. For example, imagine a colorful fish, darting through the water: few would claim that processing of the motion of the fish should be thought of as functionally dependent on the processing of its color. Similarly, eye movements and attention are influenced by a combination of sensory information and the current agenda of the observer, but they use that information to serve different functions. The evidence reviewed above suggests eye movements and spatial attention are highly interconnected systems which, in many circumstances, do depend on one another. But in some circumstances the spatial attention system needs to be independent of eye movements, and vice versa, and there are many examples of these dissociations in the literature reviewed above as well. An interesting thought experiment is to consider what the consequences for perception and behavior would be if our saccade and spatial attention systems were yoked together in an obligatory manner. In this section, with this thought experiment in mind, we will discuss some problems whose solution may depend on having eye movements and spatial attention be separable systems.

The Visual Stability Problem When the eyes move, the visual image shifts across the retina. This retinal shift has huge neural consequences, as retinotopic maps throughout the visual brain begin receiving input from a new region of space. This causes what is usually referred to as the stability problem: our perception of the world is seamless and stable, even though the retina is in almost constant motion. The stability problem is the topic of a large number of investigations, going back at least as far as Descartes (1633). Visual representations are predominantly coded in retinotopic coordinates at all levels of visual processing measured to date (Golomb and Kanwisher 2011), and profound spatial distortions occur for brief stimuli appearing just before and during eye movements (e.g., Matin and Pearce 1965; Schlag and Schlag-Rey 1995). The question of how stable visual representations are maintained when the image is displaced across the retina by an eye movement is the subject of intense debate and appears to require a complex interplay of both predictive and retrospective cross-retinal integration (for a recent review of these mechanisms, see Higgins and Rayner 2015).

Spatial attention has been proposed to play a central role in maintaining a stable and uninterrupted perception across the changes in the retinotopic locations of objects in visual scenes caused by eye movements (Cavanagh et al. 2010). It has long been known that visual cells within priority maps of the visual system (FEF, SC, and the lateral intraparietal area or LIP) respond predictively, that is, the cells begin firing in response to visual stimulation in their receptive fields before that stimulation has been shifted there by an impending eye movement (e.g., Duhamel et al. 1992; Umeno and Goldberg 1997; Walker et al. 1995). This stimulation is thought to reflect the effects of *efference copy*; that is, information about the planned eye movement is fed back into sensory areas, so they can begin compensating for the expected change in visual input that will result from execution of that eye movement (e.g., Sommer and Wurtz 2008). But what exactly does "compensation" entail? As reviewed above, FEF, SC, and LIP are thought to represent a network of priority maps (e.g., Bisley and Goldberg 2010), in which peaks of activity reflect the current retinotopic positions of behaviorally relevant objects in the environment. Based on this, Cavanagh et al. (2010) suggest that these peaks of activity reflect *attentional pointers*, which predictively shift to the expected new retinotopic location of attended objects just ahead of each eye movement. These shifts allow attention to be maintained on the correct retinotopic location of a selected number of objects across saccades. This framework for understanding the visual stability problem suggests that it is not the visual information itself that is maintained in spatiotopic coordinates, but attention to that information.

This proposal for the role of attention in visual stability implies that attention and eye movements are independent, closely coordinated systems. The eye movement system shifts the retina, and spatial attention predictively realigns its spatial coordinates with each saccade, in an equal and opposite direction to the vector of the saccade itself. In other words, the eye movements cause the problem, and spatial attention solves it. This critical role for attention in maintaining stability across eye movements is inconsistent with the proposal that attention is itself an eye movement plan and instead suggests complementary, but different, functions.

The Object Correspondence Problem Spatial attention has also been proposed to play a key role in object correspondence. This is the term usually used to refer to the problem of how we combine information from multiple visual subsystems specializing in different aspects of information processing (e.g., color, motion, shape, depth) into single, unitary objects with continuity over space and time (Robertson 2003). In Treisman's classic feature integration theory (FIT, Treisman and Gelade 1980), spatial attention is the mechanism through which basic features in a shared location are integrated into coherent objects. Although many of the details of the theory have changed over the last three decades, the core idea that spatial attention is involved in feature binding has been widely accepted. The accurate binding of features across multiple maps in the brain has been thought to be accomplished through spatial attention directed to the retinotopic location those features occupy.

As discussed in the preceding section, eye movements shift visual input on the retina, disrupting retinotopic maps. Eye movements are therefore more likely to hinder object binding than to promote it. Indeed, features can be masked by stimuli on another part of the retina when an eye movement is about to be executed that will retinotopically align them (Hunt and Cavanagh 2011). Similarly, color information is erroneously mixed for objects that appear in different physical locations, but share a location on the retina because an eye movement aligned them (Golomb et al. 2014).

For us to understand the visual world, and to track meaningful chunks of information around in a dynamic environment, we need to bind information to form object representations, and attention has been proposed to serve this function. Eye movements, in contrast, do not serve this binding function, though the saccade selection process likely profits from it: eye movements that are guided by objects would presumably be able to target more task-relevant sources of information than those guided by unbound features. For example, while tracking three moving objects among moving distractors, participants tend to fixate the center of the imaginary triangle formed by the three targets, rather than directly fixating the targets themselves (Fehd and Seiffert 2008). This suggests participants are using spatial attention to bind the three objects into a group, and they are simultaneously using eye movements to maintain the positions of all three elements as close to central vision as possible. The overall picture this paints is an eye movement system that is informed by the product of spatial attention, but that does not depend on, or determine it.

Concealing the Focus of Attention The theme in this chapter is that eye movements and spatial attention are separate systems that can behave independently. Yet the functional link between gaze direction and attention is so strong that humans have developed "joint visual attention" (Moore et al. 2014), a somewhat automatic routine that, in social situations, directs a viewer's attention to locations and objects in space toward which their counterpart is looking (e.g., Friesen and Kingstone 1998). Indeed, for this reason, eye movements are sometimes referred to as a "window into the mind." The control of attention by the gaze direction of others appears so early in development that some consider it to be innate rather than acquired (Hood et al. 1998; but see Gregory et al. 2016).

There are many social situations, however, in which an individual may not want their object of attention to be so easily determined from their direction of gaze. This would be a problem if gaze direction were an infallible cue to where and what we were attending. Our ability to volitionally control spatial attention independently from where we are looking may have evolved, in part, to solve this problem. Some sports provide an everyday context that is commonly used to exemplify the significance of concealing the focus of one's attention. An example would be the "no look pass" in hockey, where an offensive player may not want a defender to know where the pass is aimed or, better yet, where an offensive player may want to trick the defender into believing that the play is heading in a different direction. But the dissociation between eye movements and spatial attention didn't evolve for sports. Rather the selection pressure contributing to the dissociation probably arose from two social situations. In an actual aggressive encounter, the combattant will be more effective if their target cannot be anticipated from their eye movements. In groups characterized by a social hierarchy, subordinates will often want to attend to a dominant individual, but signaling this by direct gaze may trigger an unwanted aggressive responses.

Answer to Question 3 Are eye movements and spatial attention independent? Eye movements and attention often shift together, but the evidence reviewed in previous sections calls into question whether their relationship is obligatory. When and how their functioning is independent is an important but unanswered question. As a starting point, we have speculated in this section about the reasons why spatial attention and eye movements may need to be untethered from one another, at least some of the time.

5 Conclusions

We have evaluated the relationship between spatial attention and eye movements separately for two different directions of dependency. First, we considered whether spatial attention might depend on saccade preparation processes. The weight of the evidence here is quite clearly against this idea. Second, we considered whether saccade execution depends on spatial attention. Here, the answer was clearly yes, when eye movements are directed by instructions or cues, as they tend to be in the experimental setting in which their relationship has been measured. However, we pointed out that these settings encouraged a particular kind of saccade, one which has been explicitly instructed in both time and space. Eye movements in natural circumstances may not be so tightly bound to attention.

Goldberg and Wurtz, who published one of the first studies of attention in the superior colliculus in 1972, conclude their discussion with this point:

> [Attention] may not have a single physiological mechanism. We have described a phenomenon – the enhancement of a response to a stimulus when that stimulus is used by the monkey as data for an obvious voluntary act – and this condition clearly fits the description of attention. But we have performed our experiments in a very constrained manner: we have measured changes in the response to an object extrafoveally when it is the target for an eye movement. The attention associated with the analysis of complex visual stimuli may be totally different. (p. 573)

In this quote, they were recognizing an issue we continue to grapple with, which is that we tend to study attention in a stripped-down version of the environment. Attention may play quite a different role in the complex environment we navigate in our day-to-day lives. From the large literature reviewed in this chapter, we can conclude that spatial attention can be very tightly linked to eye movement preparation and execution in circumstances in which this linkage is beneficial. In more complex and dynamic natural environments, however, the visual attention system and eye movements may perform complementary, rather than linked, roles in sampling information and guiding behavior. A more fruitful discussion about the nature of the relationship between attention and eye movements going forward, therefore, may be *when*, rather than *whether*, they are coupled.

References

Becker W, Juergens R (1979) An analysis of the saccadic system by means of double step stimuli. Vis Res 19(9):967–983

Belopolsky AV, Theeuwes J (2009) When are attention and saccade preparation dissociated? Psychol Sci 20(11):1340–1347

Belopolsky AV, Theeuwes J (2012) Updating the premotor theory: the allocation of attention is not always accompanied by saccade preparation. J Exp Psychol Hum Percept Perform 38(4):902–914

Berger A, Henik A (2000) The endogenous modulation of IOR is nasal-temporal asymmetric. J Cogn Neurosci 12(3):421–428

Bisley JW, Goldberg ME (2010) Attention, intention, and priority in the parietal lobe. Annu Rev Neurosci 33:1–21

Boon PJ, Theeuwes J, Belopolsky AV (2017) Eye abduction reduces but does not eliminate competition in the oculomotor system. J Vis 17(5):1–10

Born S, Mottet I, Kerzel D (2014) Presaccadic perceptual facilitation effects depend on saccade execution: evidence from the stop-signal paradigm. J Vis 14(3):1–10

Briand KA, Klein RM (1987) Is Posner's beam the same as Treisman's glue?: on the relation between visual orienting and feature integration theory. J Exp Psychol Hum Percept Perform 13(2):228–241

Broadbent DE (1957) A mechanical model for human attention and immediate memory. Psychol Rev 64(3):205–215
Bundesen C (1990) A theory of visual attention. Psychol Rev 97(4):523–547
Cavanagh P, Hunt AR, Afraz A, Rolfs M (2010) Visual stability based on remapping of attention pointers. Trends Cogn Sci 14(4):147–153
Clarke AD, Hunt AR (2016) Failure of intuition when choosing whether to invest in a single goal or split resources between two goals. Psychol Sci 27(1):64–74
Clarke AD, Green P, Chantler MJ, Hunt AR (2016) Human search for a target on a textured background is consistent with a stochastic model. J Vis 16(7):1–16
Clarke AD, Stainer MJ, Tatler BW, Hunt AR (2017a) The saccadic flow baseline: accounting for image-independent biases in fixation behavior. J Vis 17(11):1–19
Clarke AD, Mahon A, Irvine A, Hunt AR (2017b) People are unable to recognize or report on their own eye movements. Q J Exp Psychol 70(11):2251–2270
Corbetta M, Shulman GL (2002) Control of goal-directed and stimulus-driven attention in the brain. Nat Rev Neurosci 3:201–215
Craighero L, Nascimben M, Fadiga L (2004) Eye position effects orienting of visuospatial attention. Curr Biol 14(4):331–333
Descartes, R (1633) Treatise on man (trans: Steele Hall T, 1972). Harvard University Press and Oxford University Press, Cambridge, MA
Deubel H, Schneider WX (1996) Saccade target selection and object recognition: evidence for a common attentional mechanism. Vis Res 36(12):1827–1837
Dorris MC, Par M, Munoz DP (1997) Neuronal activity in monkey superior colliculus related to the initiation of saccadic eye movements. J Neurosci 17(21):8566–8579
Duhamel JR, Colby CL, Goldberg ME (1992) The updating of the representation of visual space in parietal cortex by intended eye movements. Science 255(5040):90–92
Fehd HM, Seiffert AE (2008) Eye movements during multiple object tracking: where do participants look? Cognition 108(1):201–209
Findlay JM, Walker R (1999) A model of saccade generation based on parallel processing and competitive inhibition. Behav Brain Sci 22(4):661–621
Friesen CK, Kingstone A (1998) The eyes have it! Reflexive orienting is triggered by nonpredictive gaze. Psychon Bull Rev 5(3):490–495
Gabay S, Henik A, Gradstein L (2010) Ocular motor ability and covert attention in patients with Duane Retraction Syndrome. Neuropsychologia 48(10):3102–3109
Garey LJ, Powell TPS (1968) The projection of the retina in the cat. J Anat 102(2):189–222
Godijn R, Theeuwes J (2002) Parallel programming of saccades: evidence for a competitive integration model. J Exp Psychol Hum Percept Perform 28(5):1039–1054
Godijn R, Theeuwes J (2004) The relationship between inhibition of return and saccade trajectory deviations. J Exp Psychol Hum Percept Perform 30(3):538–554
Goldberg ME, Wurtz RH (1972) Activity of superior colliculus in behaving monkeys. II. Effect of attention on neuronal responses. J Neurophysiol 35(4):560–574
Golomb JD, Kanwisher N (2011) Higher level visual cortex represents retinotopic, not spatiotopic, object location. Cereb Cortex 22(12):2794–2810
Golomb JD, L'Heureux ZE, Kanwisher N (2014) Feature-binding errors after eye movements and shifts of attention. Psychol Sci 25(5):1067–1078
Gregory NJ, Hermens F, Facey R, Hodgson TL (2016) The developmental trajectory of attentional orienting to socio-biological cues. Exp Brain Res 234(4):1351–1362
Hall N, Colby C (2014) S-cone visual stimuli activate superior colliculus neurons in old world monkeys: implications for understanding blindsight. J Cogn Neurosci 26(6):1234–1256
Hayhoe M, Ballard D (2005) Eye movements in natural behavior. Trends Cogn Sci 9(4):188–194
Hayhoe M, Ballard D (2014) Modeling task control of eye movements. Curr Biol 24(13):R622–R628
Higgins E, Rayner K (2015) Transaccadic processing: stability, integration, and the potential role of remapping. Atten Percept Psychophys 77(1):3–27

Hodgson TL, Müller HJ (1999) Attentional orienting in two-dimensional space. Q J Exp Psychol 52 (3):615–623
Hodgson TL, Müller HJ, O'Leary MJ (1999) Attentional localization prior to simple and directed manual responses. Percept Psychophys 61(2):308–312
Hoffman JE, Subramanian B (1995) The role of visual attention in saccadic eye movements. Percept Psychophys 57(6):787–795
Hood BM, Willen JD, Driver JD (1998) Adult's eyes trigger shifts of visual attention in human infants. Psychol Sci 9(2):131–134
Hunt AR, Cavanagh P (2011) Remapped visual masking. J Vis 11(1):1–8
Hunt AR, Kingstone A (2003a) Covert and overt voluntary attention: linked or independent? Cogn Brain Res 18(1):102–105
Hunt AR, Kingstone A (2003b) Inhibition of return: dissociating attentional and oculomotor components. J Exp Psychol Hum Percept Perform 29(5):1068–1074
Hunt AR, Olk B, von Mühlenen A, Kingstone A (2004) Integration of competing saccade programs. Cogn Brain Res 19(2):206–208
Ignashchenkova A, Dicke PW, Haarmeier T, Thier P (2004) Neuron-specific contribution of the superior colliculus to overt and covert shifts of attention. Nat Neurosci 7(1):56–64
Jonides J (1981) Voluntary versus automatic control over the mind's eye movements. In: Long AD (ed) Attention and performance IX. Lawrence Erlbaum, Hillsadale, pp 187–203
Juan CH, Shorter-Jacobi SM, Schall JD (2004) Dissociation of spatial attention and saccade preparation. Proc Natl Acad Sci 101(43):15541–15544
Khan AZ, Blohm G, McPeek RM, Lefevre P (2009) Differential influence of attention on gaze and head movements. J Neurophysiol 101:198–206
Kingstone A, Klein RM (1993) Visual offsets facilitate saccadic latency: does pre disengagement of attention mediate this gap effect? J Exp Psychol Hum Percept Perform 19(6):1251–1265
Klein RM (1980) Does oculomotor readiness mediate cognitive control of visual attention? In: Nickerson R (ed) Attention and performance VIII. Academic Press, New York, pp 259–276
Klein RM (2004) On the control of visual orienting. In: Posner MI (ed) Cognitive neuroscience of attention. Guilford Press, New York, pp 29–44
Klein R (2009) On the control of attention. Can J Exp Psychol 63(3):240
Klein RM, Pontefract A (1994) Does oculomotor readiness mediate cognitive control of visual attention? Revisited! In: Nickerson R (ed) Attention and performance XV: conscious and nonconscious information processing. Erlbaum, Hillsdale, pp 333–350
Kopesz K (1995) Saccadic reaction time in gap/overlap paradigm: a model based on integration of intentional and visual information on neural, dynamic fields. Vis Res 35:2911–2925
Kowler E, Anderson E, Dosher B, Blaser E (1995) The role of attention in the programming of saccades. Vis Res 35(13):1897–1916
Lambert A, Norris A, Naiker N, Aitken V (2000) Effects of informative peripheral cues on eye movements: revisiting William James' "derived attention". Vis Cogn 7(5):545–569
Land MF (2018) The evolution of gaze shifting eye movements. In: Hodgson T (ed) Processes of visuo-spatial attention and working memory. Current topics in behavioural neurosciences. Springer, Cham
Land MF, McLeod P (2000) From eye movements to actions: how batsmen hit the ball. Nat Neurosci 3(12):1340–1345
Land MF, Mennie N, Rusted J (1999) The roles of vision and eye movements in the control of activities of daily living. Perception 28(11):1311–1328
Lu ZL, Dosher BA (2000) Spatial attention: different mechanisms for central and peripheral temporal precues? J Exp Psychol Hum Percept Perform 26(5):1534–1548
MacInnes WJ, Krüger HM, Hunt AR (2015) Just passing through: IOR in preplanned saccade sequences. Q J Exp Psychol 68:402–416
MacLean GH, Klein RM, Hilchey MD (2015) Does oculomotor readiness mediate exogenous capture of visual attention? J Exp Psychol Hum Percept Perform 41(5):1260–1270

Marrocco RT, Li RH (1977) Monkey superior colliculus: properties of single cells and their afferent inputs. J Neurophysiol 40(4):844–860
Matin L, Pearce DG (1965) Visual perception of direction for stimuli flashed during voluntary saccadic eye movements. Science 148(3676):1485–1488
McCoy B, Theeuwes J (2018) Overt and covert attention to location-based reward. Vis Res 142:27–39
Moore T, Fallah M (2001) Control of eye movements and spatial attention. Proc Natl Acad Sci 98(3):1273–1276
Moore T, Fallah M (2004) Microstimulation of the frontal eye fields and its effect on covert spatial attention. J Neurophysiol 91(1):152–162
Moore C, Dunham PJ, Dunham P (2014) Joint attention: its origins and role in development. Psychology Press, New York
Morvan C, Maloney LT (2012) Human visual search does not maximize the post-saccadic probability of identifying targets. PLoS Comput Biol 8(2):e1002342
Müller JR, Philiastides MG, Newsome WT (2005) Microstimulation of the superior colliculus focuses attention without moving the eyes. Proc Natl Acad Sci 102(3):524–529
Murthy A, Thompson KG, Schall JD (2001) Dynamic dissociation of visual selection from saccade programming in frontal eye field. J Neurophysiol 86(5):2634–2637
Najemnik J, Geisler WS (2005) Optimal eye movement strategies in visual search. Nature 434(7031):387–390
Nowakowska A, Clarke AD, Hunt AR (2017) Human visual search behaviour is far from ideal. Proc R Soc B Biol Sci 284(1849):1–6
Peterson MS, Kramer AF, Irwin DE (2004) Covert shifts of attention precede involuntary eye movements. Percept Psychophys 66(3):398–405
Posner MI (1980) Orienting of attention. Q J Exp Psychol 32(1):3–25
Posner MI, Cohen Y (1984) Component of visual orienting. In: Bouma H, Bonwhuis D (eds) Attention and performance X. Erlbaum, Hillsdale, pp 551–556
Rafal RD, Calabresi PA, Brennan CW, Sciolto TK (1989) Saccade preparation inhibits reorienting to recently attended locations. J Exp Psychol Hum Percept Perform 15(4):673–685
Rafal RD, Henik A, Smith J (1991) Extrageniculate contributions to reflex orienting in normal humans: a temporal hemifield advantage. J Cogn Neurosci 3(4):322–328
Remington RW (1980) Attention and saccadic eye movements. J Exp Psychol Hum Percept Perform 6(4):726–744
Reuter-Lorenz PA, Fendrich R (1992) Oculomotor readiness and covert orienting: differences between central and peripheral cues. Percept Psychophys 52(3):336–344
Rizzolati G, Riggio L, Dascola I, Umilta C (1987) Reorienting attention across the horizontal and vertical meridians: evidence in favor of a premotor theory of attention. Neuropsychologia 25(1):31–40
Robertson L (2003) Binding, perceptual attention and spatial awareness. Nat Rev Neurosci 4(2):93–103
Sapir A, Rafal RD, Henik A (2002) Attending to the thalamus: inhibition of return and nasal-temporal asymmetry in the pulvinar. Neuroreport 13(5):693–697
Saslow MG (1967) Effects of components of displacement-step stimuli upon latency for saccadic eye movement. J Opt Soc Am 57(8):1024–1029
Schiller PH (2013) Neural control of visually guided eye movements. In: Fafrowisc M, Marek T, Karwowski W, Schmorrow D (eds) Neuroadaptive systems: theory and applications. CRC Press, Boca Raton
Schiller PH, True SD, Conway JL (1980) Deficits in eye movements following superior colliculus and frontal eye field ablations. J Neurophysiol 44(6):1175–1189
Schiller PH, Sandell JH, Maunsell JHR (1987) The effect of frontal eye field and superior colliculus lesions on saccadic latencies in the rhesus monkey. J Neurophysiol 57(4):1033–1049
Schlag J, Schlag-Rey M (1995) Illusory localization of stimuli flashed in the dark before saccades. Vis Res 35(16):2347–2357

Schneider WX (1995) VAM: a neuro-cognitive model for visual attention control of segmentation, object recognition, and space-based motor action. Vis Cogn 2(2–3):331–376
Sheliga BM, Riggio L, Rizzolatti G (1994) Orienting of attention and eye movements. Exp Brain Res 98(3):507–522
Sheperd M, Findlay JM, Hockey RJ (1986) The relationship between eye movements and spatial attention. Q J Exp Psychol 38(3):475–491
Shulman GL (1984) An asymmetry in the control of eye movements and shifts of attention. Acta Psychol 55(1):53–69
Simion F, Valenza E, Umilta C, Barbra BD (1998) Inhibition of return in newborns is temporo-nasal asymmetrical. Infant Behav Dev 18(2):189–194
Smith DT, Schenk T (2012) The premotor theory of attention: time to move on? Neuropsychologia 50(6):1104–1114
Smith DT, Rorden C, Jackson SR (2004) Exogenous orienting of attention depends upon the ability to execute eye movements. Curr Biol 14(9):792–795
Smith DT, Ball K, Ellison A, Schenk T (2010) Deficits of reflexive attention induced by abduction of the eye. Neuropsychologia 48(5):1269–1276
Smith DT, Schenk T, Rorden C (2012) Saccade preparation is required for exogenous attention but not endogenous attention or IOR. J Exp Psychol Hum Percept Perform 38(6):1438–1447
Smith DT, Ball K, Ellison A (2014) Covert visual search within and beyond the effective oculomotor range. Vis Res 95:11–17
Sommer MA, Wurtz RH (2008) Brain circuits for the internal monitoring of movements. Annu Rev Neurosci 31:317–338
Stelmach LB, Campsall JM, Herdman CM (1997) Attentional and ocular movements. J Exp Psychol Hum Percept Perform 23(3):823–844
Sterling D (1973) Quantitative mapping with electron microscope: retinal terminals in the superior colliculus. Brain Res 54:347–354
Sumner P, Adamjee T, Mollon JD (2002) Signals invisible to the collicular and magnocellular pathways can capture visual attention. Curr Biol 12(15):1312–1316
Tatler BW (2007) The central fixation bias in scene viewing: selecting an optimal viewing position independently of motor biases and image feature distributions. J Vis 7(14):1–17
Tatler BW, Hayhoe MM, Land MF, Ballard DH (2011) Eye guidance in natural vision: reinterpreting salience. J Vis 11(5):1–23
Taylor TL, Klein RM (1998) On the causes and effects of inhibition of return. Psychon Bull Rev 5(4):625–643
Taylor T, Kingstone AF, Klein RM (1998) The disappearance of foveal and non-foveal stimuli: decomposing the gap effect. Can J Exp Psychol 52(4):192–200
Theeuwes J, Kramer AF, Hahn S, Irwin DE (1998) Our eyes do not always go where we want them to go: capture of the eyes by new objects. Psychol Sci 9:379–385
Thompson KG, Biscoe KL, Sato TR (2005) Neuronal basis of covert spatial attention in the frontal eye fields. J Neurosci 25(41):9479–9487
Trappenberg TP, Dorris MD, Munoz DP, Klein RM (2001) A model of saccade initiation based on competitive integration of endogenous and exogenous signals within the superior colliculus. J Cogn Neurosci 13(2):256–271
Treisman A, Gelade G (1980) A feature-integration theory of attention. Cogn Psychol 12(1):97–136
Umeno MM, Goldberg ME (1997) Spatial processing in the monkey frontal eye field. I. Predictive visual responses. J Neurophysiol 78(3):1373–1383
Van der Stigchel S, De Vries JP (2015) There is no attentional global effect: attentional shifts are independent of the saccade endpoint. J Vis 15(15):1–12

Võ MLH, Aizenman AM, Wolfe JM (2016) You think you know where you looked? You better look again. J Exp Psychol Hum Percept Perform 42(10):1477–1481

von Helmholtz H (1866/1962) Treatise on physiological optics, vol 2 (trans: Southall JPL). Dover, New York

Walker MF, Fitzgibbon EJ, Goldberg ME (1995) Neurons in the monkey superior colliculus predict the visual result of impending saccadic eye movements. J Neurophysiol 73(5):1988–2003

White BJ, Munoz DP (2017) Neural mechanisms of saliency, attention, and orienting. In: Computational and cognitive neuroscience of vision. Springer, Singapore, pp 1–23

Williams C, Azzopardi P, Cowey A (1995) Nasal and temporal retinal ganglion cells projecting to the midbrain: implications for "blindsight". Neuroscience 65(2):577–586

Wischnewski M, Belardinelli A, Schneider WX, Steil JJ (2010) Where to look next? Combining static and dynamic proto-objects in a TVA-based model of visual attention. Cogn Comput 2(4):326–343

Zackon DH, Casson EJ, Stelmach L, Faubert J, Racette L (1997) Distinguishing subcortical and cortical influences in visual attention: subcortical attentional processing. Invest Opthamol Vis Sci 38(2):364–371

Zackon DH, Casson EJ, Zafar A, Stelmach L, Racette L (1999) The temporal order judgment paradigm: subcortical attentional contributions under exogenous and endogenous cueing conditions. Neuropsychologia 37(5):511–520

Functional Imaging of Visuospatial Attention in Complex and Naturalistic Conditions

Emiliano Macaluso

Contents

Abstract One of the ultimate goals of cognitive neuroscience is to understand how the brain works in the real world. Functional imaging with naturalistic stimuli provides us with the opportunity to study the brain in situations similar to the everyday life. This includes the processing of complex stimuli that can trigger many types of signals related both to the physical characteristics of the external input and to the internal knowledge that we have about natural objects and environments. In this chapter, I will first outline different types of stimuli that have been used in naturalistic imaging studies. These include static pictures, short video clips, full-length movies, and virtual reality, each comprising specific advantages and disadvantages. Next, I will turn to the main issue of visual-spatial orienting in naturalistic conditions and its neural substrates. I will discuss different classes of internal signals, related to objects, scene structure, and long-term memory. All of these, together with external signals about stimulus salience, have been found to modulate the activity and the connectivity of the frontoparietal attention networks. I will conclude by pointing out some promising future directions for functional imaging with naturalistic stimuli. Despite this field of research is still in its early days, I consider that it will play a major role in bridging the gap between standard laboratory paradigms and mechanisms of brain functioning in the real world.

Keywords Attention · fMRI · Knowledge · Naturalistic · Salience · Space

E. Macaluso (✉)
Lyon Neuroscience Research Center, Bron, France
e-mail: emiliano.macaluso@univ-lyon1.fr

Curr Topics Behav Neurosci (2019) 41: 279–302
DOI 10.1007/7854_2018_73
Published Online: 15 December 2018

1 Introduction

Extensive behavioral and neuroscientific investigations of attention control provide us with advanced understanding of the many processes underlying this fundamental brain function. Nonetheless the vast majority of the experiments which have sought to uncover these processes made use of simple stimuli and stereotyped paradigms. Here, I review several neuroimaging studies that sought to understand how attention works in more complex and naturalistic conditions. More specifically, the chapter focuses on mechanisms of visuospatial attention and how different factors influence the allocation of attention in naturalistic conditions.

An important idea in visuospatial attention control concerns the distinction between processes that are triggered by stimuli in the external world (stimulus-driven control) and signals that are generated internally, as a consequence of current goals and/or expectations (endogenous control). Early studies using various versions of the Posner's spatial cueing task (Posner 1980) mapped endogenous control of visuospatial attention in dorsal frontoparietal regions (IPS, intra-parietal sulcus and, FEF, frontal eye field, in the dorsal premotor cortex) and stimulus-driven control in ventral frontoparietal regions (Corbetta et al. 1993; Arrington et al. 2000; see Corbetta and Shulman 2002; Shomstein 2012, for reviews). Nonetheless, far from being a complete dissociation, it soon became apparent that dorsal regions are also involved in the processing of external signals and that the activation of ventral frontoparietal regions depends on various endogenous constraints. On the one hand, the dorsal parietal cortex contains "sensory" maps of space, and mere visual stimulation can activate these spatial representations (e.g., Swisher et al. 2007). On the other hand, the activation of the ventral parietal cortex is strongly modulated according to the behavioral relevance of the stimuli (Indovina and Macaluso 2007; see also Corbetta et al. 2008, for review). A large number of imaging studies have made use of sophisticated designs in order to characterize the factors that influence the activation and the connectivity of the dorsal and ventral frontoparietal attention systems. These include manipulations of the low-level physical characteristics of the stimuli (e.g., contrast, luminance, Geng and Mangun 2011; Indovina and Macaluso 2007), as well as many endogenous factors concerning the task demands and task-related expectations (Nobre et al. 1999; Scalf et al. 2014). The specific constraints governing endogenous and stimulus-driven contributions to attention control are still debated, but there is a growing consensus that the inter-regional connectivity between the dorsal and ventral systems plays a pivotal role in regulating the allocation of attention resources, as a function of both the stimuli and the current task demands and expectations (Corbetta et al. 2008; Vossel et al. 2015).

While experiments employing simple and highly controlled conditions permit studying in a systematic manner the influence of endogenous and stimulus-driven factors, it should be asked to what extent these manipulations accurately reflect processes that govern attention control in everyday situations. Among the many differences between attention protocols used in the laboratory and attention control in the real world, there are two interrelated aspects that I will be focusing on here: the first concerns the complexity of the external signals that are simple and stereotyped in laboratory paradigms, while complex and highly structured in real life; the second concerns the type of internal signals that guide attention that relate primarily to task-specific goals in standard paradigms but arise from knowledge and memory in naturalistic conditions.

Concerning the characteristics of external input, standard attention paradigms make use of a few simple stimuli that are repeated over hundreds of trials. An example of this is the spatial cueing paradigm introduced by Posner (1980) that, in many different versions, has been used in a vast number of behavioral and imaging studies of spatial attention. The paradigm includes target stimuli presented in the left or the right visual field, preceded by attentional cues shown either centrally or on the left/right side. Thus, throughout the entire experiment, the participants are presented with only a few types of stimuli (typically, simple geometrical shapes), and the stimuli are shown only from two spatial locations, plus the center of the screen. This is very different from the sensory input that the brain receives in any real-life situation, when the visual input is complex and heterogeneous and includes tens to hundreds of objects/object parts occupying the entire visual field. This difference in mere "sensory load" is most likely to have an influence on the control of spatial attention.

A second important difference between standard laboratory paradigms and spatial orienting in real life involves the type of internal signals that the brain can use to guide attention. In traditional attention paradigms, internal signals typically arise from some explicit task instruction that is given to the participants. Most commonly, studies on endogenous, goal-directed attention manipulate the task relevance of specific stimuli, locations, and/or features. This may involve the instruction of directing attention toward a particular location (e.g., spatial cueing paradigms with central cues) or the instruction to detect stimuli with some specific sensory features (e.g., in visual search tasks). In real life we may also sometimes allocate attention based on a specific goal, e.g., when looking for the pepper mill in the kitchen. However, this seems more an exception than a rule, as we typically do not go around with the sole scope of finding a specific "target" object. Instead, internal signals can arise from the huge amount of knowledge that we have about the world. This includes knowledge about the identity and the spatial layout of objects in natural environments and – more generally – semantic and episodic information related to the current sensory input. All this prior knowledge can in principle contribute to the processing of the complex and structured sensory input that characterize naturalistic situations and, in turn, determining how we allocate attentional resources.

The points outlined above suggest that standard protocols for the study of attention may, on the one hand, miss relevant factors likely to play a role in real life (e.g., internal signaling based on pre-existing knowledge) and, on the other hand, inflate aspects of control that may be relatively specific for situations involving stereotyped trials (i.e., goal-directed control). Experimentation with naturalistic stimuli provides us with an opportunity to start addressing these issues. In the next section, I will review several types of naturalistic stimuli that have been used in functional imaging studies of spatial attention, discussing some of the advantages and disadvantages of different types of stimulus material. Then, I will address the main issue concerning the role of stimulus-driven (Sect. 3) and internal signals (Sect. 4) for the allocation of attention in naturalistic conditions and their neural correlates. Section 5 will point out some promising research directions for future imaging studies with naturalistic stimuli.

2 Naturalistic Stimuli

In the attempt to simulate the complexity and richness of the sensory input that the brain has to process in everyday life situations, researchers have used various types of complex visual stimuli. These include static images of natural scenes, short video clips, full-length movies, as well as virtual reality settings. All of these have also been combined with concurrent auditory stimuli that further increase the richness, and possibly the realism, of the sensory experience. Overall, it can be argued that for each experimental setting, there will be some trade-off between realism and experimental control (see Fig. 1). On the one hand, the simplest stimuli (e.g., static pictures) will allow characterizing and manipulating specific experimental variables, but the participant's experience will be still much poorer compared to everyday life. On the other hand, more complex material (e.g., full-length movies and virtual reality) will result in a more realistic experience, but the identification and characterization of specific processes will be much more challenging.

The vast majority of the imaging studies seeking to investigate naturalistic attention made use of static pictures. The main advantage of static pictures is that they can be used in "standard" fMRI protocols, including a number of experimental conditions each repeated over multiple trials. For each condition, it is possible to choose a set of unique pictures, but all sharing the relevant feature characterizing the condition. For example, one may hypothesize the existence of a visual field bias for the processing objects belonging to one vs. another category (e.g., people vs. cars). The stimulus set may include many different pictures, each containing either people on the left or on the right side of the image, or cars on the left or the right side. Each picture would be unique, with cars/peoples located at different eccentricities, with different sizes, colors, and perspectives and – more generally – embedded within different contexts. This heterogeneity will make the experiment more similar to everyday life experience, with complex and unrepeated stimuli, and may reduce strategic processes compared with paradigms employing stereotyped stimuli (i.e., few items, shown at fixed locations). Nonetheless, the different trials can still be grouped in four experimental conditions, and the data can be analyzed as a "2 categories x 2 visual hemifields" factorial design. Furthermore, the fMRI data analysis can follow a standard general linear model (GLM) approach, fitting the BOLD time series with condition-specific predictors and performing comparisons between the conditions. This straightforward link between the experimental conditions and the imaging data also facilitates the implementation of more complex analyses, including modulatory influences of single-trial activity (Nardo et al. 2014),

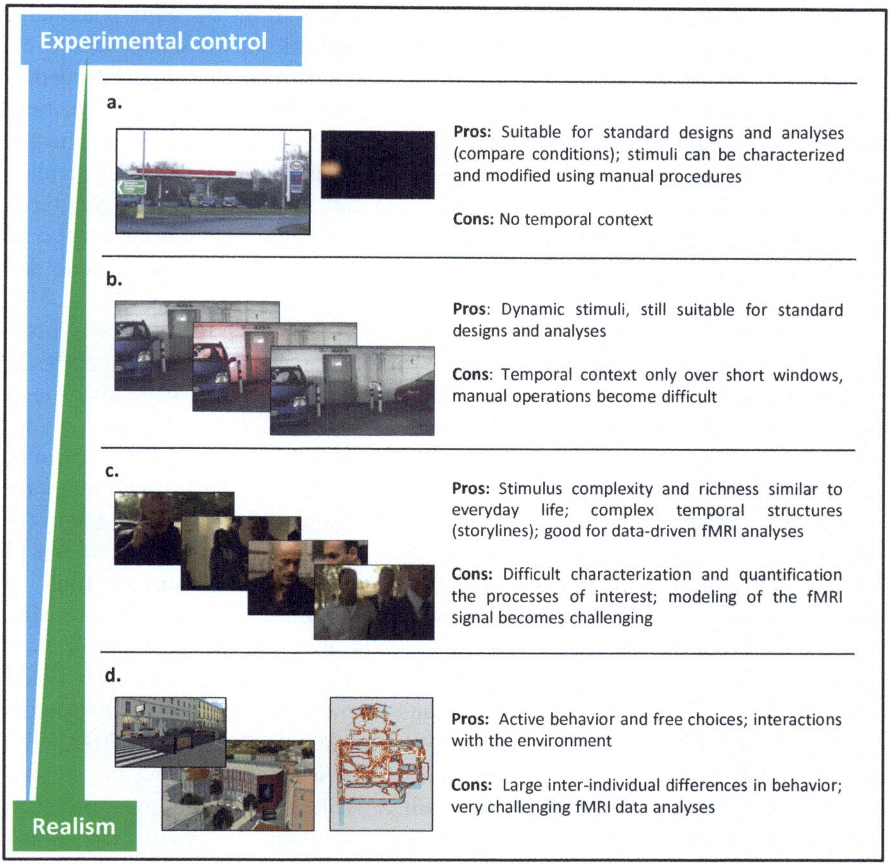

Fig. 1 Stimuli for naturalistic fMRI studies. Serval types of complex stimuli have been used to simulate brain functioning in real life. These comprise (**a**) static images of real scenes (e.g., Santangelo and Macaluso 2013); (**b**) short video clips, typically with durations of a few seconds (e.g., Nardo et al. 2016); (**c**) movies or videos with longer durations, ranging from a few minutes to up to 1 h (e.g., Nardo et al. 2011; Bordier et al. 2013); and (**d**) virtual environments allowing interactive behavior (e.g., Burgess et al. 2001; Macaluso and Ogawa 2018). Each type of stimulus has advantages ("pros") and disadvantages ("cons"), generally with a trade-off between the possibility to control experiential factors and the realism/similarity with the real life; see Sect. 2

multivariate analyses (Seidl et al. 2012), and analyses of interregional connectivity (Santangelo and Macaluso 2013; Howard et al. 2013; Sabatinelli et al. 2014).

Aside these advantages in terms of experimental protocol and data analysis, static pictures imply that only a limited number of stimuli are used for any given experiment. This makes it possible to manually characterize or modify the stimuli. Manual object annotation can be used to label the identity of objects in natural images and to quantify specific parameters such as number, size, and position of the objects (see also Sect. 5, for computational approaches that may permit to make these procedures automatic). This information can serve, for example, to verify that the pictures belonging to different conditions do not systematically differ by some potentially confounding parameter (e.g., eccentricity of people/cars, in the example above). Even with a limited set of images, manual annotation is time consuming. Nonetheless, several databases of naturalistic pictures with objects' annotations are now available (e.g., "LabelMe" database, http://labelme.csail.mit.edu/). Static images permit also manual image editing. Image editing enables creating sets of pictures with specific characteristics. For example, one can take images with congruent layouts (e.g., plates on a table set for dining) and substitute some of the objects to obtain object-context incongruent images (e.g., pairs of shoes on the same table; see Santangelo et al. 2015; see also Sect. 4, below).

In sum, static images have many advantages related to the possibility of controlling and manipulating several aspects of the stimuli that enables specific hypotheses to be addressed with standard experimental paradigms. However, complex though they may be, static images lack of any temporal dynamics. This is a characterizing aspect of any natural situation, and it is necessary in order to generate any coherent unfolding of events over time (temporal context). One first step to add a temporal dimension to naturalistic stimuli is to utilize video clips. When these are relatively short (e.g., 1–3 s), many of the advantages described above for static pictures will hold. In particular, it is possible to consider each video as an experimental "trial," to associate different videos with some specific condition and, therefore, to design experiments based on comparisons between conditions (including randomization, counterbalancing, etc.; e.g., Nardo et al. 2016). The temporal dimension will add richness and realism to the sensory input and permits, for example, to investigate interactions between the visual and the auditory modalities (Nardo et al. 2014). The increased realism of video clips comes with the downside that any manual annotation, or editing, becomes far more demanding. While manually drawing objects' contours in 100 pictures is painstaking but feasible, doing the same in 100 videos becomes almost impossible (but see, e.g., Lahnakoski et al. 2012; for manual annotation of a few movie features). Even with a short duration of 2 s and a standard frame rate of 60 fps, the total number of frames to modify corresponds to 12,000 single images! Furthermore, while short videos can include some structured temporal context, this will be very limited compared with real-life situations. The processing of temporal context in short videos is most likely to rely on information stored in short-term memory and may not be suitable to study the contribution of signals stored in long-term memory.

Full-length movies and virtual reality can provide us with experimental settings that not only comprise complex and dynamic input but also contain structured temporal contexts relatively similar to those that we experience in everyday life. Commercial movies consist of sequences of events that unfold coherently over time and that enable us to form and update knowledge and expectations about the story that develops during the movie. This contextual information includes multiple levels, encompassing short and long temporal windows (Hasson et al. 2008), as well as different levels of semantic and episodic complexity (Zacks et al. 2010; see also Kwok and Macaluso 2015). Tracking these multilevel processes, understanding their contribution for the allocation of attentional resources, and, ultimately, mapping these to changes in the brain are extraordinary challenges. Unlike studies based on pictures or short videos, the use of long and continuous stimuli often entails that experimental hypotheses cannot be addressed in the traditional way, that is, by constructing two conditions that differ for the process of interest and that can be compared to reveal the corresponding neurophysiological correlates. Indeed, this type of stimuli has been often analyzed using data-driven techniques (Bartels and Zeki 2005; Hasson et al. 2004; Lahnakoski et al. 2012), although data fitting with the GLM is also possible (e.g., Bartels et al. 2008; Bordier et al. 2013; Macaluso and Ogawa 2018; Ogawa et al. 2013).

One last point to mention is that even highly realistic full-length movies lack a key aspect of everyday life experience, that is, the possibility to interact with the environment. Within the constraints of an MR scanner, one option is to use virtual reality. Functional imaging during active exploration of virtual environments has been initially introduced in studies on navigation and memory (Burgess et al. 2001; Iaria et al. 2008). From the perspective of visuospatial attention control, little data is available to date. Conceptually, it may be suggested that active exploration will add new degrees of freedom to spatial orienting: rather than focusing processing resources only via shifts of attention/eye movements, now whole-body displacements – and active interactions with the environments – can be used to collect the maximal amount information about relevant objects and events (Gottlieb 2012). A main challenge of virtual reality concerns the fact that participant-driven behavior implies large interindividual heterogeneity of the data. Nonetheless it may still be possible to identify and track some process of interest and to use this information to analyze imaging data with GLM data fitting procedures (Macaluso and Ogawa 2018).

In sum, different types of stimuli can be used with the aim of triggering attention-related processes analogous to those taking place in real life. This broad range of stimuli is associated with variable levels of realism and experimental control (Fig. 1). Most often, the specific question one wants to address will guide the choice of the stimuli. Concurrently, it can be informative to ask the same question using different stimuli. Consistent findings across different stimuli/setups can contribute to a better understanding of attention control in naturalistic conditions. In the next sections, I will present several examples of imaging studies that have investigated the role of stimulus-driven and endogenous signals for spatial orienting. These employed a large variety of naturalistic stimuli that overall provide us with converging evidence about the involvement of dorsal and ventral frontoparietal regions for spatial orienting in naturalistic conditions.

3 Stimulus-Driven Signals

Standard imaging studies concerned with the influence of stimulus-driven signals on visual-spatial attention made use of spatial cueing or visual search tasks. To generate stimulus-driven signals, spatial cueing paradigms rely on abrupt and non-predictive physical changes, such as the appearance of a new stimulus in the visual field or some physical change of an existing stimulus (e.g., brightening). The influence of these signals can be assessed by comparing the participants response to stimuli (targets) presented at the location of the change (cued/valid location) vs. targets presented somewhere else in the visual field (uncued/invalid location). In visual search tasks, the display contains many stimuli/items, and the time it takes to find a specific stimulus (target) increases with increasing number of items. However, if the target differs from the other stimuli along one specific sensory dimension, the search is fast and independent of the number of items (e.g., search for a red-square in a display that otherwise contains only green items). The target grabs attention in a stimulus-driven manner, because it has a physical property (i.e., color, in the example) that differs from all the surrounding stimuli. Broadly, stimulus-driven signals are triggered by some physical discontinuity of the sensory input over feature space (cf. search tasks) or over time (cf. abrupt onsets, in spatial cueing tasks).

Several algorithms have been proposed to characterize physical discontinuities in naturalistic stimuli (Borji et al. 2013). A most influential model is that proposed by Itti and Koch (2001). The model is inspired by the organization of the visual system and extracts local discontinuities separately for different features (e.g., intensity, color, orientation) and at different spatial scales. The corresponding feature-specific maps are then combined in a single saliency map that highlights where in the original image there is a physical discontinuity, irrespective of the specific feature that generated the local contrast (see Fig. 2a). The model has been extended to dynamic stimuli by adding motion and flicker features (Carmi and Itti 2006) and to 3D stimuli by including local disparity (Ogawa et al. 2013, see also Jansen et al. 2009). Behavioral studies that used exploratory eye movements as an index of spatial attention demonstrated that the saliency maps can predict patterns of eye movements during the viewing of pictures (e.g., Elazary and Itti 2008) and dynamic videos (Carmi and Itti 2006). While many other factors are known to affect eye movements during free viewing of naturalistic stimuli (cf. Sect. 4; see also Hunt et al. "The Relationship Between Spatial Attention and Eye Movements"), the saliency maps are thought to well represent the contribution of stimulus-driven signals.

Saliency maps have been used in imaging studies to identify brain regions involved in the processing of stimulus-driven signals. Using cinematographic material (approx. 20 min, extracted from an episode of a TV series), we tested for brain regions where activity covaried with the saliency of the sensory input (Bordier et al. 2013). We fit the imaging data with continuous predictors based on the saliency of each single frame of the video. A conceptually analogous model was used to quantify the saliency of the auditory input, considering discontinuities in the time-frequency dimensions (see Kayser et al. 2005). The imaging results associated the processing of visual saliency with activity in extra-striate occipital cortex, plus the posterior parietal cortex (PPC), while activity in the superior temporal gyrus correlated with auditory salience. The relationship between overall stimulus salience and the activation of occipital and posterior parietal regions for vision and superior temporal areas for audition was confirmed in several other studies using different types of stimuli (e.g., static pictures in Santangelo and Macaluso 2013; short video clips in Nardo et al. 2014).

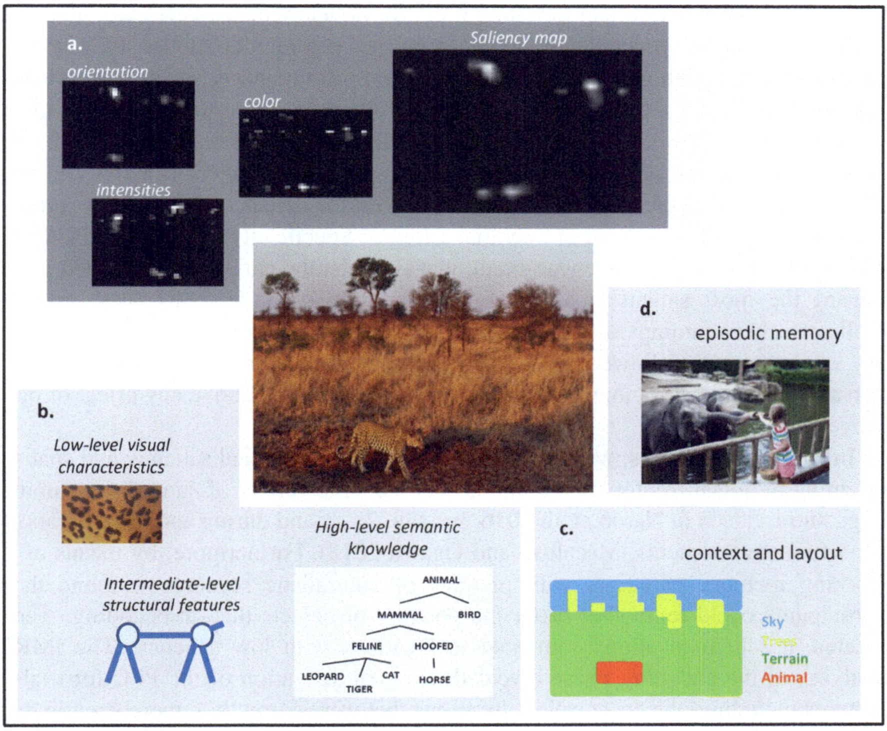

Fig. 2 Stimulus-driven and internal signals contribute to visuospatial orienting in naturalistic conditions. (**a**) Computational models of visual salience identify locations of the image that include some physical contrast, irrespective of the specific low-level feature that generates the contrast, cf. feature-specific maps of orientation, color, and intensities. Saliency maps can predict eye movements during viewing of naturalistic stimuli and are thought to well characterize the stimulus-driven component of attention orienting (see Sect. 3). Panels (**b–d**) illustrate different types of internal signals that have been found to affect the allocation of attention resources (see Sect. 4). (**b**) Object-level signals comprise knowledge about relatively low-level visual features (e.g., the pattern of leopards' skin) and intermediate-level structural aspects of familiar objects, as well as higher-level semantic information about objects and object categories. (**c**) Scene-level signals embody knowledge about the organization of real scenes, including the identity and spatial layout of objects in natural scenes. (**d**) Episodic memory for personally experienced events can also affect how we allocate resources in natural scenes. In the example here, the picture may trigger the retrieval of memories about a recent visit to the zoo that, in turn, can affect to what extent specific elements of natural scenes (e.g., the leopard) grab attention

The finding that activity in occipital cortex and PPC covaries with visual saliency provides us with some initial indication about the possible involvement of these regions in stimulus-driven attention (cf. behavioral work linking salience and spatial orienting, e.g., Carmi and Itti 2006). More direct evidence came from studies that measured eye movements during fMRI to obtain information about the allocation of spatial attention. In Nardo et al. (2011), we asked participants to freely view a 5 min video, recorded in first-person perspective and including navigation through a complex virtual environment (see also Fig. 3d). Saliency maps were computed for each frame, and the participants' eye position was recorded continuously. The eye position data allowed us to evaluate, on a frame-by-frame basis, whether the visual location with higher saliency did or did not grab attention. This information was used to construct predictors for the analysis of the fMRI data. The results showed that while activity in the occipital cortex covaried with overall saliency, activity in the PPC and the dorsal premotor cortex covaried with the distance between the current eye position and the location of maximal salience. Specifically, activity in PPC/FEF increased when this distance was small, that is, when the participants oriented gaze toward the most salient location of the scene. In separate fMRI sessions, we replicated these findings during presentation of the same stimuli, but now without any eye movement allowed. The latter permitted demonstrating the role of these dorsal frontoparietal regions in attention orienting over and above any effect of eye movements.

In subsequent studies, we confirmed the link between visual saliency and spatial orienting in dorsal frontoparietal regions using a large variety of naturalistic stimuli (e.g., short videos in Nardo et al. 2016, see Fig. 2a–c; and during active exploration of virtual environments, Macaluso and Ogawa 2018). Furthermore, by means of a working memory paradigm with pictures of naturalistic scenes, we found that participants could remember better the position of objects that, at encoding, were located at salient locations compared to locations with low saliency. The fMRI analyses of the encoding phase revealed maximal activation of the PPC for trials/pictures including objects at salient locations that were correctly remembered in the subsequent retrieval phase (Santangelo and Macaluso 2013). These results support the proposal that saliency effects in the PPC reflect the contribution of stimulus-related signals to attentional selection, including orienting of spatial attention and successful storage of spatial information in working memory.

Additional evidence for the role of dorsal frontoparietal regions in stimulus-driven attentional selection come from the study of Bogler et al. (2011). Using multivariate decoding analyses, the authors dissociated regions involved in the representation of graded salience (i.e., the overall salience of the image) and regions involved in winner-takes-all mechanisms linked to visuospatial attentional selection. Graded salience was found to covary with the signal amplitude in the occipital visual cortex and the posterior part of the parietal cortex, while the output of the winner-takes-all mechanism (i.e., position of the saliency peak in the image) was decoded in a more anterior region of the intra-parietal sulcus, plus the dorsal premotor cortex (FEF). These results are in line with the notion that stimulus-driven control entails sequential operations that start in the occipital visual cortex and proceed through the visual hierarchy via bottom-up, feed-forward signaling. Accordingly, competitive interactions at the feature-specific level would take place in specialized visual areas, these would be then integrated in feature-independent saliency representations in parietal cortex (Itti and Koch 2001) and, finally, selection processes needed to guide attentional orienting take place in a network comprising the IPS and the FEF (Bogler et al. 2011, see also Ptak 2012; Sprague et al. 2018). However, it should be noticed that other paths are also likely to be involved in stimulus-driven control of spatial attention (Veale et al. 2017).

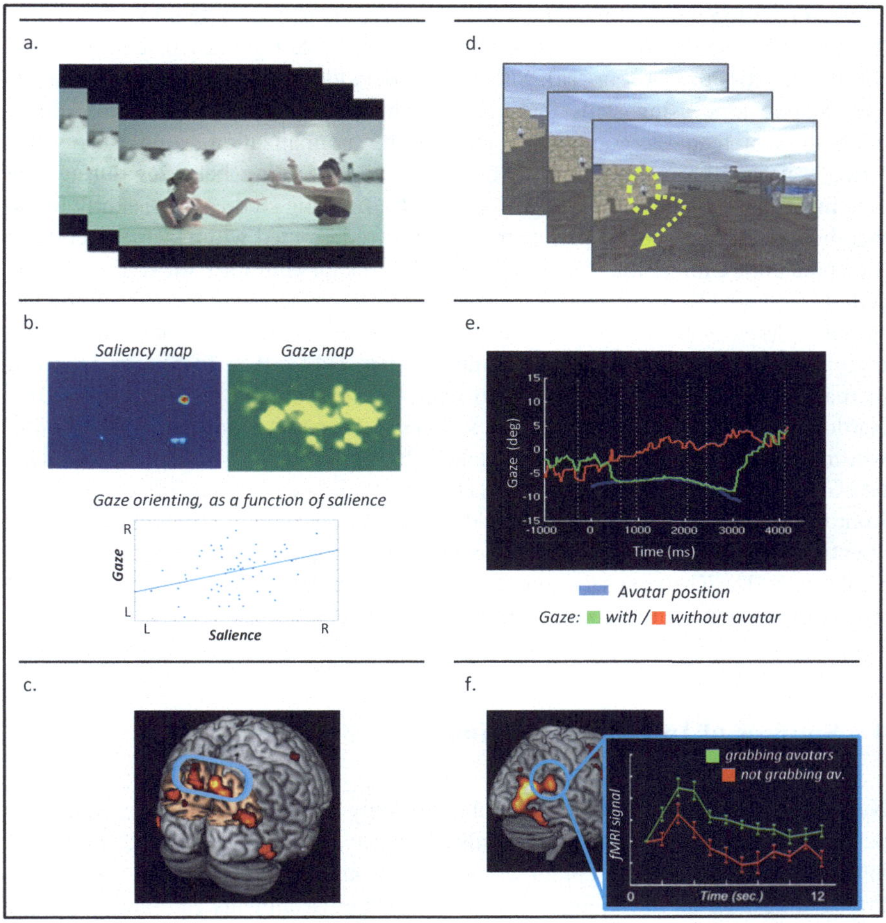

Fig. 3 Stimulus-driven and internal signals in the dorsal and ventral parietal cortex during free viewing of naturalistic stimuli. Panels (**a–c**) refer to the influence of visual salience in the posterior parietal cortex (adapted from Nardo et al. 2016), while panels (**d–f**) report the impact of internal knowledge about distinctive events in the temporoparietal junction (adapted from Nardo et al. 2011). (**a**) Example of a short video comprising visual events on both sides and entraining an effect of salience on spatial orienting. (**b**) Saliency and gaze maps (i.e., the locations fixated by a group of subjects, who freely viewed the video in panel (**a**); and the linear relationship between the level of saliency and gaze lateralization over 60 different videos (L/R: silence, gaze lateralized on the left/right side). (**c**) Regions of the brain where activity correlated with the amount of gaze lateralization in videos with events on both sides, highlighting the contribution of salience to gaze orienting in the posterior parietal cortex (PPC, circled in blue). (**d**) A few frames of the video of first-person navigation through a virtual environment, at the moment when a moving avatar enters in the visual field. The yellow dotted line indicates the approximate path of the avatar. (**e**) Horizontal position of the avatar (blue) and average gaze position of participants who watched the video with the avatar (green) versus a version of the same video but without the avatar (red). The gaze data demonstrate that the appearance of this avatar affected spatial orienting. (**f**) Activation of the temporoparietal junction (TPJ, circled in blue) time locked to the appearance of the avatars. The analyses showed greater activation of the TPJ for avatars that grabbed the participants' attention/gaze (green, in the

In sum, computational models of saliency provide us with a tool to characterize the strength and the spatial position of stimulus-related signals in complex and naturalistic stimuli (see Fig. 2a). Saliency models identify physical discontinuities and include processing steps that can be related to attentional operations (e.g., integration of features, competitive interactions, normalization, winner-takes-all). Visual saliency has been found to predict spatial orienting behavior during free viewing of naturalistic stimuli (e.g., see Fig. 3b). Imaging studies consistently found that the overall salience of naturalistic stimuli is represented within sensory cortices (occipital cortex for vision, superior temporal gyrus for audition). Most importantly, activity in dorsal frontoparietal regions has been associated with the contribution of stimulus-driven salience for the selection of relevant spatial locations. Activity in regions belonging to the dorsal attention system (PPC, IPS, FEF) was found to increase when saliency contributed to overt eye movements (Nardo et al. 2011; Nardo et al. 2016, see Fig. 3c), during successful encoding of spatial information in working memory (Santangelo and Macaluso 2013), and was associated with winner-takes-all mechanisms (Bogler et al. 2011). The finding of saliency effects in dorsal frontoparietal regions may appear at odd with the classical view that stimulus-driven attentional control depends primarily on ventral frontoparietal regions (Corbetta and Shulman 2002). The next section will address this in the broader context of internal vs. external signals during orienting in naturalistic conditions.

4 Sources of Internal Signaling

Naturalistic stimuli are rich and complex not only from a purely sensory/physical perspective (cf. saliency, above) but also in terms of their content and links with information stored in the brain (Malcolm et al. 2016a; Groen et al. 2017). Pre-existing internal knowledge about real-world environments can play a role for attentional selection (Torralba 2003; Henderson 2017). This knowledge entails multiple domains, including object-level structural and semantic information (Reeder and Peelen 2013; Malcolm et al. 2016b; Cukur et al. 2013), as well as scene-level signals about global scene properties and objects-in-context (DeCesarei et al. 2017; Hutchinson and Turk-Browne 2012; Wolfe et al. 2011; Bar 2004) (see Fig. 2b, c). Moreover, personal experience concerning objects and events in real environments can generate additional long-term (episodic) memory signals affecting attention control (e.g., Summerfield et al. 2006; Hutchinson and Turk-Browne 2012) (see Fig. 2d). The influence of these different types of internal signals on spatial attention has been studied primarily using goal-directed tasks (e.g., search for a

Fig. 3 (continued) fMRI signal plot) compared with avatars that did not affect spatial orienting (plotted red). These data indicate that the ventral parietal cortex mediates the contribution of distinctive visual events (here, the avatars) to spatial orienting in complex and dynamic environments

specific object in the scene, e.g., Reeder and Peelen 2013), but also in more natural situations entailing free viewing of naturalistic stimuli in the absence of any explicit task (Hwang et al. 2011; Nardo et al. 2016).

Behavioral experiments using goal-directed search tasks highlight that the time it takes to find a target object in a cluttered real scene is remarkably fast (Thorpe et al. 1996; Le-Hoa and Wolfe 2015). This fits with the hypothesis that prior information about natural scenes contributes to attentional selection. One interpretation of these findings is that goal-directed attention can be focused on "features" that are characteristic of the to-be-found object. Accordingly, only objects containing these relevant features would compete for attentional resources, while objects that do not share any target feature would be filtered out in an efficient manner (Peelen and Kastner 2014, but see Li et al. 2002). Object-level features may entail relatively low-level sensory characteristic, such as the typical color of the target object ("bananas are yellow"), intermediate-level structural aspects ("giraffes are made of five long segments, spatially arranged in a specific manner"), as well as higher-level semantic knowledge ("cars are motorized vehicles") (Fig. 2b). Behavioral evidence indicate that each of these levels can contribute to the deployment of spatial attention (Reeder and Peelen 2013; Seidl-Rathkopf et al. 2015; Wu et al. 2014b; Malcolm et al. 2016b; but see also Daffron and Davis 2016).

Early imaging studies investigated the neural correlates of attention to objects' categories and showed that goal-directed attention modulates activity in associative visual regions representing the relevant category (O'Craven et al. 1999, for attention to faces vs. buildings). More recently, using multivariate analyses methods and pictures of real-world scenes, Seidl et al. (2012) showed that patterns of activity in the visual cortex not only encode the currently relevant category but also reflect the suppression of irrelevant distractor categories (see also Kaiser et al. 2016, for related findings using magneto-encephalography). Together with these modulatory influences on visual representations, goal-directed attention has been found to engage frontoparietal regions, as a function of the task at hand. Nastase et al. (2017) presented participants with short videos of different animal categories (e.g., birds, reptiles) performing different types of actions (e.g., fighting, eating). The participants were asked to discriminate either the animal category (taxonomy task) or the animal behavior (action task). Multivariate analyses revealed that attending to the animals' behavior resulted in an increased discrimination of action categories in ventral occipital cortex, but also in premotor and parietal cortices. By contrast, the taxonomy task affected activity in visual cortex but not in the frontoparietal regions (see also Guo et al. 2012).

Concerning the interplay between category-based attention and the allocation of goal-directed spatial attention, we recently asked whether these two forms of endogenous control rely on common or independent brain substrates (Fagioli and Macaluso 2016). The experiment made use of pictures of real scenes containing people or cars, on the left or the right side. In different blocks, the participants were instructed to divide attention either between the two categories ("attend both to cars and people on the same side"), the two visual hemifields ("attend to one category only, but monitor both sides"), or to monitor different categories at different locations (e.g., "cars on the left and people on the right"). The imaging results showed increased activation of dorsal frontoparietal regions in the left hemisphere, when the participants monitored different categories in the two hemifields. This statistical interaction suggests that common processes underlie the control of divided spatial- and category-based attention. The results also revealed that the joint effect of space and category was most pronounced on distractors trials, that is, when the image contained an object of a task-relevant category at a currently relevant location, but the current instruction required responding to the opposite combination of category and location (e.g., scene containing a car on the left, when the participant was instructed to attend to cars on the right and people on the left). We suggested that these distractor conditions implied the strongest requirement of top-down control to suppress the processing of distractor objects matching multiple aspects of the current target template (see also, Seidl et al. 2012).

The studies described above focused on the contribution of internal knowledge about the objects and object categories (Fig. 2b). However, a central aspect of naturalistic environments is that objects do not exist in isolation, but they are embedded within meaningful contexts. Objects tend to have specific semantic and spatial relationships with each other, and these relationships are thought to be represented internally in schemas about the structure of natural environments (Bar 2004) (see Fig. 2c). The role of contextual information for object processing has been studied extensively both in memory and attention (e.g., Henderson and Hollingworth 1999; Chun 2000; van Kesteren et al. 2012). In the attention domain, behavioral studies show that out-of-context objects are easy to detect (Hollingworth and Henderson 1998), possibly because violations of schema-related information contribute to making them "attention grabbing" (Stirk and Underwood 2007). Concurrently, schemas can facilitate the search for objects located at congruent spatial locations (e.g., plates on a table) by providing goal-directed control with prior information about the likely location of the target (Mack and Eckstein 2011).

Using a working memory (WM) paradigm, we investigated the fMRI correlates of object-context congruency and the interaction between these knowledge-based internal signals and stimulus-driven external salience (Santangelo et al. 2015). The participants were presented with pictures of naturalistic scenes that, on half of the trials, contained an out-of-context object (e.g., a vacuum cleaner inside a fridge). After a short delay, they were asked to report the position of the one of the objects of the scene. The object could be either congruent or incongruent with the context of the scene and was extracted from an image location with high or low salience. The behavioral data showed better retrieval performance for out-of-context objects compared with in-context objects and that the presence of out-of-context objects reduced the facilitatory effect of salience. The latter finding is in line with behavioral studies showing a reduction of the effect of stimulus-driven salience when internal signals (e.g., task-related instructions) are available to guide attentional selection and spatial orienting (Einhauser et al. 2008; see also, Wolfe and Horowitz 2004). The imaging results showed that the encoding of context-incongruent objects was associated with a deactivation of the right temporal-parietal junction in the ventral attention system, plus the left middle frontal gyrus. By contrast, the encoding context-congruent scenes activated the posterior parietal cortex in dorsal attention system. We linked the deactivation of the ventral parietal cortex with filtering mechanisms that may come into play following the detection of the task-relevant out-of-context object that, in this experiment, was always the to-be-remembered target. Object-context congruence does not only affect activity in the frontoparietal attention systems but can also modulate the activity of the occipital visual cortex. For example, Preston et al. (2013) were able to decode the likely location of everyday-life objects based on the pattern of fMRI activity in the lateral occipital cortex and, to a lesser degree, in the intra-parietal sulcus and the retrosplenial cortex. The authors interpreted their findings as suggesting that the visual cortex represents contextually relevant information about the current search target (see also Kaiser et al. 2014, for the influence of object-context relationship on distractor processing).

The effects of object-category and object-context congruence relate primarily to semantics, but episodic information concerning personal experiences has also been reported to influence attention. Some objects may be strongly associated with episodes that we have personally experienced and, thus, attract attention based on this internally stored information. An example of this could be to see the picture of a leopard that we can associate with personal memories about a recent visit to the zoo (Fig. 2d). The mechanisms governing the interplay between attention and episodic memory are still debated (cf Ciaramelli et al. 2008; Sestieri et al. 2010), but extensive evidence indicates that long-term memory can contribute to guiding attention (e.g., Chun 2000; Chun and Turk-Browne 2007; Draschkow et al. 2014; Baldassano et al. 2016).

Functional imaging studies of memory-guided attention using naturalistic stimuli reveal the engagement of several networks, including frontoparietal attention systems, memory regions in the medial temporal cortex, and visual areas involved in scene representation. In a pioneering study, Summerfield et al. (2006) highlighted the role of the hippocampus which was found to activate when participants made use of memorized information to search for targets in natural scenes, together with dorsal frontoparietal regions (Rosen et al. 2016, for related finding and also highlighting the role the precuneus: another key region involved in episodic memory, Wagner et al. 2005). Several studies highlighted increased inter-regional connectivity between the hippocampus and the posterior partial cortex during tasks involving repeated presentation of simple displays (Manelis and Reder 2012; Goldfarb et al. 2016) or complex scenes (Howard et al. 2013). It should be noticed that the relationship between episodic memory and other contextual effects related to prior knowledge about scene organization is not fully resolved. The engagement of the hippocampus has been found to be independent of explicit memory (Manelis and Reder 2012; Goldfarb et al. 2016, see also Draschkow et al. 2014) and task relevance (Howard et al. 2013). Thus, medial temporal regions may provide the dorsal frontoparietal attention system with information about spatial layouts via implicit, pre-attentive mechanisms (Hannula and Greene 2012; Li et al. 2002, but see also Wu et al. 2014b). By contrast, episodic memory is ought to involve the explicit retrieval of the details of the encoded episode. In turns, this difference may lead to different mechanisms mediating the influence of long-term memory on attention. Future work should seek to better delineate the role of the different memory systems in memory-guided attention (Goldfarb et al. 2016; Todd and Manaligod 2018). In the framework of explicit episodic memory, this would comprise asking how specific elements of complex and naturalistic episodes (e.g., what vs. where) contribute to memory-guided attention (Kwok and Macaluso 2015; Kauttonen et al. 2018).

The imaging studies discussed up to here highlighted that different types of internal signals can affect the allocation of spatial attention, emphasizing the role dorsal frontoparietal regions and the modulation of visual representations in the occipital cortex. The vast majority of these studies made use of explicit goal-directed search tasks that prompt the participants to voluntarily focus on elements that characterize the current search target. Nonetheless, as noted at the beginning of this section, one characteristic of spatial orienting in the real life is that this rarely involves any explicit task-related goal. Free viewing paradigms provide us with one way of asking how internal signals contribute to attention in the absence of any goal-directed control. A main assumption here is that patterns of eye movements and fixations can provide us with a relevant measure of what elements/objects are selected for in-depth processing (Hayhoe and Ballard 2005; Henderson 2017; but see also Hunt et al. in this volume).

Free viewing studies have highlighted the role of stimulus-driven salience for gaze orienting in naturalistic conditions (see Sect. 3). However, behavioral studies show the mere presence of objects and/or semantically meaningful events reduce the gaze predictivity of saliency maps based on low-level sensory features only. For example, Einhauser et al. (2008) showed that spatial maps based on object location outperformed saliency maps (see also Henderson and Hayes 2017; Stoll et al. 2015; but see Elazary and Itti 2008). Furthermore, the semantic relationship between objects in natural scenes has been found to contribute to fixation patterns. Hwang and colleagues presented participants with pictures of natural scenes and examined the sequences of fixated objects, as a function of their semantic relationship (Hwang et al. 2011). The results revealed that participants were more likely to shift gaze between semantically related objects than unrelated objects, even when the task did not entail any relevant object category (i.e., scene inspection vs. explicit object search, see also Wu et al. 2014a).

Using fMRI, we investigated the contribution of semantically distinctive events for the allocation of visuospatial attention and their interaction with stimulus-driven salience (Nardo et al. 2016). The experiment involved free viewing of complex and dynamic naturalistic stimuli (1.5 s video clips). Each video contained either one semantically distinctive event, lateralized on the left or the right side, or multiple distinctive events on both sides of space (Fig. 3a). The participants were asked to simply watch the stimuli without any explicit task and were allowed to move their eyes. As expected, when the video included a single distinctive event, the participants consistently looked on the side of the event, and stimulus-driven salience did not affect the time spent on each side. However, when the videos included events on both sides, there was a linear relationship between the level of saliency bias (ratio of saliency on the left/right side) and spatial orienting (ratio of time spent looking toward the left/right side) (see Fig. 3b). This indicates that semantic signals largely dominate spatial orienting in free viewing conditions, but when multiple semantically distinctive signals compete for attention guidance, lower-level stimulus-related signals start playing a role. The imaging data linked the effect of stimulus-driven saliency on gaze orienting with the activity of the dorsal posterior parietal cortex (Fig. 3c, and Sect. 3). By contrast, the presence of multiple meaningful events was associated with processing in the temporal-parietal junction (TPJ), in the ventral attention system. We suggested that the TPJ makes use of semantic knowledge to detect distinctive elements within natural scenes and uses these internal signals to guide spatial orienting. In agreement with this, our study entailing passive navigation through a virtual environment (Nardo et al. 2011) highlighted that the TPJ activated when human-like characters (moving avatars) appeared in the visual field of the participants (see Fig. 3d, e). Critically, the results revealed larger activation of the right TPJ when the avatars grabbed participants' gaze/attention compared with avatars did not affect spatial orienting (Fig. 3f). These findings support the proposal that in free viewing conditions semantic knowledge contributes to spatial orienting via the ventral attention system.

In sum, many sources of external and internal information can contribute to guiding attention in natural conditions (Fig. 2). External signals can be quantified using computational models of stimulus-driven silence that identify image locations with a high local contrast (see Fig. 2a). Internal signals include multiple levels of knowledge, comprising object-level information about the characteristics and meaning of familiar objects, scene-level signals concerning the organization of natural scenes (e.g., object-context relationship), as well as knowledge associated with personal experiences (episodic memory) (see Fig. 2b–d). Each of these factors has been found to influence processing in the frontoparietal attention networks, but the specific contribution of the different signals – and their interactions – is not fully understood yet.

In goal-directed search tasks, each level of internal knowledge may provide the attention system with relevant "features" to distinguish the target from the surrounding stimuli (target template). In this framework, attention control may operate in a similar manner as in tasks with simple stimuli (Kastner and Ungerleider 2000). This would involve the modulation of processing in sensory areas that represent the task-relevant category/location (O'Craven et al. 1999, but see also Cukur et al. 2013; for category-based attention in higher-order associative regions). These top-down modulatory influences would arise from control regions in the dorsal frontoparietal cortex and prefrontal cortex (see Buschman and Kastner 2015; Battistoni et al. 2017). Long-term memory may further contribute via connectivity between the medial temporal cortex and the dorsal attention system (Goldfarb et al. 2016; Howard et al. 2013; see also Bar 2004; Howard et al. 2011). The ventral attention system would perform template match/mismatch operations, based on the current sensory input and task-dependent goals/expectations (Nobre et al. 1999; Doricchi et al. 2010; Macaluso and Doricchi 2013; Corbetta et al. 2008).

By contrast, studies that made use of free viewing paradigms consistently associated the dorsal frontoparietal system with the processing of stimulus-driven signals (Nardo et al. 2011; Bogler et al. 2011), and we linked the engagement of the ventral frontoparietal cortex with the processing of semantically meaningful events (e.g., Nardo et al. 2011, 2016). During free viewing, control of spatial orienting prioritizes feed-forward connectivity between sensory regions and the dorsal frontoparietal network (stimulus-driven control); and knowledge about the content and structure of natural environments triggers detection-related processes in the ventral system (see Fig. 3).

5 Perspectives and Conclusions

The introduction of naturalistic stimuli in neuroimaging studies permitted narrowing the gap between traditional studies of attention and possible mechanisms of attention control in real life. Nonetheless, the use of naturalistic stimuli pose a major challenge, in particular regarding the many processes and variables that are likely to be at play (Calhoun and Pearlson 2012; Huth et al. 2012; Maguire 2012; Hasson and Honey 2012). Most of the studies that investigate attention control in naturalistic conditions still make use of static pictures, which allow using relatively standard experimental designs and data analyses (e.g., data fitting with GLM and direct comparisons between conditions, see Sect. 2). Alternatively, imaging data acquired with dynamic stimuli (e.g., movies) can be analyzed with data-driven methods. For example, independent component analysis (ICA) has been used to tease apart regions/networks showing differential responses during movie watching, highlighting that visual areas specialized for processing different visual low-level features (motion vs. colors) also operate in a segregated manner during viewing of dynamic naturalistic stimuli (Bartels and Zeki 2005; and see Cong et al. 2014, for a discussion of ICA application in the framework of fMRI with naturalistic stimuli). ICA has the advantage that it can reveal not only areas responding to the stimuli but will highlight areas with temporally correlated signals that are most likely to form functional networks (see also, Kim et al. 2018). Hasson et al. (2004) proposed an alternative data-driven method for the analysis of fMRI data acquired with dynamic natural stimuli (ISS, inter-subject synchronization analysis; see also Poulsen et al. 2017, for a related method with electroencephalographic data). The main idea of ISS is that when two different participants are presented with the same stimuli, brain regions processing the stimuli should show similar changes of signal over time (see Hasson et al. 2010, for review). A main limitation of standard applications of ICA and ISS is that both methods assume signal stationarity, that is, the level coupling between areas (ICA) or between subjects (ISS) is assumed to be fixed during the whole time series. One option to address this limitation is to compute the relevant parameters using sliding windows (Hutchison et al. 2013; Allen et al. 2014; Bordier and Macaluso 2015). This allows gaining some temporal information about the transient engagement of specific areas or networks. Examining a posteriori the stimuli at the relevant time points can help interpreting what aspect of the stimuli drove the transient networks' coupling or increased ISS (see also Hasson et al. 2004, but note that in this seminal work considered peaks of BOLB activity – rather than temporally resolved ISS). Further developments of these data-driven methods are likely to provide us with relevant ways of analyzing imaging data acquired with naturalistic stimuli.

Several other approaches have been recently employed to identify brain regions involved in the processing of dynamic naturalistic stimuli, including artificial neural networks (Jang et al. 2017; Guclu and van Gerven 2017; Martin et al. 2017). These computational methods are highly promising, as they enable formalizing links between the low-/high-level aspects of the stimuli and changes of brain activity (Wen et al. 2017; DeCesarei et al. 2017; Groen et al. 2017, see also DiCarlo et al. 2012; Khaligh-Razavi et al. 2017, for reviews). In particular, some of the layers of trained convolutional neural networks can represent scene characteristics that have been found to contribute to attention selection processes, bridging from low-level physical aspects (salience) to high-level semantics. Thus, without any need for manual labeling, these computational approaches can be used to characterize relevant signals in complex naturalistic stimuli and to investigate possible correspondences between stimulus features represented in these networks and neuronal representations in the brain (Cichy and Teng 2017). These innovative approaches are opening a new window of opportunity for the study of brain functions using naturalistic stimuli.

In sum, traditional fMRI studies of spatial attention may fail to capture relevant characteristics of attention control in real life and overemphasize the role of other processes. Real-life situations imply the co-existence of multiple signals that can jointly contribute to the allocation of processing resources, and attention is rarely controlled based on specific task rules. Naturalistic stimuli and free viewing paradigms can help us to bridge these gaps. Naturalistic stimuli comprise low-level physical aspects that contribute to stimulus-driven attention control (cf. Sect. 3), as well as many different types of internal signals related to prior knowledge. This includes information about familiar objects, content and layout of natural scenes, and personal experiences associated with objects and real environments (Fig. 2, and Sect. 4). Tracking and quantifying all these signals is an extraordinary challenge. The combination of behavioral measures (e.g., eye movements and task performance) with computational approaches (e.g., saliency maps and neural networks) provides us with valuable tools to guide the analyses of imaging data acquired with naturalistic stimuli. Current results emphasize the role of the frontoparietal attention networks for the processing of these various signals, but further methodological developments are needed to fully seize the potential of naturalistic stimuli to reveal mechanisms of real-life brain functioning.

References

Allen EA, Damaraju E, Plis SM, Erhardt EB, Eichele T, Calhoun VD (2014) Tracking whole-brain connectivity dynamics in the resting state. Cereb Cortex 24(3):663–676

Arrington CM, Carr TH, Mayer AR, Rao SM (2000) Neural mechanisms of visual attention: object-based selection of a region in space. J Cogn Neurosci 12(Suppl 2):106–117

Baldassano C, Esteva A, Fei-Fei L, Beck DM (2016) Two distinct scene-processing networks connecting vision and memory. eNeuro 3(5):1–14

Bar M (2004) Visual objects in context. Nat Rev Neurosci 5(8):617–629

Bartels A, Zeki S (2005) Brain dynamics during natural viewing conditions--a new guide for mapping connectivity in vivo. NeuroImage 24(2):339–349

Bartels A, Zeki S, Logothetis NK (2008) Natural vision reveals regional specialization to local motion and to contrast-invariant, global flow in the human brain. Cereb Cortex 18(3):705–717

Battistoni E, Stein T, Peelen MV (2017) Preparatory attention in visual cortex. Ann N Y Acad Sci 1396(1):92–107

Bogler C, Bode S, Haynes JD (2011) Decoding successive computational stages of saliency processing. Curr Biol 21(19):1667–1671

Bordier C, Macaluso E (2015) Time-resolved detection of stimulus/task-related networks, via clustering of transient intersubject synchronization. Hum Brain Mapp 36(9):3404–3425

Bordier C, Puja F, Macaluso E (2013) Sensory processing during viewing of cinematographic material: computational modeling and functional neuroimaging. NeuroImage 67:213–226

Borji A, Sihite DN, Itti L (2013) Quantitative analysis of human-model agreement in visual saliency modeling: a comparative study. IEEE Trans Image Process 22(1):55–69

Burgess N, Maguire EA, Spiers HJ, O'Keefe J (2001) A temporoparietal and prefrontal network for retrieving the spatial context of lifelike events. NeuroImage 14(2):439–453

Buschman TJ, Kastner S (2015) From behavior to neural dynamics: an integrated theory of attention. Neuron 88(1):127–144

Calhoun VD, Pearlson GD (2012) A selective review of simulated driving studies: combining naturalistic and hybrid paradigms, analysis approaches, and future directions. NeuroImage 59(1):25–35

Carmi R, Itti L (2006) Visual causes versus correlates of attentional selection in dynamic scenes. Vis Res 46(26):4333–4345

Chun MM (2000) Contextual cueing of visual attention. Trends Cogn Sci 4(5):170–178

Chun MM, Turk-Browne NB (2007) Interactions between attention and memory. Curr Opin Neurobiol 17(2):177–184

Ciaramelli E, Grady CL, Moscovitch M (2008) Top-down and bottom-up attention to memory: a hypothesis (AtoM) on the role of the posterior parietal cortex in memory retrieval. Neuropsychologia 46(7):1828–1851

Cichy RM, Teng S (2017) Resolving the neural dynamics of visual and auditory scene processing in the human brain: a methodological approach. Philos Trans R Soc Lond Ser B Biol Sci 372(1714):20160108

Cong F, Puolivali T, Alluri V, Sipola T, Burunat I, Toiviainen P et al (2014) Key issues in decomposing fMRI during naturalistic and continuous music experience with independent component analysis. J Neurosci Methods 223:74–84

Corbetta M, Shulman GL (2002) Control of goal-directed and stimulus-driven attention in the brain. Nat Rev Neurosci 3(3):215–229

Corbetta M, Miezin FM, Shulman GL, Petersen SE (1993) A PET study of visuospatial attention. J Neurosci 13(3):1202–1226

Corbetta M, Patel G, Shulman GL (2008) The reorienting system of the human brain: from environment to theory of mind. Neuron 58(3):306–324

Cukur T, Nishimoto S, Huth AG, Gallant JL (2013) Attention during natural vision warps semantic representation across the human brain. Nat Neurosci 16(6):763–770

Daffron JL, Davis G (2016) Target templates specify visual, not semantic, features to guide search: a marked asymmetry between seeking and ignoring. Atten Percept Psychophys 78(7):2049–2065

DeCesarei A, Loftus GR, Mastria S, Codispoti M (2017) Understanding natural scenes: contributions of image statistics. Neurosci Biobehav Rev 74(Pt A):44–57

DiCarlo JJ, Zoccolan D, Rust NC (2012) How does the brain solve visual object recognition? Neuron 73(3):415–434

Doricchi F, Macci E, Silvetti M, Macaluso E (2010) Neural correlates of the spatial and expectancy components of endogenous and stimulus-driven orienting of attention in the Posner task. Cereb Cortex 20(7):1574–1585

Draschkow D, Wolfe JM, Vo ML (2014) Seek and you shall remember: scene semantics interact with visual search to build better memories. J Vis 14(8):10

Einhauser W, Spain M, Perona P (2008) Objects predict fixations better than early saliency. J Vis 8(14):18–26

Elazary L, Itti L (2008) Interesting objects are visually salient. J Vis 8(3):3–15

Fagioli S, Macaluso E (2016) Neural correlates of divided attention in natural scenes. J Cogn Neurosci 28(9):1392–1405

Geng JJ, Mangun GR (2011) Right temporoparietal junction activation by a salient contextual cue facilitates target discrimination. NeuroImage 54(1):594–601

Goldfarb EV, Chun MM, Phelps EA (2016) Memory-guided attention: independent contributions of the hippocampus and striatum. Neuron 89(2):317–324

Gottlieb J (2012) Attention, learning, and the value of information. Neuron 76(2):281–295

Groen II, Silson EH, Baker CI (2017) Contributions of low- and high-level properties to neural processing of visual scenes in the human brain. Philos Trans R Soc Lond Ser B Biol Sci 372(1714):20160102

Guclu U, van Gerven MAJ (2017) Increasingly complex representations of natural movies across the dorsal stream are shared between subjects. NeuroImage 145.(Pt B:329–336

Guo F, Preston TJ, Das K, Giesbrecht B, Eckstein MP (2012) Feature-independent neural coding of target detection during search of natural scenes. J Neurosci 32(28):9499–9510

Hannula DE, Greene AJ (2012) The hippocampus reevaluated in unconscious learning and memory: at a tipping point? Front Hum Neurosci 6:80

Hasson U, Honey CJ (2012) Future trends in neuroimaging: neural processes as expressed within real-life contexts. NeuroImage 62(2):1272–1278

Hasson U, Nir Y, Levy I, Fuhrmann G, Malach R (2004) Intersubject synchronization of cortical activity during natural vision. Science 303(5664):1634–1640

Hasson U, Yang E, Vallines I, Heeger DJ, Rubin N (2008) A hierarchy of temporal receptive windows in human cortex. J Neurosci 28(10):2539–2550

Hasson U, Malach R, Heeger DJ (2010) Reliability of cortical activity during natural stimulation. Trends Cogn Sci 14(1):40–48

Hayhoe M, Ballard D (2005) Eye movements in natural behavior. Trends Cogn Sci 9(4):188–194

Henderson JM (2017) Gaze control as prediction. Trends Cogn Sci 21(1):15–23

Henderson JM, Hayes TR (2017) Meaning-based guidance of attention in scenes as revealed by meaning maps. Nat Hum Behav 1:743–747

Henderson JM, Hollingworth A (1999) High-level scene perception. Annu Rev Psychol 50:243–271

Hollingworth A, Henderson JM (1998) Does consistent scene context facilitate object perception? J Exp Psychol Gen 127(4):398–415

Howard LR, Kumaran D, Olafsdottir HF, Spiers HJ (2011) Double dissociation between hippocampal and parahippocampal responses to object-background context and scene novelty. J Neurosci 31(14):5253–5261

Howard LR, Kumaran D, Olafsdottir HF, Spiers HJ (2013) Dissociation between dorsal and ventral posterior parietal cortical responses to incidental changes in natural scenes. PLoS One 8(7): e67988

Hutchinson JB, Turk-Browne NB (2012) Memory-guided attention: control from multiple memory systems. Trends Cogn Sci 16(12):576–579

Hutchison RM, Womelsdorf T, Allen EA, Bandettini PA, Calhoun VD, Corbetta M et al (2013) Dynamic functional connectivity: promise, issues, and interpretations. NeuroImage 80:360–378

Huth AG, Nishimoto S, Vu AT, Gallant JL (2012) A continuous semantic space describes the representation of thousands of object and action categories across the human brain. Neuron 76(6):1210–1224

Hwang AD, Wang HC, Pomplun M (2011) Semantic guidance of eye movements in real-world scenes. Vis Res 51(10):1192–1205
Iaria G, Fox CJ, Chen JK, Petrides M, Barton JJ (2008) Detection of unexpected events during spatial navigation in humans: bottom-up attentional system and neural mechanisms. Eur J Neurosci 27(4):1017–1025
Indovina I, Macaluso E (2007) Dissociation of stimulus relevance and saliency factors during shifts of visuospatial attention. Cereb Cortex 17(7):1701–1711
Itti L, Koch C (2001) Computational modelling of visual attention. Nat Rev Neurosci 2(3):194–203
Jang H, Plis SM, Calhoun VD, Lee JH (2017) Task-specific feature extraction and classification of fMRI volumes using a deep neural network initialized with a deep belief network: evaluation using sensorimotor tasks. NeuroImage 145.(Pt B:314–328
Jansen L, Onat S, Konig P (2009) Influence of disparity on fixation and saccades in free viewing of natural scenes. J Vis 9(1):29–19
Kaiser D, Stein T, Peelen MV (2014) Object grouping based on real-world regularities facilitates perception by reducing competitive interactions in visual cortex. Proc Natl Acad Sci U S A 111(30):11217–11222
Kaiser D, Oosterhof NN, Peelen MV (2016) The neural dynamics of attentional selection in natural scenes. J Neurosci 36(41):10522–10528
Kastner S, Ungerleider LG (2000) Mechanisms of visual attention in the human cortex. Annu Rev Neurosci 23:315–341
Kauttonen J, Hlushchuk Y, Jaaskelainen IP, Tikka P (2018) Brain mechanisms underlying cue-based memorizing during free viewing of movie Memento. NeuroImage 172:313–325
Kayser C, Petkov CI, Lippert M, Logothetis NK (2005) Mechanisms for allocating auditory attention: an auditory saliency map. Curr Biol 15(21):1943–1947
Khaligh-Razavi SM, Henriksson L, Kay K, Kriegeskorte N (2017) Fixed versus mixed RSA: explaining visual representations by fixed and mixed feature sets from shallow and deep computational models. J Math Psychol 76.(Pt B:184–197
Kim D, Kay K, Shulman GL, Corbetta M (2018) A new modular brain organization of the BOLD signal during natural vision. Cereb Cortex 28:3065–3081
Kwok SC, Macaluso E (2015) Exogenous features versus prior experiences modulate different subregions of the right IPL during episodic memory retrieval. Sci Rep 5:11248
Lahnakoski JM, Salmi J, Jaaskelainen IP, Lampinen J, Glerean E, Tikka P et al (2012) Stimulus-related independent component and voxel-wise analysis of human brain activity during free viewing of a feature film. PLoS One 7(4):e35215
Le-Hoa VM, Wolfe JM (2015) The role of memory for visual search in scenes. Ann N Y Acad Sci 1339:72–81
Li FF, VanRullen R, Koch C, Perona P (2002) Rapid natural scene categorization in the near absence of attention. Proc Natl Acad Sci U S A 99(14):9596–9601
Macaluso E, Doricchi F (2013) Attention and predictions: control of spatial attention beyond the endogenous-exogenous dichotomy. Front Hum Neurosci 7:685
Macaluso E, Ogawa A (2018) Visuo-spatial orienting during active exploratory behavior: processing of task-related and stimulus-related signals. Cortex 102:26–44
Mack SC, Eckstein MP (2011) Object co-occurrence serves as a contextual cue to guide and facilitate visual search in a natural viewing environment. J Vis 11(9):1–16
Maguire EA (2012) Studying the freely-behaving brain with fMRI. NeuroImage 62(2):1170–1176
Malcolm GL, Groen IIA, Baker CI (2016a) Making sense of real-world scenes. Trends Cogn Sci 20(11):843–856
Malcolm GL, Rattinger M, Shomstein S (2016b) Intrusive effects of semantic information on visual selective attention. Atten Percept Psychophys 78(7):2066–2078
Manelis A, Reder LM (2012) Procedural learning and associative memory mechanisms contribute to contextual cueing: evidence from fMRI and eye-tracking. Learn Mem 19(11):527–534
Martin CR, Khosla A, Pantazis D, Oliva A (2017) Dynamics of scene representations in the human brain revealed by magnetoencephalography and deep neural networks. NeuroImage 153:346–358

Nardo D, Santangelo V, Macaluso E (2011) Stimulus-driven orienting of visuo-spatial attention in complex dynamic environments. Neuron 69(5):1015–1028
Nardo D, Santangelo V, Macaluso E (2014) Spatial orienting in complex audiovisual environments. Hum Brain Mapp 35(4):1597–1614
Nardo D, Console P, Reverberi C, Macaluso E (2016) Competition between visual events modulates the influence of salience during free-viewing of naturalistic videos. Front Hum Neurosci 10:320
Nastase SA, Connolly AC, Oosterhof NN, Halchenko YO, Guntupalli JS, Visconti di Oleggio CM et al (2017) Attention selectively reshapes the geometry of distributed semantic representation. Cereb Cortex 27(8):4277–4291
Nobre AC, Coull JT, Frith CD, Mesulam MM (1999) Orbitofrontal cortex is activated during breaches of expectation in tasks of visual attention. Nat Neurosci 2(1):11–12
O'Craven KM, Downing PE, Kanwisher N (1999) fMRI evidence for objects as the units of attentional selection. Nature 401(6753):584–587
Ogawa A, Bordier C, Macaluso E (2013) Audio-visual perception of 3D cinematography: an fMRI study using condition-based and computation-based analyses. PLoS One 8(10):e76003
Peelen MV, Kastner S (2014) Attention in the real world: toward understanding its neural basis. Trends Cogn Sci 18(5):242–250
Posner MI (1980) Orienting of attention. Q J Exp Psychol 32(1):3–25
Poulsen AT, Kamronn S, Dmochowski J, Parra LC, Hansen LK (2017) EEG in the classroom: synchronised neural recordings during video presentation. Sci Rep 7:43916
Preston TJ, Guo F, Das K, Giesbrecht B, Eckstein MP (2013) Neural representations of contextual guidance in visual search of real-world scenes. J Neurosci 33(18):7846–7855
Ptak R (2012) The frontoparietal attention network of the human brain: action, saliency, and a priority map of the environment. Neuroscientist 18(5):502–515
Reeder RR, Peelen MV (2013) The contents of the search template for category-level search in natural scenes. J Vis 13(3):13
Rosen ML, Stern CE, Michalka SW, Devaney KJ, Somers DC (2016) Cognitive control network contributions to memory-guided visual attention. Cereb Cortex 26(5):2059–2073
Sabatinelli D, Frank DW, Wanger TJ, Dhamala M, Adhikari BM, Li X (2014) The timing and directional connectivity of human frontoparietal and ventral visual attention networks in emotional scene perception. Neuroscience 277:229–238
Santangelo V, Macaluso E (2013) Visual salience improves spatial working memory via enhanced parieto-temporal functional connectivity. J Neurosci 33(9):4110–4117
Santangelo V, Di Francesco SA, Mastroberardino S, Macaluso E (2015) Parietal cortex integrates contextual and saliency signals during the encoding of natural scenes in working memory. Hum Brain Mapp 36(12):5003–5017
Scalf PE, Ahn J, Beck DM, Lleras A (2014) Trial history effects in the ventral attentional network. J Cogn Neurosci 26(12):2789–2797
Seidl KN, Peelen MV, Kastner S (2012) Neural evidence for distracter suppression during visual search in real-world scenes. J Neurosci 32(34):11812–11819
Seidl-Rathkopf KN, Turk-Browne NB, Kastner S (2015) Automatic guidance of attention during real-world visual search. Atten Percept Psychophys 77(6):1881–1895
Sestieri C, Shulman GL, Corbetta M (2010) Attention to memory and the environment: functional specialization and dynamic competition in human posterior parietal cortex. J Neurosci 30(25):8445–8456
Shomstein S (2012) Cognitive functions of the posterior parietal cortex: top-down and bottom-up attentional control. Front Integr Neurosci 6:38
Sprague TC, Itthipuripat S, Vo VA, Serences JT (2018) Dissociable signatures of visual salience and behavioral relevance across attentional priority maps in human cortex. J Neurophysiol 119(6):2153–2165
Stirk JA, Underwood G (2007) Low-level visual saliency does not predict change detection in natural scenes. J Vis 7(10):3–10

Stoll J, Thrun M, Nuthmann A, Einhauser W (2015) Overt attention in natural scenes: objects dominate features. Vis Res 107:36–48

Summerfield JJ, Lepsien J, Gitelman DR, Mesulam MM, Nobre AC (2006) Orienting attention based on long-term memory experience. Neuron 49(6):905–916

Swisher JD, Halko MA, Merabet LB, McMains SA, Somers DC (2007) Visual topography of human intraparietal sulcus. J Neurosci 27(20):5326–5337

Thorpe S, Fize D, Marlot C (1996) Speed of processing in the human visual system. Nature 381(6582):520–522

Todd RM, Manaligod MGM (2018) Implicit guidance of attention: the priority state space framework. Cortex 102:121–138

Torralba A (2003) Modeling global scene factors in attention. J Opt Soc Am A Opt Image Sci Vis 20(7):1407–1418

van Kesteren MT, Ruiter DJ, Fernandez G, Henson RN (2012) How schema and novelty augment memory formation. Trends Neurosci 35(4):211–219

Veale R, Hafed ZM, Yoshida M (2017) How is visual salience computed in the brain? Insights from behaviour, neurobiology and modelling. Philos Trans R Soc Lond Ser B Biol Sci 372(1714):20160113

Vossel S, Mathys C, Stephan KE, Friston KJ (2015) Cortical coupling reflects Bayesian belief updating in the deployment of spatial attention. J Neurosci 35(33):11532–11542

Wagner AD, Shannon BJ, Kahn I, Buckner RL (2005) Parietal lobe contributions to episodic memory retrieval. Trends Cogn Sci 9(9):445–453

Wen H, Shi J, Zhang Y, Lu KH, Cao J, Liu Z (2017) Neural encoding and decoding with deep learning for dynamic natural vision. Cereb Cortex:1–25

Wolfe JM, Horowitz TS (2004) What attributes guide the deployment of visual attention and how do they do it? Nat Rev Neurosci 5(6):495–501

Wolfe JM, Vo ML, Evans KK, Greene MR (2011) Visual search in scenes involves selective and nonselective pathways. Trends Cogn Sci 15(2):77–84

Wu CC, Wang HC, Pomplun M (2014a) The roles of scene gist and spatial dependency among objects in the semantic guidance of attention in real-world scenes. Vis Res 105:10–20

Wu CC, Wick FA, Pomplun M (2014b) Guidance of visual attention by semantic information in real-world scenes. Front Psychol 5:54

Zacks JM, Speer NK, Swallow KM, Maley CJ (2010) The Brain's cutting-room floor: segmentation of narrative cinema. Front Hum Neurosci 4:168

Part III
Neuropsychology and Neuropsychiatry

Visual Search in Progressive Supranuclear Palsy

Daniel T. Smith and Neil Archibald

Contents

Abstract Progressive supranuclear palsy is often considered a disease of the motor system and is characterised by a profound oculomotor impairment. The oculomotor system is also known to be fundamentally important in cognitive processes such as attention and working memory, but the way in which these functions are affected by PSP is not well understood. In this chapter we outline the pathology and typical presentation of PSP, with a focus on the oculomotor impairment, briefly outline the role of the oculomotor system in spatial cognition and discuss some key studies examining spatial attention and memory in PSP. We then present new data from a study that specifically examined the effect of PSP on visual search. Our results demonstrated a profound impairment of visual search which is most severe for feature search along the vertical axis. These findings are interpreted with respect to

D. T. Smith (✉)
Department of Psychology, Durham University, Durham, UK
e-mail: daniel.smith2@durham.ac.uk

N. Archibald
South Tees Hospitals NHS Foundation Trust, The James Cook University Hospital, Middlesbrough, UK

Curr Topics Behav Neurosci (2019) 41: 305–324
DOI 10.1007/7854_2018_72
Published Online: 21 December 2018

the biased-competition theory of attention, and we discuss possible clinical applications of our results.

Keywords Attention · Oculomotor · Parkinson's-basal ganglia · PSP · Vision

1 Progressive Supranuclear Palsy

Progressive supranuclear palsy (PSP) or Steele-Richardson-Olszewski syndrome (Steele et al. 1964) is a progressive neurodegenerative disorder with a population prevalence of 5–6 cases per 100,000 and an incidence of 1–2 per 100,000 per year (Bower et al. 1997; Nath et al. 2001; Schrag et al. 1999). PSP is characterised by progressive gait disturbance, including frequent backward falls and a vertical gaze palsy. Other clinical features include bradykinesia, axial and limb rigidity, speech and swallowing problems and both cognitive and behavioural changes (Golbe 2014; Höglinger et al. 2017). There is considerable clinical heterogeneity, however, making diagnosis challenging (Boxer et al. 2017). Postmortem studies suggest misdiagnosis rates are high, with many patients either incorrectly diagnosed in life or dying without an explanation for their symptoms (Williams et al. 2005; Yoshida et al. 2017). Patients typically present in their fifth to seventh decade, and mean survival is 6–9 years (Glasmacher et al. 2017).

Although PSP is often described as an 'Atypical Parkinsonian' disorder, and may be misdiagnosed as Parkinson's disease (PD), it is a neuropathologically distinct entity. PSP is associated with the deposition of hyperphosphorylated tau as neurofibrillary tangles, neuropil threads and fibrillary gliosis in the cortex, basal ganglia and brainstem (Dickson et al. 2010). Tau proteins are involved in the development of the microtubules that maintain the structure of cells in the brain. Hyperphosphorylation of tau occurs when the phosphorylation sites become saturated, causing the protein to dissociate from the microtubules. This dissociation results in the destabilisation of microtubule, leading to the disruption of signalling between neurons. This disruption is likely to account for the neurological problems experienced by people with PSP. The dissociated tau protein also accumulates into neurofibrillary tangles that disrupt intracellular functions and may contribute to the pathology of PSP. Different patterns of tau distribution may contribute to the clinical variability in presentation and progression. For example, prominent cortical pathology may explain the broad range of cognitive impairments seen in PSP, including problems with executive functions (Gerstenecker et al. 2013; Robbins et al. 1994), apathy (Brown et al. 2010), impulsivity (Zhang et al. 2016) and social cognition (Ghosh et al. 2012; see Burrell et al. 2014 for a review). In contrast, the oculomotor problems observed in PSP are more likely to be explained by degeneration of premotor centres in the midbrain and brainstem responsible for the production of eye movements, rather than the cortical oculomotor network. PSP is characterised by the vertical paralysis of gaze, but as the disease progresses, horizontal saccades are also lost, along with other oculomotor behaviours such as vergence (Chen et al. 2010). The vertical ophthalmoplegia is likely caused by degeneration of two key midbrain structures: the medial longitudinal fasciculus (riMLF), which contains the premotor neurons that drive vertical eye movements, and the interstitial nucleus of Cajal (INC), which controls the maintenance of stable fixation. It seems likely that vertical saccades are lost before horizontal saccades because the riMLF is more rostral than the parapontine reticular formation (which controls horizontal saccades) and succumbs earlier in disease progression (Chen et al. 2010; Steele et al. 1964).

2 Visuospatial Attention

Visuospatial attention allows us to select information from task-relevant locations at the expense of signals from irrelevant locations. The focus of attention is often compared to a spotlight on a theatre stage (Posner 1980), such that the region of space within the spotlight 'illuminated' by attention and processed preferentially to region outside the spotlight (although in fact attending to a location appears to increase the perceived contrast and improve spatial resolution, rather than enhancing perceived luminance per se (Carrasco et al. 2004; Yeshurun and Carrasco 1998)). Like a spotlight on a stage, the focus of attention can be moved around the environment, and there are two different mechanisms that control this orienting of attention. Attention can be guided endogenously, in response to the observers' goals and desires, or exogenously, in response to salient events in the environment. These two modes of orienting appear somewhat independent of one another. For example, orienting of exogenous attention is rapid but transient, such that attentional facilitation is maximal ~150 ms after the onset of a cue but then rapidly decays. In contrast endogenous orienting is relatively slow, reaching its peak ~300 ms after cue onset, but sustained (Muller and Rabbitt 1989). The exogenous mode of orienting is also involuntary, to the extent that certain types of stimuli, such as sudden onsets, summon attention even when the participant is aware that the target is unlikely to appear at the cued location (Posner 1980). Endogenous orienting is largely under conscious control. Consistent with the behavioural evidence that these two modes of orienting are dissociable, exogenous and endogenous orienting appear to activate different networks of brain areas (Corbetta and Shulman 2002).

Both modes of attention are typically associated with an actual movement of the eyes, which brings the location of interest into the fovea. This mode of orienting is often referred to as an overt shift of attention. However, attentional orienting can occur covertly, such that the spotlight of attention is moved around while the eyes remain fixated (Posner 1978). These exogenous and endogenous modes of attentional facilitation are complemented by an inhibitory mechanism (Inhibition of Return) that biases the visual system away from previously explored parts of a display (Posner et al. 1985). This inhibitory effect has a perceptual component, which suppresses visual input from salient parts of the scene and a motor component, which biases the eye movement system against programming movements towards parts of the scene that have already been gazed at (Hilchey et al. 2014; Klein 2000).

At a theoretical level, the cognitive mechanisms mediating visuospatial attention are generally agreed to be tightly coupled to the systems that control eye movements (Awh et al. 2006), see also "The Relationship Between Spatial Attention and Eye Movements" by A. R. Hunt et al., this volume. For example, the brain areas involved in saccade control are spatially overlapping with those implicated in spatial attention (Corbetta et al. 1998; de Haan et al. 2008), and temporary neurodisruption of these brain areas with transcranial magnetic stimulation leads to impairments of both spatial attention and saccade control (Zangemeister et al. 1995), attentional orienting (Chambers et al. 2004; Fierro et al. 2000; Grosbras and Paus 2002; Smith et al. 2005, 2009b) and visual search (Lane et al. 2012a, b; Muggleton et al. 2003). Furthermore, there is a mandatory shift of attention to the goal of a saccadic eye movement (Shepherd et al. 1986), even when the observer knows a target is more likely to appear at a different location (Deubel and Schneider 1996), and there are several examples of neuropsychological associations between impaired oculomotor function and disrupted covert orienting of attention (Craighero et al. 2001; Gabay et al. 2010; Smith et al. 2004, 2009a; Smith and Schenk 2012). The tight coupling between oculomotor control and spatial attention can also be observed in healthy participants whose eye movements have been experimentally constrained. These participants show a deficit of covert attention in regions of space that cannot be reached with a saccadic eye movement (Craighero et al. 2004; Morgan et al. 2014; Smith et al. 2010, 2012, 2014), which is particularly severe for exogenous orienting. Given that PSP is associated with a profound oculomotor impairment, and oculomotor impairments are associated with deficits of covert attention (Smith and Schenk 2012), it seems reasonable to predict that people with PSP will have a deficit of covert attention, and this deficit will be more severe for tasks that engage exogenous attention.

Consistent with this idea, Rafal and colleagues conducted a series of experiments using Posner cueing task (Posner 1980) to examine the extent to which the vertical paralysis of gaze affected orienting of attention in PSP (Posner et al. 1982, 1985; Rafal et al. 1988). Exogenous orienting was assessed by presenting a cue in the periphery which, after a variable delay (10,150,350,550 ms), was followed by a target. This target appeared at the cued location on 50% of trials, thus removing the incentive for participants to voluntarily attend to the cued location. Critically, on half the trials stimuli appeared along the horizontal axis, and on the other half they were aligned vertically. In healthy participants, one typically observes faster reaction times at the cued location but only when the delay between cue and target onset is short (<150 ms). In contrast, when the delay is longer than ~300 ms, reaction times to targets at the cued location are slowed by IOR (Posner and Cohen 1984). This is exactly the result observed by Rafal et al. (1988) in a group of patients with Parkinson's disease (PD), and the pattern was the same for horizontally and vertically presented arrays. In contrast, patients with PSP showed a very marked horizontal/vertical asymmetry. Specifically, the facilitatory effect of peripheral cues appearing along the vertical axis was significantly weaker than those along the horizontal axis and was delayed until 350 ms. Furthermore, patients with PSP did not present the normal IOR effect at longer cue-target delays when stimuli were presented along the vertical axis, whereas IOR was normal along the horizontal axis

(Posner et al. 1985). Endogenous orienting was assessed using an arrow-cueing task, in which a centrally presented arrow correctly indicated the location of the upcoming target on 80% of trials. In this task, the cues elicited endogenous shifts of attention that had a similar time course irrespective of the axis along which attention was oriented, although the benefit was slightly smaller along the vertical axis. Rafal et al. conclude that PSP is associated with deficits of visuospatial attention and link these deficits to the ophthalmoplegia.

These data seem to suggest that people with PSP have a deficit of exogenous, covert attention that is specific to the vertical axis and is not present in PD. It is plausible to interpret this deficit of exogenous orienting as a potentially unique, cognitive marker that can differentiate PSP from PD. Given the issues surrounding the problems of differentiating between PSP and PD, particularly early in disease progression, a specific cognitive marker for PSP might have important practical applications for the early and accurate diagnosis of PSP. However, we believe this interpretation may be somewhat premature, as there are several reasons to be cautious when interpreting the data from these studies. Firstly the delay between cue and target onset was short, the cue and target overlapped, such that the target was superimposed on the cue, whereas at longer cue-target onset asynchronies (CTOAs), the target appeared on the background. Thus, there were differences in the visibility of the target at cued and uncued locations at short CTOAs which might account for some of the effects. Secondly, PSP is associated with akinesia which makes the interpretation of raw RT, which depends on the initiation of a button press, rather problematic. Unfortunately the study by Rafal makes no mention of the variance of the RT data in their sample, but the very high RT cutoff (3,500 ms) suggests a considerable variation in response times. Finally, using an arrow-cueing task is not optimal for studying endogenous orienting because arrow cues are known to engage both endogenous and exogenous attentional processes, (e.g. Ristic et al. 2002; Tipples 2002), so cannot be considered a 'pure' measure of endogenous orienting.

An alternative approach to examining visuospatial attention in PSP is to use a visual search task rather than the Posner cueing task. Visual search requires participants to detect the presence or absence of a target item among distracters. Varying parameters such as the salience of the target and the number of distracters allows the measurement of search efficiency, which refers to the amount of change in reaction time caused by adding additional distracter items (Wolfe 2003). When the target of the search is defined by a single, salient feature such as orientation or colour, search can be highly efficient, such that the target seems to 'pop-out' and additional distracters have a negligible effect on search time (<10 ms per additional item). This type of search engages more reflexive exogenous attentional processes. In contrast, when the target is non-salient or characterised by a conjunction of features, participants must engage more endogenous, effortful attentional processes to search through each item in a serial fashion until they detect the target (Wolfe 1998). This type of search is inefficient, such that additional distracter items have a substantial effect on search time (>30 ms/item).

There are several other advantages to using search tasks over cueing tasks to measure attention in PSP. Firstly, it is easier for patients to understand the task as there is a single visual event for them to detect (the target), and they do not have to worry about accidentally responding to cues. Secondly, because there is no salient pre-cue, the effects of covert attention are unlikely to be confounded with the effects of IOR. Thirdly, by examining both the overall RTs and search efficiency, it may be possible to partly control for the high levels of variance in RT. Fourthly, visual search tasks are known to be sensitive to disruptions to the oculomotor system in healthy participants, such that feature search is impaired when targets appear at locations that cannot become the goal of a saccadic eye movement, whereas conjunction search is unaffected (Smith et al. 2010, 2014). In contrast, the effect of disrupting the oculomotor system on cueing tasks is less consistent, with some authors reporting disruption to exogenous orienting but preserved endogenous orienting, social attention and perceptual IOR (Morgan et al. 2014; Smith et al. 2012), others arguing for disrupted endogenous orienting (Craighero et al. 2004) and a recent report of disrupted perceptual IOR (Michalczyk et al. 2018). Finally, visual search tasks are closer to the 'real world' than cueing tasks so offer a somewhat more ecologically valid measure of visuospatial attention than cueing tasks.

Several studies have examined visual search in PSP, typically as part of a larger battery of neuropsychological tests. These studies suggest that visual search is indeed impaired in PSP compared to PD. For example, Kimura et al. (1981) presented 20 line drawings of familiar objects on a vertical surface and examined how long it took for patients to locate a target picture. Patients with PSP were significantly slower than patients with PD, patients with frontal lesions and patients with occipital lesions. This last result is notable, given that patients with occipital lesions are likely to suffer some form of cortical blindness so might be expected to struggle with visual search tasks (e.g. Lane et al. 2010). Two other studies also report that people with PSP perform worse than PD patients on the Visual Search Test (Monza et al. 1998; Soliveri et al. 2000). However, these latter two studies are hard to interpret in terms of visuospatial attention, as the Visual Search Test used by Monza et al. actually measures the ability to name pictures of objects, rather than the ability to locate a target stimulus among distracters. Furthermore, the description of the impairment is rather incomplete. Specifically, the tasks all require patients to identify complex drawings of objects, which are likely to engage endogenous attentional processes, but none of the studies explicitly examined feature search in PSP. The extent to which feature search is also disrupted in PSP therefore remains unknown. Neither do the studies report performance on target-absent trials, which is an important indicator of participants' ability to correctly terminate a search in the absence of a target. Furthermore, none of the studies varied set size, making it impossible to know whether PSP impairs only the speed of search or whether it also affects the efficiency of the search. With respect to the impact of ophthalmoplegia on visuospatial attention, unlike the cueing task of Rafal et al., these studies did not directly compare search times for targets on the horizontal and vertical axis. So, although previous studies offer some evidence that visual search is relatively impaired in PSP, they do not offer a very complete characterisation of the visual search impairment, nor do they speak to the theoretical claim that the attentional impairment in PSP is linked to the oculomotor impairment. The work described in the next section attempts to address some of the issues by examining visual search in groups of patients with PSP.

3 Methods

3.1 *Participants*

Seventeen participants with progressive supranuclear palsy (PSP) were approached via the Movement Disorder Service at the James Cook University Hospital, Middlesbrough. Twelve agreed to participate (six male aged 63–80, mean age 71) and seven with PD (age 64–69, mean 66). All participants with PSP met the National Institute of Neurological Disorders and Stroke and Society for PSP, Inc. (NINDS-SPSP) (Litvan et al. 2003) criteria for clinically probable or definite PSP. The study was approved by the North East – Newcastle and North Tyneside 1 Research Ethics Committee (15/NE/0254) and Durham University Department of Psychology Research Ethics Committee.

3.2 *Stimuli and Apparatus*

The experimental stimuli were generated using a Cambridge Research Systems ViSaGe graphics card and displayed on a 17-in. Sony Trinitron CRT monitor with a refresh rate of 100 Hz. Responses were collected using a two-button button box. Saccadometry was performed using a BioPac 150 recording EOG at 500 Hz. The visual search target was a blue 'c' shape oriented at 45°. In the feature search task, the all distracter items were also blue 'c's, oriented at 215°. In the conjunction search task, distracters could also be either blue 'c's, oriented at 215°, or yellow 'c's, oriented at 45° (see Fig. 1). Array items were presented at 10° from the centre of the screen on a black background. In 4-item arrays, the stimuli appeared on the cardinal compass directions (N, E, S, W). In 8-item arrays, stimuli appeared at cardinal directions and intermediate points (N, NE, E, SE, S, SW, W, NW). Participants sat 50 cm from the display with the head supported by a chinrest.

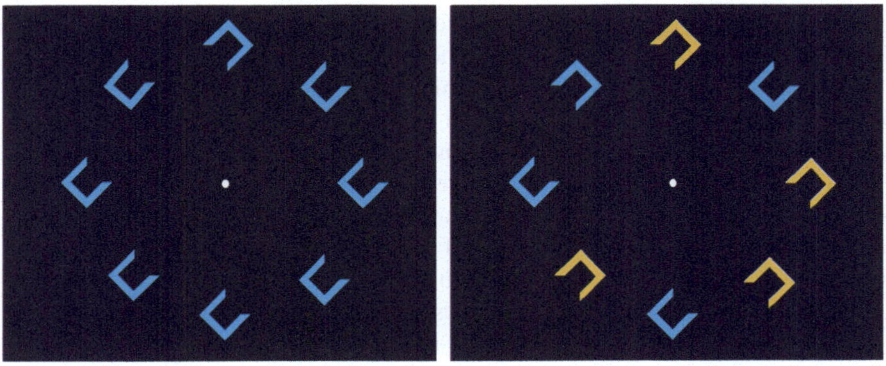

Fig. 1 Examples of 8-item visual search arrays. The left panel shows the feature search and the right the conjunction search

3.3 Procedure

Eight participants with PSP completed a saccadometric test to establish the extent of their ocular motility. In this test participants were presented with a black spot at fixation. After 2,000 ms, the spot jumped into the periphery. Participants were instructed to follow the spot with their eyes and press a button when they were fixating it. Following the button press, the spot returned to the centre, and the next trial began. Each run consisted of ten jumps that increased in magnitude in 1° steps, starting with a 1.5° jump. Participants completed four runs (Centre-Up, Centre-Left, Centre-Down, Centre-Right).

Following the saccadometry, patients completed the visual search tasks. All AMC and seven patients with PSP completed both tasks. Two patients with PSP completed the feature search task but not the conjunction search task, and three patients with PSP completed the conjunction but not the feature search tasks. Thus eight patients completed the feature search task, and nine patients completed the conjunction search task. Both tasks began with the appearance of a fixation point for 1,000 ms, followed by the appearance of a search array comprising 4 or 8 items. This array remained presented until a response was made. Participants were instructed to press one button when a target was present and the other the target was absent. They were also instructed fixate the centre of the array and try not to make eye movements. There was a 2:1 ratio of 8-item arrays to 4-item arrays and a 2:1 ratio of target-present to target-absent trials. On target-present trials, the target appeared at each location in the array with equal probability. All participants completed one block of 36 practice trials and two blocks of 108 experimental trials.

4 Results

4.1 Saccadometry

All PSP patients presented with supranuclear ophthalmoplegia that was more severe for vertical than horizontal saccadic eye movements (this is a key diagnostic criteria for PSP and was established during clinical examination by a neurologist). The extent of the ophthalmoplegia was more formally assessed with saccadometry in 8/12 PSP patients. Vertical saccadic eye movements were absent in all participants with PSP. Horizontal eye movements were present in all patients. Table 1 shows the oculomotor ranges of the eight participants for whom saccadometric data were available. The maximum amplitude of leftward saccades was significantly reduced compared to that of rightward saccades than (8.2° vs 9.6°; $t_{(12)} = 3.42$, $p < 0.01$). The main sequence for horizontal saccades was also somewhat impaired in the PSP group but was similar for left- and rightward saccades (Chen et al. 2010).

Table 1 Saccadometry results

		Max amplitude				Main seq		
ID	Group	Left	Right	Up	Down	Left	Right	Meds
1	PSP	11.5	11.5	0	0	0.16	0.1	–
2	PSP	6.8	7.8	0	0	0.73	0.5	L
3	PSP	7	8	0	0	0.70	0.89	–
4	PSP	8.8	9.9	3	0	0.68	0.34	–
5	PSP	6.3	8.9	0	0	0.38	0.64	A
6	PSP	7.5	11	0	0	0.8	0.84	L
7	PSP	10.1	10.5	0	0	0.3	0.6	D, L
8	PSP	7.5	9	0	0	0.5	0.85	–
Mean		8.2	9.6			0.53	0.59	

The oculomotor range was defined as the point at which participants ceased to scale their eye movements with increasing target eccentricity. The main sequence describes the correlation coefficient (Pearson's R) between saccade amplitude and peak velocity
Meds: *A* amantadine, *D* donepezil, *L* levodopa

4.2 Feature Search

One patient in the PSP group had a median reaction time of >10 s, so was excluded from the analysis. The remaining data were filtered to remove anticipations (<1%), incorrect responses (6% PSP), trials where the target appeared on at one of the intercardinal compass points (i.e. NE, SE, SW, NW) and target-absent trials. At each set size, we collapsed trials where the probe appeared the N and S positions into a 'vertical' condition and trials where the probe appeared the E and W positions into a 'horizontal' condition.

The median reaction times for target-present trials were subjected to a 2 × 2 × 3 factor ANOVA with within-subjects factors of set size (4, 8) and TargetAxis (horizontal, vertical) and a between-subjects factor of group (PSP, AMC). We observed a significant two-way interaction between group and TargetAxis ($F_{(1,14)} = 5.28$, $P < 0.05$, $\eta 2 = 0.27$). Inspection of Fig. 2 (top row) suggests that this effect was driven by an increase in reaction time for targets appearing on the vertical axis that was specific to the PSP group. This impression was confirmed with a t-test ($t_{(7)} = 2.56$, $p = 0.038$). Overall the PSP group was significantly slower than the AMC group (2,951 ms vs 936 ms; $F_{(1,14)} = 10.75$, $P < 0.01$, $\eta 2 = 0.26$).

We also examined the response accuracy using ANOVA. The two groups had similar overall accuracy levels (94% for both groups), and there were no significant main effects of set size or group and no interactions.

Figure 3 shows feature search times when no target was present. There was a main effect of set size, such that both groups were slower on the 8-item than the 4-item arrays ($F_{(1,14)} = 4.91$, $P = 0.044$, $\eta 2 = 0.26$). However, there was no interaction between set size and group ($F_{(1,14)} = 1.99$, $P = 0.18$, $\eta 2 = 0.12$). As with the target-present trials, there was a main effect of group, such that PSP group were slower to respond than AMC group (3,225 ms vs 1,036 ms; $F_{(1,14)} = 6.81$, $P = 0.021$, $\eta 2 = 0.33$). Accuracy on target-absent trials was also examined with a

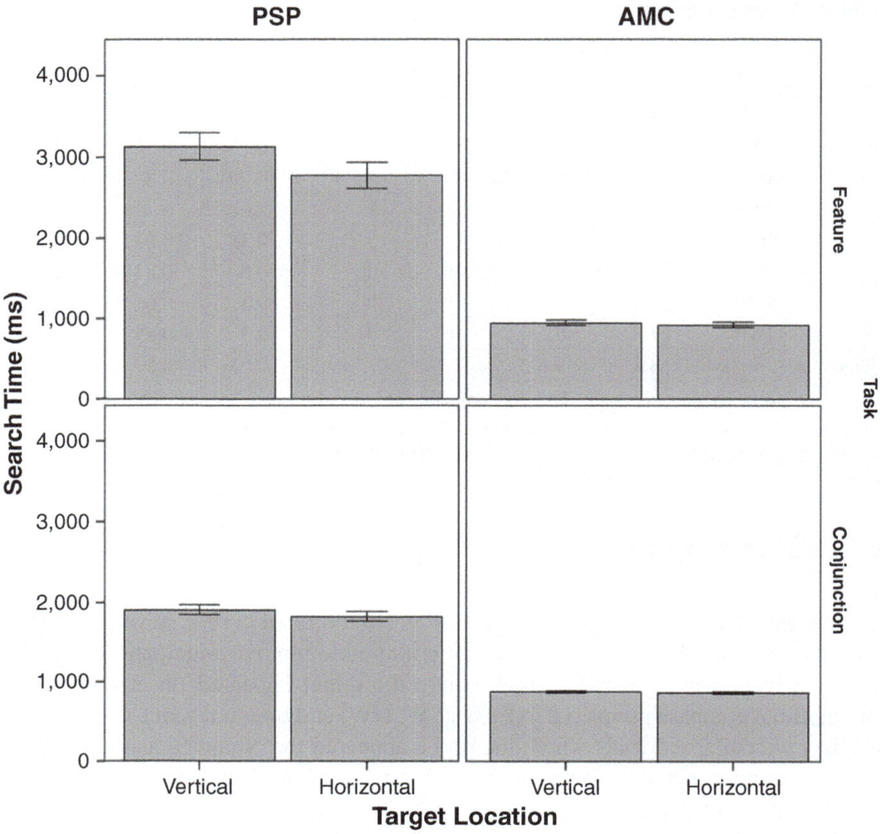

Fig. 2 Feature (upper row) and conjunction (lower row) search times on target-present trials in the PSP and age-matched control groups. Error bars show 95% within subject confidence intervals (Cousineau 2005)

2 (set size) × 2 (group) ANOVA. The analysis revealed a significant interaction, such that the PSP group tended to produce more false positives as the array size increased from 4 to 8, whereas the AMC group produced fewer false positives as the array size increased (Fig. 3).

4.3 Conjunction Search

The pattern of data in conjunction search was rather different to that in the feature search task (see Fig. 2 left column). Firstly, there was a significant main effect of set size, such that search was slower when there were 8 rather than 4 items in the search array ($F_{(1,15)} = 18.01$, $P < 0.01$, $\acute{\eta}2 = 0.55$). Secondly, there was no interaction between TargetAxis and group ($F_{(1,15)} = 1.39$, $P = 0.26$, $\acute{\eta}2 = 0.09$). However, there was a main effect of group ($F_{(1,15)} = 13.37$, $P < 0.1$, $\acute{\eta}2 = 0.48$), such that the PSP

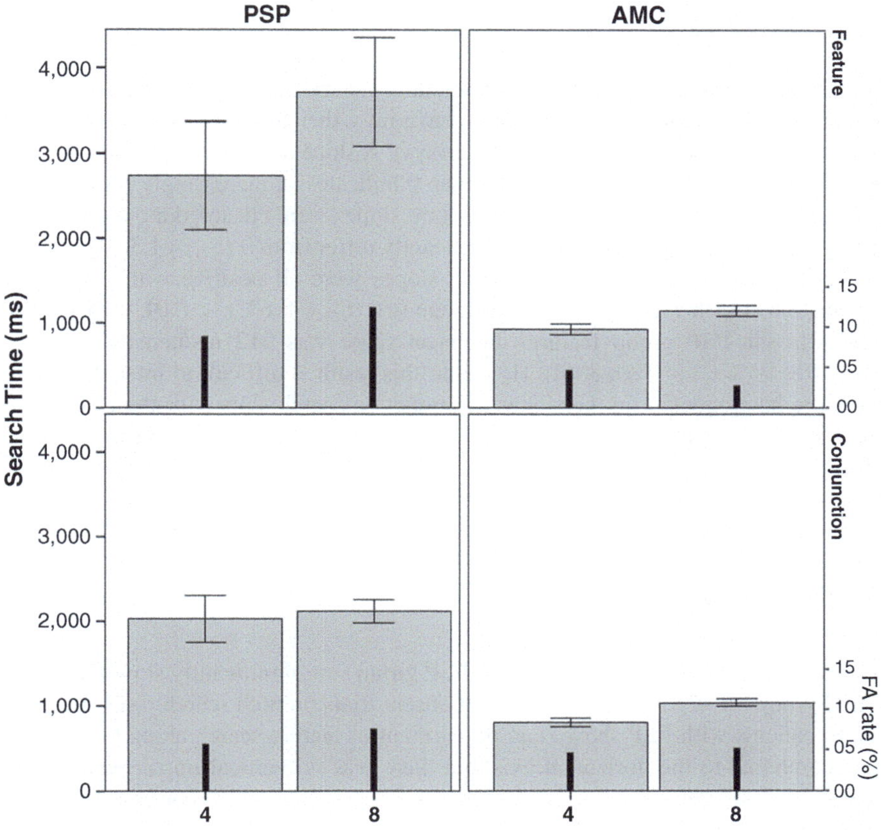

Fig. 3 Mean RTs on target-absent trials at each set size. Solid black bars show false alarm rates (%). Error Bars show 95% within-subject confidence intervals

group were slower than the AMC group (1,870 ms vs 873 ms). The response accuracy was examined using ANOVA. As with the feature search, accuracy was similar in both groups (PSP 96% correct, AMC 97% correct), and there were no significant main effects or interactions.

Figure 3 (bottom row) shows conjunction search performance on target-absent trials. There was no main effect of set size ($F_{(1,15)} = 3.84$, $P = 0.07$, $\acute{\eta}2 = 0.20$) and no interaction between set size and group ($F_{(1,12)} = 1.16$, $P = 0.229$, $\acute{\eta}2 = 0.07$), but there was a main effect of group ($F_{(1,12)} = 5.55$, $P = 0.033$, $\acute{\eta}2 = 0.27$), such that the PSP group were slower than the AMC group (2,092 ms vs 916 ms). The false alarm rate was also analysed, but there were no main effect or interactions.

4.4 Search Efficiency

Search slopes were calculated by subtracting the median search time for 4-item arrays from the median search time in 8-item arrays, then dividing the result by 4 (the difference in the number of items in the arrays). A slope of 0 ms/item is indicative of entirely parallel search; a slope greater than 0 indicates an increasingly serial search (Wolfe 2003). In the AMC group, the feature slopes were clustered around 0 with a mean of 30 ms/item, which did not significantly differ from 0 ($t_{(7)} = 1.33, p = 0.22$, 95% CI [−23, 85]). Conjunction search slopes were all positive with a mean of 91 ms/item, which was significantly different to 0 ($t_{(6)} = 8.63, p < 0.01$, 95% CI [66, 116]). In the PSP group feature, the mean slope was 312 ms/item ($t_{(7)} = 1.5$, $p = 0.18$, 95% CI [−180, 806]). However, this result is difficult to interpret due to the large heterogeneity in the slopes. Conjunction search slopes for the PSP group were all above zero, with a mean of 203 ms/item ($t_{(8)} = 3.2$, $p = 0.014$, 95% CI [54, 351]).

5 Discussion

To briefly summarise these results, the PSP group was significantly slower than the AMC group on target-present and target-absent trials on both search tasks. Furthermore, patients with PSP showed an impairment of feature search along the vertical axis compared to the horizontal axis, but there was no vertical impairment during conjunction search. In contrast, search times of age-matched controls were not affected by the location of the target. Accuracy of target detection was similar in the PSP and AMC groups. However, increasing set size during feature search led to more false alarms in the PSP group but fewer false alarms in the AMC group.

The finding that patients with PSP are significantly slower than age-matched control on both feature and conjunction search tasks indicates the presence of a problem with and is consistent with previous reports that PSP impairs visual search in complex scenes (Kimura et al. 1981). However, we have extended the previous findings by demonstrating that very simple feature search tasks are also slowed in PSP. This is an important result because feature search tasks engage relatively rapid, automatic attentional processes that do not require a serial search through the scene. The fact that PSP disrupts feature search indicates that PSP interferes with the earliest stages of visual selection and that the visual search impairment observed in PSP is not simply due to an inability to effectively explore the scene with eye movements. Furthermore, it was also found that patients with PSP were slower than age-matched controls to correctly reject no-target trials during the feature search task and more likely to report false positives as set size increased. In contrast, the AMC group became slightly slower but more accurate. The two groups therefore appear to have adopted different speed-accuracy trade-offs in response to increasing task difficulty, with the PSP group becoming less efficient as task difficulty increased. It is not entirely clear what causes this slowing, but one possibility is that the impaired search processes on target-present trials means that patients become very conservative on target-absent trials, perhaps even engaging in a very slow and effortful serial search through all the items to be sure that no target is present (Chun and Wolfe 1996), or that they engaged in more guessing than the AMC group.

Overall our findings are in agreement with previous observations by Rafal et al. (1988) and Posner et al. (1982) of impaired exogenous attentional orienting to peripheral cues presented along the vertical axis. These data appear to offer confirmation that a fully functioning oculomotor system is required for normal exogenous orienting (Jackson et al. 2005; Smith et al. 2004). Furthermore, because PSP affects eye movements very close to the output end of processing, it is tempting to conclude that the ability to make overt eye movements is *necessary* to observe exogenous orienting. Such a conclusion would certainly be consistent with neuropsychological evidence that defective saccade execution caused by problems with the extraocular muscles is associated with impaired exogenous attention (Gabay et al. 2010; Smith et al. 2004).

On first inspection, the observation of impaired attention along the vertical axis appears broadly in line with the Premotor Theory of Attention (Rizzolatti et al. 1994), which argues that covert attention depends upon the intention to make a saccadic eye movement. In this view, both endogenous and exogenous attention depend on the selection of a saccade goal by the cortical and subcortical centres responsible for eye movements, principally the frontal eye field (FEF), lateral intraparietal area (LIP) and superior colliculus (SC) (Fecteau and Munoz 2006; Schall and Cohen 2011). Indeed, there is a great weight of neurophysiological evidence that these areas are active during both saccade programming and covert orienting (Corbetta et al. 1998; de Haan et al. 2008; Nobre et al. 2000), and damage to these areas causes problems with eye movements, covert orienting and visual search (Lane et al. 2012b; Muggleton et al. 2003; Smith and Schenk 2012). However, on closer examination the interpretation is not quite so straightforward. The first potential problem with interpreting the results in terms of Premotor Theory is that the core oculomotor deficit in PSP is in the riMLF, a premotor structure in the brainstem that drives vertical eye movements (Chen et al. 2010; Steele et al. 1964). Given that riMLF does not have any feedback connections to central oculomotor structures (Sparks and Mays 1990) and is unlikely to be involved in the planning of a saccade, it isn't entirely clear by what mechanism damage to riMLF could disrupt saccade *programming*. Secondly, Premotor Theory predicts that an oculomotor impairment should be associated with deficits in both exogenous and endogenous orienting, but the PSP group were not slower to detect targets on the vertical compared to horizontal axis on the conjunction search task. This latter finding is hard to reconcile with the notion of a strict coupling between endogenous orienting and the oculomotor system posited by Premotor Theory.

The first problem may be resolved if one considers that the direct, feedforward connections between SC and riMLF leave SC vulnerable to retrograde transneuronal degeneration (Dinkin 2017; Pinching and Powell 1971), which describes a loss of presynaptic neurons as a consequence of loss of trophic support from the postsynaptic cells. This degeneration has been observed in other parts of the visual system, notably in the degeneration of the optic tract and LGN following lesions to striate cortex (Cowey et al. 2011; Kisvarday et al. 1991). If the loss of cells in riMLF leads to transneuronal degeneration in the SC, this damage would be greater for the parts of the SC that code eye movements with a more vertical component. This damage would have a profound impact on the ability of the patient to programme vertical eye movements and thus potentially to orient attention in vertical space. In this way, transneuronal degeneration could explain the problems patients with PSP experience with feature search. However, resolution of the first problem still leaves the question of why there is no difference between horizontal and vertical target detection during conjunction in PSP, an issue which may only be resolved by appealing to an alternative theoretical framework, that of biased competition.

The core idea of biased competition is that signals relating to stimulus salience (e.g. their brightness, size, contrast, orientation) compete with each other in a topographic map of space, called a salience map, in a winner-takes-all competition (Desimone 1998). The signal that wins the competition can be read out by the visual system to guide a shift of attention or by the oculomotor system to guide a saccade eye movement. The oculomotor signals are self-reinforcing, such that activation in the oculomotor system is fed back into the salience map, thus further biasing activity in favour of the activated location (Bisley et al. 2011). Competition in the salience map is also influenced by the current goals of the observer, such that the locations of stimuli that possess a feature known to be relevant to the current task are also prioritised. In this way the observer can bias the competition towards the stimuli that are most likely to be behaviourally relevant (Wolfe 1994). At a conceptual level, Fecteau and others have argued that because this biased competition relies on integrating low-level stimulus salience with top-down modulations, the map that represents the competition should be referred to as a priority map, rather than a salience map (Fecteau and Munoz 2006). Bisley and colleagues have convincingly argued that the priority map is instantiated in LIP (Bisley and Goldberg 2010; Goldberg et al. 2002), as neural activation in this area correlates with both the onset of saccadic eye movements and the covert selection of targets during visual search (Thomas and Pare 2007). This region has reciprocal connections with early visual areas, which allows for the modulation of visual signals that characterises visuospatial attention (e.g. Hillyard et al. 1998; Luck et al. 1997), and with key oculomotor centres such as the frontal eye field, which is critical for the generation of endogenous orienting signals, and the Superior Colliculus, which integrates visual and motor signals from cortical and subcortical pathways to specify a saccade goal (Munoz and Everling 2004; White and Munoz 2011). LIP therefore seems ideally placed to integrate signals from across a network of brain areas into a single priority map.

For simple tasks where one item is highly salient (e.g. in a feature search task), the competitive interactions in the priority map converge to select a single location, which is then powerfully reinforced by the re-entrant activation from the oculomotor system (Barash et al. 1991). This feedback loop allows the visual system to very rapidly select the location of the unique feature, irrespective of the number of distracters, which is the signature of efficient visual search (Wolfe 1998). This selection is driven by bottom-up factors, is largely automatic and corresponds to Posner's 'exogenous' mode of attention. A lesion to the oculomotor system that prevented it from reinforcing activation in the priority map would have the effect of slowing down the selection of the salient location, but not abolishing it completely. This would manifest as impaired feature search when targets appeared at locations that could not be represented in the oculomotor system, exactly as we observed in the PSP group (see also Smith et al. 2014). For search tasks where the target has relatively low salience, as was the case in our conjunction search, there may be multiple peaks of activation in the priority map, and the competition between the signals takes time to resolve. This in turn reduces the capacity for the oculomotor system to influence the competition, as the reinforcing signal from the oculomotor system is distributed across multiple locations. Instead, the competition is resolved by top-down processes that reflect to the observers' strategic and conscious decisions about which potential target locations should be selected. These processes correspond to Posner's 'endogenous' mode of orienting. Thus, a lesion that disrupts oculomotor selection but spares top-down processes, such as the lesion that affects people with PSP, should not affect search that relies primarily on endogenous attentional processes. Consistent with this prediction, we have previously shown that disrupting the oculomotor system does not impair conjunction search (Smith et al. 2014) or covert orienting to symbolic cues (Morgan et al. 2014; Smith et al. 2004, 2012). This interpretation also complements other evidence that disruption of the oculomotor system is associated with impaired spatial working memory in PSP (Smith and Archibald 2018) and healthy participants (Ball et al. 2013; Pearson et al. 2014), as spatial working memory and spatial attention are hypothesised to rely on the same priority map (Ikkai and Curtis 2011). Overall the data from our PSP patients fits with a model of an attention system in which orienting of attention is realised via a process of biased competition in a priority map. Activation in the oculomotor system provides an input that biases this competition towards the saccade goal, but this bias may not necessarily be sufficient to orient attention to the saccade goal. This conclusion is contrary to the central predictions of Premotor Theory and adds to the growing body of evidence that Premotor Theory does not offer an adequate account of the relationship between attention and oculomotor control.

The observation that the deficit in exogenous spatial attention in PSP observed by Rafal et al. (1988) generalises to a feature visual search task may also have clinical value. An important issue in clinical practice is the diagnosis of PSP, which is often misdiagnosed as Parkinson's disease. This is problematic because, although superficially PSP and PD often appear to be similar, the underlying pathology is very different. Misdiagnosis is upsetting for the patient and, because many patients with PSP have a poor response to standard treatments for Parkinson's disease, may lead to the use of inappropriate or ineffective treatments. If PSP can be differentiated from PSP using simple cognitive tasks such as visual search, there seems to be the possibility of developing a cost-effective tool that would make it easier to ensure that patient with PSP are correctly diagnosed at an early stage of the disease. With the advent of anti-tau immunotherapies as potential disease-modifying agents, earlier and more accurate clinical diagnosis will become increasingly important (Sigurdsson 2016). These data may also help patients and carers better understand the cognitive effects of PSP. Specifically, spatial attention permits the efficient selection of task-relevant information from the environment. When these processes go wrong, it becomes much more effortful for the patient to select relevant information in the environment. As a consequence, patients may find it harder to find everyday objects such as glasses, books and mugs, etc. particularly in cluttered environments and may also become more distracted as they find it hard to ignore irrelevant objects. Better understanding of the cognitive impacts of PSP may help patients and carers develop more effective strategies for coping with PSP.

To summarise, we set out to replicate and extend Rafal et al.'s claim that covert spatial attention is disrupted in PSP. We examined feature and conjunction search in patients with PSP and compared their performance to that of age-matched controls. Consistent with the results of Rafal et al., our patients with PSP had an impairment of feature search that was particularly severe when targets appeared along the vertical axis and not observed during conjunction search. The visual search of age-matched controls was not affected by the location of the target. We argue that that PSP interferes with the ability of the oculomotor system to contribute to competitive interactions in the priority map, thus reducing the efficacy of feature search. From a theoretical perspective, the results are consistent with our weak 'exogenous only' version of Premotor Theory (Smith and Schenk 2012), and from a clinical standpoint, these findings may have relevance for developing new tools to assist with the early diagnosis of PSP.

References

Awh E, Armstrong KM, Moore T (2006) Visual and oculomotor selection: links, causes and implications for spatial attention. Trends Cogn Sci 10(3):124–130. https://doi.org/10.1016/j.tics.2006.01.001

Ball K, Pearson DG, Smith DT (2013) Oculomotor involvement in spatial working memory is task-specific. Cognition 129(2):439–446. https://doi.org/10.1016/j.cognition.2013.08.006

Barash S, Bracewell RM, Fogassi L, Gnadt JW, Andersen RA (1991) Saccade-related activity in the lateral intraparietal area. I. Temporal properties; comparison with area 7a. J Neurophysiol 66(3):1095–1108. https://doi.org/10.1152/jn.1991.66.3.1095

Bisley JW, Goldberg ME (2010) Attention, intention, and priority in the parietal lobe. Annu Rev Neurosci 33:1–21

Bisley JW, Mirpour K, Arcizet F, Ong WS (2011) The role of the lateral intraparietal area in orienting attention and its implications for visual search. Eur J Neurosci 33(11):1982–1990. https://doi.org/10.1111/j.1460-9568.2011.07700.x

Bower JH, Maraganore DM, McDonnell SK, Rocca WA (1997) Incidence of progressive supranuclear palsy and multiple system atrophy in Olmsted County, Minnesota, 1976 to 1990. Neurology 49(5):1284–1288

Boxer AL, Yu JT, Golbe LI, Litvan I, Lang AE, Hoglinger GU (2017) Advances in progressive supranuclear palsy: new diagnostic criteria, biomarkers, and therapeutic approaches. Lancet Neurol 16(7):552–563. https://doi.org/10.1016/s1474-4422(17)30157-6

Brown RG, Lacomblez L, Landwehrmeyer BG, Bak T, Uttner I, Dubois B et al (2010) Cognitive impairment in patients with multiple system atrophy and progressive supranuclear palsy. Brain 133(Pt 8):2382–2393. https://doi.org/10.1093/brain/awq158

Burrell JR, Hodges JR, Rowe JB (2014) Cognition in corticobasal syndrome and progressive supranuclear palsy: a review. Mov Disord 29(5):684–693. https://doi.org/10.1002/mds.25872

Carrasco M, Ling S, Read S (2004) Attention alters appearance. Nat Neurosci 7(3):308–313

Chambers CD, Stokes MG, Mattingley JB (2004) Modality-specific control of strategic spatial attention in parietal cortex. Neuron 44(6):925–930. https://doi.org/10.1016/j.neuron.2004.12.009

Chen AL, Riley DE, King SA, Joshi AC, Serra A, Liao K et al (2010) The disturbance of gaze in progressive supranuclear palsy: implications for pathogenesis. Front Neurol 1:147. https://doi.org/10.3389/fneur.2010.00147

Chun MM, Wolfe JM (1996) Just say no: how are visual searches terminated when there is no target present? Cogn Psychol 30(1):39–78. https://doi.org/10.1006/cogp.1996.0002

Corbetta M, Shulman GL (2002) Control of goal-directed and stimulus-driven attention in the brain. Nat Neurosci Rev 3:201–215

Corbetta M, Akbudak E, Conturo TE, Snyder AZ, Ollinger JM, Drury HA et al (1998) A common network of functional areas for attention and eye movements. Neuron 21(4):761–773

Cousineau D (2005) Confidence intervals in within-subject designs: a simpler solution to Loftus and Masson's method. Tutor Quant Methods Psychol 1(1):4–45

Cowey A, Alexander I, Stoerig P (2011) Transneuronal retrograde degeneration of retinal ganglion cells and optic tract in hemianopic monkeys and humans. Brain 134:2149–2157. https://doi.org/10.1093/brain/awr125

Craighero L, Carta A, Fadiga L (2001) Peripheral oculomotor palsy affects orienting of visuospatial attention. Neuroreport 12(15):3283–3286

Craighero L, Nascimben M, Fadiga L (2004) Eye position affects orienting of visuospatial attention. Curr Biol 14(4):331–333

de Haan B, Morgan PS, Rorden C (2008) Covert orienting of attention and overt eye movements activate identical brain regions. Brain Res 1204:102–111. https://doi.org/10.1016/j.brainres.2008.01.105

Desimone R (1998) Visual attention mediated by biased competition in extrastriate visual cortex. Philos Trans R Soc Lond Ser B Biol Sci 353(1373):1245–1255

Deubel H, Schneider WX (1996) Saccade target selection and object recognition: evidence for a common attentional mechanism. Vis Res 36(12):1827–1837

Dickson DW, Ahmed Z, Algom AA, Tsuboi Y, Josephs KA (2010) Neuropathology of variants of progressive supranuclear palsy. Curr Opin Neurol 23(4):394–400. https://doi.org/10.1097/WCO.0b013e32833be924

Dinkin M (2017) Trans-synaptic retrograde degeneration in the human visual system: slow, silent, and real. Curr Neurol Neurosci Rep 17(2):15. https://doi.org/10.1007/s11910-017-0725-2

Fecteau JH, Munoz DP (2006) Salience, relevance, and firing: a priority map for target selection. Trends Cogn Sci 10(8):382–390

Fierro B, Brighina F, Oliveri M, Piazza A, La Bua V, Buffa D, Bisiach E (2000) Contralateral neglect induced by right posterior parietal rTMS in healthy subjects. Neuroreport 11(7):1519–1521

Gabay S, Henik A, Gradstein L (2010) Ocular motor ability and covert attention in patients with Duane Retraction Syndrome. Neuropsychologia 48(10):3102–3109

Gerstenecker A, Mast B, Duff K, Ferman TJ, Litvan I (2013) Executive dysfunction is the primary cognitive impairment in progressive supranuclear palsy. Arch Clin Neuropsychol 28(2):104–113. https://doi.org/10.1093/arclin/acs098

Ghosh BCP, Calder AJ, Peers PV, Lawrence AD, Acosta-Cabronero J, Pereira JM et al (2012) Social cognitive deficits and their neural correlates in progressive supranuclear palsy. Brain 135:2089–2102. https://doi.org/10.1093/brain/aws128

Glasmacher SA, Leigh PN, Saha RA (2017) Predictors of survival in progressive supranuclear palsy and multiple system atrophy: a systematic review and meta-analysis. J Neurol Neurosurg Psychiatry 88(5):402–411. https://doi.org/10.1136/jnnp-2016-314956
Golbe LI (2014) Progressive supranuclear palsy. Semin Neurol 34(2):151–159. https://doi.org/10.1055/s-0034-1381736
Goldberg ME, Bisley J, Powell KD, Gottlieb J, Kusunoki M (2002) The role of the lateral intraparietal area of the monkey in the generation of saccades and visuospatial attention. Neurobiol Eye Mov Mol Behav 956:205–215
Grosbras M-H, Paus T (2002) Transcranial magnetic stimulation of the human frontal eye field: effects on visual perception and attention. J Cogn Neurosci 14(7):1109–1120
Hilchey MD, Klein RM, Satel J (2014) Returning to "inhibition of return" by dissociating long-term oculomotor ior from short-term sensory adaptation and other nonoculomotor "inhibitory" cueing effects. J Exp Psychol Hum Percept Perform 40(4):1603–1616. https://doi.org/10.1037/a0036859
Hillyard SA, Vogel EK, Luck SJ (1998) Sensory gain control (amplification) as a mechanism of selective attention: electrophysiological and neuroimaging evidence. Philos Trans R Soc Lond B Biol Sci 353(1373):1257–1270. https://doi.org/10.1098/rstb.1998.0281
Höglinger GU, Respondek G, Stamelou M, Kurz C, Josephs KA, Lang AE, Movement Disorder Society-endorsed PSP Study Group et al (2017) Clinical diagnosis of progressive supranuclear palsy: the movement disorder society criteria. Mov Disord 32:853–864. https://doi.org/10.1002/mds.26987
Ikkai A, Curtis CE (2011) Common neural mechanisms supporting spatial working memory, attention and motor intention. Neuropsychologia 49(6):1428–1434
Jackson SR, Newport R, Osborne F, Wakely R, Smith DT, Walsh V (2005) Saccade-contingent spatial and temporal errors are absent for saccadic head movements. Cortex 41(2):205–212
Kimura D, Barnett HJM, Burkhart G (1981) The psychological test pattern in progressive supranuclear palsy. Neuropsychologia 19(2):301–306. https://doi.org/10.1016/0028-3932(81)90113-5
Kisvarday ZF, Cowey A, Stoerig P, Somogyi P (1991) Direct and indirect retinal input into degenerated dorsal lateral geniculate-nucleus after striate cortical removal in monkey - implications for residual vision. Exp Brain Res 86(2):271–292
Klein RM (2000) Inhibition of return. Trends Cogn Sci 4(4):138–147
Lane AR, Smith DT, Ellison A, Schenk T (2010) Visual exploration training is no better than attention training for treating hemianopia. Brain 133(6):1717–1728. https://doi.org/10.1093/brain/awq088
Lane AR, Smith DT, Schenk T, Ellison A (2012a) The involvement of posterior parietal cortex and frontal eye fields in spatially primed visual search. Brain Stimul 5(1):11–17
Lane AR, Smith DT, Schenk T, Ellison A (2012b) The involvement of posterior parietal cortex in feature and conjunction visuomotor search. J Cogn Neurosci 23(8):1964–1972. https://doi.org/10.1162/jocn.2010.21576
Litvan I, Bhatia KP, Burn DJ, Goetz CG, Lang AE, McKeith I et al (2003) Movement disorders society scientific issues committee report: SIC Task Force appraisal of clinical diagnostic criteria for Parkinsonian disorders. Mov Disord 18(5):467–486. https://doi.org/10.1002/mds.10459
Luck SJ, Chelazzi L, Hillyard SA, Desimone R (1997) Neural mechanisms of spatial selective attention in areas V1, V2, and V4 of macaque visual cortex. J Neurophysiol 77(1):24–42
Michalczyk Ł, Paszulewicz J, Bielas J, Wolski P (2018) Is saccade preparation required for inhibition of return (IOR)? Neurosci Lett 665:13–17. https://doi.org/10.1016/j.neulet.2017.11.035
Monza DD, Soliveri PP, Radice DD et al (1998) Cognitive dysfunction and impaired organization of complex motility in degenerative parkinsonian syndromes. Arch Neurol 55(3):372–378. https://doi.org/10.1001/archneur.55.3.372

Morgan EJ, Ball K, Smith DT (2014) The role of the oculomotor system in covert social attention. Atten Percept Psychophys 76(5):1265–1270. https://doi.org/10.3758/s13414-014-0716-1
Muggleton NG, Juan CH, Cowey A, Walsh V (2003) Human frontal eye fields and visual search. J Neurophysiol 89(6):3340–3343
Muller HJ, Rabbitt PMA (1989) Reflexive and voluntary orienting of visual-attention - time course of activation and resistance to interruption. J Exp Psychol Hum Percept Perform 15(2):315–330
Munoz DP, Everling S (2004) Look away: the anti-saccade task and the voluntary control of eye movement. Nat Rev Neurosci 5(3):218–228
Nath U, Ben-Shlomo Y, Thomson RG, Morris HR, Wood NW, Lees AJ, Burn DJ (2001) The prevalence of progressive supranuclear palsy (Steele–Richardson–Olszewski syndrome) in the UK. Brain 124(7):1438–1449. https://doi.org/10.1093/brain/124.7.1438
Nobre AC, Gitelman DR, Dias EC, Mesulam MM (2000) Covert visual spatial orienting and saccades: overlapping neural systems. NeuroImage 11(3):210–216
Pearson DG, Ball K, Smith DT (2014) Oculomotor preparation as a rehearsal mechanism in spatial working memory. Cognition 132(3):416–428. https://doi.org/10.1016/j.cognition.2014.05.006
Pinching AJ, Powell TPS (1971) Ultrastructural features of transneuronal cell degeneration in olfactory system. J Cell Sci 8(1):253
Posner MI (1978) Chronometric explorations of mind. Lawrence Erlbaum Associates, Hillsdale
Posner MI (1980) Orienting of attention. Q J Exp Psychol 32:3–25
Posner MI, Cohen Y (1984) Components of visual orienting. In: Bouma H, Bouwhuis D (eds) Attention and performance X. Lawrence Erlbaum Associates, London, pp 531–556
Posner MI, Cohen Y, Rafal RD (1982) Neural systems control of spatial orienting. Philos Trans R Soc Lond Ser B Biol Sci 298(1089):187–198
Posner MI, Rafal RD, Choate LS, Vaughan J (1985) Inhibition of return - neural basis and function. Cogn Neuropsychol 2(3):211–228
Rafal RD, Posner MI, Friedman JH, Inhoff AW, Bernstein E (1988) Orienting of visual-attention in progressive supranuclear palsy. Brain 111:267–280
Ristic J, Friesen CK, Kingstone A (2002) Are eyes special? It depends on how you look at it. Psychon Bull Rev 9(3):507–513. https://doi.org/10.3758/bf03196306
Rizzolatti G, Riggio L, Sheliga BM (1994) Space and selective attention. Atten Perform 15:231–265
Robbins TW, James M, Owen AM, Lange KW, Lees AJ, Leigh PN et al (1994) Cognitive deficits in progressive supranuclear palsy, Parkinson's disease, and multiple system atrophy in tests sensitive to frontal lobe dysfunction. J Neurol Neurosurg Psychiatry 57(1):79–88
Schall JD, Cohen JY (2011) The neural basis of saccade target selection. In: Liversedge SP, Gilchrist I, Everling S (eds) The Oxford handbook of eye movements. Oxford University Press, Oxford, pp 357–381
Schrag A, Ben-Shlomo Y, Quinn NP (1999) Prevalence of progressive supranuclear palsy and multiple system atrophy: a cross-sectional study. Lancet 354(9192):1771–1775
Shepherd M, Findlay JM, Hockey RJ (1986) The relationship between eye-movements and spatial attention. Q J Exp Psychol A Hum Exp Psychol 38(3):475–491
Sigurdsson EM (2016) Tau immunotherapy. Neurodegener Dis 16:34–38. https://doi.org/10.1159/000440842
Smith DT, Archibald N (2018) Spatial working memory in Progressive Supranuclear Palsy. Cortex. https://doi.org/10.1016/j.cortex.2018.07.004
Smith DT, Schenk T (2012) The premotor theory of attention: time to move on? Neuropsychologia 50(6):1104–1114
Smith DT, Rorden C, Jackson SR (2004) Exogenous orienting of attention depends upon the ability to execute eye movements. Curr Biol 14(9):792–795
Smith DT, Jackson SR, Rorden C (2005) Transcranial magnetic stimulation of the left human frontal eye fields eliminates the cost of invalid endogenous cues. Neuropsychologia 43(9):1288–1296

Smith DT, Jackson SR, Rorden C (2009a) An intact eye-movement system is not required to generate inhibition of return. J Neuropsychol 3:267–271
Smith DT, Jackson SR, Rorden C (2009b) Repetitive TMS over frontal eye fields disrupts visually cued auditory attention. Brain Stimul 2:81–87
Smith DT, Ball K, Ellison A, Schenk T (2010) Deficits of reflexive attention induced by abduction of the eye. Neuropsychologia 48:1269–1276
Smith DT, Rorden C, Schenk T (2012) Saccade preparation is required for exogenous attention but not endogenous attention or IOR. J Exp Psychol Hum Percept Perform 36(6):1438–1447. https://doi.org/10.1037/a0027794
Smith DT, Ball K, Ellison A (2014) Covert visual search within and beyond the effective oculomotor range. Vis Res 95:11–17. https://doi.org/10.1016/j.visres.2013.12.003
Soliveri P, Monza D, Paridi D, Carella F, Genitrini S, Testa D, Girotti F (2000) Neuropsychological follow up in patients with Parkinson's disease, striatonigral degeneration-type multisystem atrophy, and progressive supranuclear palsy. J Neurol Neurosurg Psychiat 69(3):313–318. https://doi.org/10.1136/jnnp.69.3.313
Sparks DL, Mays LE (1990) Signal transformations required for the generation of saccadic eye movements. Annu Rev Neurosci 13:309–336
Steele JC, Richardson JC, Olszewski J (1964) Progressive Supranuclear Palsy. A heterogeneous degeneration involving the brain stem, basal ganglia and cerebellum with vertical gaze and pseudobulbar palsy, nuchal dystonia and dementia. Arch Neurol 10:333–359
Thomas NW, Pare M (2007) Temporal processing of saccade targets in parietal cortex area LIP during visual search. J Neurophysiol 97(1):942–947. https://doi.org/10.1152/jn.00413.2006
Tipples J (2002) Eye gaze is not unique: automatic orienting in response to uninformative arrows. Psychon Bull Rev 9(2):314–318
White BJ, Munoz DP (2011) The superior colliculus. In: Liversedge SP, Gilchrist I, Everling S (eds) The Oxford handbook of eye movements. Oxford University Press, Oxford, pp 195–215
Williams DR, de Silva R, Paviour DC, Pittman A, Watt HC, Kilford L et al (2005) Characteristics of two distinct clinical phenotypes in pathologically proven progressive supranuclear palsy: Richardson's syndrome and PSP-parkinsonism. Brain 128(Pt 6):1247–1258. https://doi.org/10.1093/brain/awh488
Wolfe JM (1994) Guided search 2.0 A revised model of visual search. Psychon Bull Rev 1(2):202–238. https://doi.org/10.3758/bf03200774
Wolfe JM (1998) Visual search. In: Pashler H (ed) Attention. Psychology Press, Hove, pp 13–56
Wolfe JM (2003) Moving towards solutions to some enduring controversies in visual search. Trends Cogn Sci 7(2):70–76
Yeshurun Y, Carrasco M (1998) Attention improves or impairs visual performance by enhancing spatial resolution. Nature 396(6706):72–75
Yoshida K, Hata Y, Kinoshita K, Takashima S, Tanaka K, Nishida N (2017) Incipient progressive supranuclear palsy is more common than expected and may comprise clinicopathological subtypes: a forensic autopsy series. Acta Neuropathol 133(5):809–823. https://doi.org/10.1007/s00401-016-1665-7
Zangemeister WH, Canavan AGM, Hoemberg V (1995) Frontal and parietal transcranial magnetic stimulation (TMS) disturbs programming of saccadic eye-movements. J Neurol Sci 133(1–2):42–52
Zhang J, Rittman T, Nombela C, Fois A, Coyle-Gilchrist I, Barker RA et al (2016) Different decision deficits impair response inhibition in progressive supranuclear palsy and Parkinson's disease. Brain 139(1):161–173. https://doi.org/10.1093/brain/awv331

Working Memory in Alzheimer's Disease and Parkinson's Disease

Nahid Zokaei and Masud Husain

Contents

Abstract Working memory impairments are frequently observed in patients with Alzheimer's disease (AD) and Parkinson's disease (PD). Recent research suggests that the mechanisms underlying these deficits might be dissociable using sensitive tasks, specifically those that rely on the reproduction of the exact quality of features held in memory.

In patients with AD, working memory impairments are mainly due to an increase in *misbinding errors*. They arise when patients misremember which features (e.g., color, orientation, shape, and location) belong to different objects held in memory. Hence, they erroneously report features that belong to items in memory other than the one they are probed on. This misbinding of features that belong to different objects in memory can be considered a form of interference between stored items. Such binding errors are evident even in presymptomatic individuals with familial AD (due to gene mutations) who do not have AD yet. Overall, these findings are in line with the role of the medial temporal lobes, and specifically the hippocampus, in

N. Zokaei (✉)
Oxford Centre for Human Brain Activity, University of Oxford, Oxford, UK

Department of Experimental Psychology, University of Oxford, Oxford, UK
e-mail: nahid.zokaei@psy.ox.ac.uk

M. Husain
Department of Experimental Psychology, University of Oxford, Oxford, UK

Nuffield Department of Clinical Neuroscience, University of Oxford, Oxford, UK

Curr Topics Behav Neurosci (2019) 41: 325–344
DOI 10.1007/7854_2019_103
Published Online: 26 July 2019

retention of feature bindings, regardless of retention duration, i.e., in both short- or long-term memory.

Patients with PD, on the other hand, do not show increased misbinding. Their working memory deficits are associated with making more *random errors* or guesses. These random responses are not modulated by manipulations of their dopaminergic medication and hence may reflect involvement of non-dopaminergic neurotransmitters in this deficit. In addition, patients with PD demonstrate impairments in gating of information into relevant vs. irrelevant items in memory, a cognitive operation that is modulated by dopaminergic manipulation in line with a frontal executive effect of this neurotransmitter. Thus, although AD and PD are both associated with working memory impairments, these surface manifestations appear to be underpinned by very different mechanisms.

Keywords Alzheimer's disease · Parkinson's disease · Working memory

Working memory (WM) impairments are a common feature of many neurodegenerative disorders, such as Alzheimer's disease (AD) and Parkinson's disease (PD) (Muslimovic et al. 2005; Verbaan et al. 2007; Wolk et al. 2010). The importance of understanding the exact nature of these deficits in disease is twofold. On the one hand, WM underlies and constrains our abilities for a variety of cognitive processes including language, problem-solving, and executive functions (Baddeley 2003) which impact greatly on functional independence for patients with neurodegenerative disorders (Lau et al. 2015). Isolation of the precise nature of WM deficits in disease may assist in providing a comprehensive cognitive fingerprint of these disorders and also aid in the development of training regimens. On the other hand, given that WM impairments can be one of the earliest cognitive impairments in these disorders (Owen et al. 1993, 1997; Blackwell et al. 2004), a clearer distinction from normal healthy aging might aid in early diagnosis as well as monitoring disease progression and the impact of new treatments.

1 Binding Deficits in Working Memory in Alzheimer's Disease

Classically, long-term memory (LTM) dysfunction has been widely linked to early stages of AD (Linn et al. 1995; Fox et al. 1998; Yau et al. 2015). Medial temporal lobe (MTL) subregions, including the hippocampus, are known to contribute to LTM and are key brain areas associated with AD pathology, with respect to both neurofibrillary tangle disposition and as an early site of atrophy (Braak and Braak 1998; Ridha et al. 2006; Brier et al. 2016). Some models of memory argue that although other parts of the MTL may be involved in retention of information regarding an event, or contextual information, the hippocampus binds features belonging to a memory episode, for example, an item to its context (Davachi 2006; Eichenbaum et al. 2007; Konkel and Cohen 2009; Nadel and Peterson 2013).

More specifically, it has been hypothesized that the hippocampus does not play a role in all aspects of memory but rather performs a specific computation: high-resolution binding of features in perception as well as working and long-term memories, *regardless of duration* (Pertzov et al. 2013; Yonelinas 2013). For example, maintenance of complex scenes or tasks that require precise retention of bound information appears to require hippocampal contributions (Hartley et al. 2007; Koen et al. 2016). Recent neurophysiological studies provide further evidence for the role of MTL – and specifically the hippocampus – in integration and coordination of disparate cortical representations, supporting relational binding of features that belong to a specific episode in LTM (Cashdollar et al. 2011; Watrous et al. 2013). There is also evidence from lesion studies that the hippocampus and/or MTL structures play a similar role in binding feature information together over far shorter periods (seconds) in WM (Hannula et al. 2006; Olson et al. 2006).

Consistent with these models, an important line of research has provided evidence that the ability to retain bound features in WM is also critically affected in AD, a disorder that, as mentioned above, is classically associated with hippocampal and MTL dysfunction. Traditionally, WM has been measured using neuropsychological tasks that index span. In such tests, participants are presented with increasing digit, word, or spatial location sequences and have to recall the sequence of the array, in forward or backward order. The maximum sequence length successfully recalled by the participants is taken as a measure of WM capacity.

Several studies using digit and word spans have found reduced WM capacity in patients with AD compared to healthy controls (Miller 1973; Spinnler et al. 1988; Becker 1988; Hulme et al. 1993). However, these findings are not always reliably replicable (Martin et al. 1985; Carlesimo et al. 1994; Perry and Hodges 2000) and importantly are not sensitive enough to detect subtle variations in performance in the healthy populations, for example, between elderly and young healthy participants (Belleville et al. 1996). Moreover, span measures of WM do not provide any further detail on the exact nature of the deficits, that is, whether any reduced WM capacity observed in a patient group such as AD is due to encoding, retention, or retrieval aspects of this cognitive process.

1.1 Change-Detection Tasks

More in-depth experimental studies however provide a means to better quantify the nature of WM impairments in AD patients. In a series of studies, Parra and colleagues demonstrated that maintenance of bound features, for example, color and shape, is selectively impaired in AD (Parra et al. 2009, 2010, 2011; Della Sala et al. 2012). In one such report (Della Sala et al. 2012), patients with AD and non-AD dementias as well as healthy controls performed a visual WM task in which they were presented with memory arrays consisting of either single features (color squares or objects), multiple features bound together in a single object (colored objects), or unbound features (colored squares and black and white line drawing objects presented separately). Participants were asked to keep in mind and recall as many features or bound objects following a brief delay interval.

Importantly, performance for single-feature conditions was matched between patients and healthy controls. This avoided ceiling- and floor-level performance in healthy controls and patients, respectively, and also attempted to minimize the contribution of task difficulty in the two conditions of interest: binding and unbound multiple-feature conditions. AD patients performed significantly worse compared to non-AD dementia patients only in the binding condition (Della Sala et al. 2012).

Binding deficits have also been observed in both sporadic and familial AD (FAD) patients and importantly in asymptomatic carriers of genes that have been shown to cause FAD (Parra et al. 2010, 2011). Mutations in three genes associated with amyloid processing – amyloid precursor protein (APP), presenilin-1, and presenilin-2 – have been identified in FAD; all are inherited in an autosomal dominant fashion. That is, inheritance of the abnormal gene from only one parent will result in the development of the disease. Furthermore, mutation carriers have increased hippocampal and whole-brain atrophy which is evident prior to a clinical diagnosis of AD (Ridha et al. 2006).

WM binding deficits have now been reported in both patients with FAD and asymptomatic carriers of these genetic mutations. In one study (Parra et al. 2011), participants were presented with a memory array and were asked to keep in mind either single features (colors) or bound features of objects (colored objects). Following a short-delay interval, a probe array appeared, and participants had to indicate whether the probe array was the same or different to the memory array. Patients with FAD and clinically asymptomatic carriers performed similar to sporadic AD patients, demonstrating impaired performance in only the bound condition. Impaired retention of bound features in WM even in clinically asymptomatic carriers highlights the sensitivity of this deficit in detecting AD-related changes even in the absence of other cognitive deficits.

The studies employed by Parra and his colleagues employed a change-detection paradigm, which can be used to measure the capacity of WM or the *number of items* which participants can recall correctly from the WM array in a binary manner (Luck 1997; Zhang and Luck 2008). Although these "quantal" measures of WM performance have been essential to our understanding of WM and its impairments in disease, they might not be sensitive to detect changes in memory resolution. In other words, change-detection paradigms require participants to make a binary response: either the item was remembered correctly or it was not remembered at all, in an all-or-none manner.

However, it is important to highlight that just because an individual fails to recall an item, it does not necessarily mean that information regarding that item is completely lost from memory. Conversely, a correct response does not inform us regarding the quality with which an item was retained in memory. Furthermore, the condition of interest (i.e., the binding condition) required an additional operation compared to single-feature trials. In this condition, participants have to remember both single features and their associations with one another. Thus, direct comparison of such trials to unimpaired single-feature trials should be considered with caution, especially since the single-feature capacity of patients differed from healthy controls in some of these studies (Parra et al. 2011).

1.2 Delayed Reproduction Tasks

A more recent theoretical and empirical approach aims to examine the resolution with which items are retained in memory, in a continuous manner (for a review, see Fallon et al. 2016; Ma et al. 2014). Instead of asking participants to report whether a change occurred between the memory and the probe array, they are requested to reproduce the exact quality of the remembered feature in an analogue response space (e.g., Bays et al. 2009; Gorgoraptis et al. 2011). These delayed reproduction tasks (Fig. 1) measure the resolution or precision of recall by calculating the difference in the response from the true value of the feature from the memory array and hence provide an index of the quality of memory representation.

The delayed reproduction tasks have now been successfully used to study memory for a range of features, including color, orientation, motion direction, location, and even auditory pitch (Bays et al. 2009; Zokaei et al. 2011; Gorgoraptis et al. 2011; Joseph et al. 2016). Using this methodology, it has been found that memory does not reach a capacity limit, but rather the precision with which these features are retained, as measured by variability in recall error, increases gradually

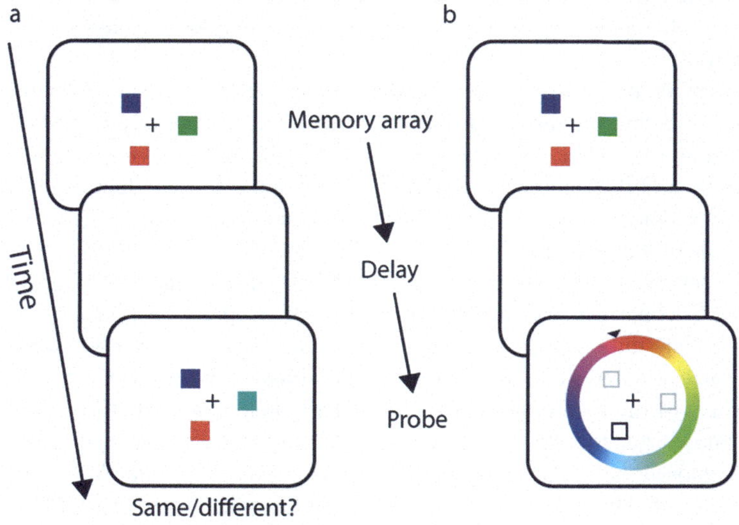

Fig. 1 Example of a color change-detection and delayed reproduction tasks. In the change-detection task, participants have to identify whether the probe array was the same or different to the memory array. In the delayed reproduction tasks, observers must report the color of the item probed by location by selecting it from the color wheel. (**a**) Change-detection. (**b**) Delayed reproduction

and continuously with memory array set size. Importantly, these tasks have been shown to be more sensitive in detecting subtle changes in WM compared to conventional span measures of WM, such as digit or spatial spans (Zokaei et al. 2014a).

On a delayed reproduction task, FAD patients with pathological mutations in presenilin-1 or APP genes demonstrated significant WM deficits for object locations compared to healthy controls (Liang et al. 2016). Participants were presented with one or three colored shapes (fractals) on a touchscreen computer and were asked to keep in mind both their identity and locations (Fig. 2a). Following a delay, they were presented with two shapes, one from the memory array and a novel foil. They first had to select the shape that appeared in the memory array by tapping on it (identification memory) and then drag it to its remembered location (localization memory).

Localization memory error, i.e., the difference in the reported location and the true location of the object from the memory array, was larger in patients with FAD (Fig. 2b). Importantly, an advantage of these delayed reproductions tasks, such as the one employed in this study, is that it is possible to dissect out different sources of error that contribute to impaired WM. Specifically, error can arise from misreporting features of "other" none-probed items that were presented in the memory array, instead of reporting the features of the probed item. These *binding errors* are often labelled as "swap errors."

For example, in the object-location WM task, a swap occurs if participants pick the correct item at identification but drags it to the location of one of the other objects on the screen (Fig. 2c). Importantly, unlike change-detection paradigms, the proportion of binding errors – or swaps – can be estimated from the same trial rather than requiring participants to actively retain bound vs. unbound features in WM on different trials.

In line with previous research, FAD patients showed greater binding errors compared to healthy controls and hence mislocalized a correctly identified object to the location of other items held in WM. Significantly, asymptomatic carriers (people with an FAD gene who were not yet showing cognitive deficits on standard neuropsychological tests) also made increased binding errors (Fig. 2c). Consistent with the role of hippocampus in feature-binding (Yonelinas 2013; Libby et al. 2014), decreased hippocampal volume across familial AD participants was associated with deficits of WM performance and the proportion of swap errors (Fig. 2d).

In a similar study, using an identical object-location memory task, WM performance was examined in another group of individuals with a genetic risk factor of AD: carriers of the apolipoprotein-E (APOE) ε4 allele gene. APOE ε4 confers the highest-known genetic risk for developing AD in the older age, with 30–60% of those diagnosed with AD carrying one or two copies of the APOE ε4 allele (Saunders et al. 1993; Myers et al. 1996; Sando et al. 2008). This high risk of conversion makes individuals with ε4 allele ideal candidates for investigating cognitive impairments that may point to early signs of AD. Importantly, variants of the APOE allele occur normally in the population and are not causative of AD, but rather the APOE ε4 allele results in a higher-than-normal risk of developing this disorder.

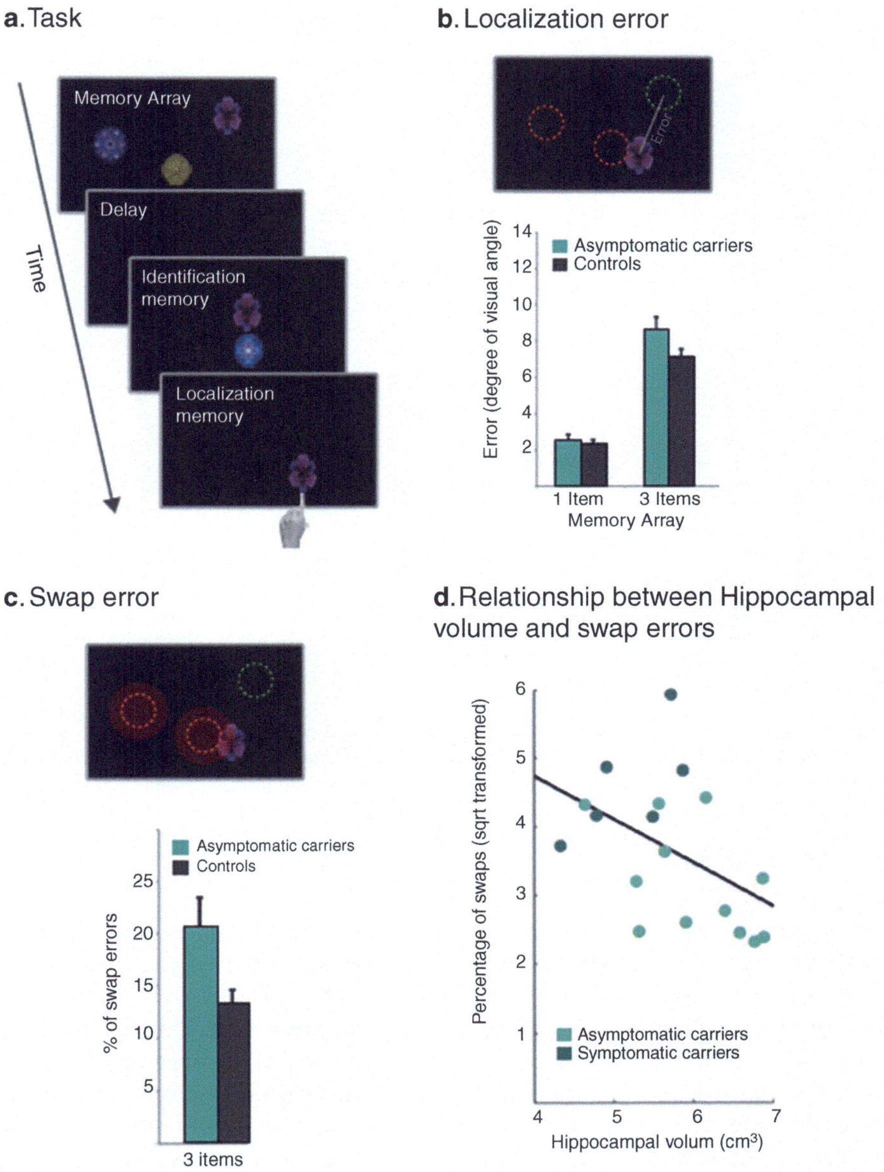

Fig. 2 Object-location memory task and performance. (**a**) Schematic of the working memory task. Participants are presented with a memory array followed by a delay. They are then presented with two items, one from the memory array and a foil. On a touchscreen computer, participants first have to touch the object they have seen before in the memory array and drag it to its remembered location. (**b**) Different types of error that can occur in this task. Localization error is measured by a difference in response location compared to the true location of the item in the memory array. Swap errors are proportion of trials where the object is dragged to a location near the location of one of the other, non-probed, items from the memory array. (**c**) Asymptomatic patients with familial AD make significantly larger errors in memory and make significantly more swap errors compared to healthy controls. (**d**) Proportion of swap errors correlated with the hippocampus volume for both symptomatic and asymptomatic carriers (figure from Liang et al. 2016)

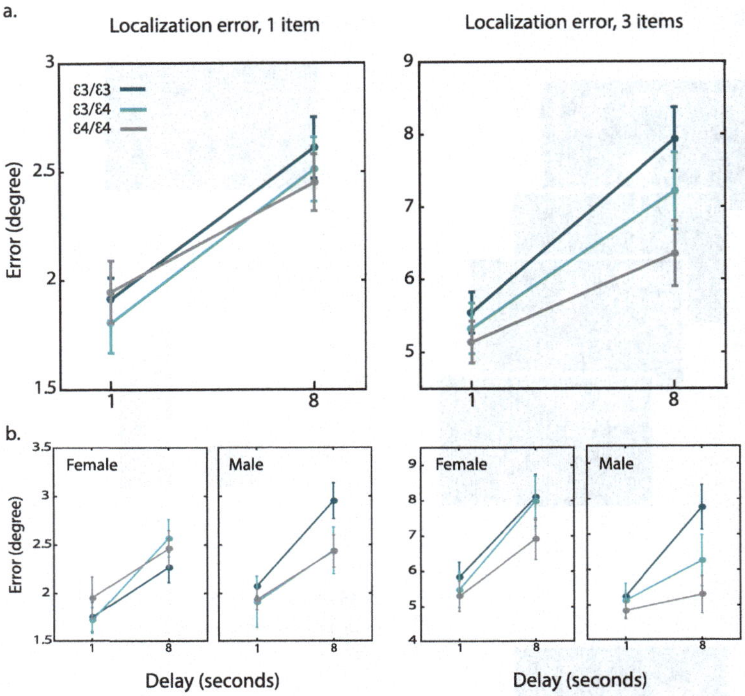

Fig. 3 Localization performance according to genotype and gender in middle age Localization error, indexed by the difference in angular degrees between the target location and the response, for different memory set sizes and delays (**a**) for all participants and (**b**) broken down by sex. Location memory was more precise and decayed less in male APOE ε4/ε4 carriers than male ε4/ε3 or ε3/ε3 carriers, particularly, when more items had to be retained in memory for longer durations (figure from Zokaei et al. 2017)

Interestingly, middle-aged carriers of the APOE ε4 allele demonstrated enhanced WM performance, indexed by both reduced memory decay and a decrease in number of swap errors, in a gene dosage-dependent manner (Zokaei et al. 2017, 2019a). That is, individuals with two copies of the ε4 allele performed better than those with a single copy compared to noncarriers, specifically in male compared to female carriers (Fig. 3). These results are consistent with an antagonistic pleiotropy hypothesis (Williams 1957), which proposes that carrying the APOE ε4 allele might actually confer a WM advantage earlier in life. Detrimental effects associated with this allele become evident only at a point beyond normal reproductive age and would not have been a factor until only relatively recently in human history when people have begun to live far longer than in previous centuries. Although the presence of the APOE ε4 allele reduced swap errors in middle age, the exact effects of the APOE gene on neural mechanisms related to AD pathology and its interaction with cognition are not yet well understood but have been subject to vigorous research in the past decade (Wolk et al. 2010; Suri et al. 2013; Heise et al. 2014; Zimmermann and Butler 2018).

In other pathologies like AD that affects MTL structures, WM binding deficits have also been reported (Pertzov et al. 2013; Koen et al. 2016; Zokaei et al. 2019b). For example, individuals with focal MTL damage due to voltage-gated potassium channel antibody (VGKC-Ab)-mediated limbic encephalitis showed the binding deficit in WM, measured by an increase in proportion of swap errors in the object-location WM task, while crucially their memory for object identities and locations remained intact (Pertzov et al. 2013). Similarly, patients with hippocampal lesions also appear to have a binding deficit in memory, regardless of retention duration (Olson et al. 2006; van Geldorp et al. 2014; Esfahani-Bayerl et al. 2016; Zokaei et al. 2018), although some have contested these conclusions (Jeneson et al. 2010). Together, these findings point to a specific impairment in retention of bound objects in WM in MTL disorders, such as sporadic and familial AD.

2 Working Memory in Parkinson's Disease

Parkinson's disease (PD) is a progressive neurodegenerative movement disorder characterized by depletion of nigrostriatal and mesocortical dopamine, although it is now established that other neurotransmitter systems such as noradrenaline (Zarow et al. 2003), serotonin (Kish et al. 2008), GABA (Emir et al. 2012), or acetylcholine (Dubois et al. 1983) are also affected in PD. The neurotransmitter dopamine nevertheless is thought to play a crucial role in many complex cognitive functions such as WM. Considering the link between dopamine depletion and PD, cognitive deficits, including those in WM, are an important feature of PD, often apparent at very early stages of the disease (Owen et al. 1993, 1997; Dujardin et al. 1999; Muslimovic et al. 2005; Verbaan et al. 2007; Savica et al. 2010).

The influence of disease progression on WM performance, however, is highly complex, and some studies have failed to identify impairments in unmedicated PD patients at earlier stages of the illness using variants of span tasks (e.g., Owen et al. 1997). For example, in the previously described study by Della Sala et al. (2012), WM performance for memory arrays consisting of single features or multiple features – bound or unbound – was intact in patients with PD compared to healthy controls. More sensitive delayed reproduction tasks, however, have been successful in identifying the pattern of impairment in PD patients and in at-risk cohorts.

2.1 Delayed Reproduction Tasks

In one study, unmedicated PD patients (those who had not yet been started on any drug therapy by their clinician) were tested on a sequential delayed reproduction task within a few months of being diagnosed in order to identify the nature of WM impairments in early PD (Zokaei et al. 2014a). Participants were presented with a sequence of four colored oriented bars and were asked to keep in mind both the color and the orientation of these bars. At the end of the sequence, they were presented with a probe orientation in the color of one of the orientations from the memory array. They were asked to “dial up” the orientation of the probed item to match the one from memory (Fig. 4a). WM orientation precision, as measured by the difference in angular deviation between the response orientation and the target angle, was significantly impaired in these unmedicated patients with PD compared to healthy controls.

But what is the source of error of the impaired WM precision in patient with PD? As mentioned earlier, delayed reproduction tasks allow us to dissect out sources of error contributing to impaired performance (Fig. 4b). In addition to misbinding errors, participants may make random errors because, on some trials, they may be simply guessing because of failure to encode or retrieve the probed item. Patients with PD made significantly more *random responses* when performing the described delayed reproduction task (Zokaei et al. 2014b). Moreover, in this study, patients with PD – regardless of the presence or absence of a genetic risk factor for PD: glucocerebrosidase (*GBA*) mutation – were making significantly more guesses (Fig. 4c).

Mutations in the gene encoding the lysosomal enzyme GBA, classically associated with Gaucher’s disease (Pastores and Hughes 1993), have been identified as the highest genetic risk factor for developing PD (Clark et al. 2009; Neumann et al. 2009; Sidransky et al. 2009). However, it is important to note that this gene also constitutes a rare susceptibility factor for PD (Sato et al. 2005). Individuals with GBA mutation, regardless of whether they had PD, showed increased *binding errors* (Zokaei et al. 2014b). Interestingly, both mouse models of *GBA* mutation (Sardi et al. 2011) and human Gaucher’s disease patients with dementia or PD have pathological changes in MTL regions (Wong et al. 2004), in line with the role of MTL in binding of information.

Unlike people with *GBA* mutations, individuals with idiopathic rapid eye movement (REM) sleep behavior disorder (RBD) provide an “enriched” at-risk cohort to examine whether WM impairments, specifically increased random responses, are present in prodromal PD – the period between the onset of neurodegeneration and the diagnosis. In healthy people, the body is atonic and motionless during dreaming during REM sleep. RBD is a parasomnia typified by simple or complex motor behaviors associated with vivid dreams occurring during REM sleep (Boeve 2010). Prospective cohort studies have reported a very strong association between RBD and subsequent development of neurodegenerative disorders, with up to 80% of cases affected (Iranzo et al. 2006; Postuma et al. 2009; Boot et al. 2012; Schenck et al. 2013). While some patients with RBD may develop dementia with Lewy bodies, most will eventually develop PD (Boeve 2010).

Moreover, dysfunction within the dopaminergic systems has also been reported in RBD prior to the onset of clinically defined neurodegenerative disorders. For example, in one study using single-photon emission computed tomography (SPECT), a decrease in ^{123}I-FP-CIT uptake in the striatum of RBD patients was demonstrated with approximately 40% of cases having an abnormal scan (Selikhova et al. 2009; Lawton et al. 2015). Similarly, loss of dopaminergic neurons as

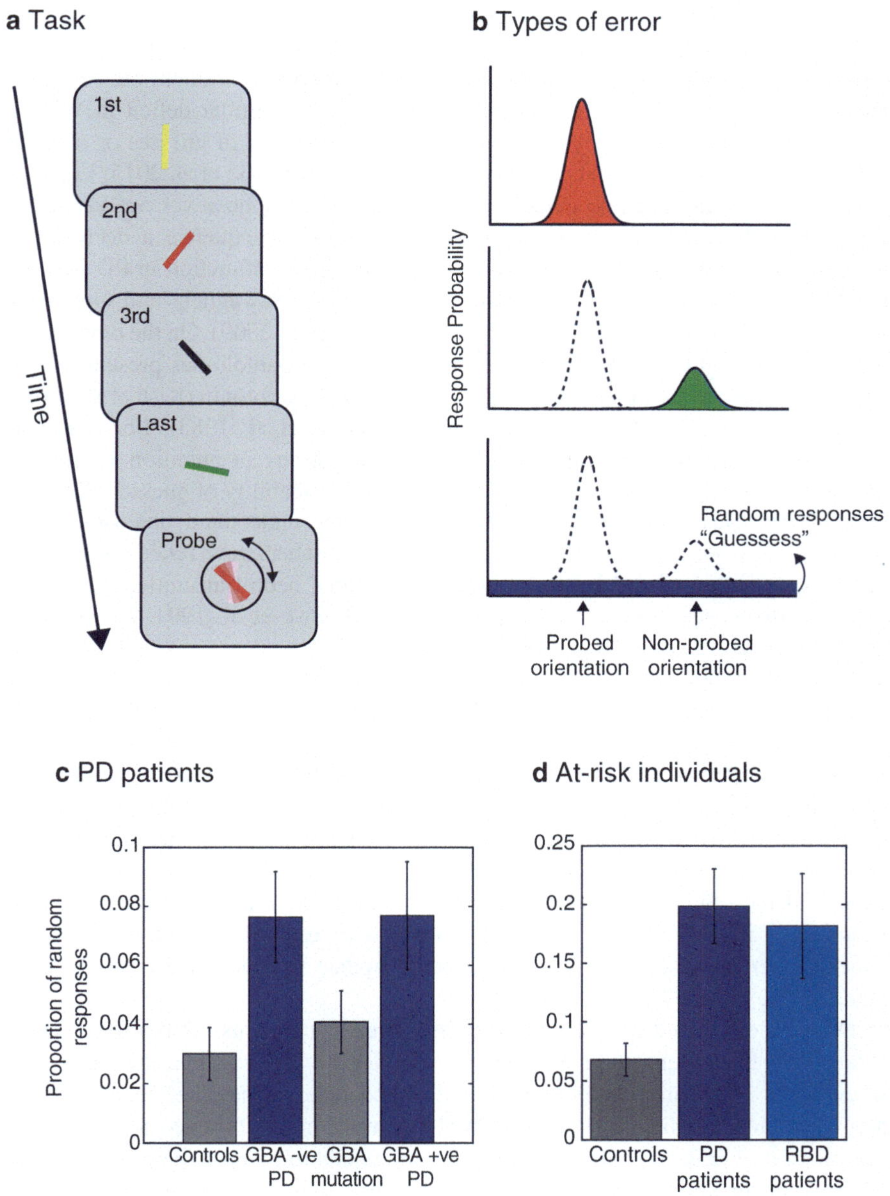

Fig. 4 Sequential orientation task and performance. (**a**) Sequences of four colored oriented bars were presented. Participants had to adjust the orientation of the probe stimuli to the orientation of the bar with the same color in the memory array. (**b**) Sources of error in memory. Error can arise due to increased variability in responses to the probed orientation, due to responses toward the non-probed items or misbinding error, and lastly due to random responses or guesses. Patients with PD with or without GBA mutation (**c**) as well as those at risk of developing PD due to RBD (**d**) made significantly more random responses/guesses compared to healthy controls (figures from Zokaei et al. 2014a, b; Rolinski et al. 2015)

measured by a decrease in ^{11}C-dihydrotetrabenazine (^{11}C-DTBZ) striatal binding on positron emission tomography (PET) scanning has also been reported in RBD patients without PD (Ferini-Strambi et al. 2004). Importantly though, patients with RBD, who were not yet diagnosed with PD, also had a similar deficit in WM to patients with PD, that is, they made an increased proportion of guesses or *random responses* as measured by delayed reproduction tasks (Rolinski et al. 2015; Fig. 4d).

The precise mechanisms underlying this type of error however are yet to be established. Increased random responses could be a consequence a decrease in signal-to-noise ratio due to deficits in the dopaminergic dysfunction in PD patients resulting in responses to fall within the random range (Sawaguchi and Goldman-Rakic 1991; Winterer and Weinberger 2004; Kroener et al. 2009). On the other hand, WM deficits might arise because of non-dopaminergic pathologies present in PD, such as disruptions in noradrenaline (Zarow et al. 2003), serotonin (Kish et al. 2008), GABA (Emir et al. 2012), or acetylcholine (Dubois et al. 1983). For example, cholinergic disruptions in PD may result in fluctuations in attention leading to encoding or retrieval failure and therefore increased probability of guesses (Kehagia et al. 2010; Hasselmo and Sarter 2011). To better understand and isolate the mechanisms associated with WM deficits in PD, researchers have relied on studying the effects of drugs that selectively alter an individual neurotransmitter by investigating performance both on and off medication (Poewe et al. 1991; Lange et al. 1992; Lewis et al. 2005; Hughes et al. 2013).

Using this approach, studies investigating WM performance both on and off dopaminergic medication have resulted in conflicting findings. Dopaminergic medication has been reported to both improve (Lange et al. 1992) and impair WM performance (Poewe et al. 1991; Cools et al. 2010; Uitvlugt et al. 2016), sometimes depending on modality of the memory array (Owen et al. 1997; Postle et al. 1997; Gruszka et al. 2016), in tasks employing span or change-detection methodology. To better quantify the effect of dopaminergic medication on WM, a recent study (Fig. 5) examined the effect of dopaminergic medication on a delayed reproduction task examining retention over time (simple maintenance), as well as recall when people either had to ignore distracting information or update the contents of WM (Fallon et al. 2017).

Dopaminergic medication selectively improved PD patients' ability to precisely recall items from memory *in the presence of irrelevant information*, irrespective of whether the information had to be ignored or updated. On the other hand, dopaminergic medication did not influence WM retention in the absence to competing information, that is, in trials that required participants to retain a memory array for either short- or long-delay intervals – maintenance conditions. Simple retention of information however was impaired in PD patients compared to healthy controls, pointing to possible involvement of non-dopaminergic neurotransmitters involved in WM maintenance processes.

Together these findings point to the possibility of two distinct mechanisms that may contribute to WM deficits in PD. Firstly, a frontal dysexecutive effect which is modulated by dopamine and deals with gating of information into relevant vs. irrelevant information in WM (Dalrymple-Alford et al. 1994; Gruszka et al.

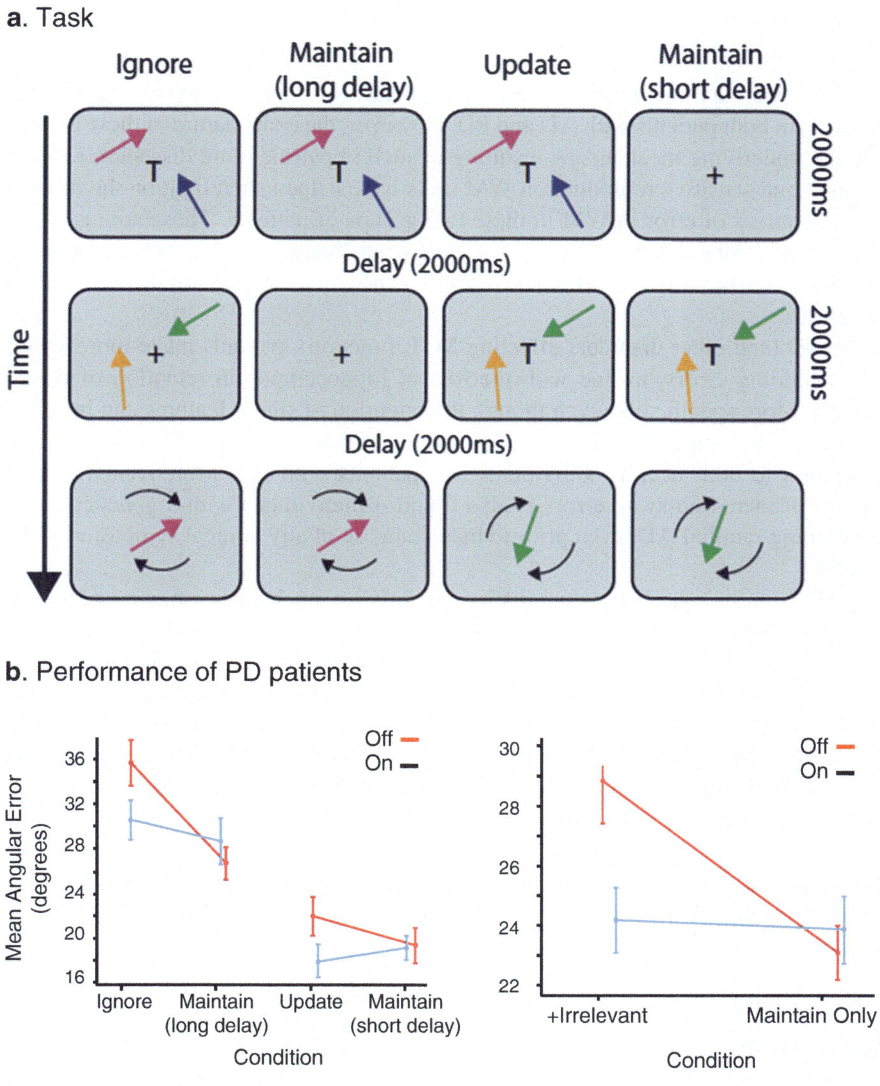

Fig. 5 Working memory task with and without irrelevant information. (**a**) Participants are presented with two colored arrows to keep in mind either for short or long duration (conditions maintain long and short delay) in half of the trials. In the remaining half, they will be presented with a second array of arrows that they either have to ignore (ignore condition) or replace with the array in mind (update). (**b**) Dopaminergic medication only influenced by the trials with irrelevant information, that is, ignore or update conditions (figures from Fallon et al. 2017)

2016). Secondly, studies reported here also support the possibility of a non-dopaminergic syndrome that influences the successful retention of information and may result in increased forgetting over short periods of time (Kehagia et al. 2010; Gratwicke et al. 2015).

3 Summary

Working memory impairment is a central feature of cognitive dysfunction, often reported in both patients with AD and PD. However, the exact nature of these deficits and the underlying mechanisms resulting in such impairment are dissociable. More detailed and sensitive reproduction WM tasks have helped shed light on the distinction in sources of error in WM in these two groups of patients. These are tasks that allow us to distinguish between errors arising from changes in resolution of memory, increased misbinding of retained features, and increased guessing or random responses.

In AD (and other disorders affecting MTL function), patients make more swaps or misbinding errors, in line with the role of hippocampus in retention of bound objects. More specifically, even though the retention of single features can be intact in these individuals, the binding of features into objects is significantly affected compared to both healthy individuals and patients with PD. Moreover, a similar pattern of increased swap errors is also found in individuals with a genetic risk of developing familial AD, even prior to manifestation of any clinical symptoms of the disease.

In PD, on the other hand, two distinct underlying mechanisms may contribute to WM impairments. Firstly, there is an increased proportion of random responses or guesses. This deficit is not influenced by dopaminergic medication, highlighting the role of a non-dopaminergic contributions to impairments associated with simple retention of information over brief periods of time. However, impairments in gating of information in WM in PD can be attributed to a frontal dysfunction and are modulated by dopamine, in studies that manipulate the effects of drug to selectively alter this neurotransmitter system.

Acknowledgments This work is supported by the Wellcome Trust, British Academy, NIHR Oxford Biomedical Research Centre, and Wellcome Centre for Integrative Neuroscience.

References

Baddeley A (2003) Working memory: looking back and looking forward. Nat Rev Neurosci 4:829–839. https://doi.org/10.1038/nrn1201

Bays PM, Catalao RFG, Husain M (2009) The precision of visual working memory is set by allocation of a shared resource. J Vis 9:7. https://doi.org/10.1167/9.10.7

Becker JT (1988) Working memory and secondary memory deficits in Alzheimer's disease. J Clin Exp Neuropsychol 10:739–753. https://doi.org/10.1080/01688638808402811

Belleville S, Peretz I, Malenfant D (1996) Examination of the working memory components in normal aging and in dementia of the Alzheimer type. Neuropsychologia 34:195–207

Blackwell AD, Sahakian BJ, Vesey R, Semple JM, Robbins TW, Hodges JR (2004) Detecting dementia: novel neuropsychological markers of preclinical Alzheimer's disease. Dement Geriatr Cogn Disord 17:42–48. https://doi.org/10.1159/000074081

Boeve BF (2010) REM sleep behavior disorder: updated review of the core features, the REM sleep behavior disorder-neurodegenerative disease association, evolving concepts, controversies, and future directions. Ann N Y Acad Sci 1184:15–54. https://doi.org/10.1111/j.1749-6632.2009.05115.x

Boot BP, Boeve BF, Roberts RO, Ferman TJ, Geda YE, Pankratz VS, Ivnik RJ, Smith GE, McDade E, Christianson TJH, Knopman DS, Tangalos EG, Silber MH, Petersen RC (2012) Probable rapid eye movement sleep behavior disorder increases risk for mild cognitive impairment and Parkinson disease: a population-based study. Ann Neurol 71:49–56. https://doi.org/10.1002/ana.22655

Braak H, Braak E (1998) Evolution of neuronal changes in the course of Alzheimer's disease. J Neural Transm Suppl 53:127–140

Brier MR, Gordon B, Friedrichsen K, McCarthy J, Stern A, Christensen J, Owen C, Aldea P, Su Y, Hassenstab J, Cairns NJ, Holtzman DM, Fagan AM, Morris JC, Benzinger TLS, Ances BM (2016) Tau and Aβ imaging, CSF measures, and cognition in Alzheimer's disease. Sci Transl Med 8:338ra66. https://doi.org/10.1126/scitranslmed.aaf2362

Carlesimo GA, Fadda L, Lorusso S, Caltagirone C (1994) Verbal and spatial memory spans in Alzheimer's and multi-infarct dementia. Acta Neurol Scand 89:132–138

Cashdollar N, Duncan JS, Duzel E (2011) Challenging the classical distinction between long-term and short-term memory: reconsidering the role of the hippocampus. Future Neurol 6:351–362. https://doi.org/10.2217/fnl.11.12

Clark LN, Kartsaklis LA, Wolf Gilbert R, Dorado B, Ross BM, Kisselev S, Verbitsky M, Mejia-Santana H, Cote LJ, Andrews H, Vonsattel J-P, Fahn S, Mayeux R, Honig LS, Marder K (2009) Association of glucocerebrosidase mutations with dementia with lewy bodies. Arch Neurol 66:578–583. https://doi.org/10.1001/archneurol.2009.54

Cools R, Miyakawa A, Sheridan M, D'Esposito M (2010) Enhanced frontal function in Parkinson's disease. Brain J Neurol 133:225–233. https://doi.org/10.1093/brain/awp301

Dalrymple-Alford JC, Kalders AS, Jones RD, Watson RW (1994) A central executive deficit in patients with Parkinson's disease. J Neurol Neurosurg Psychiatry 57:360–367

Davachi L (2006) Item, context and relational episodic encoding in humans. Curr Opin Neurobiol 16:693–700. https://doi.org/10.1016/j.conb.2006.10.012

Della Sala S, Parra MA, Fabi K, Luzzi S, Abrahams S (2012) Short-term memory binding is impaired in AD but not in non-AD dementias. Neuropsychologia 50:833–840. https://doi.org/10.1016/j.neuropsychologia.2012.01.018

Dubois B, Ruberg M, Javoy-Agid F, Ploska A, Agid Y (1983) A subcortico-cortical cholinergic system is affected in Parkinson's disease. Brain Res 288:213–218

Dujardin K, Degreef JF, Rogelet P, Defebvre L, Destee A (1999) Impairment of the supervisory attentional system in early untreated patients with Parkinson's disease. J Neurol 246:783–788

Eichenbaum H, Yonelinas AP, Ranganath C (2007) The medial temporal lobe and recognition memory. Annu Rev Neurosci 30:123–152. https://doi.org/10.1146/annurev.neuro.30.051606.094328

Emir UE, Tuite PJ, Öz G (2012) Elevated pontine and putamenal GABA levels in mild-moderate Parkinson disease detected by 7 tesla proton MRS. PLoS One 7:e30918

Esfahani-Bayerl N, Finke C, Braun M, Düzel E, Heekeren HR, Holtkamp M, Hasper D, Storm C, Ploner CJ (2016) Visuo-spatial memory deficits following medial temporal lobe damage: a comparison of three patient groups. Neuropsychologia 81:168–179. https://doi.org/10.1016/j.neuropsychologia.2015.12.024

Fallon SJ, Zokaei N, Husain M (2016) Causes and consequences of limitations in visual working memory. Ann N Y Acad Sci 1369:40–54. https://doi.org/10.1111/nyas.12992

Fallon SJ, Mattiesing RM, Muhammed K, Manohar S, Husain M (2017) Fractionating the neurocognitive mechanisms underlying working memory: independent effects of dopamine and Parkinson's disease. Cereb Cortex 27:5727–5738. https://doi.org/10.1093/cercor/bhx242

Ferini-Strambi L, Di Gioia MR, Castronovo V, Oldani A, Zucconi M, Cappa SF (2004) Neuropsychological assessment in idiopathic REM sleep behavior disorder (RBD): does the idiopathic form of RBD really exist? Neurology 62:41–45

Fox NC, Warrington EK, Seiffer AL, Agnew SK, Rossor MN (1998) Presymptomatic cognitive deficits in individuals at risk of familial Alzheimer's disease. A longitudinal prospective study. Brain J Neurol 121(Pt 9):1631–1639

Gorgoraptis N, Catalao RFG, Bays PM, Husain M (2011) Dynamic updating of working memory resources for visual objects. J Neurosci 31:8502–8511. https://doi.org/10.1523/JNEUROSCI.0208-11.2011

Gratwicke J, Jahanshahi M, Foltynie T (2015) Parkinson's disease dementia: a neural networks perspective. Brain J Neurol 138:1454–1476. https://doi.org/10.1093/brain/awv104

Gruszka A, Bor D, Barker RR, Necka E, Owen AM (2016) The role of executive processes in working memory deficits in Parkinson's disease. Pol Psychol Bull 47:123–130. https://doi.org/10.1515/ppb-2016-0013

Hannula DE, Tranel D, Cohen NJ (2006) The long and the short of it: relational memory impairments in amnesia, even at short lags. J Neurosci 26:8352–8359

Hartley T, Bird CM, Chan D, Cipolotti L, Husain M, Vargha-Khadem F, Burgess N (2007) The hippocampus is required for short-term topographical memory in humans. Hippocampus 17:34–48. https://doi.org/10.1002/hipo.20240

Hasselmo ME, Sarter M (2011) Modes and models of forebrain cholinergic neuromodulation of cognition. Neuropsychopharmacology 36:52–73. https://doi.org/10.1038/npp.2010.104

Heise V, Filippini N, Trachtenberg AJ, Suri S, Ebmeier KP, Mackay CE (2014) Apolipoprotein E genotype, gender and age modulate connectivity of the hippocampus in healthy adults. NeuroImage 98:23–30. https://doi.org/10.1016/j.neuroimage.2014.04.081

Hughes LE, Altena E, Barker RA, Rowe JB (2013) Perseveration and choice in Parkinson's disease: the impact of progressive frontostriatal dysfunction on action decisions. Cereb Cortex 23:1572–1581

Hulme C, Lee G, Brown GD (1993) Short-term memory impairments in Alzheimer-type dementia: evidence for separable impairments of articulatory rehearsal and long-term memory. Neuropsychologia 31:161–172

Iranzo A, Molinuevo JL, Santamaría J, Serradell M, Martí MJ, Valldeoriola F, Tolosa E (2006) Rapid-eye-movement sleep behaviour disorder as an early marker for a neurodegenerative disorder: a descriptive study. Lancet Neurol 5:572–577. https://doi.org/10.1016/S1474-4422(06)70476-8

Jeneson A, Mauldin KN, Squire LR (2010) Intact working memory for relational information after medial temporal lobe damage. J Neurosci 30:13624–13629. https://doi.org/10.1523/JNEUROSCI.2895-10.2010

Joseph S, Teki S, Kumar S, Husain M, Griffiths TD (2016) Resource allocation models of auditory working memory. Brain Res 1640:183–192. https://doi.org/10.1016/j.brainres.2016.01.044

Kehagia AA, Barker RA, Robbins TW (2010) Neuropsychological and clinical heterogeneity of cognitive impairment and dementia in patients with Parkinson's disease. Lancet Neurol 9:1200–1213. https://doi.org/10.1016/S1474-4422(10)70212-X

Kish SJ, Tong J, Hornykiewicz O, Rajput A, Chang L-J, Guttman M, Furukawa Y (2008) Preferential loss of serotonin markers in caudate versus putamen in Parkinson's disease. Brain 131:120–131

Koen JD, Borders AA, Petzold MT, Yonelinas AP (2016) Visual short-term memory for high resolution associations is impaired in patients with medial temporal lobe damage. Hippocampus 27:184–193. https://doi.org/10.1002/hipo.22682

Konkel A, Cohen NJ (2009) Relational memory and the hippocampus: representations and methods. Front Neurosci 3:166–174. https://doi.org/10.3389/neuro.01.023.2009

Kroener S, Chandler LJ, Phillips PEM, Seamans JK (2009) Dopamine modulates persistent synaptic activity and enhances the signal-to-noise ratio in the prefrontal cortex. PLoS One 4: e6507. https://doi.org/10.1371/journal.pone.0006507

Lange KW, Robbins TW, Marsden CD, James M, Owen AM, Paul GM (1992) L-dopa withdrawal in Parkinson's disease selectively impairs cognitive performance in tests sensitive to frontal lobe dysfunction. Psychopharmacology 107:394–404

Lau KM, Parikh M, Harvey DJ, Huang C-J, Farias ST (2015) Early cognitively based functional limitations predict loss of independence in instrumental activities of daily living in older adults. J Int Neuropsychol Soc 21:688–698. https://doi.org/10.1017/S1355617715000818

Lawton M, Baig F, Rolinski M, Ruffman C, Nithi K, May MT, Ben-Shlomo Y, Hu MTM (2015) Parkinson's disease subtypes in the Oxford Parkinson disease Centre (OPDC) discovery cohort. J Park Dis 5:269–279. https://doi.org/10.3233/JPD-140523

Lewis SJG, Foltynie T, Blackwell AD, Robbins TW, Owen AM, Barker RA (2005) Heterogeneity of Parkinson's disease in the early clinical stages using a data driven approach. J Neurol Neurosurg Psychiatry 76:343–348

Liang Y, Pertzov Y, Nicholas JM, Henley SMD, Crutch S, Woodward F, Leung K, Fox NC, Husain M (2016) Visual short-term memory binding deficit in familial Alzheimer's disease. Cortex 78:150–164. https://doi.org/10.1016/j.cortex.2016.01.015

Libby LA, Hannula DE, Ranganath C (2014) Medial temporal lobe coding of item and spatial information during relational binding in working memory. J Neurosci 34:14233–14242. https://doi.org/10.1523/JNEUROSCI.0655-14.2014

Linn RT, Wolf PA, Bachman DL, Knoefel JE, Cobb JL, Belanger AJ, Kaplan EF, D'Agostino RB (1995) The "preclinical phase" of probable Alzheimer's disease. A 13-year prospective study of the Framingham cohort. Arch Neurol 52:485–490

Luck V (1997) The capacity of visual working memory for features and conjunctions. Nature 390:279–281. https://doi.org/10.1038/36846

Ma WJ, Husain M, Bays PM (2014) Changing concepts of working memory. Nat Neurosci 17:347–356. https://doi.org/10.1038/nn.3655

Martin A, Brouwers P, Cox C, Fedio P (1985) On the nature of the verbal memory deficit in Alzheimer's disease. Brain Lang 25:323–341

Miller E (1973) Short- and long-term memory in patients with presenile dementia (Alzheimer's disease). Psychol Med 3:221–224

Muslimovic D, Post B, Speelman JD, Schmand B (2005) Cognitive profile of patients with newly diagnosed Parkinson disease. Neurology 65:1239–1245. https://doi.org/10.1212/01.wnl.0000180516.69442.95

Myers RH, Schaefer EJ, Wilson PW, D'Agostino R, Ordovas JM, Espino A, Au R, White RF, Knoefel JE, Cobb JL, McNulty KA, Beiser A, Wolf PA (1996) Apolipoprotein E epsilon4 association with dementia in a population-based study: the Framingham study. Neurology 46:673–677

Nadel L, Peterson MA (2013) The hippocampus: part of an interactive posterior representational system spanning perceptual and memorial systems. J Exp Psychol Gen 142:1242–1254. https://doi.org/10.1037/a0033690

Neumann J, Bras J, Deas E, O'Sullivan SS, Parkkinen L, Lachmann RH, Li A, Holton J, Guerreiro R, Paudel R, Segarane B, Singleton A, Lees A, Hardy J, Houlden H, Revesz T, Wood NW (2009) Glucocerebrosidase mutations in clinical and pathologically proven Parkinson's disease. Brain 132:1783–1794. https://doi.org/10.1093/brain/awp044

Olson IR, Moore KS, Stark M, Chatterjee A (2006) Visual working memory is impaired when the medial temporal lobe is damaged. J Cogn Neurosci 18:1087–1097. https://doi.org/10.1162/jocn.2006.18.7.1087

Owen AM, Beksinska M, James M, Leigh PN, Summers BA, Marsden CD, Quinn NP, Sahakian BJ, Robbins TW (1993) Visuospatial memory deficits at different stages of Parkinson's disease. Neuropsychologia 31:627–644

Owen AM, Iddon JL, Hodges JR, Summers BA, Robbins TW (1997) Spatial and non-spatial working memory at different stages of Parkinson's disease. Neuropsychologia 35:519–532

Parra MA, Abrahams S, Fabi K, Logie R, Luzzi S, Della Sala S (2009) Short-term memory binding deficits in Alzheimer's disease. Brain J Neurol 132:1057–1066. https://doi.org/10.1093/brain/awp036

Parra MA, Abrahams S, Logie RH, Méndez LG, Lopera F, Della Sala S (2010) Visual short-term memory binding deficits in familial Alzheimer's disease. Brain J Neurol 133:2702–2713. https://doi.org/10.1093/brain/awq148

Parra MA, Sala SD, Abrahams S, Logie RH, Méndez LG, Lopera F (2011) Specific deficit of colour-colour short-term memory binding in sporadic and familial Alzheimer's disease. Neuropsychologia 49:1943–1952. https://doi.org/10.1016/j.neuropsychologia.2011.03.022

Pastores GM, Hughes DA (1993) Gaucher disease. In: Pagon RA, Adam MP, Bird TD, Dolan CR, Fong C-T, Stephens K (eds) GeneReviews™. University of Washington, Seattle

Perry RJ, Hodges JR (2000) Fate of patients with questionable (very mild) Alzheimer's disease: longitudinal profiles of individual subjects' decline. Dement Geriatr Cogn Disord 11:342–349. https://doi.org/10.1159/000017264

Pertzov Y, Miller TD, Gorgoraptis N, Caine D, Schott JM, Butler C, Husain M (2013) Binding deficits in memory following medial temporal lobe damage in patients with voltage-gated potassium channel complex antibody-associated limbic encephalitis. Brain J Neurol 136:2474–2485. https://doi.org/10.1093/brain/awt129

Poewe W, Berger W, Benke TH, Schelosky L (1991) High-speed memory scanning in Parkinson's disease: adverse effects of levodopa. Ann Neurol 29:670–673

Postle BR, Jonides J, Smith EE, Corkin S, Growdon JH (1997) Spatial, but not object, delayed response is impaired in early Parkinson's disease. Neuropsychology 11:171–179

Postuma RB, Gagnon JF, Vendette M, Montplaisir JY (2009) Markers of neurodegeneration in idiopathic rapid eye movement sleep behaviour disorder and Parkinson's disease. Brain J Neurol 132:3298–3307. https://doi.org/10.1093/brain/awp244

Ridha BH, Barnes J, Bartlett JW, Godbolt A, Pepple T, Rossor MN, Fox NC (2006) Tracking atrophy progression in familial Alzheimer's disease: a serial MRI study. Lancet Neurol 5:828–834. https://doi.org/10.1016/S1474-4422(06)70550-6

Rolinski M, Zokaei N, Baig F, Giehl K, Quinnell T, Zaiwalla Z, Mackay CE, Husain M, Hu MTM (2015) Visual short-term memory deficits in REM sleep behaviour disorder mirror those in Parkinson's disease. Brain 139:47–53. https://doi.org/10.1093/brain/awv334

Sando SB, Melquist S, Cannon A, Hutton ML, Sletvold O, Saltvedt I, White LR, Lydersen S, Aasly JO (2008) APOE epsilon 4 lowers age at onset and is a high risk factor for Alzheimer's disease; a case control study from Central Norway. BMC Neurol 8:9. https://doi.org/10.1186/1471-2377-8-9

Sardi SP, Clarke J, Kinnecom C, Tamsett TJ, Li L, Stanek LM, Passini MA, Grabowski GA, Schlossmacher MG, Sidman RL, Cheng SH, Shihabuddin LS (2011) CNS expression of glucocerebrosidase corrects α-synuclein pathology and memory in a mouse model of Gaucher-related synucleinopathy. Proc Natl Acad Sci 108:12101–12106. https://doi.org/10.1073/pnas.1108197108

Sato C, Morgan A, Lang AE, Salehi-Rad S, Kawarai T, Meng Y, Ray PN, Farrer LA, St George-Hyslop P, Rogaeva E (2005) Analysis of the glucocerebrosidase gene in Parkinson's disease. Mov Disord 20:367–370. https://doi.org/10.1002/mds.20319

Saunders AM, Strittmatter WJ, Schmechel D, George-Hyslop PHS, Pericak-Vance MA, Joo SH, Rosi BL, Gusella JF, Crapper-MacLachlan DR, Alberts MJ, Hulette C, Crain B, Goldgaber D, Roses AD (1993) Association of apolipoprotein E allele ϵ4 with late-onset familial and sporadic Alzheimer's disease. Neurology 43:1467–1467. https://doi.org/10.1212/WNL.43.8.1467

Savica R, Rocca WA, Ahlskog J (2010) When does Parkinson disease start? Arch Neurol 67:798–801. https://doi.org/10.1001/archneurol.2010.135

Sawaguchi T, Goldman-Rakic P (1991) D1 dopamine receptors in prefrontal cortex: involvement in working memory. Science 251:947–950. https://doi.org/10.1126/science.1825731

Schenck CH, Boeve BF, Mahowald MW (2013) Delayed emergence of a parkinsonian disorder or dementia in 81% of older men initially diagnosed with idiopathic rapid eye movement sleep behavior disorder: a 16-year update on a previously reported series. Sleep Med 14:744–748. https://doi.org/10.1016/j.sleep.2012.10.009

Selikhova M, Williams DR, Kempster PA, Holton JL, Revesz T, Lees AJ (2009) A clinico-pathological study of subtypes in Parkinson's disease. Brain J Neurol 132:2947–2957. https://doi.org/10.1093/brain/awp234

Sidransky E, Nalls MA, Aasly JO, Aharon-Peretz J, Annesi G, Barbosa ER, Bar-Shira A, Berg D, Bras J, Brice A, Chen C-M, Clark LN, Condroyer C, de Marco EV, Dürr A, Eblan MJ, Fahn S, Farrer MJ, Fung H-C, Gan-Or Z, Gasser T, Gershoni-Baruch R, Giladi N, Griffith A, Gurevich T, Januario C, Kropp P, Lang AE, Lee-Chen G-J, Lesage S, Marder K, Mata IF, Mirelman A, Mitsui J, Mizuta I, Nicoletti G, Oliveira C, Ottman R, Orr-Urtreger A, Pereira LV, Quattrone A, Rogaeva E, Rolfs A, Rosenbaum H, Rozenberg R, Samii A, Samaddar T, Schulte C, Sharma M, Singleton A, Spitz M, Tan E-K, Tayebi N, Toda T, Troiano AR, Tsuji S, Wittstock M, Wolfsberg TG, Wu Y-R, Zabetian CP, Zhao Y, Ziegler SG (2009) Multicenter analysis of glucocerebrosidase mutations in Parkinson's disease. N Engl J Med 361:1651–1661. https://doi.org/10.1056/NEJMoa0901281

Spinnler H, Sala SD, Bandera R, Baddeley A (1988) Dementia, ageing, and the structure of human memory. Cogn Neuropsychol 5:193–211. https://doi.org/10.1080/02643298808252933

Suri S, Heise V, Trachtenberg AJ, Mackay CE (2013) The forgotten APOE allele: a review of the evidence and suggested mechanisms for the protective effect of APOE ε2. Neurosci Biobehav Rev 37:2878–2886. https://doi.org/10.1016/j.neubiorev.2013.10.010

Uitvlugt MG, Pleskac TJ, Ravizza SM (2016) The nature of working memory gating in Parkinson's disease: a multi-domain signal detection examination. Cogn Affect Behav Neurosci 16:289–301

van Geldorp B, Bouman Z, Hendriks MPH, Kessels RPC (2014) Different types of working memory binding in epilepsy patients with unilateral anterior temporal lobectomy. Brain Cogn 85:231–238. https://doi.org/10.1016/j.bandc.2013.12.009

Verbaan D, Marinus J, Visser M, van Rooden SM, Stiggelbout AM, Middelkoop HAM, van Hilten JJ (2007) Cognitive impairment in Parkinson's disease. J Neurol Neurosurg Psychiatry 78:1182–1187. https://doi.org/10.1136/jnnp.2006.112367

Watrous AJ, Tandon N, Conner CR, Pieters T, Ekstrom AD (2013) Frequency-specific network connectivity increases underlie accurate spatiotemporal memory retrieval. Nat Neurosci 16:349–356. https://doi.org/10.1038/nn.3315

Williams GC (1957) Pleiotropy, natural selection, and the evolution of senescence. Evolution 11:398–411. https://doi.org/10.2307/2406060

Winterer G, Weinberger DR (2004) Genes, dopamine and cortical signal-to-noise ratio in schizophrenia. Trends Neurosci 27:683–690. https://doi.org/10.1016/j.tins.2004.08.002

Wolk DA, Dickerson BC, Alzheimer's Disease Neuroimaging Initiative (2010) Apolipoprotein E (APOE) genotype has dissociable effects on memory and attentional-executive network function in Alzheimer's disease. Proc Natl Acad Sci U S A 107:10256–10261. https://doi.org/10.1073/pnas.1001412107

Wong K, Sidransky E, Verma A, Mixon T, Sandberg GD, Wakefield LK, Morrison A, Lwin A, Colegial C, Allman JM, Schiffmann R (2004) Neuropathology provides clues to the pathophysiology of Gaucher disease. Mol Genet Metab 82:192–207. https://doi.org/10.1016/j.ymgme.2004.04.011

Yau W-YW, Tudorascu DL, McDade EM, Ikonomovic S, James JA, Minhas D, Mowrey W, Sheu LK, Snitz BE, Weissfeld L, Gianaros PJ, Aizenstein HJ, Price JC, Mathis CA, Lopez OL, Klunk WE (2015) Longitudinal assessment of neuroimaging and clinical markers in autosomal dominant Alzheimer's disease: a prospective cohort study. Lancet Neurol 14:804–813. https://doi.org/10.1016/S1474-4422(15)00135-0

Yonelinas AP (2013) The hippocampus supports high-resolution binding in the service of perception, working memory and long-term memory. Behav Brain Res 254:34–44. https://doi.org/10.1016/j.bbr.2013.05.030

Zarow C, Lyness SA, Mortimer JA, Chui HC (2003) Neuronal loss is greater in the locus coeruleus than nucleus basalis and substantia nigra in Alzheimer and Parkinson diseases. Arch Neurol 60:337–341

Zhang W, Luck SJ (2008) Discrete fixed-resolution representations in visual working memory. Nature 453:233–235

Zimmermann JF, Butler CR (2018) Accelerated long-term forgetting in asymptomatic APOE ε4 carriers. Lancet Neurol 17:394–395. https://doi.org/10.1016/S1474-4422(18)30078-4

Zokaei N, Gorgoraptis N, Bahrami B, Bays PM, Husain M (2011) Precision of working memory for visual motion sequences and transparent motion surfaces. J Vis 11:2. https://doi.org/10.1167/11.14.2

Zokaei N, Burnett Heyes S, Gorgoraptis N, Budhdeo S, Husain M (2014a) Working memory recall precision is a more sensitive index than span. J Neuropsychol 9:319–329. https://doi.org/10.1111/jnp.12052

Zokaei N, McNeill A, Proukakis C, Beavan M, Jarman P, Korlipara P, Hughes D, Mehta A, Hu MTM, Schapira AHV, Husain M (2014b) Visual short-term memory deficits associated with GBA mutation and Parkinson's disease. Brain J Neurol 137:2303–2311. https://doi.org/10.1093/brain/awu143

Zokaei N, Giehl K, Sillence A, Neville MJ, Karpe F, Nobre AC, Husain M (2017) Sex and APOE: a memory advantage in male APOE ε4 carriers in midlife. Cortex 88:98–105. https://doi.org/10.1016/j.cortex.2016.12.016

Zokaei N, Nour MM, Sillence A, Drew D, Adcock J, Stacey R, Voets N, Sen A, Husain M (2018) Binding deficits in visual short-term memory in patients with temporal lobe lobectomy. Hippocampus 29:63–67. https://doi.org/10.1002/hipo.22998

Zokaei N, Čepukaitytė G, Board AG, Mackay CE, Husain M, Nobre AC (2019a) Dissociable effects of the apolipoprotein-E (APOE) gene on short- and long-term memories. Neurobiol Aging 73:115–122. https://doi.org/10.1016/j.neurobiolaging.2018.09.017

Zokaei N, Nour MN, Sillence A, Drew D, Adcock J, Stacey R, Voets N, Sen A, Husain M (2019b) Binding deficits in visual short-term memory in patients with temporal lobe lobectomy. Hippocampus 29:63–67

Mechanisms Underlying Visuospatial Working Memory Impairments in Schizophrenia

Megan Ichinose and Sohee Park

Contents

Abstract Working memory deficits are observed in the vast majority of individuals diagnosed with schizophrenia and those at risk for the disorder. Working memory impairments are present during the prodromal stage and persist throughout the course of schizophrenia. Given the importance of cognition in functional outcome, working memory deficits are an important therapeutic target for schizophrenia. This chapter examines mechanisms underlying working memory deficits in schizophrenia, focusing on the roles of perception and attention in the encoding process. Lastly, we present a comprehensive discussion of neural oscillation and internal noise in the context of the etiology of working memory deficits in schizophrenia and introduce noninvasive treatment strategies that could improve encoding processes.

Keywords Encoding · Neural oscillation · Psychosis · Visual perception · Working memory

---

M. Ichinose and S. Park (✉)
Department of Psychology, Vanderbilt University, Nashville, TN, USA
e-mail: sohee.park@vanderbilt.edu

Curr Topics Behav Neurosci (2019) 41: 345–368
DOI 10.1007/7854_2019_99
Published Online: 13 August 2019

1 Introduction

Working memory deficits are a core cognitive feature of schizophrenia (Silver et al. 2003), observed in the majority of individuals diagnosed with schizophrenia and a significant proportion of those at risk for the disorder (see Park and Gooding 2014). Importantly, persistent cognitive deficits such as working memory impairments are better candidates for an endophenotypic marker of schizophrenia than the fluctuating surface manifestations of psychotic symptoms and thus can be more precisely targeted for intervention (Park and Gooding 2014; Saperstein et al. 2005). Given the lifelong personal and societal burden of this condition (Salomon et al. 2015), it is of utmost importance to develop effective treatments for those with schizophrenia that target these cognitive deficits. However, the etiology of working memory deficits in schizophrenia is not fully understood. The main goals of this chapter are to present a selective review of visuospatial working memory deficits in schizophrenia spectrum and to identify mechanisms underlying these deficits, with a specific focus on visual encoding difficulties. Ultimately, we hope this synthesis will facilitate discussions concerning future treatment and intervention strategies.

2 Definitions and a Brief Summary of Visuospatial Working Memory Deficits in Schizophrenia

Working memory is a limited-capacity, active short-term memory system that maintains information to guide and control behavior (Baddeley 2007). This capacity to encode, generate, and maintain mental representations allows us to guide our behavior adaptively by generating mental representations or predictions to plan for future actions. According to Goldman-Rakic (1994), the etiology of schizophrenia stems from a problem of internal representations, such that "a defect in working memory—the ability to guide behavior by representations—may be the fundamental impairment leading to schizophrenic thought disorder." In this framework, cognitive deficits of executive control and related problems of increased distractibility, perseveration, and failure to inhibit irrelevant responses reflect an inability to utilize working memory to adaptively guide behavior. These working memory impairments are thought to arise from abnormal regulation of multiple prefrontal-parietal circuits that are recruited to hold information "on line" while updating past and current information moment to moment.

There is robust evidence to indicate that stable and reliable working memory deficits are present in individuals with schizophrenia across diverse paradigms, methods, and techniques (e.g., Lee and Park 2005; Park and Gooding 2014). Moreover, those at risk for schizophrenia such as unaffected first-degree relatives (Conklin et al. 2005; Glahn et al. 2003; Myles-Worsley and Park 2002; Park et al. 1995; Zhang et al. 2016) and healthy individuals with elevated psychosis proneness (Park and McTigue 1997; Tallent and Gooding 1999; Xie et al. 2018) exhibit similar

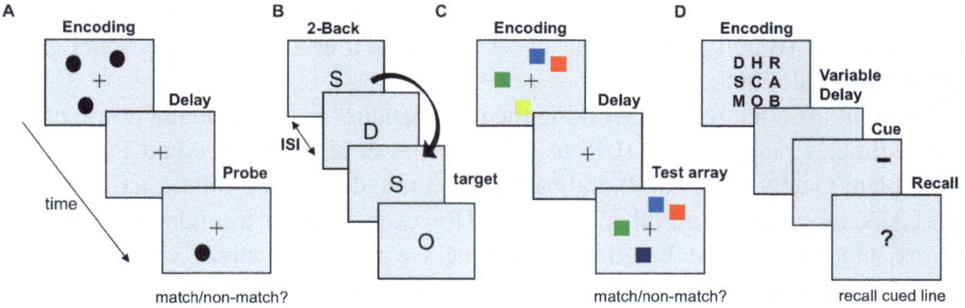

Fig. 1 Common experimental working memory task paradigms. (**a**) Delayed response. (**b**) N-back. (**c**) Change detection. (**d**) Sperling partial report

working memory impairment. Overall, in individuals with schizophrenia, working memory deficits appear to be permanent and stable even when clinical symptoms are in remission (see Park et al. 1999).

A wide range of experimental paradigms are currently used to assess working memory (See Fig. 1). Depending on the task parameters, different subcomponents of working memory and associated cognitive and perceptual functions are recruited. Online maintenance of information can be tested using delayed response or delayed discrimination tasks (e.g., Badcock et al. 2008; Cannon et al. 2005; Gooding and Tallent 2003; Haenschel et al. 2007; Haenschel et al. 2009; Park and Holzman 1992). Maintenance and manipulation of information can be assessed using N-back tasks (e.g., Barch et al. 2002; Callicott et al. 2003), backward digit span (Glahn et al. 2006; Pirkola et al. 2005), and letter-number sequencing tasks (e.g., Gold et al. 1997). Change detection tasks (Gold et al. 2003; Mayer et al. 2012) and the Sternberg paradigm (Manoach et al. 1999) among others have been used to estimate the capacity of working memory.

Neuroimaging studies of schizophrenia patients performing working memory tasks have typically demonstrated task-related hypo-frontality (Barch et al. 2002; Callicott et al. 2003). However, there are also reports of working memory-associated hyper-frontality (Manoach et al. 2000) and anomalous frontal functional asymmetry (Kim et al. 2010; Lee et al. 2008; Walter et al. 2003). These results suggest that the task demand, manipulation of task parameters, and individual differences must always be considered when comparing the neural activation patterns in individuals with schizophrenia with those of healthy participants (Callicott et al. 2003; Manoach 2003). Recruitment of a functional network may not necessarily correlate with the behavioral performance. Neural activation also depends on the degree of expertise; inactivity of a network sometimes reflects a certain level of automaticity for that particular task and therefore mastery rather than failure. Given the many possible routes to working memory errors and nonlinear nature of the brain-behavior relationship, it is important to analyze the specific mechanisms underlying memory errors in neuroimaging studies. Therefore, group comparisons must be interpreted in specific behavioral contexts in which different types of errors arise (Lee et al. 2008). A network approach to understanding working memory deficits suggests that a broad network of frontal, parietal, temporal, and subcortical systems are involved in schizophrenia (Barch and Csernansky 2007; Driesen et al. 2008; Haenschel et al. 2009; Lee et al. 2008).

Overall, meta-analyses of working memory deficits in schizophrenia reveal effect size estimates ranging from 0.45 to 1.29 (Forbes et al. 2009; Lee and Park 2005; Mesholam-Gately et al. 2009), indicating that these deficits are robust across different tasks, modalities, and other variables. However, in order to understand mechanisms underlying these broad impairments, we need to analyze components of working memory systems in disease and health. When components of working memory deficits are examined, encoding and maintenance problems seem specific to the schizophrenia spectrum and do not necessarily extend to bipolar disorder (Mayer and Park 2012). Individuals with schizophrenia seem to be affected by poor encoding processes that may originate from reduced fidelity or quality of sensory input ("bottom-up" processes) as well as those that require more "top-down" control that involve attention. Difficulties with encoding could lead to either having limited memory or "false memory" of what has just occurred. In this chapter, we focus on the encoding components of working memory because without accurate encoding, it is impossible to achieve accurate memory retrieval.

3 Analyzing Components of Working Memory: An Emphasis on Encoding

A growing body of research indicates that deficient encoding and early maintenance processes are largely culpable for working memory deficits in schizophrenia (Haenschel et al. 2007; Hahn et al. 2010; Hartman et al. 2003; Lee and Park 2005; Mayer and Park 2012; Tek et al. 2002). An emphasis on encoding as the major deficient stage of working memory processing in schizophrenia is driven by a collection of findings from behavioral and neural studies aimed at differentiating failures during the encoding of target stimuli versus those taking place once the stimuli are already in working memory and must be maintained or retrieved (Haenschel et al. 2007). However, the cause of this encoding deficit remains unclear, particularly because encoding itself is not a unitary process. In the broadest sense, encoding describes the mechanism(s) by which sensory input is transferred into working memory (Bays et al. 2011). Thus, not only does proper encoding involve intact sensory and perceptual processing, it also entails intact top-down attentional mechanisms that direct which sensory information is to be maintained in working memory and those involved in the inhibition of distractors.

4 Visual Processing Deficits

Because any subsequent stage of processing in working memory beyond encoding is constrained by the quality of sensory information available during encoding, deficient or degraded sensory and perceptual information will inevitably lead to a reduction in memory accuracy or fidelity. Delineating the role of early sensory and perceptual processing abnormalities in higher-order cognitive abilities in schizophrenia has become a popular area of research (e.g., Butler et al. 2005, 2006, 2009; Dias et al. 2011; Javitt et al. 1999), particularly as the complex interactions between early sensory regions and higher cognitive regions continue to be flushed out in the basic cognitive and neuroscience literatures. Regarding visual perception, individuals with schizophrenia exhibit visual processing deficits at nearly every level of visual system processing, from the sensory organs to primary sensory cortices and associated areas (for reviews, see Silverstein and Rosen 2015; Gagné et al. 2015; Lavoie et al. 2014). Even the earliest stages of retinal processing seem to be affected, including abnormal photoreceptor functioning (Balogh et al. 2008) and neuroanatomical differences of the retina (Chu et al. 2012; Lee et al. 2013), such as decreased thickness of the retinal nerve fiber layer and reduced macular volume (Lee et al. 2013). Further neuroanatomical abnormalities are observed as visual signals are relayed from the retina to primary visual cortex (V1) via the lateral geniculate nucleus. For instance, some research finds that individuals with schizophrenia exhibit reduced white matter integrity of optic radiations – tracts connecting the lateral geniculate nucleus (LGN) to areas of visual cortex (Butler et al. 2006) – while other studies have found reduced neuron number and volume of the pulvinar (Byne et al. 2007), a thalamic hub which receives inputs from LGN and is thought to play a functional, integrative role in visual attention (Fischer and Whitney 2012).

Individuals with schizophrenia also exhibit differences at the level of processing in visual cortex. For instance, there is evidence that individuals with schizophrenia have reduced cell number and volume in primary visual cortex (V1), suggesting an overall smaller cortical area for primary visual perception (Dorph-Petersen et al. 2007; Narr et al. 2005). Early visual processing abnormalities in schizophrenia are further captured with a number of behavioral measures thought to involve low-level visual areas, including reduced contrast sensitivity (Skottun and Skoyles 2007), orientation discrimination (Rokem et al. 2011), motion and velocity perception (Chen 2011; Kim et al. 2006), lack of preference for global versus local visual processing (Brittain et al. 2010; Coleman et al. 2009), spatial integration (Silverstein and Keane 2011), and reduced center-surround suppression (Serrano-Pedraza et al. 2014; Tadin et al. 2006; Tibber et al. 2013). Surround suppression is a property of visual processing in which neural activity is dampened in response to adjacent stimuli and/or contextual stimuli (e.g., perceiving a target is of higher contrast when presented in isolation versus within a high-contrast surround). Accordingly, reduced center-surround suppression in individuals with schizophrenia has been linked to impaired inhibitory connections in early visual cortex (Serrano-Pedraza et al. 2014). This interpretation is further supported by findings that show an approximately 10% reduction in the concentration of GABA, the primary neurotransmitter involved in cortical inhibition, in the visual cortex of individuals with schizophrenia (Yoon et al. 2010).

Neuroimaging studies have also captured abnormalities in visual processes in schizophrenia. One of the most consistent findings is that of reduced steady-state visual evoked potentials (ssVEPs), electrophysiological responses entrained to the frequency, and phase of a periodic visual stimulus. These ssVEPS are thought to be generated primarily by occipital cortex, with regional variation related to the frequency at which the stimulus is presented (Di Russo et al. 2007). Additionally, individuals with schizophrenia exhibit differences in the generation of event-related potentials (ERPs) thought to reflect early visual processing, such as reduced P1 (~90–120 ms) and N1 (~170 ms) amplitudes across a range of visual paradigms (Javitt 2009). Such P1 reductions do not seem to correlate with age, illness duration, or medication (Yeap et al. 2008) and have also been captured in first-degree relatives (Yeap et al. 2006) and those at higher genetic risk for schizophrenia (Donohoe et al. 2008), pointing to a relatively stable aspect of schizophrenia pathophysiology (See Fig. 2). While functional magnetic resonance imaging (fMRI) studies of visual perception may lack the temporal specificity to capture very early visual processing deficits in schizophrenia, reduced activation in primary and secondary visual cortical regions during basic visual perception tasks is consistent with dysfunction in primary visual areas (Martinez et al. 2008). Such functional deficits may be driven by neuroanatomical differences of these visual regions, as more recent work has found overall reduced cortical thickness of low-level visual areas in individuals with schizophrenia (lateral occipital cortex, retinotopic cortex), which in turn is related to their visual processing abnormalities (Reavis et al. 2017).

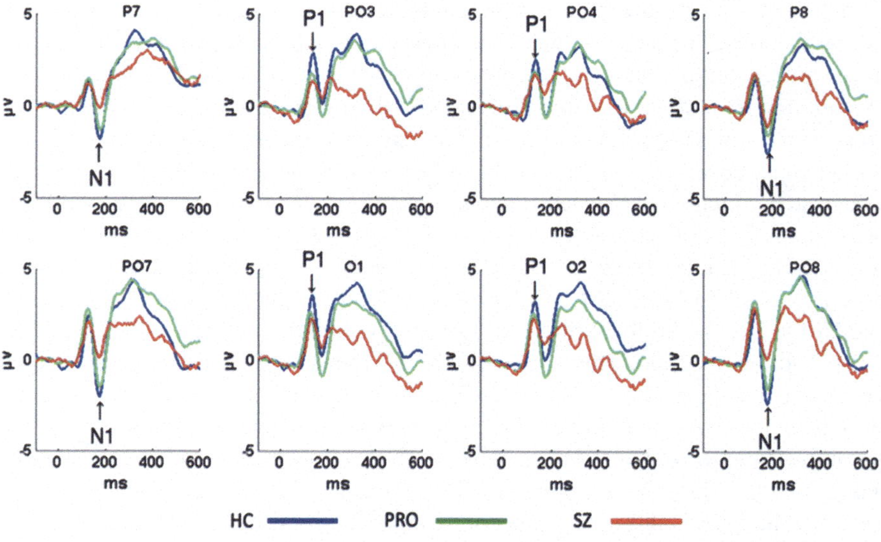

Fig. 2 Indices of reduced early visual activity in schizophrenia. Figure from Starc et al. (2017). *HC* healthy controls, *PRO* prodromal individuals, *SZ* first-episode schizophrenia patients

5 Visual Processing and Working Memory Performance

Several groups have more directly examined how basic sensory and perceptual deficits during encoding contribute to working memory impairment. Such work suggests that early visual processing, as reflected by sensory ERP components, are often deficient during visual encoding in those with schizophrenia and at-risk or vulnerable groups compared to controls (Dias et al. 2011; Haenschel et al. 2007; Koychev et al. 2010). However, the same visual ERPs are not always reduced in schizophrenia during encoding (Bachman et al. 2009; Green et al. 2017; Zhao et al. 2011), and thus further research is necessary to more definitively demonstrate this connection. Some of these mixed findings may be a function of task stimuli and conditions. For instance, using abstract visual stimuli as stimuli, Haenschel et al. (2007) found that adolescents with a recent onset of schizophrenia exhibited reduced P1 amplitudes over central occipital sites during encoding and additionally did not show the same increase in P1 amplitude with an increase in working memory load as the control group. Interestingly, fMRI analyses from this study largely complemented ERP findings of impaired early visual processing during encoding, such that individuals with schizophrenia exhibited less activation in extrastriatal visual areas relative to controls. Additionally, this reduced activation was most pronounced for higher working memory loads, suggesting a failure in recruitment of visual areas during the perception of more complex visual stimuli.

On the other hand, Zhao et al. (2011) found that P1 amplitudes of individuals with schizophrenia were no different from that of controls during visual encoding of simple digits, though P2 and N1 amplitudes over frontal and parietal areas were reduced. Similarly, Green et al. (2017) did not find that early sensory processing played a significant role in successful memory encoding of words in schizophrenia. Event-related potential (ERP) analyses revealed similar N1 amplitudes and latencies across groups (measured at occipital sites) but a reduction in P2 amplitudes (at frontal, central, and parietal midline sites) in the schizophrenia group compared to controls. Though the P2 component is not functionally well characterized due to the range of factors that affect it, it is generally thought to reflect later-stage perceptual processing that is modulated by attention (Luck and Hillyard 1994). Amplitude reductions in sensory components thus may be more frequently found in conditions that tax perceptual processing (e.g., more complex visual stimuli), such that early sensory indices contribute more to working memory performance in schizophrenia.

One important note across these tasks is the degree to which certain early visual ERPs such as P1 and N1 and oscillatory metrics index attentional impairments, as both of these components are also thought to be modulated by attention (e.g., Eimer and Schlaghecken 1998). However, it is unlikely that patient or high-risk groups' amplitude reductions of these early sensory components mainly reflect greater lapses in attention. First, these studies only examined ERPs across correct trials in order to establish adequate task engagement. Additionally, results within studies often revealed select deficits in the schizophrenia group for either P1 or N1 but rarely both, limiting the likelihood that amplitude reductions reflected generally poor attention or engagement throughout the task. Nonetheless, attention is critical throughout encoding, in that it can amplify the activation (or processing) of sensory information that should be remembered and the degree to which it is consolidated as a memory representation. Attention's influence can be observed prior to when the stimulus appears, while it appears, as well as once the stimulus has disappeared, such that its focus or its diversion from target information at any of these stages contributes to the degree to which that information is encoded. Given that impairments in attention are considered a core cognitive deficit in schizophrenia (Nuechterlein and Dawson 1984), it is likely that dysfunctional attentional processing also plays a role in working memory encoding deficits.

6 Selective Attention, Distraction, and Encoding

Deficits in attention may contribute to reduced working memory encoding in several ways: an inability to select the right target stimulus (or stimulus features) to be encoded, overcome distractors during encoding, or some combination of both. Several of these hypotheses have been tested with various working memory paradigms employing attention-directing cues or manipulating target feature that either promote or hinder memory performance. Such studies find that reduced working memory capacity is not typically due to a failure in the ability to selectively attend to the relevant or salient aspects of target stimuli in schizophrenia (Erickson et al. 2015; Gold et al. 2006; Smith et al. 2011), though novelty may serve as an intriguing exception (see Mayer et al. 2013). For example, Gold et al. (2006) utilized cues that preceded and appeared simultaneously with the memory array and cues that were presented in a variety of locations (e.g., centrally, peripherally), both indicating where attention should be directed to best detect a change after the delay. While the schizophrenia group exhibited reduced working memory capacity in all conditions relative to controls, they were still able to utilize all spatial cue types to direct attention toward or away from certain array items. Furthermore, the authors showed that individuals with schizophrenia did not recall un-cued items to a relatively greater extent than controls, suggesting that attention was not diverted to the un-cued items during initial encoding. Individuals with schizophrenia also appear to be able to selectively encode targets from a memory array based on perceptual features of shape (Gold et al. 2006) and color (Erickson et al. 2015; Mayer et al. 2012; Smith et al. 2011), again supporting intact deployment of attention to relevant targets during encoding.

However, selective attention in individuals with schizophrenia *does* seem to break down during encoding when distractors are highly salient, and in these cases, attentional deficits have been shown to relate to reduced working memory capacity in schizophrenia (e.g., Hahn et al. 2010; Mayer et al. 2012). Distractor salience may be determined by the degree to which they share similar features or properties with the target (Mayer et al. 2012) or the manner in which they are presented (e.g., static vs. flickering) (Hahn et al. 2010). Highly salient distractors are generally thought to require significantly more top-down attentional control, while inhibition of less salient or less relevant distractors to ongoing task demands requires less top-down control. Thus, more broadly, heightened distractibility in schizophrenia may be more problematic under task conditions in which attentional control becomes challenging. Indeed, Ducato et al. (2008) found a similar degree of task interference by a moving lateral distractor between patients and controls when task conditions were less demanding but increased distraction in patients when task conditions were more demanding. While visual masking paradigms indicate heightened visual distractor interference in those with schizophrenia, it is important to note that the extent of this interference does not sufficiently account for the working memory capacity reduction in patients, suggesting that poor distractor inhibition may magnify an already existing working memory deficit (Erickson et al. 2014). Indeed, visual working memory deficits in schizophrenia are well documented in the absence of any task distractors, suggesting that overt distraction cannot fully explain encoding impairments and reduced working memory capacity in schizophrenia.

7 Imprecision of Working Memory Representations

Early visual processing dysfunction in schizophrenia may also contribute to working memory deficit by prolonging the overall encoding process (Hartman et al. 2003). In neural network models of working memory, encoding of memory representations has been characterized as strong, recurrent excitation within cortical networks that sustain activity corresponding to specific target stimulus features (Bays 2015; Compte et al. 2000; Murray et al. 2014). The efficiency with which this specific pattern of recurrent excitation is established would likely depend on the ease with which perceptual representations were formed. Indeed, some studies have shown that working memory performance can be facilitated in those with schizophrenia by increasing stimulus presentation time (Badcock et al. 2008; Hartman et al. 2003; Tek et al. 2002) or manipulating target features to ease their perceptual processing burden (Javitt et al. 2007; Lee and Park 2006). However, slowed perceptual encoding in individuals with schizophrenia may not be entirely culpable for the level and breadth of working memory deficits typically observed, as increasing stimulus duration does not seem to yield significant capacity gains in tasks requiring memory of spatial location (Tek et al. 2002) or orientation (Gold et al. 2003). Thus, early visual processing deficits involved in perception of certain low-level stimuli features (e.g., orientation, location) also likely reduce the precision ("tuning") of neuron population activity involved in forming of the perceptual working memory representation, resulting in less precise visuospatial working memory representations (Bays 2015).

Though imprecise working memory representations have often been discussed in the auditory domain (Javitt et al. 1997, 1999), few studies have utilized continuous measures of visual working memory retrieval to examine altered precision in schizophrenia. Interestingly, evidence for such visual imprecision during working memory recall is somewhat mixed depending on the visual modality assessed. For instance, Gold et al. (2010) did not find increased imprecision in working memory recall for color in schizophrenia. On the other hand, studies assessing visuospatial working memory have found reduced visuospatial precision during recall in schizophrenia (Badcock et al. 2008; Starc et al. 2017). Starc et al. (2017) not only tested group differences in recall precision across varying delays, they also examined differences in the impact of visual distractors on recall precision. The schizophrenia group exhibited higher response variability (poorer precision) relative to controls at every delay interval and additionally exhibited a steeper increase in variability with delay duration versus controls. Badcock et al. (2008) found that individuals with schizophrenia only exhibited greater imprecision during manual recall of a visual target at a 4 s delay, not a 0 s delay (participants utilized a stylus to locate target stimuli). However, the authors effectively controlled for potential between-group differences in perceptual processing by using individually derived stimulus durations necessary for participants to achieve at least 79% accurate detection of targets' spatial locations. Consistent with earlier work, the schizophrenia group required, on average, a longer stimulus duration to obtain similar target detection accuracy rates as controls, implicating a prominent role of perceptual processing deficits in working memory precision. Nonetheless, this study points to the instability of working memory representations in schizophrenia even during improved perceptual encoding conditions, such that representations themselves are more susceptible to drift or disruption during later encoding and maintenance stages. Indeed, in addition to capturing increased recall variability over time in schizophrenia, Starc et al. (2017) also found that recall precision was more contaminated by near-target visuospatial distractors in schizophrenia compared to controls, again pointing to overall reduced stability of visuospatial working memory representations (See Fig. 3). Finally, reduced visuospatial working memory precision in college students has also been related to increased schizotypy, but not depressive symptoms (Xie et al. 2018), suggesting a somewhat specific vulnerability associated with the schizophrenia spectrum as opposed to overall working memory deficits.

Interestingly, some prominent models of working memory (Cowan 1988, 1999; Chun and Turk-Brown 2007) propose that attention facilitates the encoding of stimulus features within long-term memory: without the sustained focus of visual attention alongside intact long-term memory processing, the working memory trace is more susceptible to interference and temporal decay. While encoding begins when the stimulus is presented, the readiness of the brain system at that moment determines the efficiency of encoding. In other words, context, past experience, and predictions matter. To that end, recent findings suggest that an inability to utilize long-term memory (i.e., stored regularities) during working memory encoding is the major source of visual attention deficits in schizophrenia (Reinhart et al. 2018). Thus, reduced recall precision in schizophrenia may also reflect a failure in the temporal coordination of long-term memory activity with incoming perceptual information, resulting in inefficient, potentially compensatory recruitment of top-down attentional control systems during encoding. Degraded sensory information, in this situation, may further destabilize network coordination by increasing the

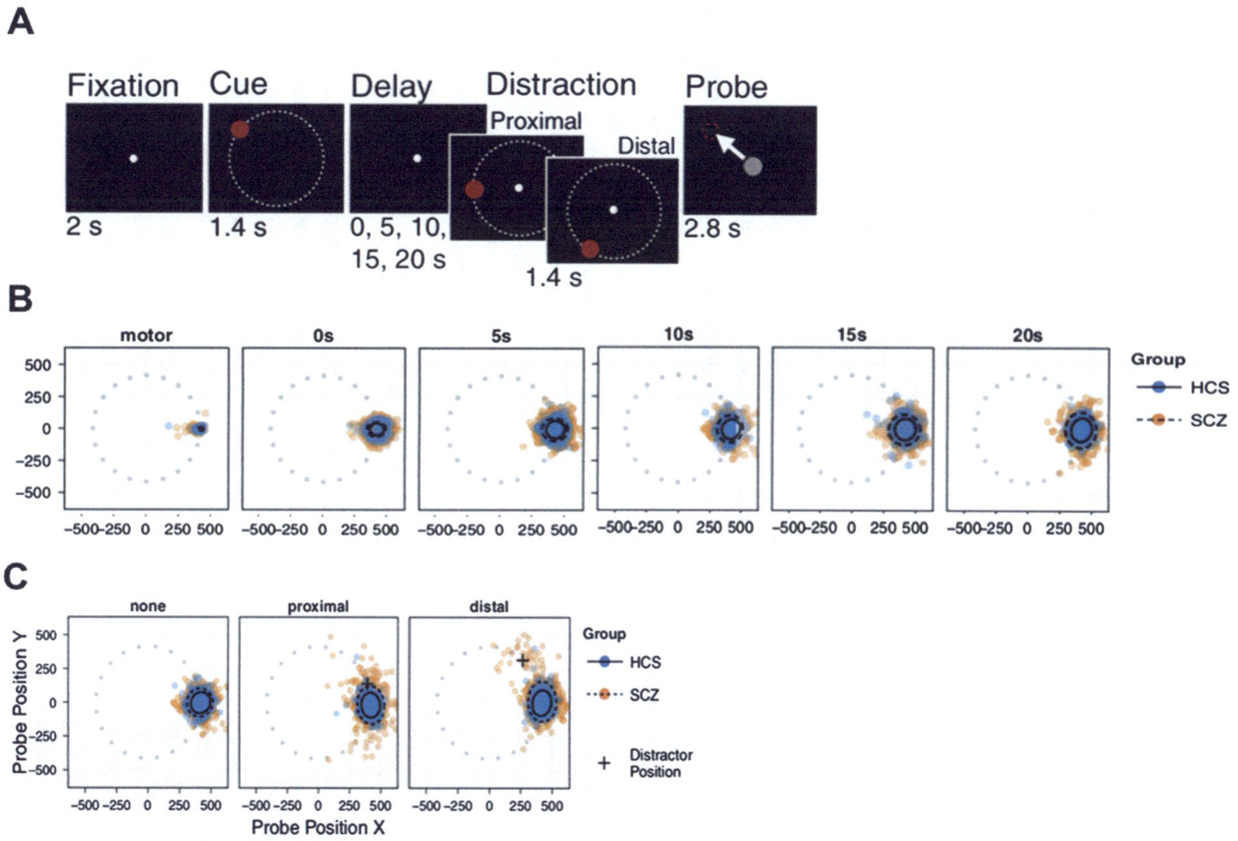

Fig. 3 Instability and imprecision of visuospatial working memory representations in schizophrenia (Adapted from Oribe et al. 2013). **(a)** Participants completed a visuospatial delayed response task requiring manual replication of a cue's location. A distractor condition contained a distractor during the delay. **(b)** Performance on the no-distractor condition indicates less precise recall of the cue in those with schizophrenia. **(c)** Performance on the distractor condition indicates greater influence of the distractor on recall precision in the schizophrenia group

variability of neural activity involved in sensory and perceptual coding. For example, fMRI studies have found that individuals with schizophrenia exhibit reduced recruitment of subcortical and cortical sensory regions during explicit sensory processing tasks (e.g., reduced activation of V2–V4 during visual integration (Silverstein et al. 2009) and motion-sensitive areas (MT) during motion perception (Chen et al. 2008)) and that these reductions in sensory regions are often accompanied by abnormal responses in frontal regions associated with higher level cognitive processing – a pattern interpreted as compensatory activity in response to impaired sensory processing (Chen et al. 2008; Silverstein et al. 2009).

8 Neural Oscillations and Internal "Noise"

The complex coordination of attention, perception, and memory processes that underlie working memory encoding can also be indexed by neural oscillatory activity, thought to enable large-scale communication and integration of information across the brain through synchronous firing of neural populations at various frequencies (Varela et al. 2001). Because oscillations also reflect fluctuations in cortical excitability contributing to the ongoing variability in stimulus processing (Bates et al. 2009; Becker et al. 2011; Coon et al. 2016; Ergenoglu et al. 2004), it can also provide insights into the degree to which encoding is a more variable, "noisier" process in those with schizophrenia (Javitt et al. 1999; Krystal et al. 2017; Starc et al. 2017). The rapidly growing literature on neural oscillations and working memory in healthy individuals largely implicates theta (4–7 Hz), alpha (8–12 Hz), and gamma range (>30 Hz) activity as functionally relevant for the encoding and maintenance of working memory representations (for a review, see Roux and Uhlhaas 2014). Unsurprisingly, studies of working memory-related oscillatory activity in schizophrenia point to patient differences across all of these frequency bands (for reviews, see Haenschel and Linden 2011; Senkowski and Gallinat 2015; Uhlhaas and Singer 2010, 2013).

Perhaps the most commonly reported oscillatory findings related to both perceptual deficits and working memory dysfunction in schizophrenia is abnormal gamma-band activity (Senkowski and Gallinat 2015; Uhlhaas and Singer 2010, 2013, 2015). According to Roux and Uhlhaas (2014), gamma-band activity seems to support the top-down, active maintenance of working memory items. In a visual working memory partial response paradigm requiring memory for letter arrays, Chen et al. (2014) found reduced task-induced frontal gamma during all stages of working memory (encoding, retention, retrieval), though only gamma activity during encoding related to individuals' working memory performance. Similar patient reductions in gamma have been found during a preparatory cognitive control task (Minzenberg et al. 2010), suggesting that gamma abnormalities in schizophrenia may impair working memory encoding and maintenance under task conditions requiring heightened top-down control. Individuals with schizophrenia also exhibit weaker theta-gamma cross-frequency coupling (CFC) compared to controls during visual working memory (Barr et al. 2017). In healthy adults, theta-gamma CFC has been implicated in working memory encoding (Friese et al. 2013) and maintenance (Axmacher et al. 2010), with recent findings showing causal links between theta-gamma CFC enhancement and visuospatial working memory task performance (Alekseichuk et al. 2016). Similar to gamma activity, frontal theta synchronization has been associated with executive control processes, consolidation, and the maintenance-related period of working memory (Sauseng et al. 2010; Roux and Uhlhaas 2014). While some work has found evidence of aberrant theta activity during visual working memory in schizophrenia, it is unclear whether these deficits are most apparent during the encoding versus maintenance phases (Griesmayr et al. 2014; Haenschel et al. 2009).

Studies also find abnormal posterior alpha activity during visual working memory encoding in those with schizophrenia, including adolescents with early-onset schizophrenia (Haenschel et al. 2009, 2010), individuals with chronic schizophrenia, and, to a lesser degree, their unaffected co-twins (Bachman et al. 2009). Alpha oscillations were initially interpreted as cortical idling or noise. Prompted by Berger's (1929) classic findings of a task-related *decrease* in alpha amplitude (power) during visual stimulation and *increase* in alpha power during visual suppression (e.g., closing one's eyes), this viewpoint purports that a large increase in alpha power characterizes a state of task disengagement. Research has since demonstrated that alpha may play a more functional role during visual perception, attention, and working memory (Klimesch et al. 2007; Roux and Uhlhaas 2014). For instance, lower pre-stimulus posterior alpha power is related to more accurate visual detection and discrimination (Ergenoglu et al. 2004; Hanslmayr et al. 2007). Additionally, the magnitude of alpha desynchronization (suppression) over parieto-occipital regions during visual working memory has been shown to index working memory set size (Fukuda et al. 2015, 2016, 2017), such that greater suppression is related to better encoding and higher working memory capacity. Some hypothesize this phenomenon to reflect an attentional gating mechanism that attunes attention for perceptual detection and inhibits task-irrelevant activity (Jensen and Mazaheri 2010). At least one study has found that patients with schizophrenia failed to suppress alpha activity to the same extent as controls during a visual working memory task, which was more apparent at higher working memory loads (Bachman et al. 2009), though a similar pattern of weakened alpha suppression in schizophrenia has also been found for auditory stimulus detection (Krause 2006). Further research is needed to clarify whether aberrant alpha activity in schizophrenia contributes to specific failures in perceptual encoding of stimuli or more broadly to deficits in inhibition of task-irrelevant activity.

Given evidence of abnormal amplitudes of ERP components and EEG frequency band activity in schizophrenia during working memory tasks and at rest, substantial efforts have been made to better characterize the contributions of "noise" and "variability" to behavioral and neural indices of information processing dysfunction in schizophrenia. Influential work by Winterer and colleagues (Winterer et al. 2000, 2004, 2006; Winterer and Weinberger 2003, 2004) provided weight to the theory that individuals with schizophrenia have "noisy" cognitive processing by showing that a measure of cortical noise during event-related activity from an auditory oddball task was elevated in schizophrenia patients, unaffected siblings, and schizotypal individuals, indicating that heightened cortical noise may convey core features of schizophrenia pathophysiology. Cortical noise power was computed to capture the spontaneous background activity and variability of the event-related signal (e.g., Winterer et al. 2004). Individuals with schizophrenia and at-risk groups showed both a reduction in signal power and an increase in background spectral noise power across all frequency bands, observed mainly over prefrontal and temporoparietal cortices (Winterer and Weinberger 2003). Interestingly, the largest reduction in signal and increase in noise were observed in a relatively early time interval following stimulus presentation (100–450 ms), supporting the notion that background noise impairs stimulus encoding. Additionally, reduced SNR in frontal alpha and delta frequency bands was related to working memory performance on a visual N-back task (Winterer et al. 2004). Taken together, these studies of oscillatory activity during working memory may more broadly reflect a failure in the temporal precision with which neural activity responds to incoming sensory information, driven by heightened, task-unrelated activity (noise). Aberrant oscillations during working memory encoding is observed over both sensory and frontal areas, providing support to the idea that sensory region involvement is pivotal for working memory alongside parietal and frontal regions. However, because this study utilized an auditory paradigm, it remains to be seen whether such cortical noise is a more global or local feature of perceptual and memory processing.

Computational models of perception and working memory that incorporate the hypothesized neurochemical and neural network dysfunction in schizophrenia also implicate noisier encoding. For example, the impact of glutamatergic dysfunction in schizophrenia on visuospatial working memory has been examined (Murray et al. 2014; Starc et al. 2017). The glutamate hypothesis proposes that hypofunction of *N*-methyl-D-aspartate receptors (NMDARs), particularly in frontal cortex, causes the development of positive, negative, and cognitive symptoms of schizophrenia. In this working memory model, NMDAR hypofunction led to increased cortical excitability and GABAergic disinhibition of PFC circuits coding and maintaining stimulus features in working memory. Behaviorally, this manipulation resulted in less precise working memory representations that were more contaminated by nearby distractors during maintenance, which matched the performance of patients with schizophrenia (Starc et al. 2017) as well as healthy adults who had been administered a sub-anesthetic dose of the NMDAR antagonist, ketamine (Murray et al. 2014). This model importantly demonstrates that a cortical excitation/inhibition (E/I) imbalance can account for the less precise and more variable working memory encoding observed in individuals with schizophrenia. Earlier models have also tested the hypothesized effects of dopaminergic abnormalities on working memory performance in schizophrenia (Braver et al. 1999; Braver and Cohen 1999; Cohen and Servan-Schreiber 1992). Such work showed that increasing the noise level in a parameter simulating DA's role in the PFC led to an increase in tonic, baseline DA activity as well as reduced gain in phasic activity related to encoding new information (signal), resulting in a failure to update or encode new information in working memory.

While these models certainly point to dysfunctional working memory encoding, they both focus on the impact of neurochemical dysfunction in PFC without incorporating sensory and thalamic regions, the latter of which is believed to process and integrate sensory information relevant to ongoing demands through the regulation of oscillatory activity across multiple cortical areas (Cronenwett and Csernansky 2010). In addition to neuroanatomical abnormalities, there's evidence of abnormal dopaminergic, glutamatergic, and GABAergic expression in the thalamus (e.g., Clinton and Meador-Woodruff 2004) and reduced GABA concentration in the visual cortex of those with schizophrenia (Yoon et al. 2010). Yoon et al. (2010) further showed that GABA concentration in visual cortex was associated with deficits in a behavioral measure of visual inhibition. Future research may benefit from the inclusion of abnormal neurochemical transmission involved at earlier stages of sensory and perceptual processing in primarily PFC-based models of schizophrenia working memory deficits.

9 Summary

This chapter examined potential key mechanisms underlying visuospatial working memory deficits of schizophrenia, with a focus on aberrant encoding. Encoding encompasses the period during which sensory information is transformed into relatively stable working memory representations and thus involves the interaction of visual perceptual, attention, and memory processes. The collective results across these behavioral and neuroimaging studies indicate that multiple aspects of these encoding processes, as well as their coordination and integration, are disrupted in those with schizophrenia and contribute to the observed visuospatial working memory deficits in this population.

Interestingly, sensory and perceptual processing dysfunction appears quite central to downstream impairments (See Fig. 4). There is ample evidence that those with schizophrenia exhibit neuroanatomical and/or functional abnormalities at nearly every level of visual processing, with electrophysiological findings implicating visual dysfunction as early as 100–200 ms of processing a visual stimulus. Deficits in the actual consolidation of sensory memory in individuals with schizophrenia may be driven by a general failure in the temporal coordination of memory-related oscillatory activity with incoming perceptual information, which is likely exacerbated by poor-quality sensory information. Tasks requiring high levels of attentional control may further result in an inability of top-down executive control networks to sufficiently inhibit highly salient distractors and focus on targets during encoding. Both of these disruptions would render the working memory representation less stable during maintenance and more susceptible to distraction, as seen in the heightened variability (poorer precision) of working memory recall in schizophrenia (Badcock et al. 2008; Park et al. 1999; Starc et al. 2017). In other words, greater levels of internal noise seem to interfere with visual perceptual (e.g., Krishnan et al. 2005) as well as higher-order cognitive processing (e.g., Winterer and Weinberger 2003) in those with schizophrenia, though the degree to which these noise

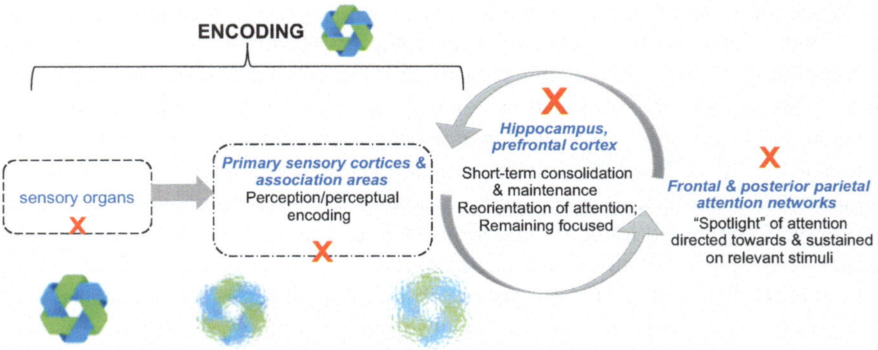

Fig. 4 Inefficient working memory encoding in schizophrenia. Initial degradation begins with less precise perception and perceptual encoding of a stimulus. Disrupted top-down processing adds further variability ("noise") to the representation, as its instability is impacted by sustained attention (red crosses indicate possible sources of errors)

mechanisms are driven by the same neuroanatomical and neurochemical pathways remains to be seen.

10 Opportunities for Therapeutic Interventions

The current overview of visual encoding dysfunction in schizophrenia has several implications for cognitive interventions. First, given the downstream consequences of visual perceptual abnormalities, performance benefits may be better realized when early sensory and perceptual processing is targeted. Indeed, "bottom-up" cognitive remediation strategies targeting auditory processing have shown to improve working memory with moderate-to-large effect sizes in first-episode schizophrenia patients (Fisher et al. 2014) and patients with chronic schizophrenia (Adcock et al. 2009; Fisher et al. 2009). Some studies also show increased engagement of prefrontal regions following training, suggesting that a focus on sensory processing may facilitate top-down control as well (Dale et al. 2015). One potential drawback is the intensive nature of training (40+ h of computerized exercises). However, ongoing work to improve existing protocols and establish novel sensory-based interventions may address this limitation. For example, an innovative study by Appelbaum et al. (2012) found that one 15–45-min session of stroboscopic visual training, in which one practices an activity (e.g., throwing and catching a ball) while only experiencing snapshots of visual stimuli, enhanced visual working memory in healthy adults.

The application of noninvasive electrical stimulation over specific brain regions may offer an alternative approach for enhancing sensory and perceptual processing, which should theoretically ease the burden of encoding during working memory. Indeed, at least one study has found that anodal transcranial direct current stimulation (tDCS) over occipital cortex during visual working memory improved performance in healthy adults (Makovski and Lavidor 2014). Alternatively, anodal tDCS over frontal and/or parietal regions may stabilize existing working memory representations, though findings of stimulation-related working memory improvements in controls and patient populations (including schizophrenia) have been mixed (for reviews, see Andrews et al. 2011; Hill et al. 2016; Mancuso et al. 2016; Mulquiney et al. 2011). Such interventions may be more broadly reducing cortical noise or enhancing signal power by improving synchronization of task-related networks. Given the importance of working memory impairment to functional outcome in schizophrenia, improving encoding processes has the potential to ultimately improve the prognosis and well-being of individuals with schizophrenia.

References

Adcock RA, Dale C, Fisher M et al (2009) When top-down meets bottom-up: auditory training enhances verbal memory in schizophrenia. Schizophr Bull 35(6):1132–1141

Alekseichuk I, Turi Z, de Lara GA et al (2016) Spatial working memory in humans depends on theta and high gamma synchronization in the prefrontal cortex. Curr Biol 26(12):1513–1521

Andrews SC, Hoy KE, Enticott PG et al (2011) Improving working memory: the effect of combining cognitive activity and anodal transcranial direct current stimulation to the left dorsolateral prefrontal cortex. Brain Stimul 4(2):84–89

Appelbaum LG, Cain MS, Schroeder JE et al (2012) Stroboscopic visual training improves information encoding in short-term memory. Atten Percept Psychophys 74(8):1681–1691

Axmacher N, Henseler MM, Jensen O et al (2010) Cross frequency coupling supports multi-item working memory in the human hippocampus. Proc Natl Acad Sci U S A 107(7):3228–3233

Bachman P, Kim J, Yee CM et al (2009) Efficiency of working memory encoding in twins discordant for schizophrenia. Psychiatry Res 174(2):97–104

Badcock JC, Badcock DR, Read C et al (2008) Examining encoding imprecision in spatial working memory in schizophrenia. Schizophr Res 100(1–3):144–152

Baddeley A (2007) Working memory, thought, and action. Oxford University Press, Oxford

Balogh Z, Benedek G, Kéri S (2008) Retinal dysfunctions in schizophrenia. Prog Neuro-psychopharmacol Biol Psychiatry 32(1):297–300

Barch DM, Csernansky JG (2007) Abnormal parietal cortex activation during working memory in schizophrenia: verbal phonological coding disturbances versus domain-general executive dysfunction. Am J Psychiatry 164(7):1090–1098

Barch DM, Csernansky JG, Conturo T et al (2002) Working and long-term memory deficits in schizophrenia: is there a common prefrontal mechanism? J Abnorm Psychol 111(3):478–494

Barr MS, Rajji TK, Zomorrodi R et al (2017) Impaired theta-gamma coupling during working memory performance in schizophrenia. Schizophr Res 189:104–110

Bates AT, Kiehl KA, Laurens KR et al (2009) Low-frequency EEG oscillations associated with information processing in schizophrenia. Schizophr Res 115(2–3):222–230

Bays PM (2015) Spikes not slots: noise in neural populations limits working memory. Trends Cogn Sci 19(8):431–438

Bays PM, Gorgoraptis N, Wee N et al (2011) Temporal dynamics of encoding, storage, and reallocation of visual working memory. J Vis 11(10):6. https://doi.org/10.1167/11.10.6

Becker R, Reinacher M, Freyer F et al (2011) How ongoing neuronal oscillations account for evoked fMRI variability. J Neurosci 31(30):11016–11027

Berger H (1929) Über das elektrenkephalogramm des menschen. Arch Psychiatr Nervenkr 87(1):527–570

Braver TS, Cohen JD (1999) Dopamine, cognitive control, and schizophrenia: the gating model. Prog Brain Res 121:327–349

Braver TS, Barch DM, Cohen JD (1999) Cognition and control in schizophrenia: a computational model of dopamine and prefrontal function. Biol Psychiatry 46(3):312–328

Brittain PJ, Surguladze S, McKendrick AM et al (2010) Backward and forward visual masking in schizophrenia and its relation to global motion and global form perception. Schizophr Res 124(1–3):134–141

Butler PD, Zemon V, Schechter I et al (2005) Early-stage visual processing and cortical amplification deficits in schizophrenia. Arch Gen Psychiatry 62(5):495–504

Butler PD, Martinez A, Foxe JJ, Kim D, Zemon V, Silipo G, Mahoney J, Shpaner M, Jalbrzikowski M, Javitt DC (2006) Subcortical visual dysfunction in schizophrenia drives secondary cortical impairments. Brain 130(2):417–430

Butler PD, Abeles IY, Weiskopf NG et al (2009) Sensory contributions to impaired emotion processing in schizophrenia. Schizophr Bull 35(6):1095–1107

Byne W, Fernandes J, Haroutunian V et al (2007) Reduction of right medial pulvinar volume and neuron number in schizophrenia. Schizophr Res 90(1–3):71–75

Callicott JH, Mattay VS, Verchinski BA, Marenco S, Egan MF, Weinberger DR (2003) Complexity of prefrontal cortical dysfunction in schizophrenia: more than up or down. Am J Psychiatry 160 (12):2209–2215

Cannon TD, Glahn DC, Kim J et al (2005) Dorsolateral prefrontal cortex activity during maintenance and manipulation of information in working memory in patients with schizophrenia. Arch Gen Psychiatry 62(10):1071–1080

Chen Y (2011) Abnormal visual motion processing in schizophrenia: a review of research progress. Schizophr Bull 37(4):709–715

Chen Y, Grossman ED, Bidwell LC et al (2008) Differential activation patterns of occipital and prefrontal cortices during motion processing: evidence from normal and schizophrenic brains. Cogn Affect Behav Neurosci 8(3):293–303

Chen CMA, Stanford AD, Mao X et al (2014) GABA level, gamma oscillation, and working memory performance in schizophrenia. NeuroImage Clin 4:531–539

Chu EMY, Kolappan M, Barnes TR et al (2012) A window into the brain: an in vivo study of the retina in schizophrenia using optical coherence tomography. Psychiatry Res 203(1):89–94

Chun M, Turk-Brown NB (2007) Interactions between attention and memory. Curr Opin Neurobiol 17(2):177–184

Clinton SM, Meador-Woodruff JH (2004) Abnormalities of the NMDA receptor and associated intracellular molecules in the thalamus in schizophrenia and bipolar disorder. Neuropsychopharmacology 29(7):1353–1362

Cohen JD, Servan-Schreiber D (1992) Context, cortex, and dopamine: a connectionist approach to behavior and biology in schizophrenia. Psychol Rev 99(1):45–77

Coleman MJ, Cestnick L, Krastoshevsky O et al (2009) Schizophrenia patients show deficits in shifts of attention to different levels of global-local stimuli: evidence for magnocellular dysfunction. Schizophr Bull 35(6):1108–1116

Compte A, Brunel N, Goldman-Rakic PS et al (2000) Synaptic mechanisms and network dynamics underlying spatial working memory in a cortical network model. Cereb Cortex 10(9):910–923

Conklin HM, Curtis CE, Calkins ME et al (2005) Working memory functioning in schizophrenia patients and their first-degree relatives: cognitive functioning shedding light on etiology. Neuropsychologia 43(6):930–942

Coon WG, Gunduz A, Brunner P et al (2016) Oscillatory phase modulates the timing of neuronal activations and resulting behavior. NeuroImage 133:294–301

Cowan N (1988) Evolving conceptions of memory storage, selective attention, and their mutual constraints within the human information-processing system. Psychol Bull 104(2):163–191

Cowan N (1999) An embedded-processes model of working memory. In: Miyake A, Shah P (eds) Models of working memory: mechanisms of active maintenance and executive control. Cambridge University Press, Cambridge, pp 62–101

Cronenwett WJ, Csernansky J (2010) Thalamic pathology in schizophrenia. In: Swerdlow NR (ed) Behavioral neurobiology of schizophrenia and its treatment. Springer, Berlin, pp 509–528
Dale CL, Brown EG, Fisher M et al (2015) Auditory cortical plasticity drives training-induced cognitive changes in schizophrenia. Schizophr Bull 42(1):220–228
Di Russo F, Pitzalis S, Aprile T et al (2007) Spatiotemporal analysis of the cortical sources of the steady-state visual evoked potential. Hum Brain Mapp 28(4):323–334
Dias EC, Butler PD, Hoptman MJ, Javitt DC (2011) Early sensory contributions to contextual encoding deficits in schizophrenia. Arch Gen Psychiatry 68(7):654–664
Donohoe G, Morris DW, De Sanctis P et al (2008) Early visual processing deficits in dysbindin-associated schizophrenia. Biol Psychiatry 63(5):484–489
Dorph-Petersen KA, Pierri JN, Wu Q et al (2007) Primary visual cortex volume and total neuron number are reduced in schizophrenia. J Comp Neurol 501(2):290–301
Driesen NR, Leung HC, Calhoun VD et al (2008) Impairment of working memory maintenance and response in schizophrenia: functional magnetic resonance imaging evidence. Biol Psychiatry 64(12):1026–1034
Ducato MG, Michael GA, Thomas P et al (2008) Attentional capture in schizophrenia: failure to resist interference from motion signals. Cogn Neuropsychiatry 13(3):185–209
Eimer M, Schlaghecken F (1998) Effects of masked stimuli on motor activation: behavioral and electrophysiological evidence. J Exp Psychol Hum Percept Perform 24(6):1737–1747
Ergenoglu T, Demiralp T, Bayraktaroglu Z et al (2004) Alpha rhythm of the EEG modulates visual detection performance in humans. Brain Res Cogn Brain Res 20(3):376–383
Erickson M, Hahn B, Leonard CJ et al (2014) Enhanced vulnerability to distraction does not account for working memory capacity reduction in people with schizophrenia. Schizophr Res Cogn 1(3):149–154
Erickson MA, Hahn B, Leonard CJ et al (2015) Impaired working memory capacity is not caused by failures of selective attention in schizophrenia. Schizophr Bull 41(2):366–373
Fischer J, Whitney D (2012) Attention gates visual coding in the human pulvinar. Nat Commun 3:1051. https://doi.org/10.1038/ncomms2054
Fisher M, Holland C, Merzenich MM et al (2009) Using neuroplasticity based auditory training to improve verbal memory in schizophrenia. Am J Psychiatry 166(7):805–811
Fisher M, Loewy R, Carter C et al (2014) Neuroplasticity-based auditory training via laptop computer improves cognition in young individuals with recent onset schizophrenia. Schizophr Bull 41(1):250–258
Forbes NF, Carrick LA, McIntosh AM, Lawrie SM (2009) Working memory in schizophrenia: a meta-analysis. Psychol Med 39(6):889–905
Friese U, Köster M, Hassler U et al (2013) Successful memory encoding is associated with increased cross-frequency coupling between frontal theta and posterior gamma oscillations in human scalp-recorded EEG. NeuroImage 66:642–647
Fukuda K, Mance I, Vogel EK (2015) α power modulation and event-related slow wave provide dissociable correlates of visual working memory. J Neurosci 35(41):14009–14016
Fukuda K, Kang MS, Woodman GF (2016) Distinct neural mechanisms for spatially lateralized and spatially global visual working memory representations. J Neurophysiol 116(4):1715–1727
Fukuda K, Sundby C, Woodman G (2017) Parieto-occipital alpha power dynamics selectively code for the storage of spatial locations in visual working memory. J Vis 17(10):336. https://doi.org/10.1167/17.10.336
Gagné AM, Hebert M, Maziade M (2015) Revisiting visual dysfunctions in schizophrenia from the retina to the cortical cells: a manifestation of defective neurodevelopment. Prog Neuro-Psychopharmacol Biol Psychiatry 62:29–34
Glahn DC, Therman S, Manninen M et al (2003) Spatial working memory as an endophenotype for schizophrenia. Biol Psychiatry 53(7):624–626
Glahn DC, Bearden CE, Cakir S, Barrett JA, Najt P, Serap Monkul E, Maples N, Velligan DI, Soares JC (2006) Differential working memory impairment in bipolar disorder and schizophrenia: effects of lifetime history of psychosis. Bipolar Disord 8(2):117–123

Gold JM, Carpenter C, Randolph C et al (1997) Auditory working memory and Wisconsin Card Sorting Test performance in schizophrenia. Arch Gen Psychiatry 54(2):159–165
Gold JM, Wilk CM, McMahon RP et al (2003) Working memory for visual features and conjunctions in schizophrenia. J Abnorm Psychol 112(1):61–71
Gold JM, Fuller RL, Robinson BM et al (2006) Intact attentional control of working memory encoding in schizophrenia. J Abnorm Psychol 115(4):658–673
Gold JM, Hahn B, Zhang WW et al (2010) Reduced capacity but spared precision and maintenance of working memory representations in schizophrenia. Arch Gen Psychiatry 67(6):570–577
Goldman-Rakic PS (1994) Working memory dysfunction in schizophrenia. J Neuropsychiatry Clin Neurosci 6(4):348–357
Gooding DC, Tallent KA (2003) Spatial, object, and affective working memory in social anhedonia: an exploratory study. Schizophr Res 63(3):247–260
Green AE, Fitzgerald PB, Johnston PJ et al (2017) Evidence for a differential contribution of early perceptual and late cognitive processes during encoding to episodic memory impairment in schizophrenia. World J Biol Psychiatry 18(5):369–381
Griesmayr B, Berger B, Stelzig-Schoeler R et al (2014) EEG theta phase coupling during executive control of visual working memory investigated in individuals with schizophrenia and in healthy controls. Cogn Affect Behav Neurosci 14(4):1340–1355
Haenschel C, Linden DE (2011) Exploring intermediate phenotypes with EEG: working memory dysfunction in schizophrenia. Behav Brain Res 216(2):481–495
Haenschel C, Bittner RA, Haertling R et al (2007) Contribution of impaired early-stage visual processing to working memory dysfunction in adolescents with schizophrenia. Arch Gen Psychiatry 64:1229–1240
Haenschel C, Bittner RA, Waltz J et al (2009) Cortical oscillatory activity is critical for working memory as revealed by deficits in early-onset schizophrenia. J Neurosci 29(30):9481–9489
Haenschel C, Linden DE, Bittner RA et al (2010) Alpha phase locking predicts residual working memory performance in schizophrenia. Biol Psychiatry 68(7):595–598
Hahn B, Robinson BM, Kaiser ST et al (2010) Failure of schizophrenia patients to overcome salient distractors during working memory encoding. Biol Psychiatry 68(7):603–609
Hanslmayr S, Aslan A, Staudigl T et al (2007) Prestimulus oscillations predict visual perception performance between and within subjects. NeuroImage 37(4):1465–1473
Hartman M, Steketee MC, Silva S et al (2003) Working memory and schizophrenia: evidence for slowed encoding. Schizophr Res 59(2–3):99–113
Hill AT, Fitzgerald PB, Hoy KE (2016) Effects of anodal transcranial direct current stimulation on working memory: a systematic review and meta-analysis of findings from healthy and neuropsychiatric populations. Brain Stimul 9(2):197–208
Javitt DC (2009) When doors of perception close: bottom-up models of disrupted cognition in schizophrenia. Annu Rev Clin Psychol 5(1):249–275
Javitt DC, Strous RD, Grochowski S et al (1997) Impaired precision, but normal retention, of auditory sensory ("echoic") memory information in schizophrenia. J Abnorm Psychol 106 (2):315–324
Javitt DC, Liederman E, Cienfuegos A et al (1999) Panmodal processing imprecision as a basis for dysfunction of transient memory storage systems in schizophrenia. Schizophr Bull 25 (4):763–775
Javitt DC, Rabinowicz E, Silipo G et al (2007) Encoding vs. retention: differential effects of cue manipulation on working memory performance in schizophrenia. Schizophr Res 91 (1–3):159–168
Jensen O, Mazaheri A (2010) Shaping functional architecture by oscillatory alpha activity: gating by inhibition. Front Hum Neurosci 4:186. https://doi.org/10.3389/fnhum.2010.00186
Kim D, Wylie G, Pasternak R et al (2006) Magnocellular contributions to impaired motion processing in schizophrenia. Schizophr Res 82(1):1–8
Kim MA, Tura E, Potkin SG et al (2010) Working memory circuitry in schizophrenia shows widespread cortical inefficiency and compensation. Schizophr Res 117(1):42–51
Klimesch W, Sauseng P, Hanslmayr S (2007) EEG alpha oscillations: the inhibition–timing hypothesis. Brain Res Rev 53(1):63–88

Koychev I, El-Deredy W, Haenschel C et al (2010) Visual information processing deficits as biomarkers of vulnerability to schizophrenia: an event-related potential study in schizotypy. Neuropsychologia 48(7):2205–2214
Krause CM (2006) Cognition-and memory-related ERD/ERS responses in the auditory stimulus modality. Prog Brain Res 159:197–207
Krishnan GP, Vohs JL, Hetrick WP et al (2005) Steady state visual evoked potential abnormalities in schizophrenia. Clin Neurophysiol 116(3):614–624
Krystal JH, Anticevic A, Yang GJ et al (2017) Impaired tuning of neural ensembles and the pathophysiology of schizophrenia: a translational and computational neuroscience perspective. Biol Psychiatry 81(10):874–885
Lavoie J, Maziade M, Hébert M (2014) The brain through the retina: the flash electroretinogram as a tool to investigate psychiatric disorders. Prog Neuro-Psychopharmacol Biol Psychiatry 48:129–134. https://doi.org/10.1016/j.pnpbp.2013.09.020
Lee J, Park S (2005) Working memory impairments in schizophrenia: a meta-analysis. J Abnorm Psychol 114(4):599–611
Lee J, Park S (2006) The role of stimulus salience in CPT-AX performance of schizophrenia patients. Schizophr Res 81(2–3):191–197
Lee J, Folley BS, Gore J et al (2008) Origins of spatial working memory deficits in schizophrenia: an event-related FMRI and near-infrared spectroscopy study. PLoS One 3(3):e1760. https://doi.org/10.1371/journal.pone.0001760
Lee WW, Tajunisah I, Sharmilla K et al (2013) Retinal nerve fiber layer structure abnormalities in schizophrenia and its relationship to disease state: evidence from optical coherence tomography. Invest Ophthalmol Vis Sci 54(12):7785–7792
Luck SJ, Hillyard SA (1994) Electrophysiological correlates of feature analysis during visual search. Psychophysiology 31(3):291–308
Makovski T, Lavidor M (2014) Stimulating occipital cortex enhances visual working memory consolidation. Behav Brain Res 275:84–87. https://doi.org/10.1016/j.bbr.2014.09.004
Mancuso LE, Ilieva IP, Hamilton RH et al (2016) Does transcranial direct current stimulation improve healthy working memory? A meta-analytic review. J Cogn Neurosci 28(8):1063–1089
Manoach DS (2003) Prefrontal cortex dysfunction during working memory performance in schizophrenia: reconciling discrepant findings. Schizophr Res 60(2):285–298
Manoach DS, Press DZ, Thangaraj V, Searl MM, Goff DC, Halpern E, Saper CB, Warach S (1999) Schizophrenic subjects activate dorsolateral prefrontal cortex during a working memory task, as measured by fMRI. Biol Psychiatry 45(9):1128–1137
Manoach DS, Gollub RL, Benson ES et al (2000) Schizophrenic subjects show aberrant fMRI activation of dorsolateral prefrontal cortex and basal ganglia during working memory performance. Biol Psychiatry 48(2):99–109
Martinez A, Hillyard SA, Dias EC et al (2008) Magnocellular pathway impairment in schizophrenia: evidence from functional magnetic resonance imaging. J Neurosci 28(30):7492–7500
Mayer JS, Park S (2012) Working memory encoding and false memory in schizophrenia and bipolar disorder in a spatial delayed response task. J Abnorm Psychol 121(3):784–794
Mayer JS, Fukuda K, Vogel EK (2012) Impaired contingent attentional capture predicts reduced working memory capacity in schizophrenia. PLoS One 7(11):e48586. https://doi.org/10.1371/journal.pone.0048586
Mayer JS, Kim J, Park S (2013) Failure to benefit from target novelty during encoding contributes to working memory deficits in schizophrenia. Cogn Neuropsychiatry 19(3):268–279
Mesholam-Gately RI, Giuliano AJ, Goff KP et al (2009) Neurocognition in first-episode schizophrenia: a meta-analytic review. Neuropsychology 23(3):315–335
Minzenberg MJ, Firl AJ, Yoon JH et al (2010) Gamma oscillatory power is impaired during cognitive control independent of medication status in first-episode schizophrenia. Neuropsychopharmacology 35(13):2590–2599

Mulquiney PG, Hoy KE, Daskalakis ZJ et al (2011) Improving working memory: exploring the effect of transcranial random noise stimulation and transcranial direct current stimulation on the dorsolateral prefrontal cortex. Clin Neurophysiol 122(12):2384–2389
Murray JD, Anticevic A, Gancsos M et al (2014) Linking microcircuit dysfunction to cognitive impairment: effects of disinhibition associated with schizophrenia in a cortical working memory model. Cereb Cortex 24(4):859–872
Myles-Worsley M, Park S (2002) Spatial working memory deficits in schizophrenia patients and their first degree relatives from Palau, Micronesia. Am J Med Genet 114(6):609–615
Narr KL, Toga AW, Szeszko P et al (2005) Cortical thinning in cingulate and occipital cortices in first episode schizophrenia. Biol Psychiatry 58(1):32–40
Nuechterlein KH, Dawson ME (1984) Information processing and attentional functioning in the developmental course of schizophrenic disorders. Schizophr Bull 10(2):160–203
Oribe N, Hirano Y, Kanba S et al (2013) Early and late stages of visual processing in individuals in prodromal state and first episode schizophrenia: an ERP study. Schizophr Res 146(1–3):95–102
Park S, Gooding DC (2014) Working memory impairment as an endophenotypic marker of a schizophrenia diathesis. Schizophr Res Cogn 1(3):127–136
Park S, Holzman PS (1992) Schizophrenics show spatial working memory deficits. Arch Gen Psychiatry 49(12):975–982
Park S, McTigue K (1997) Working memory and the syndromes of schizotypal personality. Schizophr Res 26(2):213–220
Park S, Holzman PS, Goldman-Rakic PS (1995) Spatial working memory deficits in the relatives of schizophrenic patients. Arch Gen Psychiatry 52(10):821–828
Park S, Püschel J, Sauter BH et al (1999) Spatial working memory deficits and clinical symptoms in schizophrenia: a 4-month follow-up study. Biol Psychiatry 46(3):392–400
Pirkola T, Tuulio-Henriksson A, Glahn D, Kieseppä T, Haukka J, Kaprio J, Lönnqvist J, Cannon TD (2005) Spatial working memory function in twins with schizophrenia and bipolar disorder. Biol Psychiatry 58(12):930–936
Reavis EA, Lee J, Wynn JK et al (2017) Cortical thickness of functionally defined visual areas in schizophrenia and bipolar disorder. Cereb Cortex 27(5):2984–2993
Reinhart RM, Park S, Woodman GF (2018) Localization and elimination of attentional dysfunction in schizophrenia during visual search. Schizophr Bull 45(1):96–105
Rokem A, Yoon JH, Ooms RE et al (2011) Broader visual orientation tuning in patients with schizophrenia. Front Hum Neurosci 5:127. https://doi.org/10.3389/fnhum.2011.00127
Roux F, Uhlhaas PJ (2014) Working memory and neural oscillations: alpha–gamma versus theta–gamma codes for distinct WM information? Trends Cogn Sci 18(1):16–25
Salomon JA, Haagsma JA, Davis A et al (2015) Disability weights for the global burden of disease 2013 study. Lancet Glob Health 3(11):e712–e723
Saperstein AM, Fuller RL, Avila MT et al (2005) Spatial working memory as a cognitive endophenotype of schizophrenia: assessing risk for pathophysiological dysfunction. Schizophr Bull 32(3):498–506
Sauseng P, Griesmayr B, Freunberger R et al (2010) Control mechanisms in working memory: a possible function of EEG theta oscillations. Neurosci Biobehav Rev 34(7):1015–1022
Senkowski D, Gallinat J (2015) Dysfunctional prefrontal gamma-band oscillations reflect working memory and other cognitive deficits in schizophrenia. Biol Psychiatry 77(12):1010–1019
Serrano-Pedraza I, Romero-Ferreiro V, Read JC et al (2014) Reduced visual surround suppression in schizophrenia shown by measuring contrast detection thresholds. Front Psychol 5:1431. https://doi.org/10.3389/fpsyg.2014.01431
Silver H, Feldman P, Bilker W et al (2003) Working memory deficit as a core neuropsychological dysfunction in schizophrenia. Am J Psychiatry 160(10):1809–1816
Silverstein SM, Keane BP (2011) Perceptual organization impairment in schizophrenia and associated brain mechanisms: review of research from 2005 to 2010. Schizophr Bull 37(4):690–699
Silverstein SM, Rosen R (2015) Schizophrenia and the eye. Schizophr Res Cogn 2(2):46–55

Silverstein SM, Berten S, Essex B et al (2009) An fMRI examination of visual integration in schizophrenia. J Integr Neurosci 8(2):175–202
Skottun BC, Skoyles JR (2007) Contrast sensitivity and magnocellular functioning in schizophrenia. Vis Res 47(23):2923–2933
Smith EE, Eich TS, Cebenoyan D et al (2011) Intact and impaired cognitive-control processes in schizophrenia. Schizophr Res 126(1–3):132–137
Starc M, Murray JD, Santamauro N et al (2017) Schizophrenia is associated with a pattern of spatial working memory deficits consistent with cortical disinhibition. Schizophr Res 181:107–116
Tadin D, Kim J, Doop ML (2006) Weakened center-surround interactions in visual motion processing in schizophrenia. J Neurosci 26(44):11403–11412
Tallent KA, Gooding D (1999) Working memory and Wisconsin Card Sorting Test performance in schizotypic individuals: a replication and extension. Psychiatry Res 89(3):161–170
Tek C, Gold J, Blaxton T et al (2002) Visual perceptual and working memory impairments in schizophrenia. Arch Gen Psychiatry 59:146–153
Tibber MS, Anderson EJ, Bobin T et al (2013) Visual surround suppression in schizophrenia. Front Psychol 4:88. https://doi.org/10.3389/fpsyg.2013.00088
Uhlhaas PJ, Singer W (2010) Abnormal neural oscillations and synchrony in schizophrenia. Nat Rev Neurosci 11(2):100–113
Uhlhaas PJ, Singer W (2013) High-frequency oscillations and the neurobiology of schizophrenia. Dialogues Clin Neurosci 15(3):301–313
Uhlhaas PJ, Singer W (2015) Oscillations and neuronal dynamics in schizophrenia: the search for basic symptoms and translational opportunities. Biol Psychiatry 77(12):1001–1009
Varela F, Lachaux JP, Rodriguez E et al (2001) The brainweb: phase synchronization and large-scale integration. Nat Rev Neurosci 2(4):229–239
Walter H, Wunderlich AP, Blankenhorn M et al (2003) No hypofrontality, but absence of prefrontal lateralization comparing verbal and spatial working memory in schizophrenia. Schizophr Res 61 (2–3):175–184
Winterer G, Weinberger DR (2003) Cortical signal-to-noise ratio: insight into the pathophysiology and genetics of schizophrenia. Clin Neurosci Res 3(1–2):55–66
Winterer G, Weinberger DR (2004) Genes, dopamine and cortical signal-to-noise ratio in schizophrenia. Trends Neurosci 27(11):683–690
Winterer G, Ziller M, Dorn H et al (2000) Schizophrenia: reduced signal-to-noise ratio and impaired phase-locking during information processing. Clin Neurophysiol 111(5):837–849
Winterer G, Coppola R, Goldberg TE et al (2004) Prefrontal broadband noise, working memory, and genetic risk for schizophrenia. Am J Psychiatry 161(3):490–500
Winterer G, Egan MF, Kolachana BS et al (2006) Prefrontal electrophysiologic "noise" and catechol-O-methyltransferase genotype in schizophrenia. Biol Psychiatry 60(6):578–584
Xie W, Cappiello M, Park HB et al (2018) Schizotypy is associated with reduced mnemonic precision in visual working memory. Schizophr Res 193:91–97
Yeap S, Kelly SP, Sehatpour P et al (2006) Early visual sensory deficits as endophenotypes for schizophrenia: high-density electrical mapping in clinically unaffected first-degree relatives. Arch Gen Psychiatry 63(11):1180–1188
Yeap S, Kelly SP, Sehatpour P et al (2008) Visual sensory processing deficits in schizophrenia and their relationship to disease state. Eur Arch Psychiatry Clin Neurosci 258(5):305–316
Yoon JH, Maddock RJ, Rokem A et al (2010) GABA concentration is reduced in visual cortex in schizophrenia and correlates with orientation-specific surround suppression. J Neurosci 30 (10):3777–3781
Zhang R, Picchioni M, Allen P, Toulopoulou T (2016) Working memory in unaffected relatives of patients with schizophrenia: a meta-analysis of functional magnetic resonance imaging studies. Schizophr Bull 42(4):1068–1077
Zhao YL, Tan SP, De Yang F et al (2011) Dysfunction in different phases of working memory in schizophrenia: evidence from ERP recordings. Schizophr Res 133(1–3):112–119

Dopamine and Working Memory: Genetic Variation, Stress and Implications for Mental Health

Kristel Klaus and Kyla Pennington

Contents

Abstract At the molecular level, the neurotransmitter dopamine (DA) is a key regulatory component of executive function in the prefrontal cortex (PFC) and dysfunction in dopaminergic (DAergic) circuitry has been shown to result in impaired working memory (WM). Research has identified multiple common genetic variants suggested to impact on the DA system functionally and also behaviourally to alter WM task performance. In addition, environmental stressors impact on DAergic tone, and this may be one mechanism by which stressors confer vulnerability to the development of neuropsychiatric conditions. This chapter aims to evaluate the impact of key DAergic gene variants suggested to impact on both synaptic DA levels (*COMT*, *DAT1*, *DBH*, *MAOA*) and DA receptor function (*ANKK1*, *DRD2*, *DRD4*) in terms of their influence on visuospatial WM. The role of stressors and interaction with the genetic background is discussed in addition to discussion around some of the implications for precision psychiatry. This and future work in this area aim to disentangle the neural mechanisms underlying susceptibility

K. Klaus
MRC Brain and Cognition Unit, University of Cambridge, Cambridge, UK
e-mail: Kristel.Klaus@mrc-cbu.cam.ac.uk

K. Pennington (✉)
School of Psychology, University of Lincoln, Lincoln, UK
e-mail: KPennington@lincoln.ac.uk

Curr Topics Behav Neurosci (2019) 41: 369–392
DOI 10.1007/7854_2019_113
Published Online: 10 September 2019

to stress and their impact and relationship with cognitive processes known to influence mental health vulnerability.

Keywords Dopamine · Genetic variation · Stress · Working memory

1 Background

Dopamine is a modulatory monoaminergic neurotransmitter with projections to brain regions involved in motor behaviour, cognition and emotion. It is thought to contribute to the coordination of cognitive ability and is susceptible to changes in arousal state (Arnsten et al. 2010) possibly through its role in salience appraisal, reward and habit-forming processes (Lee et al. 2016; Vaessen et al. 2015). It also plays a critical role in executive functions attributed to the PFC that likely underlie these such as cognitive flexibility, response inhibition and working memory (WM) (Goldman-Rakic and Selemon 1997; Seamans and Yang 2004).

As the last monoaminergic neurotransmitter system to develop, the DAergic system has been suggested to have an important stabilising and integrative influence on brain circuitry (Grace 2016), although it may also confer increased sensitivity to environmental stressors at key developmental stages (Sinclair et al. 2014). This has relevance in the context of mental health vulnerability given the evidence for the involvement of DA in schizophrenia spectrum disorders (Howes and Kapur 2009; Meltzer and Stahl 1976), substance abuse (Nutt et al. 2015; Volkow et al. 2017) and post-traumatic stress disorder (Hoexter et al. 2012; Lee et al. 2016). Indeed, it may be that these conditions have underlying similarities that reside on the role of DA in the interpretation of and response to stressors (Vaessen et al. 2015). The combination of genetic variation impacting on DAergic function and the experience of stressors at key periods of development may result in heightened responsivity to later stressors and impact on executive functions such as WM through altered prefrontal circuitry and activity (Arnsten 2015; Howes et al. 2017; Perlstein et al. 2001).

This chapter aims to present the current understanding of the role of the DAergic system in visuospatial WM and some of the genetic variants that have been shown to contribute to differences in WM ability. The experience of stress and how this affects the DAergic system and WM will also be discussed in addition to the implications that this may have for our understanding and treatment of mental health disorders.

2 The Dopaminergic System and Its Role in Visuospatial WM

Dopaminergic neurons are predominately found in the ventral tegmental area (VTA) and substantia nigra (SN) within the midbrain. There are three major pathways that originate in these areas, the mesocortical pathway which projects from the VTA to

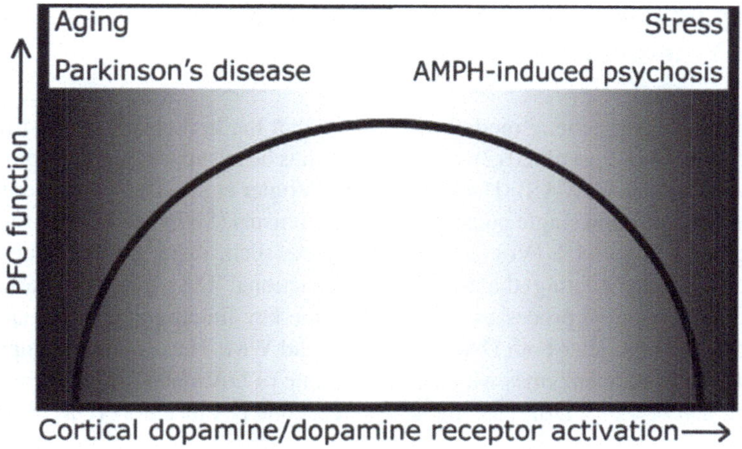

Fig. 1 Illustrates the proposed inverted U-shaped relationship between dopamine levels and prefrontal cortex (PFC) function (Goldman-Rakic et al. 2000). PFC function is impaired in states of dopamine hypofunction (e.g., aging and Parkinson's disease) and hyperfunction (e.g., stress and amphetamine-induced psychosis). Common genetic variants such as *COMT* Val158Met have been linked to altered working memory performance and/or PFC activity. Stress and aging may affect individuals differently in terms of PFC function depending on their genetic background and resulting dopamine concentrations. Adapted from Tunbridge et al. (2006b)

the PFC, the mesolimbic pathway from the VTA to the nucleus accumbens and the nigrostriatal pathway from the SN to the basal ganglia. The mesocortical pathway is the pathway most relevant when discussing the role of DA in visuospatial WM, and concentration levels of DA within the PFC have been proposed to modulate executive function according to an inverted U pattern (Fig. 1; Goldman-Rakic et al. 2000; Tunbridge et al. 2006a, b). Multiple factors can influence the concentration of DA in the PFC and consequently performance in executive function tasks with key factors being age and stress levels.

Of the five DA receptor subtypes, D1 receptors have been found to be relatively more common in the PFC compared to D2 receptors, whereas the latter are abundant in the striatum (Charuchinda et al. 1987; Lidow et al. 1991). This has relevance to cognitive function according to a model proposed by Durstewitz and Seamans (2008), whereby effective cognitive control may result from the relative balance of D1 and D2 receptor activation, with a D1-dominated state being important for the online maintenance of information and greater D2 activation important for flexible information updating. Indeed, D2 receptors have been shown to be central to mental flexibility, abstraction, response inhibition and attention in humans (Uchida et al. 2009; Volkow et al. 1998) though they have been found to influence spatial WM (SWM) and attention performance in mice (Glickstein et al. 2002, 2005; Kellendonk et al. 2006). This suggests that both D1 and D2 receptors are potentially important for WM.

3 Dopaminergic Gene Variants and Their Impact on Visuospatial WM

Twin studies using genetic modelling estimate that the heritability of WM may be around 60–65% (Lee et al. 2012), and this work has been supported by genome-wide association studies (GWAS) (Heck et al. 2014; Vogler et al. 2014), with heritability estimates for common single nucleotide polymorphisms (SNPs) in WM performance ranging from 31 to 41% (Vogler et al. 2014). As such, there is now a substantial evidence base delineating the role of key functional DAergic gene variants in WM-related cognitive processes. Table 1 lists the key functional gene variants that have been found to alter both DAergic activity and WM. These can be grouped into genes which encode enzymes which effect the rate of DA metabolism and synthesis (*COMT*, *MAOA*, *DAT*, *DBH*) and those which encode DA receptors (*DRD2*, *DRD4*).

3.1 Gene Variants Influencing Synaptic Dopamine Levels

The enzyme catechol-O-methyltransferase, encoded by *COMT*, has received most attention to date in relation to both cognitive abilities and disorders characterised by DAergic abnormalities. This enzyme is responsible for breaking down ~50–60% of DA produced in the frontal cortex (Karoum et al. 1994; Yavich et al. 2007), and studies on knockout models have shown that a homozygous *COMT* knockout results in improved SWM, particularly in male mice (Babovic et al. 2008), which suggests improved SWM in the context of higher PFC DA levels. COMT enzyme activity and protein expression in the PFC increase from neonate to adulthood, which may contribute to the attenuated PFC DA levels in the mature adult brain (Tunbridge et al. 2006a, b). Perhaps as a consequence of this, the effects of *COMT* genotype on brain activity and visuospatial WM performance become more apparent from mid-adolescence onwards (Dumontheil et al. 2011).

The Val158Met (rs4680) SNP is the most common variation in the *COMT* gene (Tunbridge et al. 2006b), whereby a single guanine to arginine (G/A) base pair substitution leads to a valine (Val) to methionine (Met) substitution at codon 158. This variation results in ~40% lower enzyme activity leading to substantially higher extracellular DA levels in Met carriers (Chen et al. 2004; Lotta et al. 1995). An association between *COMT* Val158Met genotype and PFC activation has been reported, with the Val allele associated with higher PFC activity and therefore supposedly reduced PFC efficiency (Mier et al. 2010; Opmeer et al. 2013). However, results from studies investigating the association between genetic variation in *COMT* Val158Met and PFC-mediated cognitive functions in healthy and clinical samples have been inconsistent (Barnett et al. 2007a, b; Barnett et al. 2008). Increased Val loading has been associated with slower reaction times in a visual WM task (Berryhill et al. 2013), but a number of studies have failed to find an effect of *COMT* Val158Met on SWM performance in healthy participants using a spatial

Table 1 Provides a summary of some of the studies that have shown an impact on working memory (WM) or prefrontal cortex (PFC) function by key dopaminergic gene variants

Gene	Protein encoded	Polymorphisms investigated	Molecular outcome	Research findings showing an effect of genetic variation on working memory and PFC function	References
ANKK1	Ankyrin repeat and kinase domain containing 1	rs#1800437 (Taq1a)	A1 allele associated with decreased striatal D2 receptor binding	• A1 allele associated with poorer performance on visual WM • Effect of *Taq1a* reported only when combined as a haplotype analysis with *DRD2*	Berryhill et al. (2013) and Markett et al. (2010)
COMT	Catechol-O-methyltransferase	rs#4680 (Val158Met)	Lower enzyme activity, higher DA levels in met carriers Val allele associated with greater D2 receptor activity	• Age × genotype interaction with benefit of met allele on visuospatial WM capacity from age 10 • Val allele associated with higher PFC activity; reduced PFC efficiency • Val allele associated with slower reaction times in visual WM task • Coupling of episodic and WM in older Val homozygotes	Dumontheil et al. (2011), Mier et al. (2010), Opmeer et al. (2013), Berryhill et al. (2013) and Papenberg et al. (2014)
DAT1	Dopamine transporter	VNTR 3′-untranslated region	10-repeat homozygotes have higher striatal DAT density than 9-/10-repeat carriers	• DAT knockout mice have decreased SWM • 10-repeat homozygotes had poorer performance on visuospatial WM • 9-repeat carriers larger gains at visuospatial WM training	Li et al. (2010), Dzirasa et al. (2009), Morice et al. (2007), Zilles et al. (2012) and Brehmer et al. (2009)

(continued)

Table 1 (continued)

Gene	Protein encoded	Polymorphisms investigated	Molecular outcome	Research findings showing an effect of genetic variation on working memory and PFC function	References
DBH	Dopamine beta-hydroxylase	rs#1108580 (G444A)	G allele associated with higher plasma DBH in mood and anxiety disorder patients and higher CSF DBH levels in schizophrenia patients	• G allele associated with improved accuracy in SWM	Parasuraman et al. (2005)
		rs#1611115 (C1021T)	TT homozygosity results in lower DBH enzyme activity in the blood and higher dopamine levels	• Age × genotype interaction COMT *Val/Val* and DBH *C/C* individuals showing impaired performance in SWM in higher task load conditions	Greenwood et al. (2014)
DRD2	Dopamine receptor D2	rs#6277 (C957T) Haplotype; rs#1800497, rs#6277, rs#2283265	CC homozygosity associated with striatal D2 receptor availability and affinity but higher extrastriatal D2 receptor availability	• CC homozygosity associated with poorer verbal WM • CC homozygosity associated with impaired SWM • Haplotype risk associated with visuospatial WM	Xu et al. (2007), Klaus et al. (2017) and Markett et al. (2010)
DRD4	Dopamine receptor D4	VNTR in exon 3	7-repeat carriers exhibit lower ability of DA to inhibit cAMP formation, suggesting lower functional activity in carriers of this allele	• 7-repeat carriers show higher increases in PFC activity with increasing task load • 7-repeat carriers associated with poorer SWM performance in children	Herrmann et al. (2007) and Froehlich et al. (2007)

MAOA	Monoamine oxidase A	VNTR in 3′-untranslated region	Alleles with 3.5 or 4 copies result in 2–10 times more efficient gene transcription and higher MAOA levels	• High activity allele associated with increased VLPFC activity in during high load n-back task • *MAOA* × *COMT* interaction effect on WM	Cerasa et al. (2008) and Barnett et al. (2011)
		Haplotype; rs#12343268, rs#3027400, and rs#1137070	Haplotype previously linked to ADHD	• Haplotype risk effected visuospatial WM	Rommelse et al. (2008)

For details of negative findings, see the expanded in-text summary

delayed response task (Bruder et al. 2005), spatial n-back task (Blanchard et al. 2011; Smyrnis et al. 2007), Cambridge Neuropsychological Test Automated Battery (CANTAB) SWM task (Klaus et al. 2017) and visuospatial rehearsal and pattern maintenance tasks in a mixed sample of individuals with and without a psychiatric disorder (Zilles et al. 2012). These mixed findings may be partly attributable to the effects of different aspects of DA circuitry on the cognitive domains and the close interaction between DA receptor activation and *COMT* activity. For example, it has been suggested that *COMT* Val allele is associated with greater D2 receptor activity and therefore with better cognitive flexibility but with worse performance on tests of WM (Durstewitz and Seamans 2008). In addition, some studies have failed to find a main effect of *COMT* on cognition but have identified a *COMT*-by-sex interaction. Consequently it has been suggested that *COMT* may exert sexually dimorphic effects on behavioural and cognitive outcomes (Barnett et al. 2007a; Harrison and Tunbridge 2008). It has also been found that the effect of *COMT* on the reaction times in SWM tasks is magnified by age in healthy participants, with Val homozygotes showing poorer performance in older age, and a further three-way interaction with the gene encoding the brain-derived neurotrophic factor (BDNF) has been reported (Nagel et al. 2008). Furthermore, a reported coupling between episodic and spatial WM in older Val homozygotes suggests that genetic effects may be magnified when cognitive resources decline, as in older age (Papenberg et al. 2014).

In addition to *COMT*, a set of enzymes called monoamine oxidases (MAO) initiate the breakdown of monoamine neurotransmitters, including DA, serotonin and norepinephrine in the PFC (Green and Youdim 1975). The 30-base pair variable number tandem repeat (VNTR) in the monoamine oxidase A (*MAOA*) gene promoter region has been the most commonly investigated variation in this gene in relation to cognitive function, whereby individuals with 3.5 or 4 copies have been shown to have higher levels of MAOA (Sabol et al. 1998). It has been found that in an adult male sample matched for *COMT* Val158Met, the high activity *MAOA* allele increased ventrolateral PFC activity during the high load spatial n-back task, but there were no significant differences in task performance (Cerasa et al. 2008). Furthermore, in a study on a population-representative sample of Caucasian children, no main effect of *MAOA* on cognitive performance was found, but a significant *MAOA*-by-*COMT* interaction on WM in boys was reported (Barnett et al. 2011). It has also been proposed that other loci in linkage disequilibrium with this VNTR may regulate PFC MAOA levels (Balciuniene et al. 2002). A study that investigated a *MAOA* haplotype previously associated with attention deficit hyperactivity disorder (ADHD) found a significant effect of this haplotype on visuospatial WM in girls (Rommelse et al. 2008). Together these findings suggest a limited and potentially sex-specific effect of *MAOA* on WM and provide evidence for considering the combined influence of genetic variants which are known to influence DA concentration and activity in the PFC.

Two other gene variants that have been investigated in the context of their influence on WM are *DBH* and *DAT1*. *DBH* encodes dopamine beta-hydroxylase – an enzyme that converts DA to NE. In exon 2 of the *DBH* gene is the G444A polymorphism, where carriers of the G allele have been found to have higher plasma

DBH in mood and anxiety disorder patients and higher cerebrospinal fluid levels of DBH in schizophrenia patients (Cubells et al. 1998). Increasing dose of the G allele was associated with improved accuracy in SWM in a sample of 103 healthy individuals (Parasuraman et al. 2005). Another functional polymorphism within *DBH*, C1021T (rs1611115) results in tenfold difference in the serum DBH levels (Zabetian et al. 2003), whereby TT homozygosity results in lower DBH enzyme activity in the bloodstream (Zabetian et al. 2001) which theoretically results in higher extracellular DA and lower NE levels. The authors predicted that the carriers of both *COMT* Val/Val and *DBH* CC genotypes, associated with high activity COMT and DBH enzymes and therefore lower extracellular DA levels, would show impaired SWM performance on a delayed match-to-sample test. The outcome of this study was indeed this pattern of results, but only in the older age group and under the difficult discrimination conditions of the task. The results suggest that the effects of chronically lower extracellular DA levels due to genetic variation are exacerbated in older age (Greenwood et al. 2014).

DAT1 encodes the DA transporter which is involved in the clearance of synaptic DA levels by taking the released dopamine back up to the presynaptic terminals. Unlike COMT, DAT is relatively sparsely expressed in the PFC but is abundant in the striatum where it serves as primary means of DA reuptake (Brooks 2016; Madras et al. 2005). The *DAT1* gene includes a polymorphic 40-base pair VNTR in the 3′-untranslated region of *DAT1*. The VNTR is repeated between 3 and 13 times but occurs with highest frequency in the 9- and 10-repeat forms (Mitchell et al. 2000; Vandenbergh et al. 1992). This VNTR has been associated with striatal DAT availability in both in vitro (van Ness et al. 2005) and in vivo studies (van de Giessen et al. 2009). Studies on knockout mouse models, which produce a genetically induced hyperdopaminergia, have shown that DAT knockouts show significantly decreased SWM performance compared to wild type in the Y-maze (Li et al. 2010), eight-arm radial maze task (Dzirasa et al. 2009) and Morris water maze (Morice et al. 2007). Furthermore, the knockouts had significantly decreased BDNF levels in the frontal cortex. As BDNF is one of the key proteins in neural plasticity, learning and memory formation, this may provide a potential mechanism for the DAergic effects on WM impairment (Li et al. 2010).

Human imaging studies have investigated *DAT1* VNTR in the 3′-untranslated region, whereby 10-repeat homozygotes have higher striatal DAT density than 9-/10-repeat carriers (Heinz et al. 2000), which may lead to increased synaptic DA clearance and therefore lower striatal DA levels in the 10-repeat homozygotes. *DAT1* variation has been associated with performance on a visuospatial WM task (visuospatial rehearsal), with 10-repeat homozygotes showing inferior performance compared to the 9-repeat carriers in a combined sample of schizophrenia, bipolar disorder, obsessive-compulsive disorder patients and healthy controls (Zilles et al. 2012), whereas 9-repeat carriers have showed larger gains at visuospatial WM training (Brehmer et al. 2009). However, some authors have failed to find an effect of *DAT1* (9/10) on visuospatial WM performance (Blanchard et al. 2011), and a recent meta-analysis investigating the association of *DAT1* and cognition (including WM) concluded that this gene plays a limited role in cognitive function (Ettinger et al. 2016).

3.2 Gene Variants Influencing Dopamine Receptor Function

The C957T polymorphism within the dopamine receptor D2 encoding *DRD2* (rs6277; located in exon 7 of *DRD2* on chromosome 11q22–q23) has recently gained interest due to its proposed role in the development of schizophrenia (Gonzalez-Castro et al. 2016; Liu et al. 2014). More specifically, the C allele has been associated with heightened schizophrenia risk in Caucasian samples (Liu et al. 2014), whereas the T allele has been found to confer risk towards schizophrenia in Asian populations (Fan et al. 2010). Although C957T is a synonymous substitution not resulting in an amino acid change (Pro319Pro), it has been shown to affect *DRD2* mRNA stability (Duan et al. 2003), and positron emission tomography studies have demonstrated the functional significance of this SNP in striatal (Hirvonen et al. 2004, 2005) and extrastriatal D2 receptor availability (Hirvonen et al. 2009b). CC homozygosity has been associated with lower striatal D2 receptor availability and affinity (CC < CT < TT; Hirvonen et al. 2004, 2005, 2009a) but with higher D2 receptor availability in the cortex and thalamus (CC > CT > TT; Hirvonen et al. 2009b). A recent meta-analysis confirmed the effect of C957T on striatal D2 receptor availability but failed to find an association between C957Tand D2 availability in cortical regions (Smith et al. 2017). The genetic studies looking at the association of *DRD2* C957T with cognitive function have shown that CC homozygosity is associated with poorer performance on tests of verbal WM, but not on the spatial delayed response test in the general population (Xu et al. 2007), and others have reported an association of C957T with visuospatial WM capacity only in the *DRD2* haplotype (rs1800497, rs6277, rs2283265) analyses (Markett et al. 2010). Our group has recently shown an association between CC homozygosity and impaired visuospatial WM assessed with the Cambridge Neuropsychological Test Automated Battery (CANTAB) SWM task in a sample of 122 healthy males with no current diagnoses of psychiatric disorders and mean age of 35 years, lending evidence to the inverted U-shaped DA function (Cools and D'Esposito 2011) (see Fig. 2). In age- and education-adjusted analysis, both the number of errors and a measure of WM strategy showed reduced performance in CC homozygotes in comparison to CT heterozygotes.

The Taq1A (rs1800497) SNP was initially thought to be part of the *DRD2* gene but was later shown to lie 10 kb downstream of the last exon of *DRD2*, within the ankyrin repeat and kinase domain containing 1 (*ANKK1*) gene (Neville et al. 2004). However, due to linkage disequilibrium with *DRD2*, it has been considered a reliable marker of D2 receptor availability in the striatum (Zhang et al. 2007b). Decreased D2 receptor binding in A1 allele carriers has been repeatedly demonstrated (Jönsson et al. 1999; Pohjalainen et al. 1998; Ritchie and Noble 2003), and the role of this polymorphism in striatal D2 receptor availability was recently confirmed in a meta-analysis summarising the findings from five previous studies (Gluskin and Mickey 2017). Studies investigating the association between Taq1A and WM have produced mixed findings, with some claiming that the A1 allele is associated with poorer performance on tests of visual WM (Berryhill et al. 2013) or reporting an effect of

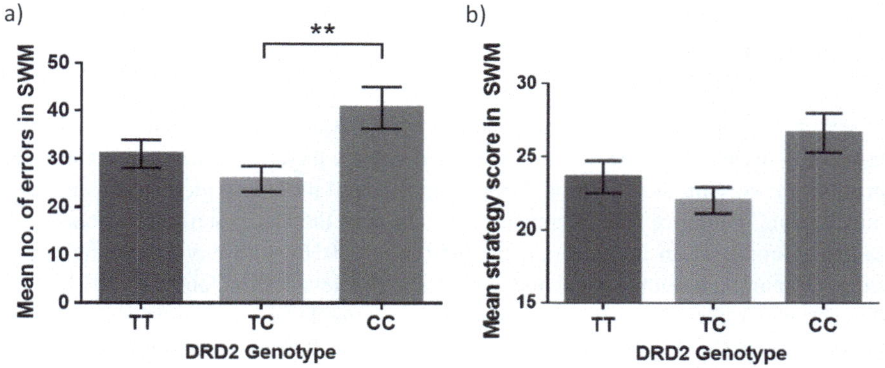

Fig. 2 (**a**) Mean number of errors (±SEM) made in the spatial working memory (SWM) test across *DRD2* genotypes. (**b**) Mean strategy score (the number of distinct boxes used to start a new search; lower is better) (±SEM) in the spatial working memory (SWM) test across *DRD2* genotypes $**p \leq 0.01$. Adapted from Klaus et al. (2017)

Taq1A on visuospatial WM performance only in *ANKK1/DRD2* haplotype analyses (Markett et al. 2010). Our recent systematic review and meta-analysis investigated the effect of Taq1a and *DRD2* C957T on phonological and visuospatial tests of WM (Taq1A, 6 samples, $n = 1{,}270$; C957T, 6 samples, $n = 977$). Although we found limited evidence for an effect of these gene variants on WM, test and sample heterogeneities are inherent issues in these types of analyses (Klaus et al. 2019a).

The D4 receptor is abundantly expressed in the PFC and therefore likely contributes to the PFC-mediated cognitive functions (Noaín et al. 2006; Oak et al. 2000) with WM task performance in mice being dose-dependently affected by D4 antagonists (Zhang et al. 2004). In humans, within exon 3 of *DRD4*, there is a 48 bp VNTR, ranging from 2 to 11-repeat alleles (van Tol et al. 1992). Carrying the 7-repeat allele of the VNTR seems to confer risk for disorders such as ADHD (Gizer et al. 2009) and has functionally been associated with reduced cyclic adenosine monophosphate (cAMP) formation when DA is receptor-bound (Asghari et al. 1995) which may result in reduced DA responsiveness. Studies in healthy participants have shown an interaction between this *DRD4* variant and n-back task difficulty, whereby 7-repeat carriers show higher increases in oxygenated haemoglobin during the task compared to the 7-repeat allele noncarriers, which the authors interpret as being indicative of ineffective brain activity in the 7-repeat allele carriers, albeit in the absence of behavioural differences (Herrmann et al. 2007). In addition, behavioural studies have shown that the *DRD4* 7-repeat is associated with poorer SWM performance in children (Froehlich et al. 2007).

4 The Impact of Environmental Stressors on Dopamine Function and WM

Whether we appraise an event as being benign or threatening is one of the initial processes involved in the activation of a complex physiological stress response, primarily involving the "fight-or-flight" response and including increased heart rate, muscle tone, vigilance and awareness. Activation of the DA system in response to a salient stimulus is an integral part of this (Grace 2016) with environmental stress shown to augment both cortical and subcortical DA levels (Sinclair et al. 2014). The downstream effect that stressors may have on the DAergic system depends on factors such as the degree and duration of stress, as well as the developmental timing of the stressors.

The duration of exposure to a stressor can be important with acute transitory stress induced under laboratory conditions resulting in the brief inhibition of DA neuron firing in affect-related areas of the medial VTA and substantia nigra and a transient increase in excitement of the lateral tegmental area which is proposed to be involved in salience (Floresco et al. 2001; Grace 2016). On the contrary, prolonged or chronic stress has been found to result in increased tonic DA activity in the medial and lateral VTA and greater concentration of DA in the PFC and nucleus accumbens (Abercrombie et al. 1989; Holly and Miczek 2016) (see Fig. 1). These changes have been proposed to result in an increase in basal activity of DA firing and greater responsivity of the system when phasic DA neuronal activity is triggered (Grace 1991).

Due to the sensitivity of the PFC and the DAergic system during childhood, the developmental timing of the stressor, in combination with the length of time the individual is exposed for, can also affect the impact on DA-related processes (Sinclair et al. 2014). Stressors in early developmental stages have been found to increase striatal DA levels in 10-day-old and juvenile rats (Kehoe et al. 1996a, b) and to be related to increased DA release in the ventral striatum in response to acute psychosocial stress in humans who have experienced poor early life maternal care (Pruessner et al. 2004). In terms of there being a direct link between the effects of stress on the DAergic system and SWM, previous research has shown that chronic stress induces impairment of SWM, along with a decrease in DA transmission and a concomitant increase in D1 receptor density in the PFC (Mizoguchi et al. 2000). In addition, prefrontal cortical D2 receptors have been proposed to regulate responsivity to stress via subcortical projections which act to attenuate DA release in the midbrain (Deutch 1993).

4.1 Environment × Gene Interactions

It may be that the mixed evidence to date in terms of the effects of genetic variation in the DAergic system on processes key to SWM may be partly attributable to variation in an individual's exposure to adverse life events and associated activation of the stress response system. There is some evidence to support this with studies investigating *COMT* Val158Met showing that Met allele carriers have greater stress responsivity and hypothalamic-pituitary-adrenal axis activity (Hernaus et al. 2013; Walder et al. 2010) and impaired WM function under acute stress (Buckert et al. 2012). Furthermore, *COMT* knockout mice exhibit decreased resilience to stress and better WM performance, whereas overexpression of the human *COMT* Val polymorphism may result in a blunted stress response and impaired attentional set-shifting ability, as well as reduced working and recognition memory (Papaleo et al. 2008).

Research has also started to explore gene methylation as an epigenetic mechanism key to the environmental regulation of gene expression and activity (Jirtle and Skinner 2007; Montano et al. 2016). There has been some research investigating the effect of *COMT* methylation on WM tasks and PFC activity and the further association with stress (Ursini et al. 2011; Walton et al. 2014). The effects of childhood malnutrition on the methylation profiles in the region of several genes including *COMT* and further on attentional abilities have also been noted (Peter et al. 2016). Furthermore, *COMT* has been shown to interact with methylenetetrahydrofolate reductase (*MTHFR* rs1801133), important in the regulation of intracellular methylation, in affecting SWM performance in both healthy groups and schizophrenia patients (Kontis et al. 2013). *DRD2* methylation has also been investigated in the context of schizophrenia (Kordi-Tamandani et al. 2013; Yoshino et al. 2016; Zhang et al. 2007a), and an interactive effect of an eating disorder diagnosis and childhood adversity on *DRD2* methylation has been noted (Groleau et al. 2014). In addition, although our group did not find evidence for an association between childhood adversity and *DRD2* methylation levels, we did find a significant positive association with number of errors made on a SWM task and increasing levels of *DRD2* methylation at two CpG sites within the promotor region in a sample of healthy adult males (Klaus et al. 2019b).

5 Precision Psychiatry and Mental Health Implications

In order to improve the efficacy of medications, increasing attention is now directed to individualised therapeutic methods. Approaches such as pharmacogenetics and pharmacogenomics, which investigate how genetic variation may influence the individual's response to medications, are gaining ground. Findings from healthy subjects have shown that the COMT inhibitor tolcapone, in adjunctive treatment of Parkinson's disease, results in impaired WM performance in *COMT* Met homozygotes (associated with higher PFC DA levels) but increased performance in Val homozygotes (associated with decreased PFC DA levels) in a visual n-back task of WM (Farrell et al. 2012). Similarly, the administration of pro-DAergic agents such as amphetamine has been found to improve PFC efficiency and the performance on an n-back task in healthy Val homozygotes by increasing DA availability in the PFC.

However, amphetamine exposure impaired n-back task performance and decreased PFC efficiency in Met homozygotes during the most taxing version, 3-back condition of the task (Mattay et al. 2003). The studies on schizophrenia patients have, rather unanimously, found that Met homozygotes show greater improvement in WM tasks after antipsychotic medication (i.e. dopamine receptor antagonist) treatment, compared to other genotype carriers (Burdick et al. 2011). These findings lend evidence once again to the inverted U-shaped DA function in the PFC and suggest that Val allele carriers benefit cognitively from pro-DAergic agents, whereas DA blockade predicts beneficial WM effects in the Met allele carriers. Indeed, in a large sample of Parkinson's disease patients ($n = 372$), *COMT* Val158Met genotype was associated with SWM performance in an inverted U-shaped manner, whereby Val158Met heterozygotes made fewer errors than either homozygote group. These results from Val homozygotes may result from insufficient DA receptor stimulation, whereas the Met homozygotes may be subject to DA receptor overstimulation (Fallon et al. 2015).

COMT Val158Met Val/Val genotype has also been associated with a higher percentage of gambling disorders compared to other genotype carriers, as well as with increased errors on the CANTAB SWM task in the same study. As lower WM capacity is associated with higher levels of automatic behaviour, the authors suggested that WM deficits may result in impaired self-regulation, making these individuals more likely to engage in gambling behaviours (Grant et al. 2015). As tolcapone has previously been shown to reverse WM deficits in healthy Val homozygotes (Farrell et al. 2012; Giakoumaki et al. 2008), it has been proposed that gamblers who are the carriers of the risk genotype (Val/Val) and exhibit cognitive dysfunction may benefit from this medication (Grant et al. 2013, 2015).

Other genes, such as *DAT1* have also recently received attention in the area of pharmacogenetics. A study on 108 drug-naive children and adolescents with the diagnosis of ADHD investigated the response to methylphenidate – a NE and DA reuptake inhibitor – treatment in relation to *DAT1* VNTR genotype. It was found that the 10/10 VNTR carriers showed significant gains after 8 and 24 weeks of treatment, compared to 9/9 and 9/10 genotype carriers as measured by the performance on the visuospatial n-back task. Furthermore, the gains on the WM performance were maintained after 8 weeks of methylphenidate discontinuation (Pasini et al. 2013). More recently, attention has been directed to DA D1 receptors, which are abundantly expressed in the PFC and the optimal function of which is needed for WM function. It is hoped that D1 agonists may provide a future avenue for treating cognitive deficits in schizophrenia (Arnsten et al. 2017). Altogether, the above results suggest that the dopaminergic agents may not benefit all patients, but rather these medications may prove an efficient cognitive enhancer for a subgroup of patients with a particular genetic background.

6 Conclusions and Future Directions

Executive function deficits, including WM, are a promising candidate for an intermediate transdiagnostic phenotype across multiple mental health disorders, and understanding the shared and unique genetic and environmental mechanisms involved is of increasing interest to the clinical and research communities (European Network of National Networks studying Gene-Environment Interactions in Schizophrenia et al. 2014; Miller and Rockstroh 2013; Reininghaus et al. 2013). Although not fully comprehensive in its scope, we hope that it is clear from the research presented in this chapter that DA plays a fundamental role in WM and that DA activity and function can be altered as a result of individual genetic variation and environmental stressors. The mechanisms by which these and other genetic variants influence processes such as salience evaluation and WM are really key to our ability to help identify individuals most at risk of mental health problems and enable more effective treatment when needed.

References

Abercrombie ED, Keefe KA, DiFrischia DS, Zigmond MJ (1989) Differential effect of stress on in vivo dopamine release in striatum, nucleus accumbens, and medial frontal cortex. J Neurochem 52(5):1655–1658. https://doi.org/10.1111/j.1471-4159.1989.tb09224.x

Arnsten AF (2015) Stress weakens prefrontal networks: molecular insults to higher cognition. Nat Neurosci 18(10):1376–1385. https://doi.org/10.1038/nn.4087

Arnsten AF, Paspalas CD, Gamo NJ, Yang Y, Wang M (2010) Dynamic network connectivity: a new form of neuroplasticity. Trends Cogn Sci 14(8):365–375. https://doi.org/10.1016/j.tics.2010.05.003

Arnsten AF, Girgis RR, Gray DL, Mailman RB (2017) Novel dopamine therapeutics for cognitive deficits in schizophrenia. Biol Psychiatry 81(1):67–77. https://doi.org/10.1016/j.biopsych.2015.12.028

Asghari V, Sanyal S, Buchwaldt S, Paterson A, Jovanovic V, Van Tol HHM (1995) Modulation of intracellular cyclic AMP levels by different human dopamine D4 receptor variants. J Neurochem 65(3):1157–1165. https://doi.org/10.1046/j.1471-4159.1995.65031157.x

Babovic D, O'Tuathaigh CM, O'Connor AM, O'Sullivan GJ, Tighe O, Croke DT et al (2008) Phenotypic characterization of cognition and social behavior in mice with heterozygous versus homozygous deletion of catechol-O-methyltransferase. Neuroscience 155(4):1021–1029. https://doi.org/10.1016/j.neuroscience.2008.07.006

Balciuniene J, Emilsson L, Oreland L, Pettersson U, Jazin E (2002) Investigation of the functional effect of monoamine oxidase polymorphisms in human brain. Hum Genet 110(1):1–7. https://doi.org/10.1007/s00439-001-0652-8

Barnett JH, Heron J, Ring SM, Golding J, Goldman D, Xu K, Jones PB (2007a) Gender-specific effects of the catechol-o-methyltransferase val 108/158 met polymorphism on cognitive function in children. Am J Psychiatr 164(1):142–149. https://doi.org/10.1176/ajp.2007.164.1.142

Barnett JH, Jones PB, Robbins TW, Müller U (2007b) Effects of the catechol-O-methyltransferase Val158Met polymorphism on executive function: a meta-analysis of the Wisconsin card sort test in schizophrenia and healthy controls. Mol Psychiatry 12(5):502–509. https://doi.org/10.1038/sj.mp.4001973

Barnett JH, Scoriels L, Munafò MR (2008) Meta-analysis of the cognitive effects of the catechol-O-methyltransferase gene Val158/108Met polymorphism. Biol Psychiatry 64(2):137–144. https://doi.org/10.1016/j.biopsych.2008.01.005

Barnett JH, Xu K, Heron J, Goldman D, Jones PB (2011) Cognitive effects of genetic variation in monoamine neurotransmitter systems: a population-based study of COMT, MAOA, and 5HTTLPR. Am J Med Genet B Neuropsychiatr Genet 156(2):158–167. https://doi.org/10.1002/ajmg.b.31150

Berryhill ME, Wiener M, Stephens JA, Lohoff FW, Coslett HB (2013) COMT and ANKK1-Taq-Ia genetic polymorphisms influence visual working memory. PLoS One 8(1):e55862. https://doi.org/10.1371/journal.pone.0055862

Blanchard MM, Chamberlain SR, Roiser J, Robbins TW, Muller U (2011) Effects of two dopamine-modulating genes (DAT1 9/10 and COMT Val/met) on n-back working memory performance in healthy volunteers. Psychol Med 41(3):611–618. https://doi.org/10.1017/s003329171000098x

Brehmer Y, Westerberg H, Bellander M, Fürth D, Karlsson S, Bäckman L (2009) Working memory plasticity modulated by dopamine transporter genotype. Neurosci Lett 467(2):117–120. https://doi.org/10.1016/j.neulet.2009.10.018

Brooks DJ (2016) Molecular imaging of dopamine transporters. Ageing Res Rev 30:114–121. https://doi.org/10.1016/j.arr.2015.12.009

Bruder GE, Keilp JG, Xu H, Shikhman M, Schori E, Gorman JM, Gilliam TC (2005) Catechol-O-methyltransferase (COMT) genotypes and working memory: associations with differing cognitive operations. Biol Psychiatry 58(11):901–907. https://doi.org/10.1016/j.biopsych.2005.05.010

Buckert M, Kudielka BM, Reuter M, Fiebach CJ (2012) The COMT Val158Met polymorphism modulates working memory performance under acute stress. Psychoneuroendocrinology 37 (11):1810–1821. https://doi.org/10.1016/j.psyneuen.2012.03.014

Burdick KE, Gopin CB, Malhotra AK (2011) Pharmacogenetic approaches to cognitive enhancement in schizophrenia. Harv Rev Psychiatry 19(3):102–108. https://doi.org/10.3109/10673229.2011.581899

Cerasa A, Gioia MC, Fera F, Passamonti L, Liguori M, Lanza P et al (2008) Ventro-lateral prefrontal activity during working memory is modulated by MAO a genetic variation. Brain Res 1201:114–121. https://doi.org/10.1016/j.brainres.2008.01.048

Charuchinda C, Supavilai P, Karobath M, Palacios J (1987) Dopamine D2 receptors in the rat brain: autoradiographic visualization using a high-affinity selective agonist ligand. J Neurosci 7(5):1352–1360. https://doi.org/10.1523/JNEUROSCI.07-05-01352.1987

Chen J, Lipska BK, Halim N, Ma QD, Matsumoto M, Melhem S et al (2004) Functional analysis of genetic variation in catechol-O-methyltransferase (COMT): effects on mRNA, protein, and enzyme activity in postmortem human brain. Am J Hum Genet 75(5):807–821. https://doi.org/10.1086/425589

Cools R, D'Esposito M (2011) Inverted-U–shaped dopamine actions on human working memory and cognitive control. Biol Psychiatry 69(12):e113–e125. https://doi.org/10.1016/j.biopsych.2011.03.028

Cubells JF, van Kammen DP, Kelley ME, Anderson GM, O'Connor DT, Price LH et al (1998) Dopamine β-hydroxylase: two polymorphisms in linkage disequilibrium at the structural gene DBH associate with biochemical phenotypic variation. Hum Genet 102(5):533–540. https://doi.org/10.1007/s004390050736

Deutch AY (1993) Prefrontal cortical dopamine systems and the elaboration of functional corticostriatal circuits: implications for schizophrenia and Parkinson's disease. J Neural Transm Gen Sect 91(2–3):197–221. https://doi.org/10.1007/BF01245232

Duan J, Wainwright MS, Comeron JM, Saitou N, Sanders AR, Gelernter J, Gejman PV (2003) Synonymous mutations in the human dopamine receptor D2 (DRD2) affect mRNA stability and synthesis of the receptor. Hum Mol Genet 12(3):205–216. https://doi.org/10.1093/hmg/ddg055

Dumontheil I, Roggeman C, Ziermans T, Peyrard-Janvid M, Matsson H, Kere J, Klingberg T (2011) Influence of the COMT genotype on working memory and brain activity changes during development. Biol Psychiatry 70(3):222–229. https://doi.org/10.1016/j.biopsych.2011.02.027

Durstewitz D, Seamans JK (2008) The dual-state theory of prefrontal cortex dopamine function with relevance to catechol-o-methyltransferase genotypes and schizophrenia. Biol Psychiatry 64(9):739–749. https://doi.org/10.1016/j.biopsych.2008.05.015

Dzirasa K, Ramsey AJ, Takahashi DY, Stapleton J, Potes JM, Williams JK et al (2009) Hyperdopaminergia and NMDA receptor hypofunction disrupt neural phase signaling. J Neurosci 29(25):8215–8224. https://doi.org/10.1523/JNEUROSCI.1773-09.2009

Ettinger U, Merten N, Kambeitz J (2016) Meta-analysis of the association of the SLC6A3 3′-UTR VNTR with cognition. Neurosci Biobehav Rev 60:72–81. https://doi.org/10.1016/j.neubiorev.2015.09.021

European Network of National Networks studying Gene-Environment Interactions in Schizophrenia, van Os J, Rutten BP, Myin-Germeys I, Delespaul P, Viechtbauer W et al (2014) Identifying gene-environment interactions in schizophrenia: contemporary challenges for integrated, large-scale investigations. Schizophr Bull 40(4):729–736. https://doi.org/10.1093/schbul/sbu069

Fallon SJ, Smulders K, Esselink RA, van de Warrenburg BP, Bloem BR, Cools R (2015) Differential optimal dopamine levels for set-shifting and working memory in Parkinson's disease. Neuropsychologia 77:42–51. https://doi.org/10.1016/j.neuropsychologia.2015.07.031

Fan H, Zhang FQ, Xu Y, Huang XZ, Sun GX, Song YQ et al (2010) An association study of DRD2 gene polymorphisms with schizophrenia in a Chinese Han population. Neurosci Lett 477 (2):53–56. https://doi.org/10.1016/j.neulet.2009.11.017

Farrell SM, Tunbridge EM, Braeutigam S, Harrison PJ (2012) COMT Val(158)met genotype determines the direction of cognitive effects produced by catechol-O-methyltransferase inhibition. Biol Psychiatry 71(6):538–544. https://doi.org/10.1016/j.biopsych.2011.12.023

Floresco SB, Blaha CD, Yang CR, Phillips AG (2001) Modulation of hippocampal and amygdalar-evoked activity of nucleus accumbens neurons by dopamine: cellular mechanisms of input selection. J Neurosci 21(8):2851–2860. https://doi.org/10.1523/JNEUROSCI.21-08-02851.2001

Froehlich TE, Lanphear BP, Dietrich KN, Cory-Slechta DA, Wang N, Kahn RS (2007) Interactive effects of a DRD4 polymorphism, Lead, and sex on executive functions in children. Biol Psychiatry 62(3):243–249. https://doi.org/10.1016/j.biopsych.2006.09.039

Giakoumaki SG, Roussos P, Bitsios P (2008) Improvement of prepulse inhibition and executive function by the COMT inhibitor tolcapone depends on COMT Val158Met polymorphism. Neuropsychopharmacology 33(13):3058. https://doi.org/10.1038/npp.2008.82

Gizer IR, Ficks C, Waldman ID (2009) Candidate gene studies of ADHD: a meta-analytic review. Hum Genet 126(1):51–90. https://doi.org/10.1007/s00439-009-0694-x

Glickstein SB, Hof PR, Schmauss C (2002) Mice lacking dopamine D2 and D3 receptors have spatial working memory deficits. J Neurosci 22(13):5619–5629. https://doi.org/10.1523/JNEUROSCI.22-13-05619.2002

Glickstein SB, DeSteno DA, Hof PR, Schmauss C (2005) Mice lacking dopamine D2 and D3 receptors exhibit differential activation of prefrontal cortical neurons during tasks requiring attention. Cereb Cortex 15(7):1016–1024. https://doi.org/10.1093/cercor/bhh202

Gluskin B, Mickey B (2017) Genetic variation and dopamine D2 receptor availability: a systematic review and meta-analysis of human in vivo molecular imaging studies. Transl Psychiatry 6(3): e747. https://doi.org/10.1038/tp.2016.22

Goldman-Rakic PS, Selemon LD (1997) Functional and anatomical aspects of prefrontal pathology in schizophrenia. Schizophr Bull 23(3):437–458. https://doi.org/10.1093/schbul/23.3.437

Goldman-Rakic PS, Muly EC 3rd, Williams GV (2000) D(1) receptors in prefrontal cells and circuits. Brain Res Brain Res Rev 31(2–3):295–301. https://doi.org/10.1016/S0165-0173(99)00045-4

Gonzalez-Castro TB, Hernandez-Diaz Y, Juarez-Rojop IE, Lopez-Narvaez ML, Tovilla-Zarate CA, Genis-Mendoza A, Alpuin-Reyes M (2016) The role of C957T, TaqI and Ser311Cys polymorphisms of the DRD2 gene in schizophrenia: systematic review and meta-analysis. Behav Brain Funct 12(1):29. https://doi.org/10.1186/s12993-016-0114-z

Grace AA (1991) Phasic versus tonic dopamine release and the modulation of dopamine system responsivity: a hypothesis for the etiology of schizophrenia. Neuroscience 41(1):1–24. https://doi.org/10.1016/0306-4522(91)90196-U

Grace AA (2016) Dysregulation of the dopamine system in the pathophysiology of schizophrenia and depression. Nat Rev Neurosci 17(8):524–532. https://doi.org/10.1038/nrn.2016.57

Grant JE, Odlaug BL, Chamberlain SR, Hampshire A, Schreiber LR, Kim SW (2013) A proof of concept study of tolcapone for pathological gambling: relationships with COMT genotype and brain activation. Eur Neuropsychopharmacol 23(11):1587–1596. https://doi.org/10.1016/j.euroneuro.2013.07.008

Grant JE, Leppink EW, Redden SA, Odlaug BL, Chamberlain SR (2015) COMT genotype, gambling activity, and cognition. J Psychiatr Res 68:371–376. https://doi.org/10.1016/j.jpsychires.2015.04.029

Green A, Youdim M (1975) Effects of monoamine oxidase inhibition by clorgyline, deprenil or tranylcypromine on 5-hydroxytryptamine concentrations in rat brain and hyperactivity following subsequent tryptophan administration. Br J Pharmacol 55(3):415–422. https://doi.org/10.1111/j.1476-5381.1975.tb06946.x

Greenwood PM, Lin M-K, Sundararajan R, Fryxell KJ, Parasuraman R (2014) Healthy aging increases the cognitive effects of two genes that influence extracellular dopamine. Psychol Aging 29(2):363. https://doi.org/10.1037/a0036109

Groleau P, Joober R, Israel M, Zeramdini N, DeGuzman R, Steiger H (2014) Methylation of the dopamine D2 receptor (DRD2) gene promoter in women with a bulimia-spectrum disorder: associations with borderline personality disorder and exposure to childhood abuse. J Psychiatr Res 48(1):121–127. https://doi.org/10.1016/j.jpsychires.2013.10.003

Harrison PJ, Tunbridge EM (2008) Catechol-O-methyltransferase (COMT): a gene contributing to sex differences in brain function, and to sexual dimorphism in the predisposition to psychiatric disorders. Neuropsychopharmacology 33(13):3037–3045. https://doi.org/10.1038/sj.npp.1301543

Heck A, Fastenrath M, Ackermann S, Auschra B, Bickel H, Coynel D et al (2014) Converging genetic and functional brain imaging evidence links neuronal excitability to working memory, psychiatric disease, and brain activity. Neuron 81(5):1203–1213. https://doi.org/10.1016/j.neuron.2014.01.010

Heinz A, Goldman D, Jones DW, Palmour R, Hommer D, Gorey JG et al (2000) Genotype influences in vivo dopamine transporter availability in human striatum. Neuropsychopharmacology 22(2):133. https://doi.org/10.1016/S0893-133X(99)00099-8

Hernaus D, Collip D, Lataster J, Ceccarini J, Kenis G, Booij L et al (2013) COMT Val 158 met genotype selectively alters prefrontal [18 F] fallypride displacement and subjective feelings of stress in response to a psychosocial stress challenge. PLoS One 8(6):e65662. https://doi.org/10.1371/journal.pone.0065662

Herrmann MJ, Walter A, Schreppel T, Ehlis AC, Pauli P, Lesch KP, Fallgatter AJ (2007) D4 receptor gene variation modulates activation of prefrontal cortex during working memory. Eur J Neurosci 26(10):2713–2718. https://doi.org/10.1111/j.1460-9568.2007.05921.x

Hirvonen M, Laakso A, Nagren K, Rinne JO, Pohjalainen T, Hietala J (2004) C957T polymorphism of the dopamine D2 receptor (DRD2) gene affects striatal DRD2 availability in vivo. Mol Psychiatry 9(12):1060–1061. https://doi.org/10.1038/sj.mp.4001561

Hirvonen M, Laakso A, Nagren K, Rinne JO, Pohjalainen T, Hietala J (2005) C957T polymorphism of the dopamine D2 receptor (DRD2) gene affects striatal DRD2 availability in vivo. Mol Psychiatry 10(9):889. https://doi.org/10.1038/sj.mp.4001707

Hirvonen MM, Laakso A, Nagren K, Rinne JO, Pohjalainen T, Hietala J (2009a) C957T polymorphism of dopamine D2 receptor gene affects striatal DRD2 in vivo availability by changing the receptor affinity. Synapse 63(10):907–912. https://doi.org/10.1002/syn.20672

Hirvonen MM, Lumme V, Hirvonen J, Pesonen U, Nagren K, Vahlberg T et al (2009b) C957T polymorphism of the human dopamine D2 receptor gene predicts extrastriatal dopamine receptor availability in vivo. Prog Neuro-Psychopharmacol Biol Psychiatry 33(4):630–636. https://doi.org/10.1016/j.pnpbp.2009.02.021

Hoexter MQ, Fadel G, Felicio AC, Calzavara MB, Batista IR, Reis MA et al (2012) Higher striatal dopamine transporter density in PTSD: an in vivo SPECT study with [(99m)Tc]TRODAT-1. Psychopharmacology 224(2):337–345. https://doi.org/10.1007/s00213-012-2755-4

Holly EN, Miczek KA (2016) Ventral tegmental area dopamine revisited: effects of acute and repeated stress. Psychopharmacology 233(2):163–186. https://doi.org/10.1007/s00213-015-4151-3

Howes OD, Kapur S (2009) The dopamine hypothesis of schizophrenia: version III – the final common pathway. Schizophr Bull 35(3):549–562. https://doi.org/10.1093/schbul/sbp006

Howes OD, McCutcheon R, Owen MJ, Murray RM (2017) The role of genes, stress, and dopamine in the development of schizophrenia. Biol Psychiatry 81(1):9–20. https://doi.org/10.1016/j.biopsych.2016.07.014

Jirtle RL, Skinner MK (2007) Environmental epigenomics and disease susceptibility. Nat Rev Genet 8(4):253. https://doi.org/10.1038/nrg2045

Jönsson E, Nöthen M, Grünhage F, Farde L, Nakashima Y, Propping P, Sedvall G (1999) Polymorphisms in the dopamine D2 receptor gene and their relationships to striatal dopamine receptor density of healthy volunteers. Mol Psychiatry 4(3):290. https://doi.org/10.1038/sj.mp.4000532

Karoum F, Chrapusta SJ, Egan MF (1994) 3-Methoxytyramine is the major metabolite of released dopamine in the rat frontal cortex: reassessment of the effects of antipsychotics on the dynamics of dopamine release and metabolism in the frontal cortex, nucleus accumbens, and striatum by a simple two pool model. J Neurochem 63(3):972–979. https://doi.org/10.1046/j.1471-4159.1994.63030972.x

Kehoe P, Clash K, Skipsey K, Shoemaker W (1996a) Brain dopamine response in isolated 10-day-old rats: assessment using D2 binding and dopamine turnover. Pharmacol Biochem Behav 53(1):41–49. https://doi.org/10.1016/0091-3057(95)00196-4

Kehoe P, Shoemaker WJ, Triano L, Hoffman J, Arons C (1996b) Repeated isolation in the neonatal rat produces alterations in behavior and ventral striatal dopamine release in the juvenile after amphetamine challenge. Behav Neurosci 110(6):1435–1444

Kellendonk C, Simpson EH, Polan HJ, Malleret G, Vronskaya S, Winiger V et al (2006) Transient and selective overexpression of dopamine D2 receptors in the striatum causes persistent abnormalities in prefrontal cortex functioning. Neuron 49(4):603–615. https://doi.org/10.1016/j.neuron.2006.01.023

Klaus K, Butler K, Durrant SJ, Ali M, Inglehearn CF, Hodgson TL et al (2017) The effect of COMT Val158Met and DRD2 C957T polymorphisms on executive function and the impact of early life stress. Brain Behav 7(5):e00695. https://doi.org/10.1002/brb3.695

Klaus K, Butler K, Curtis F, Bridle C, Pennington K (2019a) The effect of ANKK1 Taq1A and DRD2 C957T polymorphisms on executive function: a systematic review and meta-analysis. Neurosci Biobehav Rev 100:224–236. https://doi.org/10.1016/j.neubiorev.2019.01.021

Klaus K, Fachim HF, Reynolds GP, Pennington K (2019b) The effect of early life stress on human dopamine receptor D2 (DRD2) and parvalbumin (PVALB) promotor methylation and the association with cognition. Eur Neuropsychopharmacol 29:S414

Kontis D, Theochari E, Fryssira H, Kleisas S, Sofocleous C, Andreopoulou A et al (2013) COMT and MTHFR polymorphisms interaction on cognition in schizophrenia: an exploratory study. Neurosci Lett 537:17–22. https://doi.org/10.1016/j.neulet.2013.01.012

Kordi-Tamandani DM, Sahranavard R, Torkamanzehi A (2013) Analysis of association between dopamine receptor genes' methylation and their expression profile with the risk of schizophrenia. Psychiatr Genet 23(5):183–187. https://doi.org/10.1097/YPG.0b013e328363d6e1

Lee T, Mosing MA, Henry JD, Trollor JN, Ames D, Martin NG et al (2012) Genetic influences on four measures of executive functions and their covariation with general cognitive ability: the older Australian twins study. Behav Genet 42(4):528–538. https://doi.org/10.1007/s10519-012-9526-1

Lee JC, Wang LP, Tsien JZ (2016) Dopamine rebound-excitation theory: putting brakes on PTSD. Front Psych 7:163. https://doi.org/10.3389/fpsyt.2016.00163

Li B, Arime Y, Hall FS, Uhl GR, Sora I (2010) Impaired spatial working memory and decreased frontal cortex BDNF protein level in dopamine transporter knockout mice. Eur J Pharmacol 628 (1):104–107. https://doi.org/10.1016/j.ejphar.2009.11.036

Lidow M, Goldman-Rakic P, Gallager D, Rakic P (1991) Distribution of dopaminergic receptors in the primate cerebral cortex: quantitative autoradiographic analysis using [3H] raclopride, [3H] spiperone and [3H] SCH23390. Neuroscience 40(3):657–671. https://doi.org/10.1016/0306-4522(91)90003-7

Liu L, Fan D, Ding N, Hu Y, Cai G, Wang L et al (2014) The relationship between DRD2 gene polymorphisms (C957T and C939T) and schizophrenia: a meta-analysis. Neurosci Lett 583:43–48. https://doi.org/10.1016/j.neulet.2014.09.024

Lotta T, Vidgren J, Tilgmann C, Ulmanen I, Melen K, Julkunen I, Taskinen J (1995) Kinetics of human soluble and membrane-bound catechol O-methyltransferase: a revised mechanism and description of the thermolabile variant of the enzyme. Biochemistry 34(13):4202–4210. https://doi.org/10.1021/bi00013a008

Madras BK, Miller GM, Fischman AJ (2005) The dopamine transporter and attention-deficit/hyperactivity disorder. Biol Psychiatry 57(11):1397–1409. https://doi.org/10.1016/j.biopsych.2004.10.011

Markett SA, Montag C, Reuter M (2010) The association between dopamine DRD2 polymorphisms and working memory capacity is modulated by a functional polymorphism on the nicotinic receptor gene CHRNA4. J Cogn Neurosci 22(9):1944–1954. https://doi.org/10.1162/jocn.2009.21354

Mattay VS, Goldberg TE, Fera F, Hariri AR, Tessitore A, Egan MF et al (2003) Catechol O-methyltransferase val158-met genotype and individual variation in the brain response to amphetamine. Proc Natl Acad Sci 100(10):6186–6191. https://doi.org/10.1073/pnas.0931309100

Meltzer HY, Stahl SM (1976) The dopamine hypothesis of schizophrenia: a review. Schizophr Bull 2(1):19–76. https://doi.org/10.1093/schbul/2.1.19

Mier D, Kirsch P, Meyer-Lindenberg A (2010) Neural substrates of pleiotropic action of genetic variation in COMT: a meta-analysis. Mol Psychiatry 15(9):918. https://doi.org/10.1038/mp.2009.36

Miller GA, Rockstroh B (2013) Endophenotypes in psychopathology research: where do we stand? Annu Rev Clin Psychol 9:177–213. https://doi.org/10.1146/annurev-clinpsy-050212-185540

Mitchell R, Howlett S, Earl L, White N, McComb J, Schanfield M et al (2000) Distribution of the 3'VNTR polymorphism in the human dopamine transporter gene in world populations. Hum Biol 72:295–304. https://www.jstor.org/stable/41465826

Mizoguchi K, Yuzurihara M, Ishige A, Sasaki H, Chui DH, Tabira T (2000) Chronic stress induces impairment of spatial working memory because of prefrontal dopaminergic dysfunction. J Neurosci 20(4):1568–1574. https://doi.org/10.1523/JNEUROSCI.20-04-01568.2000

Montano C, Taub MA, Jaffe A, Briem E, Feinberg JI, Trygvadottir R et al (2016) Association of DNA methylation differences with schizophrenia in an epigenome-wide association study. JAMA Psychiat 73(5):506–514. https://doi.org/10.1001/jamapsychiatry.2016.0144

Morice E, Billard J-M, Denis C, Mathieu F, Betancur C, Epelbaum J et al (2007) Parallel loss of hippocampal LTD and cognitive flexibility in a genetic model of hyperdopaminergia. Neuropsychopharmacology 32(10):2108. https://doi.org/10.1038/sj.npp.1301354

Nagel IE, Chicherio C, Li SC, von Oertzen T, Sander T, Villringer A et al (2008) Human aging magnifies genetic effects on executive functioning and working memory. Front Hum Neurosci 2:8. https://doi.org/10.3389/neuro.09.001.2008

Neville MJ, Johnstone EC, Walton RT (2004) Identification and characterization of ANKK1: a novel kinase gene closely linked to DRD2 on chromosome band 11q23. 1. Hum Mutat 23 (6):540–545. https://doi.org/10.1002/humu.20039

Noaín D, Avale ME, Wedemeyer C, Calvo D, Peper M, Rubinstein M (2006) Identification of brain neurons expressing the dopamine D4 receptor gene using BAC transgenic mice. Eur J Neurosci 24(9):2429–2438. https://doi.org/10.1111/j.1460-9568.2006.05148.x

Nutt DJ, Lingford-Hughes A, Erritzoe D, Stokes PR (2015) The dopamine theory of addiction: 40 years of highs and lows. Nat Rev Neurosci 16(5):305–312. https://doi.org/10.1038/nrn3939

Oak JN, Oldenhof J, van Tol HH (2000) The dopamine D4 receptor: one decade of research. Eur J Pharmacol 405(1–3):303–327. https://doi.org/10.1016/S0014-2999(00)00562-8

Opmeer EM, Kortekaas R, van Tol M-J, van der Wee NJ, Woudstra S, van Buchem MA et al (2013) Influence of COMT val158met genotype on the depressed brain during emotional processing and working memory. PLoS One 8(9):e73290. https://doi.org/10.1371/journal.pone.0073290

Papaleo F, Crawley JN, Song J, Lipska BK, Pickel J, Weinberger DR, Chen J (2008) Genetic dissection of the role of catechol-O-methyltransferase in cognition and stress reactivity in mice. J Neurosci 28(35):8709–8723. https://doi.org/10.1523/JNEUROSCI.2077-08.2008

Papenberg G, Bäckman L, Nagel IE, Nietfeld W, Schröder J, Bertram L et al (2014) COMT polymorphism and memory dedifferentiation in old age. Psychol Aging 29(2):374–383. https://doi.org/10.1037/a0033225

Parasuraman R, Greenwood PM, Kumar R, Fossella J (2005) Beyond heritability: neurotransmitter genes differentially modulate visuospatial attention and working memory. Psychol Sci 16 (3):200–207. https://doi.org/10.1111/j.0956-7976.2005.00804.x

Pasini A, Sinibaldi L, Paloscia C, Douzgou S, Pitzianti MB, Romeo E et al (2013) Neurocognitive effects of methylphenidate on ADHD children with different DAT genotypes: a longitudinal open label trial. Eur J Paediatr Neurol 17(4):407–414. https://doi.org/10.1016/j.ejpn.2013.02.002

Perlstein WM, Carter CS, Noll DC, Cohen JD (2001) Relation of prefrontal cortex dysfunction to working memory and symptoms in schizophrenia. Am J Psychiatry 158(7):1105–1113. https://doi.org/10.1176/appi.ajp.158.7.1105

Peter CJ, Fischer LK, Kundakovic M, Garg P, Jakovcevski M, Dincer A et al (2016) DNA methylation signatures of early childhood malnutrition associated with impairments in attention and cognition. Biol Psychiatry 80(10):765–774. https://doi.org/10.1016/j.biopsych.2016.03.2100

Pohjalainen T, Rinne JO, Nagren K, Syvalahti E, Hietala J (1998) Sex differences in the striatal dopamine D2 receptor binding characteristics in vivo. Am J Psychiatry 155(6):768–773. https://doi.org/10.1176/ajp.155.6.768

Pruessner JC, Champagne F, Meaney MJ, Dagher A (2004) Dopamine release in response to a psychological stress in humans and its relationship to early life maternal care: a positron emission tomography study using [11C]raclopride. J Neurosci 24(11):2825–2831. https://doi.org/10.1523/JNEUROSCI.3422-03.2004

Reininghaus U, Priebe S, Bentall RP (2013) Testing the psychopathology of psychosis: evidence for a general psychosis dimension. Schizophr Bull 39(4):884–895. https://doi.org/10.1093/schbul/sbr182

Ritchie T, Noble EP (2003) Association of seven polymorphisms of the D2 dopamine receptor gene with brain receptor-binding characteristics. Neurochem Res 28(1):73–82. https://doi.org/10.1023/A:1021648128758

Rommelse NNJ, Altink ME, Arias-Vásquez A, Buschgens CJM, Fliers E, Faraone SV et al (2008) Differential association between MAOA, ADHD and neuropsychological functioning in boys and girls. Am J Med Genet Part B Neuropsychiatr Genet 147B(8):1524–1530. https://doi.org/10.1002/ajmg.b.30845

Sabol SZ, Hu S, Hamer D (1998) A functional polymorphism in the monoamine oxidase A gene promoter. Hum Genet 103(3):273–279. https://doi.org/10.1007/s004390050816

Seamans JK, Yang CR (2004) The principal features and mechanisms of dopamine modulation in the prefrontal cortex. Prog Neurobiol 74(1):1–58. https://doi.org/10.1016/j.pneurobio.2004.05.006

Sinclair D, Purves-Tyson TD, Allen KM, Weickert CS (2014) Impacts of stress and sex hormones on dopamine neurotransmission in the adolescent brain. Psychopharmacology 231(8):1581–1599. https://doi.org/10.1007/s00213-013-3415-z

Smith CT, Dang LC, Buckholtz JW, Tetreault AM, Cowan RL, Kessler RM, Zald DH (2017) The impact of common dopamine D2 receptor gene polymorphisms on D2/3 receptor availability: C957T as a key determinant in putamen and ventral striatum. Transl Psychiatry 7(4):e1091. https://doi.org/10.1038/tp.2017.45

Smyrnis N, Avramopoulos D, Evdokimidis I, Stefanis CN, Tsekou H, Stefanis NC (2007) Effect of schizotypy on cognitive performance and its tuning by COMT val158 met genotype variations in a large population of young men. Biol Psychiatry 61(7):845–853. https://doi.org/10.1016/j.biopsych.2006.07.019

Tunbridge E, Weickert C, Kleinman J, Herman M, Chen J, Kolachana B et al (2006a) Catechol-o-methyltransferase enzyme activity and protein expression in human prefrontal cortex across the postnatal lifespan. Cereb Cortex 17(5):1206–1212. https://doi.org/10.1093/cercor/bhl032

Tunbridge EM, Harrison PJ, Weinberger DR (2006b) Catechol-o-methyltransferase, cognition, and psychosis: Val158Met and beyond. Biol Psychiatry 60(2):141–151. https://doi.org/10.1016/j.biopsych.2005.10.024

Uchida H, Rajji TK, Mulsant BH, Kapur S, Pollock BG, Graff-Guerrero A et al (2009) D2 receptor blockade by risperidone correlates with attention deficits in late-life schizophrenia. J Clin Psychopharmacol 29(6):571–575. https://doi.org/10.1097/JCP.0b013e3181bf4ea3

Ursini G, Bollati V, Fazio L, Porcelli A, Iacovelli L, Catalani A et al (2011) Stress-related methylation of the catechol-O-methyltransferase Val158 allele predicts human prefrontal cognition and activity. J Neurosci 31(18):6692–6698. https://doi.org/10.1523/JNEUROSCI.6631-10.2011

Vaessen T, Hernaus D, Myin-Germeys I, van Amelsvoort T (2015) The dopaminergic response to acute stress in health and psychopathology: a systematic review. Neurosci Biobehav Rev 56:241–251. https://doi.org/10.1016/j.neubiorev.2015.07.008

van de Giessen EM, de Win MM, Tanck MW, van den Brink W, Baas F, Booij J (2009) Striatal dopamine transporter availability associated with polymorphisms in the dopamine transporter gene SLC6A3. J Nucl Med 50(1):45–52. https://doi.org/10.2967/jnumed.108.053652

van Ness SH, Owens MJ, Kilts CD (2005) The variable number of tandem repeats element in DAT1 regulates in vitro dopamine transporter density. BMC Genet 6(1):55. https://doi.org/10.1186/1471-2156-6-55

van Tol HHM, Wu CM, Guan H-C, Ohara K, Bunzow JR, Civelli O et al (1992) Multiple dopamine D4 receptor variants in the human population. Nature 358:149. https://doi.org/10.1038/358149a0

Vandenbergh DJ, Persico AM, Hawkins AL, Griffin CA, Li X, Jabs EW, Uhl GR (1992) Human dopamine transporter gene (DAT1) maps to chromosome 5p15. 3 and displays a VNTR. Genomics 14(4):1104–1106. https://doi.org/10.1016/S0888-7543(05)80138-7

Vogler C, Gschwind L, Coynel D, Freytag V, Milnik A, Egli T et al (2014) Substantial SNP-based heritability estimates for working memory performance. Transl Psychiatry 4:e438. https://doi.org/10.1038/tp.2014.81

Volkow ND, Gur RC, Wang GJ, Fowler JS, Moberg PJ, Ding YS et al (1998) Association between decline in brain dopamine activity with age and cognitive and motor impairment in healthy individuals. Am J Psychiatr 155(3):344–349. https://doi.org/10.1176/ajp.155.3.344

Volkow ND, Wise RA, Baler R (2017) The dopamine motive system: implications for drug and food addiction. Nat Rev Neurosci 18(12):741–752. https://doi.org/10.1038/nrn.2017.130

Walder DJ, Trotman HD, Cubells JF, Brasfield J, Tang Y, Walker EF (2010) Catechol-O-methyltransferase (COMT) modulation of cortisol secretion in psychiatrically at-risk and healthy adolescents. Psychiatr Genet 20(4):166–170. https://doi.org/10.1097/YPG.0b013e32833a1ff3

Walton E, Liu JY, Hass J, White T, Scholz M, Roessner V et al (2014) MB-COMT promoter DNA methylation is associated with working-memory processing in schizophrenia patients and healthy controls. Epigenetics 9(8):1101–1107. https://doi.org/10.4161/epi.29223

Xu H, Kellendonk CB, Simpson EH, Keilp JG, Bruder GE, Polan HJ et al (2007) DRD2 C957T polymorphism interacts with the COMT Val158Met polymorphism in human working memory ability. Schizophr Res 90(1–3):104–107. https://doi.org/10.1016/j.schres.2006.10.001

Yavich L, Forsberg MM, Karayiorgou M, Gogos JA, Mannisto PT (2007) Site-specific role of catechol-O-methyltransferase in dopamine overflow within prefrontal cortex and dorsal striatum. J Neurosci 27(38):10196–10209. https://doi.org/10.1523/JNEUROSCI.0665-07.2007

Yoshino Y, Kawabe K, Mori T, Mori Y, Yamazaki K, Numata S et al (2016) Low methylation rates of dopamine receptor D2 gene promoter sites in Japanese schizophrenia subjects. World J Biol Psychiatry 17(6):449–456. https://doi.org/10.1080/15622975.2016.1197424

Zabetian CP, Anderson G, Buxbaum S, Elston R, Ichinose H, Nagatsu T et al (2001) A quantitative-trait analysis of human plasma–dopamine β-hydroxylase activity: evidence for a major functional polymorphism at the DBH locus. Am J Hum Genet. 68:515–522. https://doi.org/10.1086/318198

Zabetian CP, Buxbaum S, Elston R, Köhnke MD, Anderson G, Joel Gelernter J, Cubells J (2003) The structure of linkage disequilibrium at the DBH locus strongly influences the magnitude of association between diallelic markers and plasma dopamine beta-hydroxylase activity. Am J Hum Genet 72:1389–1400

Zhang K, Grady CJ, Tsapakis EM, Andersen SL, Tarazi FI, Baldessarini RJ (2004) Regulation of working memory by dopamine D4 receptor in rats. Neuropsychopharmacology 29(9):1648–1655. https://doi.org/10.1038/sj.npp.1300491

Zhang AP, Yu J, Liu JX, Zhang HY, Du YY, Zhu JD et al (2007a) The DNA methylation profile within the 5′-regulatory region of DRD2 in discordant sib pairs with schizophrenia. Schizophr Res 90(1–3):97–103. https://doi.org/10.1016/j.schres.2006.11.007

Zhang Y, Bertolino A, Fazio L, Blasi G, Rampino A, Romano R et al (2007b) Polymorphisms in human dopamine D2 receptor gene affect gene expression, splicing, and neuronal activity during working memory. Proc Natl Acad Sci U S A 104(51):20552–20557. https://doi.org/10.1073/pnas.0707106104

Zilles D, Meyer J, Schneider-Axmann T, Ekawardhani S, Gruber E, Falkai P, Gruber O (2012) Genetic polymorphisms of 5-HTT and DAT but not COMT differentially affect verbal and visuospatial working memory functioning. Eur Arch Psychiatry Clin Neurosci 262(8):667–676. https://doi.org/10.1007/s00406-012-0312-0

Eye Movements in Neuropsychological Tasks

Timothy L. Hodgson, Gemma Ezard, and Frouke Hermens

Contents

Abstract This chapter reviews how recording and analysis of eye movements have been applied to understanding cognitive functioning in patients with neurological disease. Measures derived from the performance of instructed eye movement tests such as the anti-saccade and memory-guided saccade tasks have been shown to be associated with cognitive test performance and the early stages of neurodegenerative

---

T. L. Hodgson (✉) and F. Hermens
School of Psychology, University of Lincoln, Lincoln, UK
e-mail: tlhodgson@lincoln.ac.uk

G. Ezard
Lincolnshire Partnership NHS Foundation Trust, Lincoln, UK

Curr Topics Behav Neurosci (2019) 41: 393–418
DOI 10.1007/7854_2019_98
Published Online: 25 August 2019

disorders including Alzheimer's and Parkinson's disease. Other researchers have taken an ecological approach and recorded the uninstructed pattern of saccades made by patients during performance of established neuropsychological tasks. Studies that have analysed the eye movement strategies used in a number of widely used tests are reviewed, including the Corsi blocks, Tower of London, 'CANTAB' Spatial Working Memory and Brixton Spatial Anticipation test. The findings illustrate that eye movements are not purely in the service of vision, but support visuospatial working memory and forward action planning. Eye movement tests and measures also have potential for application in the assessment and diagnosis of neurological disease and cognitive impairment. Establishing large-scale normative data sets in healthy older adults and use of machine learning multivariate classifier algorithms may be key to further developing eye tracking applications in neuropsychological assessment.

Keywords Cognitive planning · Dementia · Executive function · Frontal lobe · Oculomotor · Saccades

1 Introduction

1.1 Eye Movements in Neuropsychological Assessment

Analysis of ocular motility is commonly used as a routine part of neurological clinical examinations (Bedell and Stevenson 2013; Pensyl and Benjamin 2006). Abnormalities in eye movements can indicate a wide variety of neurological conditions (Wong 2008), and sometimes the site of a brain lesion can be deduced with remarkable precision based purely on changes in the dynamic characteristics of eye movements (e.g. Schon et al. 2001). This is because the pathways involved in oculomotor control are very well understood, from the cranial nerves which innervate the extraocular muscles to the cerebral cortical centres contributing either directly or indirectly to oculomotor control via perceptual and cognitive processes (Leigh and Zee 2006).

The use of eye movement recording in neuropsychological assessment is less well developed even though many traditional forms of neuropsychological assessment implicitly demand the patient to coordinate complex patterns of eye movements even when these are not explicitly instructed. This chapter summarises some of the research which has used eye movements to assess cognitive function in healthy people and patients with neurological and psychiatric disorders. As well as enhancing understanding of the role of processes such as visuospatial memory and cognitive planning in task performance, it is the contention of the chapter that as eye tracking technology advances and the social challenges of an aging population increase, eye movement measures are likely to play a more important role in neuropsychological assessment.

1.2 From the Lab to the Real World

Two general approaches to studying eye movements and cognitive function can be identified. The first approach is referred to here as the *instructed* saccade task, where participants are given explicit instructions to make isolated saccades in response to simple visual stimuli according to a particular instruction (e.g. follow the target spot with your eyes). Traditionally these tests would have been carried out using the type of clinical laboratory illustrated in Fig. 1a. The other approach taken in eye

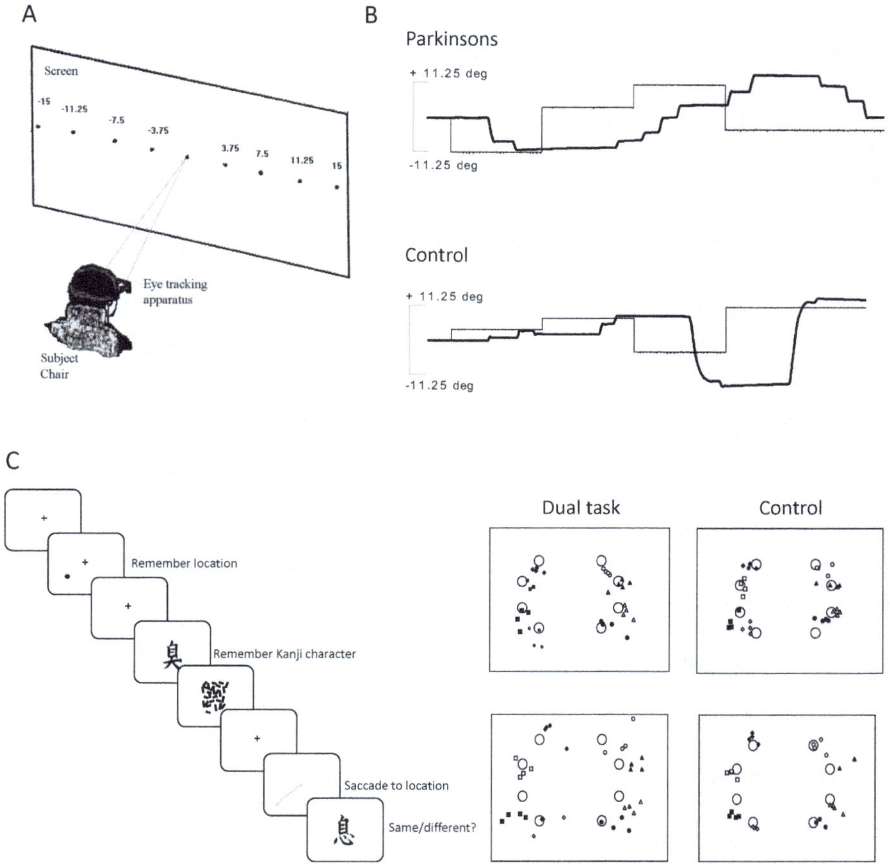

Fig. 1 (**a**) A typical eye tracking and display configuration used in a traditional clinical eye movement lab. A patient is seated in a chair wearing an infra-red eye tracking helmet and stimuli are displayed on an LED array whilst a patient performs a battery of saccade tasks based upon instructions given to them by the researcher (e.g. pro-saccade, anti-saccade and memory-guided saccades). (**b**) Parkinsonian hypometria or 'multi-stepping' of saccades in a sequential memory-guided saccade task. Example eye movement traces showing movements made in the dark by a patient and a control towards a memorised sequence of locations (Hodgson et al. 1999). (**c**) Dual task investigation of memory-guided saccades and visuospatial memory (Akdal et al. 2002). Left panel shows schematic of trial sequence in dual task condition in which participants must memorise a saccade target and a Kanji character. Right panel shows scatter plots for individual participants' memory-guided saccade endpoints in the dual task and control conditions

movement studies is referred to here as *ecological*, as it involves recording and analysis of the natural spontaneous pattern of eye movements observed during task performance, often using video-based mobile eye trackers to study eye movements in the real world outside the laboratory. These studies have shown that tasks as diverse as making a cup of tea (Land et al. 1999), reading music (Kinsler and Carpenter 1995) and driving a car (Land and Lee 1994) require the coordination of complex sequences of eye movements. Saccadic eye movements and the intervening periods of stable fixation often seem to be related to advanced planning of actions. The findings also imply an important role for trans-saccadic memory as information encoded in earlier fixations is often used to guide actions whilst the eye movement system continues to 'look ahead' towards the next goal.

The studies reviewed below show that recording eye movements during performance of neuropsychological tests used in clinical assessment can provide more detailed insights into action planning, visuospatial working memory and executive processes than is provided by accuracy and response time measures alone (Hodgson and Golding 2003; Hodgson et al. 2000, 2002; Huddy et al. 2007; Kaller et al. 2009; Mosimann et al. 2004). Research using both the instructed and ecological approach has also highlighted the potential of eye tracking measures for enhancing diagnosis and assessment of common psychiatric and neurological disorders (Benson et al. 2012; Kaufman et al. 2012; Shakespeare et al. 2015).

The rest of the chapter summarises findings using the instructed eye movement task approach and reviews studies of the natural pattern of eye movements observed during performance of established neuropsychological tests. Where possible, estimates are given for eye movement metrics' potential sensitivity and specificity for detecting the presence of a particular neurological disorder. Given their high relative prevalence and importance amongst neurodegenerative diseases, the review has a focus on Alzheimer's and Parkinson's disease, although studies involving patients with other types of dementia and schizophrenia are also included. Finally, the insights offered by the research are incorporated in an updated version of a model of oculomotor control in real-world tasks, and future directions for applications of eye movement research in neuropsychological assessment are outlined.

2 Instructed Eye Movement Tasks

2.1 Anti-saccades

The most widely used instructed saccade task to probe cognitive function is the anti-saccade task (Hallet and Adams 1980). In contrast to pro-saccades, in which the viewer must saccade towards a peripheral stimulus, in the anti-saccade task, participants are instructed to make eye movements away from the location of the target spot which steps from a central fixation to either the left or right. Increased errors, in which the first saccade after target onset is made in the direction of the target stimulus rather than in the opposite direction, occur following focal lesions to the frontal lobe in humans, and the test is considered to test the ability to exert inhibitory control over stimulus-driven saccades. Although the frontal eye fields and dorsolateral prefrontal cortex are often assumed to be the critical lesion site for deficits in oculomotor inhibitory control, structural magnetic resonance imaging analysis in patients with a variety of lesion aetiologies suggests that damage to the ventrolateral frontal cortex, rather than dorsolateral frontal cortex, is in fact the strongest predictor of increased errors in the anti-saccade task (Hodgson et al. 2007; Walker et al. 1998).

Several recent studies have also indicated that anti-saccade errors are sensitive to the early signs of Alzheimer's disease (Hellmuth et al. 2012). *Uncorrected errors* in the anti-saccade task, in which errors are directed towards the target location but not followed by a secondary saccade towards the opposite (correct) location, may be a particularly informative indicator of cognitive impairment. Crawford et al. (2005) showed that uncorrected error rates in patients with suspected Alzheimer's disease correlated significantly with the Mini-mental State Examination and Spatial Span score (see Sect. 3.2). Another study estimated that uncorrected anti-saccade errors have a sensitivity of 95% and a specificity of 65% for discriminating between patients with early stage mild Alzheimer's disease and healthy controls (Kaufman et al. 2012).

There is considerable variation in the literature concerning whether anti-saccade errors are increased in Parkinson's disease. A number of studies report significantly increased *corrected* errors in Parkinson's disease patients (Antoniades et al. 2015; Chan et al. 2005; Briand et al. 1999; Fukushima et al. 1994), whilst others do not find significant impairments relative to age-matched controls (Lueck et al. 1990; Rivaud-Pechoux et al. 2007). Several factors may influence the presence or absence of a group difference between Parkinson's patients and controls, and in our experience, there is considerable heterogeneity amongst patients with respect to anti-saccade errors. Studies of anti-saccades differ with respect to the severity of patients' symptoms as assessed by the Unified Parkinson's Disease Rating Scale Part III (UPDRS III) (e.g. mean 28.6 for Cameron et al. 2012 relative to 15.2 for Rivaud-Pechoux et al. 2007). Some studies that find a significant difference between Parkinson's and controls also report unusually low error rates for healthy control participants (4% in the case of Chan et al. 2005), whereas a more typical representative error rate for older health controls would be around 15–20% (Hutton and Ettinger 2006). Non-dopaminergic medication used in the treatment of Parkinsonian symptoms has also been shown to increase anti-saccade errors (Kitagawa et al. 1994), and studies vary with respect to the proportion of patients included who were taking anticholinergic and antipsychotic drugs at the time of testing. Finally, although most studies exclude patients with Parkinson's disease dementia, it is possible that increased anti-saccade errors in Parkinson's in a subset of patients may relate to the onset of more severe cognitive symptoms associated with more widespread neurodegenerative changes beyond the nigrostriatal dopamine system. Increased anti-saccade errors might therefore provide an early indicator of the onset of Parkinson's disease dementia although this has not been tested directly in longitudinal studies.

2.2 Memory-Guided Saccades

The memory-guided saccade task has been used to probe visuospatial memory function in healthy participants and patients. In this test the patient must refrain from making an eye movement towards a peripheral stimulus, but instead make a saccade towards the memorised location of the target. The accuracy of the memory-guided saccade can be assessed by measuring the *amplitude gain* (size) of the primary saccade made to the memorised location and the *variability* in the final eye position relative to the correct target position.

In Parkinson's disease the memory-guided saccade task can be used to elicit the hypometria (reduced amplitude) or multistepping saccades which are characteristic of eye movements in the disorder in the absence of visual cues (Crawford et al. 1989) (Fig. 1b). Hypometria is observed in Parkinson's patients on and off medication as well as in de novo patients who have never received dopaminergic medication (Crawford et al. 1997; Hodgson et al. 1999). The effect appears to be maximal when a sequence of targets has to be remembered and reproduced, i.e. a sequential memory-guided saccade task (Hodgson et al. 1999; Lueck et al. 1992), and when the testing environment is blacked out to remove any extraneous visual cues. Normal amplitude saccades are present when patients execute a memorised sequence towards an array of constantly illuminated targets, confirming that saccades are normal accepted under uncued or *open loop* conditions (Hodgson et al. 1999). Although saccades are reduced in size, patients usually reproduce the overall movement pattern presented earlier in the trial correctly, although errors where the recalled sequence is clearly incorrect are found to be slightly increased in patients. As is suggested for anti-saccades above (Sect. 2.1), it is possible that these sequence errors might reflect the onset of more severe cognitive impairments in a subset of Parkinson's patients. The magnitude of the difference in amplitude gain in memory-guided saccades between patients and controls is also such that it could have potential as a useful diagnostic marker for Parkinson's disease and also be useful in discriminating between Parkinson's and other related disorders, although this has not been tested directly.

The variability in the final eye position of memory-guided saccades may also provide an index of the accuracy of visuospatial working memory representations. Akdal et al. (2002) asked healthy participants to perform a memory-guided saccade task under dual-task conditions (Fig. 1c). In a control condition, the location of a briefly flashed target had to be maintained across a memory delay whilst holding fixation. In the dual-task condition, a secondary task was included which required participants to remember the form of an abstract shape (a Japanese Kanji character) during the delay period of the saccade task. The accuracy of memory-guided saccades under both conditions was indexed by examining the variability of the final eye position under the two conditions. The additional visuospatial memory load imposed in the Kanji condition was found to increase the variability of the final eye position of memory-guided saccades (Fig. 1c). This suggests that the target for memory-guided eye movements may be represented within the same limited capacity memory resource that stores information about the shape and form of visual objects. Variability in memory-guide saccade endpoint may represent the accuracy of mnemonic representations of spatial location.

2.3 *Stimulus-Response Rule Learning and Switching*

Other eye movement tests require participants to make saccades to the left or right not based upon memory, but on the basis of changing task rules. A series of studies have tested different patient groups on an oculomotor *rule-switching* task based upon an associative learning paradigm developed for neurophysiological studies in macaque monkeys (Asaad et al. 1998). In the original task, monkeys were given juice rewards for the correct response, but in the human version feedback is given in the form of a happy or sad face and high or low tone following execution of the saccade. Changes in rules are indicated by unexpected errors following a saccade. Both patients and controls make increased corrected saccade errors on the trial following a rule change, in which a saccade in the wrong direction is quickly followed by a secondary movement towards the correct location according to the new rule. These errors appear to reflect the influence of the old rule which still appears to exert an influence saccade programming even after the rule has changed (Hodgson et al. 2004).

Patients with Parkinson's disease are found to be impaired at learning the first rule linking the two coloured shapes with saccades to the left or right, but normal when switching between the two rules once the first rule has been acquired (Hodgson et al. 2013). This contrasts with patients with frontal lobe lesions who are able to acquire the initial association between a coloured shape and the correct direction of a saccade, but make more errors and take longer to switch between rules when the rule linking the shape with direction of response changes (Hodgson et al. 2007). This pattern of impairments in patients is remarkably consistent with neurophysiological studies using the same task which have recorded response and cue-related neural activity in both the prefrontal cortex and the caudate nucleus. Neurons in the caudate nucleus rapidly become tuned to cue-response-reward associations, whereas prefrontal cortex cells take longer to establish tuning to the conditional rules (Pasupathy and Miller 2005). One possibility is that the basal ganglia represent direct stimulus-response mappings as opposed to abstracted task rules, whereas the prefrontal cortex monitors activity within the basal ganglia and extracts information about conditional rules. Rule-based representations can then update and modify associative mappings represented within the basal ganglia when task goals change (see also Hodgson et al. 2015; Huddy et al. 2011).

Another rule-based task that has been widely used in patient assessment is the conditional pro-/anti-saccade task in which a central cue instructs the participant to execute either a pro- or an anti-saccade on each trial. Rivaud-Pechoux et al. (2007) found that patients with Parkinson's, whilst unimpaired on anti-saccades showed a differential increase in errors relative to controls in mixed task blocks where either a green (pro) or red (anti) spot at fixation instructed the correct saccade response to a target appearing to the left or right. Robert et al. (2009) used the same task in patients with Huntington's disease and presymptomatic gene carriers for the disorder. Their results showed a difference in saccade latency (response time) in the task between patient and control groups but also a relationship between the length of the disease-specific gene repeat sequence in the presymptomatic group and performance of the pro-/anti-saccade task. This suggests that saccade tests may be useful in predicting onset of Huntington's disease in presymptomatic gene carriers (see also Golding et al. 2006).

2.4 Smooth Pursuit

Although this chapter focusses on tasks and measures that assess fast saccadic eye movements, it is important to note the large number of studies showing that slow eye movements used to track continuously moving objects are a potentially useful tool in the assessment of cognitive disorders. In the smooth pursuit task, participants must track a target spot continuously moving at different velocities in either a predictable or unpredictable pattern. Key measures include *pursuit gain* (reflecting the lag between the eye position and the target position) and *time on the target* (reflecting breaks in fixation due to saccadic intrusions into the smooth tracking). Controlling continuous tracking movements in this way would seem to require a number of complex processes, including movement prediction, error detection and error correction.

Pavisic et al. (2017) report that a proportion of time on target is significantly different between patients with young onset Alzheimer's disease and age-matched controls. These authors trained an automated classifier algorithm to discriminate between the Alzheimer's and control groups based on smooth pursuit eye movement data. Using half the data set to train a Bayesian logistic regression classifier, the algorithm was then able to classify the rest of the data sets as having been derived from a control or patient with at least 95% accuracy. Smooth pursuit measures correlated with standard tests of visual cognition such as dot counting and visual cancellation tests, although correlations with other cognitive tests (e.g. digit/spatial span) were not reported. The patient group in this study also included patients with the posterior cortical atrophy variant Alzheimer's which has predominantly visual cognitive as opposed to mnemonic impairments as its defining feature.

Other work has applied similar machine learning approaches to smooth pursuit performance in patients with schizophrenia. Benson et al. (2012) found significantly increased corrective saccades in pursuit made by schizophrenic patients relative to controls, particularly when the target moved in an unpredictable 'Lissajous' pattern rather than in a predictable sine wave. A probabilistic machine learning algorithm was trained to discriminate between patients and controls based upon data from the pursuit and other eye movement tasks. When retested on a new group of patients and controls, the algorithm classified the eye movement data as belonging to either a healthy or a schizophrenic participant with close to 100% accuracy.

Other studies have confirmed the presence of smooth pursuit abnormalities in newly diagnosed schizophrenic patients, although interestingly, unlike anti-saccade errors in schizophrenia, smooth pursuit measures do not appear to correlate with tests of higher-level cognitive functioning in patients (Hutton et al. 2004). As with the memory-guided saccade task (Sect. 2.2), studies have also utilised a dual-task approach in an attempt to better understand the cognitive demands of the smooth pursuit task in health participants. Hutton and Tegally (2005) combined smooth pursuit tracking with concurrent performance of an auditory tone discrimination task and a sequential finger tapping task and found that both reduced pursuit accuracy compared to control tasks. Most recently Stubbs et al. (2018) have devised an ingenious version of the 'N-Back' working memory task, in which the pursuit target itself regularly changes colour and participants must compare the colour of the current target to the target colour 1 or 2 changes back. However, adding this concurrent working memory load was found to significantly *improve* pursuit accuracy, presumably due to an enhanced motivation to actively attend to the pursuit target to monitor colour changes.

Taken together, the evidence suggests that whilst smooth pursuit tracking relies on what appear to be high-level or even cognitive processes (e.g. prediction and error detection), pursuit may operate in parallel and independently to working memory and higher-level cognitive functions which are involved in other eye movement tests such as the anti-saccade task.

3 Ecological Studies of Eye Movements in Neuropsychological Tasks

3.1 Eye Movements and Task Schema

A complementary approach to using instructed eye movement tasks to investigate cognitive function is to measure the *scan path* of fixations and saccade during performance of cognitively demanding tasks. This ecological approach was pioneered by Land and colleagues who studied eye movements in a range of real-world tasks, most famously tea making (see Land 2019 this volume). Land and Furneaux (1997) proposed a simple but useful conceptual model of how eye movements are coordinated in real-world tasks (Fig. 2a). The model proposes that for any task, appropriate *eye movement strategies* are represented along with other actions and goals within a task *schema*. Studies by Land and others have demonstrated how eye movements play an important role in guiding (non-oculomotor) motor execution. Saccades often look ahead towards objects that are the target of actions like an 'advanced patrol' (Land and Furneaux 1997). Eye fixations can be directed towards objects which are several steps ahead in the action sequence, such

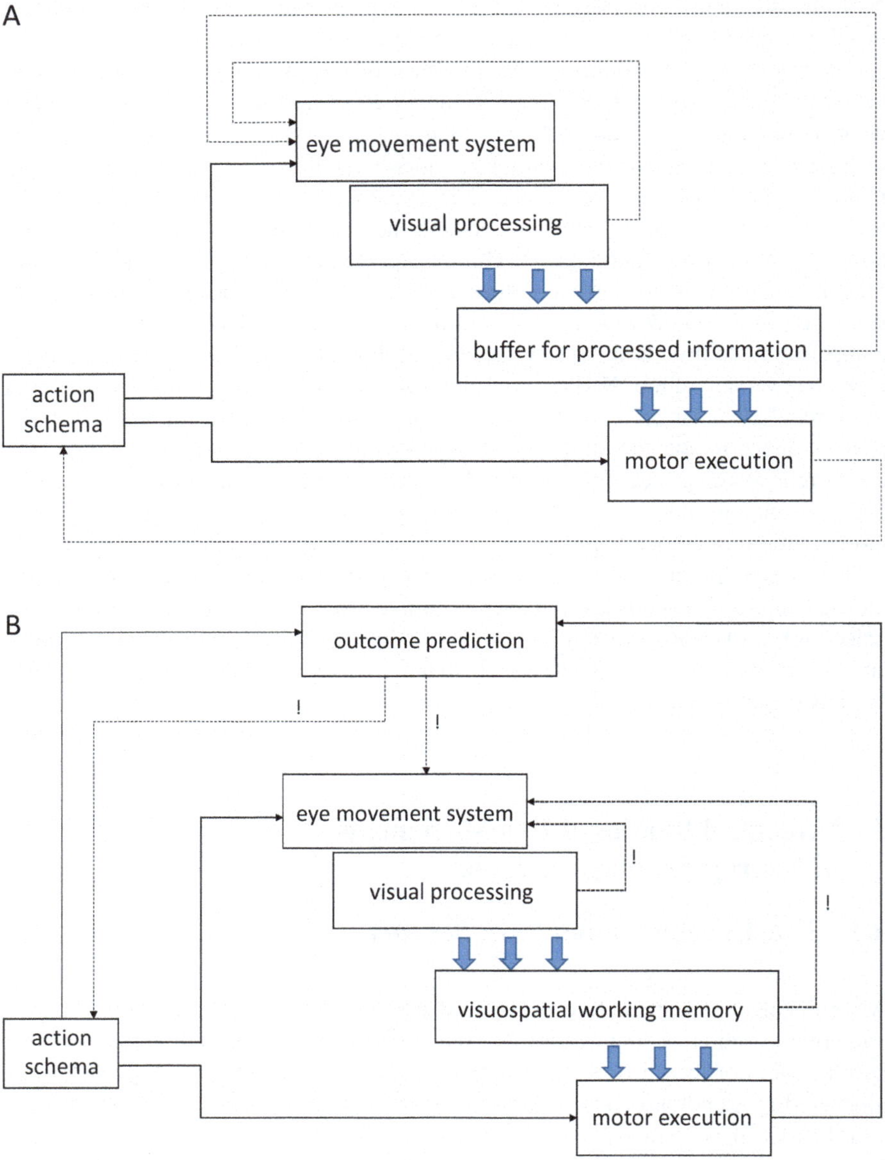

Fig. 2 (**a**) Land and Furneaux's model of eye movement control in real-world tasks. Action schema specifies eye movement strategies appropriate for extracting visual information and directing action in different contexts. A buffer for processed information maintains information across eye fixations. Feedback connections are proposed whereby sequences of eye movements and action schema can be 'chained' together. (**b**) An homage to Land and Furneaux's model with additional features added to explain eye movements in neuropsychological tasks. '!' indicates an interrupt or trigger function (see Sect. 4.1 for description)

that they may be considered as reflecting advanced cognitive planning. For this reason an 'information processing buffer' is included in the model so that information acquired via previous fixations can guide current actions. Links are also proposed between the processing buffer and visual processing by which the eye movement system can be activated and action schemas repeated.

The rest of this section reviews studies which have examined eye movements in neuropsychological tests in patients and healthy controls. As well as considering potential applications for eye tracking in neuropsychological assessment, the findings are discussed relative to the framework proposed by Land and Furneaux.

3.2 *Corsi Blocks Task*

Several studies have examined how eye movements are used in a classic neuropsychological test of spatial working memory known as the spatial span or 'Corsi blocks' test (Corsi 1972). In this test, a sequence of spatial locations is cued by the experimenter, and the participant must then point sequentially to the same locations either in the same (forward span) or reverse (backward span) order. Recordings of naturally occurring eye movements made by participants during the forward spatial span task show that an adaptive strategy taken by participants is to maintain fixation during stimulus presentation prior to reproduction of the memorised sequence (Martin et al. 2017; Patt et al. 2014). One explanation for why this might be an adaptive strategy is that shifts in fixation might disrupt retinotopically encoded visuospatial coordinates within working memory. It may be advantageous wherever possible to maintain correspondences between retinotopic and craniotopic reference frames during the spatial span task. This eye movement strategy might also facilitate configural processing of the entire sequence into a single shape or chunking of the sequence into discrete segments (see De Lillo 2019).

The role of eye movements in rehearsal processes during the period between sequence presentation and recall has also been examined. As with the sequence encoding period, Patt et al. (2014) showed that most participants tend to adopt a steady fixation strategy during the entire retention period, with only occasional looks towards target locations. Even so, Tremblay et al. (2006) observed an association between the number of sequences successfully recalled and the extent to which successive fixations mirrored the target sequence during the memory delay period. This suggests that eye movements and visuospatial working memory are closely related and that eye movement rehearsal can facilitate recall. This is also consistent with earlier work which has shown correlations between eye movement sequences during viewing and retention of random dot patterns using a visual-working memory task (Brandt and Stark 1997).

Ketcham et al. (2003) used a Corsi-like sequential memory recall task in healthy controls and patients with Parkinson's disease using a touchscreen to record pointing responses. These authors report a propensity to adopt a strategy of holding fixation, most commonly at the location of the first of the locations in the four-step sequence.

There appeared to be no significant difference in the application of this strategy or the general characteristics of saccades in the task between patients with Parkinson's and controls. However, the accuracy of the touchscreen responses made by Parkinson's patients during the recall period was found to be significantly more variable.

Recently we have recorded eye movements during a 'real-world' Corsi blocks task, using a physical block array as is commonly used in clinical practice. We categorised eye movements according to which block was currently fixated as well as the 'phase' of each task that fixation occurred in: the *Watch* phase in which the sequence was presented by the experimenter by pointing to blocks in order; a *Pre-response* phase, prior to initiation of the first movement by the participant to indicate which of the blocks they remembered in the sequence; and finally the *Response* phase during which the participant touched the blocks in the order they recalled seeing them being cued (Fig. 3a). A key question of interest was whether the eye movements made during the Watch and Pre-response phase mirrored the memorised sequence to be executed, indicative of strategic encoding and rehearsal of the eye movement sequence.

As with earlier studies (Martin et al. 2017; Patt et al. 2014), we found that although healthy control participants made an average of around 16 fixations whilst

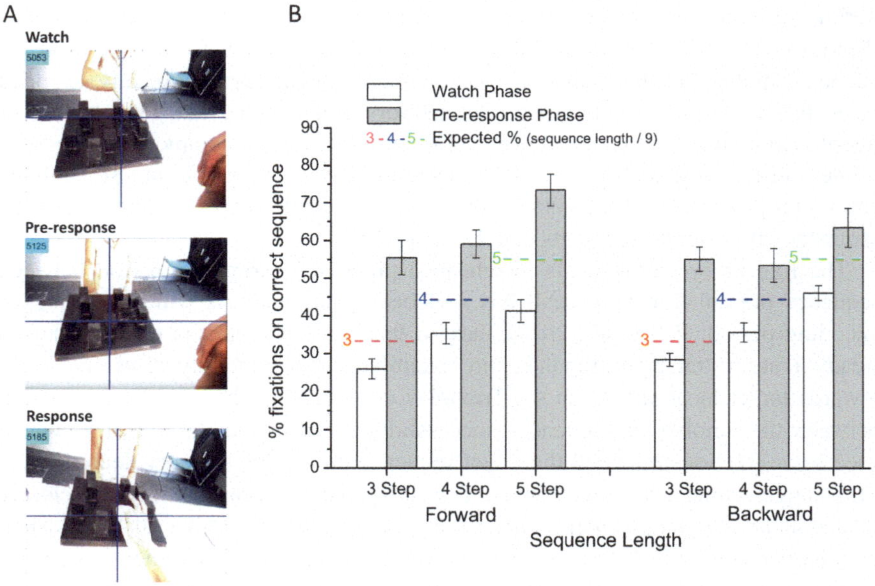

Fig. 3 'Real-world' eye tracking in the Forward and Backward Corsi Blocks task. (**a**) Indicative screen shots from a video of the participant's view overlaid with a gaze cursor during the three phases of a Corsi blocks trial. (**b**) Percentage of fixations directed towards blocks which were part of the sequence to be remembered during either the *Watch* phase (when the experimenter was pointing to the blocks) or the *Pre-response* phase (before the participant made their response). The dashed lines show the expected probability of fixating a block in the sequence if fixations were randomly distributed across blocks and unrelated to the correct sequence

viewing a four-step sequence, these fixations did not necessarily follow the sequence of blocks being pointed to by the tester. In fact, rates of fixating blocks that formed part of the correct sequence were significantly *lower* than would be expected by chance. In contrast during the *Pre-response* phase, the proportion of fixations directed towards blocks that were part of the sequence was significantly *higher* than would be expected by chance particularly in the 'backward' condition where the sequence had to be recalled in reverse (Fig. 1a). This suggests that eye movements are indeed used strategically to rehearse, maintain or plan the sequence to be executed in the spatial span task. Further studies in patients could investigate whether eye movements in the Corsi task might provide more insights into the reasons for different patients' impairments in the task than is provided by analysis of response errors alone.

3.3 Tower of London Task

Another widely used neuropsychological test which has been studied using eye movement recording is the Tower of London (TOL) task. The TOL was originally developed with the aim of testing the subtle deficits in behaviour observed following prefrontal cerebral cortex damage in humans (Shallice 1982) and is implemented in the Cambridge Neuropsychological Test Automated Battery (CANTAB[1]) as the 'Stockings of Cambridge'. In this task participants are presented with two arrangements of coloured billiard balls on top of each other in pockets. The aim is to rearrange the lower set of balls to match the upper set using the shortest possible number of moves.

In the 'one touch' variant, problems are viewed and solutions are planned but not executed, and the participant instead indicates the minimum number of moves they think are required to solve the problem once a solution has been generated through 'cognitive planning' (Owen et al. 1995). Analysis of eye movements in the one touch version is able to delineate between different cognitive operations/demands at different stages of the task. An initial period of problem *appraisal* is observed in which both the 'Goal' and the 'Start' configuration are inspected equally (in order to compare to the two arrangements of balls). This is followed by a *solution planning* period where the Start array is the focus of attention as the viewer considers the sequence of moves required to complete the problem. Towards the end of the planning period, eye movements reflect a period of *problem verification* as the Start space is reinspected (Hodgson et al. 2000).

Eye movements also revealed insights into the cognitive plan being elaborated to solve the problems. Efficient problem solvers quickly directed their gaze towards the key ball that needs to be moved in order to solve complex problems.

[1]CANTAB® [Cognitive assessment software]. Cambridge Cognition (2017). All rights reserved. www.cantab.com.

In contrast, participants that make more errors spent more time fixating balls that, whilst obviously out of place, were not the critical ones to manoeuvre first to correctly solve the puzzle (Hodgson et al. 2000). A further study by Nitschke et al. (2012) systematically varied the difficulty of problem definition and demands of cognitive planning using a series of specially designed TOL problems and measured eye movement in participants during solution. Higher demands on forming an internal problem representation were associated with an increased number of gaze alternations between Start state and Goal state (lower and upper part of the display), but did not show any effect on the durations of fixation periods. In contrast, higher demands on actual planning in terms of complexity of the sequence to be planned in working memory coincided with a significantly longer duration of the very last fixation immediately preceding movement execution.

Two studies have examined eye movements in the TOL in patient groups. Hodgson et al. (2002) recorded eye movements in a group of mild to moderately affected Parkinson's patients without dementia, whilst they completed the one touch Tower of London problem. The patients made more errors in the task but also had a qualitatively different distribution of gaze between the Goal and Start configuration. Whilst older controls directed proportionately more gaze time to the configuration of balls to be moved as problem complexity increased, patients distributed equal amounts of time to each half of the display regardless of problem difficulty and made more shifts between the upper and lower half of the problem. It is possible that patients had more difficulty in maintaining information in visuospatial memory across saccades, and this leads to a more equal distribution of gaze times between the two halves of the display.

Schizophrenic patients also make significantly increased errors in the TOL task, and whilst the spatial distribution of gaze is found to be identical to age-matched controls, a significant increase is found in the duration of eye fixations for schizophrenic patients compared to controls (Huddy et al. 2007). One interpretation of this is that schizophrenia leads to problems in encoding, rather than maintaining information in visual working memory. More processing time is required to extract relevant information, and sometimes this is not successfully achieved leading to errors where aspects of the Goal or Start space are not successfully processed and encoded. This conclusion is consistent with the idea that schizophrenic patients' deficit in working memory tasks is due to problems in perceptual encoding of stimulus features into working memory (Ichinose and Park 2019).

3.4 CANTAB Spatial Working Memory Task

Recently we have examined how eye movements contribute to another neuropsychological test which places demands on both spatial memory and cognitive planning processes (Hodgson et al. 2019). Also incorporated into CANTAB, the Spatial Working Memory (SWM) test was originally developed as a human analogue of tests of memory in animals (Olton 1982; Passingham 1985; Petrides and Milner 1982). In the SWM task, patients have to find reward tokens hidden within an array of boxes displayed on a computer screen. Patients are asked to search through the boxes by selecting them using a mouse click or touchscreen response. Following a box selection, the contents of the box are revealed as either a token (a coloured square) or as empty (a blank space). The patient is told that there is only ever one token hidden at a time, but when a token has been found, another one is immediately hidden. Crucially, a token is never hidden under a box where a token has been previously found. Compared to the Corsi blocks test, the SWM task requires patients to actively search whilst simultaneously maintaining memory for spatial locations in working memory. This makes it difficult to adopt a strategy of maintaining a constant fixation point to support retinotopic encoding of the token locations as appears to happen in the Corsi test (see Sect. 3.2).

Patients need to keep track of where they have found reward tokens in the SWM task and only search boxes where rewards have not already been found in the current set. A key measure of performance in the task are *Between search errors* (BSEs) which occur when a box where a token has already been found is reselected during a later token search within the same set. Investigations comparing patients with focal brain lesions and patients with idiopathic Parkinson's in different disease states have revealed significantly increased BSEs compared to healthy controls (Owen et al. 1995, 1997).

We found that healthy participants make around ten fixations per token search in the SWM test, and these eye movements often relate to advanced planning of the search sequence and rehearsal of the preceding search path (Hodgson et al. 2019) (Fig. 4a). Eye movements occurred at a rate higher than chance on boxes that were the targets for mouse clicks up to two selections ahead and one selection back. The extent to which participants 'look ahead' and 'look back' in this way correlated inversely with BSEs, such that individuals that look ahead and back made less errors in the task (Fig. 4b). Patients with Parkinson's disease were also found to make significantly fewer forward planning fixations consistent with a deficit in strategic planning rather than spatial working memory in these patients (Hodgson et al. 2019).

Refixations on locations where a token had been found were very rare, indicating that eye movements were guided by memory for locations where a token had already been found. Participants that made more refixations also tended to make more BSEs (although refixations did not always result in a mouse click at the refixated location). This suggests that refixations might reflect a weak memory trace or partial forgetting of locations where a token has already been found. This is consistent with a limited resource model where the quality of information may become degraded but not completely lost from memory (Aagten-Murphy and Bays 2019; Zokaei and Husain 2019). Because of this, eye movement refixations might provide a more sensitive measure of impairments in the quality of spatial memory representations than BSEs (mouse click selection errors) in the task.

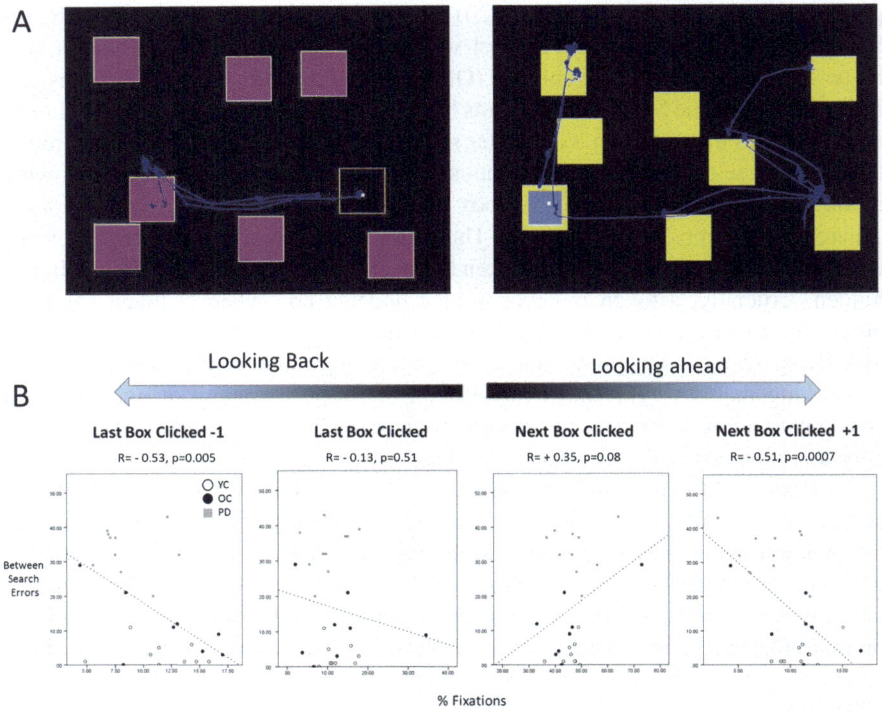

Fig. 4 (**a**) Scan paths during performance of the 'CANTAB' SWM task. Patients have to search boxes using a mouse click to find tokens. Tokens are never hidden under the same box twice such that the patient needs to remember where they have already found tokens to direct their search efficiently. Left image shows a patient with Parkinson's disease and the right image a healthy control. (**b**) Scatter plots show correlations between fixations which were directed towards boxes ahead or back along the sequence of selected boxes and errors where a box in which a token had already been found was reselected with a mouse click

3.5 *Visual Matching and Comparison Tasks*

Another class of cognitive test that requires eye movements are tasks which directly test visuospatial memory function by requiring matching or comparison between visual stimuli. Archibald et al. (2013) examined visual exploration strategies in patients with Parkinson's disease whilst they matched a cue image to four possible alternative options displayed simultaneously with the cue. These included different oriented line conjunctions (i.e. angles), the time on a clock face, shapes and overlapping animal figures. Patients with Parkinson's without cognitive impairment, with mild cognitive impairment and healthy controls were tested. Patients without cognitive impairment showed significantly increased fixation durations overall, and this effect correlated with UPDRS III scores. Cognitively impaired patients additionally showed an increase in the number of regions of interest refixated with secondary fixations. Consistent with findings described above for the TOL test (Sect. 3.3), this suggests that visuospatial working memory may be affected in the early stages of Parkinson's disease and this impairment is more pronounced in patients with measurable mild cognitive impairment. These may relate to either problems in encoding information into memory or maintaining information in memory or both.

Other studies have examined eye movements during the Visual Paired Comparison (VPC) test (Crutcher et al. 2009). This task has previously been examined in animals where hippocampal lesions have been shown to impair performance (Clark et al. 2000; Zola et al. 2000). On each trial two identical stimuli are initially presented to the left and right of a screen. Following a delay period, two stimulus objects are displayed, one of them identical to the originally presented stimulus and the other a novel object. A key measure is the amount of fixation time spent viewing the novel stimulus, with the novel stimulus receiving more fixation dwell time than the old stimulus. One recent study compared patients with mild cognitive impairment (MCI), Parkinson's disease patients and healthy controls in the task using either a 2 s or 2 min delay period. MCI patients spent a significantly reduced proportion of time viewing the novel stimulus for the 2 min but not 2 s delay period compared to both Parkinson's patients and controls. Interestingly in the context of the findings for patients in the TOL task outlined above, Parkinson's were found *not* to be impaired on the VPC task relative to controls. One possibility is that the VPC task probes the integrity of longer-term memory systems than those that facilitate the short-term integration of information across saccades.

Lagun et al. (2011) have applied automatic classification algorithms including Naïve Bayes, logistic regression and support vector machine approaches to eye movement measures derived from the VPC test in order to discriminate between patients with MCI and healthy controls. Classifier algorithms were initially trained to discriminate between healthy control data and data collected from Alzheimer's disease patients, and then its effectiveness at discriminating between data from patients with MCI and healthy controls was assessed. Proportion of viewing time on the novel stimulus remained the main predictive variable weighted by the classifier, but improved performance of the classifier algorithms was observed when other eye movement variables such as fixation duration, saccade direction and refixations were included. The best type of machine learning algorithm was found to be the support vector machine approach, which could discriminate between MCI patients and controls with 97% sensitivity and 77% specificity.

3.6 *Set Shifting*

A common feature implicit in many tasks of frontal lobe and executive function is the demand to learn, monitor and flexibly shift between task rules or 'sets' (e.g. TOL task). Section 2.3 described an *instructed* eye movement task, the oculomotor rule-switching test, which can be used to assess this function in patient groups. However, set switching is also explicitly tested by several neuropsychological tests, and the influence of task set on eye movements in these tests has been studied using eye tracking.

In the Wisconsin card sort task (Grant and Berg 1948), a card depicting a number of coloured shapes has to be matched with an array of cards based upon a single sorting dimension (colour, shape, number). The sorting dimension can change from trial to trial based either on response contingent feedback or direct instruction. Golding (2004) recorded eye movements whilst healthy participants performed a simplified version of the task in which saccades were made towards one of the three test stimuli that matched a central cue on either the colour or shape dimension. When the sorting dimension was different to that applied on the previous trial ('switch trial'), the previous sorting dimension was seen to affect the dynamics of saccades made between the Stimulus and Response cards. Saccade trajectories and endpoints appeared to be pulled towards the location of nontarget cards which shared a common feature with the Stimulus card (Hodgson and Golding 2003). The effect was particularly marked on the first two trials after a rule switch when colour was the distracting stimulus dimension (Fig. 5a). Despite this competing influence of the distracting stimulus features on eye movement dynamics, participants correctly selected the target with a mouse click on 99% of trials. This suggests that eye movement measures may be more sensitive to the effect of competition between rule sets than overt response errors. Golding (2004) also reported data in the eye movement version of the Wisconsin task in a single patient with a prefrontal cortex lesion who was found to be impaired under conditions where he had to maintain the current sorting rule in memory (rather than it being cued with an instruction on each trial), although the potential usefulness of using eye movement measures to assess impairments in set-shifting ability in patients in a clinical setting has not been otherwise explored.

Another established neuropsychological test which probes flexible switching between response sets is the Brixton Spatial Anticipation (BSA) test (Burgess and Shallice 1997). Traditionally this test is carried out by neuropsychologists using a physical booklet in which successive pages reveal an array of dots with one dot highlighted in black. As the dot moves from one location to another, it conforms to a pattern, such that after a few examples, the next location at which the highlighted dot will be shown can be predicted. It can move in a straight line, zig-zag pattern or a displaced zig-zag in which the dot jumps an extra row on every step. The task assesses a patient's ability to anticipate movement as well as to learn and switch between the rules that dictate the spot's movement pattern.

Primativo et al. (2017) adapted the BSA task into a computerised version and tracked eye movements whilst controls and patients with fronto-temporal dementia and semantic dementia performed the task. A problem with the original BSA and many of other pencil and paper tests is that they require verbal responses from a patient which can themselves be affected in many conditions, including in frontal temporal dementia. Therefore eye movement measures have the advantage of not relying on verbal communication between the patient and neuropsychologist. The researchers found that patients with behavioural variant fronto-temporal dementia made less correct anticipatory saccades in the task and more saccades to locations

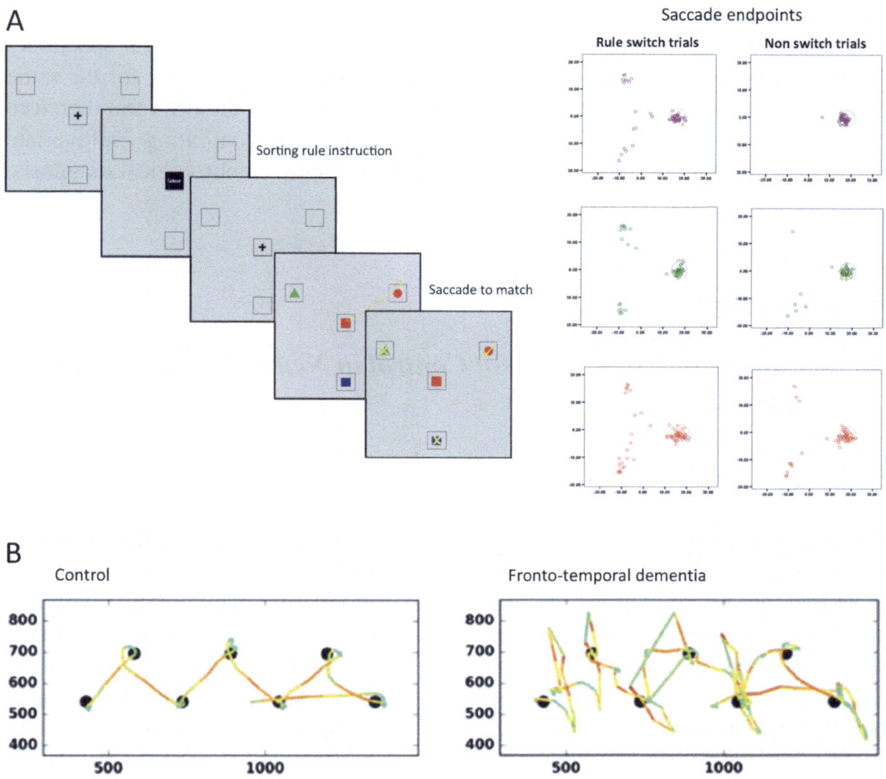

Fig. 5 (**a**) Eye movements during a modified version of the Wisconsin Card Sorting task (Golding 2004). Scatter plots show endpoints of saccades following a change in the dimensional rule which the subject had to use to match the centrally displayed cue with the correct peripheral target. The grey circle overlaid onto the scatter plots indicates the correct target location. (**b**) Examples of eye movements recorded in a computerised version of the Brixton Spatial Anticipation test in a healthy control and a fronto-temporal dementia patient. The target made a predictable zig-zag movement pattern which the patient had to track with saccadic eye movements (adapted from Primativo et al. 2017)

which were not the next step in the sequence, compared to both healthy controls and semantic dementia patients (Fig. 5b). Interestingly, these errors were not random, but were most likely to be directed towards locations that mirrored the previous movement pattern, consistent with eye movements being influenced by interference from the preceding movement rule.

As with the VPC test (Sect. 3.5; Lagun et al. 2011), automated classifier algorithms were also applied to eye movements in the BSA test. Primativo et al. (2017) used the unfiltered raw eye position time series for each subject to train a Bayesian classification algorithm. Classification based upon the eye movement metrics was found to have superior sensitivity and specificity compared to the traditional pen and paper version of the BSA test and was also more effective than standard univariate statistics for discriminating between patients with fronto-temporal dementia and those suffering from semantic dementia (91.7% sensitivity with 100% specificity for eye movements, compared to 58% sensitivity and 67% specificity for the same patients on the original BSA). Analysis of eye movements in the BSA test therefore provides further evidence of the potential usefulness of combining multivariate classifier approaches with oculomotor measures to improve neurological diagnosis.

4 Discussion

4.1 A Model of Eye Movement Control in Neuropsychological Tasks

Studies of eye movements in neuropsychological tasks are consistent with the idea proposed by Land and Furneaux (1997) that oculomotor strategies are represented alongside task goals and strategies within task schema. The studies reviewed above also emphasise the important role played by eye movements and fixations in responding to the demands of visuospatial memory and cognitive planning and show how eye movements can be sensitive to interference from recently active rules, sets or schema. As a ‘homage’ to Land and Furneaux, we propose some updates to their model based upon our own and others’ studies of eye movements in neuropsychological tasks. These are shown in Fig. 2b and described in more detail below.

The first update to the original model depiction is that the information processing buffer has been explicitly relabelled as ‘visuospatial working memory’. Rather than being a specialist memory buffer associated exclusively with the oculomotor system, it is suggested that Land and Furneaux’s information processing buffer is the same mechanism which maintains visuospatial information across delays even when the eyes are not in motion. This modification is based upon the evidence reviewed above on how eye movements are used in visual working memory tasks (e.g. Akdal et al. 2002; Brandt and Stark 1997; Hodgson et al. 2019; Tremblay et al. 2006).

The second addition is to specify an ‘interrupt’ or trigger function for the link between visuospatial memory and the oculomotor system. It is suggested that when task critical information stored within memory degrades such that it contains insufficient detail to guide action, the oculomotor system is triggered to fixate the location or object which contains the relevant information. In simple terms, visuospatial working memory can call the eye movement system for help when needed. The way eye movements are used in tests such as the TOL and SWM task is consistent with this. Under high memory load, or when visuospatial working memory is impaired, increased refixations or fixation durations are required to support working memory function. A similar triggering function is proposed for the feedback connection between the visual processing module and the oculomotor system, whereby the eye movement system can be tasked to foveate locations or objects to extract task relevant information which is not directly available within the current retinal image.

The final modification provides a mechanism via which action schema are initiated, repeated or modified. This proposes that action schema include *outcome predictions*. Once an action sequence has been completed, the outcome prediction generated in advance by the schema (what was expected to happen) is compared with the outcome as determined through sensory feedback (what actually happened). Based upon this comparison, the current action schema is then either repeated or modified or a new schema recalled from long-term memory. This modification is inspired by the studies in patients and healthy controls reviewed above which show how competing task rules and sets influences eye movements. In tasks such as oculomotor rule-switching and the BSA test (Sects. 2.3 and 3.6), detecting differences between predicted and expected sensory feedback is necessary to update rules and schema effectively, and this process can be impaired in patients with frontal lobe damage and disease.

4.2 Future of Eye Movement Metrics in Neuropsychological Assessment and Diagnosis

An aim of the chapter was to consider whether eye movement recording, tests and metrics might have value in clinical assessment and diagnosis. A common criticism of applying eye movement testing to the diagnosis of neurological and psychiatric disease is that they have high sensitivity but low specificity for particular disorders and this view appears to be supported prima facie by the findings reviewed above on both instructed eye movement tasks and ecological studies of eye movement recordings in neuropsychological tasks. A range of different neurological and psychiatric conditions are associated with increased *corrected* errors in the anti-saccade task, but even completely healthy subjects sometimes have high anti-saccade error rates (40%+). However, some measures may prove to have better specificity than others. For example, *uncorrected* errors in the anti-saccade task (Alzheimer's) or multi-stepping saccades in sequential memory-guided saccades (Parkinson's). Recording of eye movements during neuropsychological tasks may also hold more promise for discriminating between specific diseases than standard test scores especially when combined with multivariate classifier algorithms.

An increasing awareness of neurological disease amongst the general public also means that many people are concerned about their own brain health and particularly whether or not they might be developing dementia. The presence of self-reported 'subjective' complaints is in fact a very poor predictor of future diagnosis of dementia (Iliffe and Pealing 2010). Most people with subjective memory complaints do not develop dementia, whereas many without self-reported concerns do. Eye movement tests may have a number of advantages in this context. Tests with low specificity to a particular condition but high sensitivity to a wide range of disorders might in fact be useful in determining which patients require referral for detailed investigation and diagnosis and which may just need reassurance. As part of a routine 'brain health check', a short battery of simple oculomotor tests also has the benefit of not overtly testing patient's memory and cognitive function as many patients will become anxious and performance may be affected if they believe that memory and cognition is being assessed.

Nevertheless despite many decades of research, clinical applications of oculomotor tests and measures are still underdeveloped compared to pencil and paper and computerised cognitive testing. One factor which explains this is that established neuropsychological tests have benefited from the creation of standardised normative data sets which can be used as a benchmark to compare patient performance against controls (Casaletto and Heaton 2017). The development of similar normative data for eye movement tasks and measures, banded by age, IQ and other factors may therefore be key to more widespread use. Development of more robust, affordable eye tracking technology which allows reliable and accurate calibration and tracking in patients with minimal supervision and expertise on the part of the tester would also assist more widespread applications.

4.3 Final Reflections on the Relationship Between Oculomotor Control and Cognitive Function

The debate concerning whether there is a direct equivalence between oculomotor programming and covert orienting of visuospatial attention is well reviewed earlier in this volume (Hunt et al. 2019), but when one considers the integral role played by eye movements in many tests of cognitive and executive function, it is the border between oculomotor control and cognitive processing that becomes blurred. If the brain is considered as an organism for transforming sensory input into motor output, then cognitive processes must logically be reducible to highly complex sensorimotor transformations. The intricate and captivating patterns of eye movements which are observed during many neuropsychological tests[2] (apparently without conscious effort or detailed awareness on the part of the patient) offer a fascinating window onto these transformations and also offer potential practical benefits for improving neurological and neuropsychological assessment in the future.

References

Aagten-Murphy D, Bays PM (2019) Functions of memory across saccadic eye movements. In: Hodgson TL (ed) Current topics in behavioural neurosciences: processes of visuo-spatial attention and working memory. Springer, New York

Akdal G, Hodgson TL, Hill AC, Kennard C (2002) Visual object memory and memory guided saccades rely on shared mental representations. Exp Brain Res 143(4):509–514

[2]Videos showing eye movements in many of the tasks described in this chapter can be viewed online at https://www.youtube.com/user/HodgsonTim/playlists.

Antoniades CA, Demeyere N, Kennard C, Humphreys GW, Hu MT (2015) Antisaccades and executive dysfunction in early drug-naive Parkinson's disease: the discovery study. Mov Disord 30(6):843–847

Archibald NK, Hutton SB, Clarke MP, Mosimann UP, Burn DJ (2013) Visual exploration in Parkinson's disease and Parkinson's disease dementia. Brain 136:739–750

Asaad WF, Rainer G, Miller EK (1998) Neural activity in the primate prefrontal cortex during associative learning. Neuron 21:1399–1407

Bedell HE, Stevenson SB (2013) Eye movement testing in clinical examination. Vis Res 90:32–37

Benson PJ, Beedie SA, Shephard E, Giegling I, Rujescu D, St Clair D (2012) Simple viewing tests can detect eye movement abnormalities that distinguish schizophrenia cases from controls with exceptional accuracy. Biol Psychiatry 72(9):716–724

Brandt SA, Stark LW (1997) Spontaneous eye movements during visual imagery reflect the content of the visual scene. J Cogn Neurosci 9(1):27–38

Briand KA, Strallow D, Hening W, Poizner H, Sereno AB (1999) Control of voluntary and reflexive saccades in Parkinson's disease. Exp Brain Res 129:38–48

Burgess PW, Shallice T (1997) The Hayling and Brixton Test manual. Thames Valley Test Co., Bury St Edmunds

Cameron IGM, Pari G, Alahyane N, Brien DC, Coe BC, Stroman PW, Munoz DP (2012) Impaired executive function signals in motor brain regions in Parkinson's disease. NeuroImage 60(2):1156–1170

Casaletto KB, Heaton RK (2017) Neuropsychological assessment: past and future. J Int Neuropsychol Soc 23:778–790

Chan F, Armstrong IT, Pari G, Riopelle RJ, Munoz DP (2005) Deficits in saccadic eye-movement control in Parkinson's disease. Neuropsychologia 43:784–796

Clark RE, Zola SM, Squire LR (2000) Impaired recognition memory in rats after damage to the hippocampus. J Neurosci 20:8853–8860

Corsi PM (1972) Memory and the medial temporal region of the brain (unpublished doctoral dissertation). McGill University, Montreal

Crawford TJ, Henderson L, Kennard C (1989) Abnormalities of non-visual guided eye movements in Parkinson's disease. Brain 112:1573–1586

Crawford TJ, Shaunak S, OSullivan E, Blunt S, Lawden M, Kennard C (1997) A dissociation of internal representation and hypometric memory-guided saccades in de novo Parkinson's disease. Abstract presented to the European conference on eye movements, Ulm

Crawford TJ, Higham S, Renvoize T, Patel J, Dale M, Suriya A, Tetley S (2005) Inhibitory control of saccadic eye movements and cognitive impairment in Alzheimer's disease. Biol Psychiatry 57:1052–1060

Crutcher MD, Calhoun-Haney R, Manzanares CM, Lah JJ, Levey AI, Zola SM (2009) Eye tracking during a visual paired comparison task as a predictor of early dementia. Am J Alzheimers Dis Other Dement 24(3):258–266

De Lillo C (2019) Variations in the beneficial effects of spatial structure and serial organisation on working memory span in humans and other species. In: Hodgson TL (ed) Current topics in behavioural neurosciences: processes of visuo-spatial attention and working memory. Springer, New York

Fukushima J, Fukushima K, Miyasaka K, Yamashita I (1994) Voluntary control of saccadic eye movement in patients with frontal cortical lesions and Parkinsonian patients in comparison with that in schizophrenics. Biol Psychiatry 36:21–30

Golding CVP (2004) Eye movements during task switching. PhD thesis, Imperial College Medical School, University of London

Golding CVP, Danchaivijitr C, Hodgson TL, Tabrizi SJ, Kennard C (2006) Identification of an oculomotor biomarker of preclinical Huntington disease. Neurology 67(3):485–487

Grant DA, Berg E (1948) A behavioral analysis of degree of reinforcement and ease of shifting to new responses in Weigl-type card-sorting problem. J Exp Psychol 38:404–411

Hallet PE, Adams BD (1980) The predictability of saccadic latency in a novel voluntary oculomotor task. Vis Res 20:329–339

Hellmuth J, Mirsky J, Heuer HW et al (2012) Multicenter validation of a bedside antisaccade task as a measure of executive function. Neurology 78:1824–1831

Hodgson TL, Golding C (2003) Executive contributions to eye movement control. In: Hyona J, Radach R, Deubel H (eds) The mind's eye: cognitive and applied aspects of eye movement research. Elsevier, Amsterdam, pp 49–64

Hodgson TL, Dittrich W, Henderson L, Kennard C (1999) Eye movements and spatial working memory in Parkinson's disease. Neuropsychologia 37(8):927–938

Hodgson TL, Bajwa A, Owen AM, Kennard C (2000) The strategic control of gaze direction in the Tower of London task. J Cogn Neurosci 12(5):894–907

Hodgson TL, Tiesman B, Owen AM, Kennard C (2002) Abnormal gaze strategies during problem solving in Parkinson's disease. Neuropsychologia 40:411–422

Hodgson TL, Golding C, Molyva D, Rosenthal C, Kennard C (2004) Eye movements during task switching: reflexive, symbolic, and affective contributions to response selection. J Cogn Neurosci 16(2):318–330

Hodgson TL, Chamberlain M, Parris BA, James M, Gutowski NJ, Husain M, Kennard C (2007) The role of the ventrolateral frontal cortex in inhibitory oculomotor control. Brain 130:1525–1537

Hodgson TL, Sumner P, Molyva M, Sheridan R, Kennard C (2013) Learning and switching between stimulus-saccade associations in Parkinson's disease. Neuropsychologia 51 (7):1350–1360

Hodgson TL, Parris BA, Benattayallah A, Summers IR (2015) Multi-modal representation of effector modality in frontal cortex during rule switching. Front Hum Neurosci 9:486

Hodgson TL, Hermans F, Pennington K, Pickering JS, Ezard G, Clarke R, Sharma J, Owen AM (2019) Eye movements in the "Morris Maze" spatial working memory task reveal deficits in strategic planning. J Cogn Neurosci 31(4):497–509

Huddy VC, Hodgson TL, Harrison I, Kapasi M, Stanley H, Thomas M, Barnes TRE, Joyce EM (2007) Gaze strategies during performance of the Tower of London planning task in first episode schizophrenia. J Abnorm Psychol 116(3):589–598

Huddy VC, Hodgson TL, Ron MA, Barnes TRE, Joyce EM (2011) Abnormal negative feedback processing in first episode schizophrenia: evidence from an oculomotor rule switching task. Psychol Med 41(9):1805–1814

Hunt AR, Reuther J, Hilchey MD, Klein RM (2019) The relationship between spatial attention and eye movements. In: Hodgson TL (ed) Current topics in behavioural neurosciences: processes of visuo-spatial attention and working memory. Springer, New York

Hutton SB, Ettinger U (2006) The antisaccade task as a research tool in psychopathology: a critical review. Psychophysiology 43:302–313

Hutton SB, Tegally D (2005) The effects of dividing attention on smooth pursuit eye tracking. Exp Brain Res 163:306–313

Hutton SB, Huddy V, Barnes TRE, Robbins TW, Crawford TJ, Kennard C, Joyce EM (2004) The relationship between antisaccades, smooth pursuit, and executive dysfunction in first-episode schizophrenia. Biol Psychiatry 56:553–559

Ichinose M, Park S (2019) Mechanisms underlying visuospatial working memory impairments in schizophrenia. In: Hodgson TL (ed) Current topics in behavioural neurosciences: processes of visuo-spatial attention and working memory. Springer, New York

Iliffe S, Pealing L (2010) Subjective memory problems. Br Med J 340:1425

Kaller CP, Rahm B, Bolkenius K, Unterrainer JM (2009) Eye movements and visuospatial problem solving: identifying separable phases of complex cognition. Psychophysiology 46(4):818–883

Kaufman LD, Pratt J, Levine B, Black SE (2012) Executive deficits detected in mild Alzheimers disease using the antisaccade task. Brain Behav 2(1):15–21

Ketcham CJ, Hodgson TL, Kennard C, Stelmach GE (2003) Memory-motor transformations are impaired in Parkinson's disease. Exp Brain Res 149(1):30–39

Kinsler V, Carpenter RHS (1995) Saccadic eye movements while reading music. Vis Res 35:1447–1458

Kitagawa M, Fukushima J, Tashiro K (1994) Relationship between antisaccades and the clinical symptoms in Parkinson's disease. Neurology 44:2285–2289

Lagun D, Manzanares C, Zola SM, Buffalo EA, Agichtein E (2011) Detecting cognitive impairment by eye movement analysis using automatic classification algorithms. J Neurosci Methods 201:196–203

Land MF (2019) The evolution of gaze shifting eye movements. In: Hodgson TL (ed) Current topics in behavioural neurosciences: processes of visuo-spatial attention and working memory. Springer, New York

Land MF, Furneaux S (1997) The knowledge base of the oculomotor system. Philos Trans R Soc Lond B 352:1231–1239

Land MF, Lee D (1994) Where we look when we steer. Nature 369:742–744

Land MF, Mennie N, Rusted J (1999) The roles of vision and eye movements in the control of activities of daily living. Perception 28:1311–1328

Leigh RJ, Zee DS (2006) The neurology of eye movements, 4th edn. Oxford University Press, Oxford

Lueck CJ, Tanyeri S, Crawford TJ, Henderson L, Kennard C (1990) Antisaccades and remembered saccades in Parkinson's disease. J Neurol Neurosurg Psychiatry 53:284–288

Lueck CJ, Crawford TL, Henderson L, Van Gisbergen JAM, Duysens J, Kennard C (1992) Saccadic eye movements in Parkinson's disease II remembered saccades: towards a unified hypothesis. Q J Exp Psychol 45A(2):211–233

Martin L, Tapper A, Gonzalez DA, Leclerc M, Niechwiej-Szwedo E (2017) The effects of task-relevant saccadic eye movements performed during the encoding of a serial sequence on visuospatial memory performance. Exp Brain Res 235(5):1519–1529

Mosimann UP, Felblinger J, Ballinari P, Hess W, Muri RM (2004) Visual exploration behaviour during clock reading in Alzheimer's disease. Brain 127:431–438

Nitschke K, Ruh N, Kappler S, Stahl C, Kaller CP (2012) Dissociable stages of problem solving (I): temporal characteristics revealed by eye-movement analyses. Brain Cogn 80:160–169

Olton DS (1982) Spatially organised behaviours of animals: behavioural and neurological studies. In: Potegal M (ed) Spatial abilities. Pergamon, London

Owen AM, Sahakian BJ, Hodges JR, Summers BA, Polkey CE, Robbins TW (1995) Dopamine-dependent fronto-striatal planning deficits in early Parkinson's disease. Neuropsychology 9:126–140

Owen AM, Iddon JL, Hodges JR, Summers BA, Robbins TW (1997) Spatial and non-spatial working memory at different stages of Parkinson's disease. Neuropsychologia 35(4):519–532

Passingham RE (1985) Memory of monkeys (Macaca mulatta) with lesions in the prefrontal cortex. Behav Neurosci 99:3–21

Pasupathy A, Miller EK (2005) Different time courses of learning-related activity in the prefrontal cortex and striatum. Nature 433:873–876

Patt VM, Thomas ML, Minassian A, Geyer MA, Brown GG, Perry W (2014) Disentangling working memory processes during spatial span assessment: a modelling analysis of preferred eye movement strategies. J Clin Exp Neuropsychol 36(2):186–204

Pavisic IM, Firth NC, Parsons S, Martinez Rego D, Shakespeare TJ, Yong KXX, Slattery CF, Paterson RW, Foulkes AJM, Macpherson K, Carton AM, Alexander DC, Shawe-Taylor J, Fox NC, Schott JM, Crutch SJ, Primativo S (2017) Eye tracking metrics in young onset Alzheimer's disease: a window into cognitive visual functions. Front Neurol 8:377

Pensyl CC, Benjamin WJ (2006) Ocular motility. In: Benjamin WJ (ed) Borish's clinical refraction, 2nd edn. Butterworth, St Louis, pp 356–399

Petrides M, Milner B (1982) Deficits on subject-ordered tasks after frontal- and temporal-lobe lesions in man. Neuropsychologia 20:249–262

Primativo S, Clark C, Yong KXX, Firth NC, Nicholas J, Alexander D, Warren JD, Rohrer JD, Crutch SJ (2017) Eyetracking metrics reveal impaired spatial anticipation in behavioural variant frontotemporal dementia. Neuropsychologia 106:328–340

Rivaud-Pechoux S, Vidailhet M, Brandel JP, Gaymard B (2007) Mixing pro- and antisaccades in patients with parkinsonian syndromes. Brain 130:256–264

Robert MPA, Nachev PC, Hicks SL, Golding CVP, Tabrizi SJ, Kennard C (2009) Saccadometry of conditional rules in presymptomatic Huntington's disease. basic and clinical aspects of vertigo and dizziness. Ann N Y Acad Sci 1164:444–450

Schon F, Hodgson TL, Mort D, Kennard C (2001) Ocular flutter associated with a localised lesion in the paramedian pontine reticular formation. Ann Neurol 50(3):413–416

Shakespeare TJ, Kaski D, Yong KXX, Paterson RW, Slattery CF, Ryan NS, Schott JM, Crutch SJ (2015) Abnormalities of fixation, saccade and pursuit in posterior cortical atrophy. Brain 138:1976–1991

Shallice T (1982) Specific impairments of planning. Philos Trans R Soc Lond B 298:199–209

Stubbs J, Corrow S, Kiang B, Panenka W, Barton J (2018) The effects of enhanced attention and working memory on smooth pursuit eye movement. Exp Brain Res 236:485–495

Tremblay S, Saint-Aubin J, Jalbert A (2006) Rehearsal in serial memory for visual-spatial information: evidence from eye movements. Psychon Bull Rev 13(3):452–457

Walker R, Husain M, Hodgson TL, Harrison J, Kennard C (1998) Saccadic eye movements and working memory following damage to human prefrontal cortex. Neuropsychologia 36(11):1141–1159

Wong AMF (2008) Eye movement disorders. Oxford University Press, Oxford

Zokaei N, Husain M (2019) Working memory in Alzheimer's disease and Parkinson's disease. In: Hodgson TL (ed) Current topics in behavioural neurosciences: processes of visuo-spatial attention and working memory. Springer, New York

Zola SM, Squire LR, Teng E, Stefanacci L, Buffalo EA, Clark RE (2000) Impaired recognition memory in monkeys after damage limited to the hippocampal region. J Neurosci 20:451–463

The manufacturer's authorised representative in the EU is Springer Nature Customer Service Centre GmbH, Europaplatz 3, 69115 Heidelberg, Germany. If you have any concerns regarding our products, please contact ProductSafety@springernature.com

Printed and bound by CPI Group (UK) Ltd, Croydon, CR0 4YY
05/10/2026
02228130-0016